THE COSMETIC INDUSTRY

COSMETIC SCIENCE AND TECHNOLOGY SERIES

Series Editor
ERIC JUNGERMANN
Jungermann Associates, Inc.
Phoenix, Arizona

Other Volumes in Preparation

THE COSMETIC INDUSTRY

SCIENTIFIC AND REGULATORY FOUNDATIONS

Edited by

NORMAN F. ESTRIN

THE COSMETIC, TOILETRY AND FRAGRANCE
ASSOCIATION, INC.
WASHINGTON, D.C.

CRC Press
Taylor & Francis Group
Boca Raton London New York

CRC Press is an imprint of the
Taylor & Francis Group, an **informa** business

First published 1984 by Marcel Dekker, Inc.

Published 2019 by CRC Press
Taylor & Francis Group
6000 Broken Sound Parkway NW, Suite 300
Boca Raton, FL 33487-2742

© 1984 by Taylor & Francis Group, LLC
CRC Press is an imprint of Taylor & Francis Group, an Informa business

First issued in paperback 2019

No claim to original U.S. Government works

ISBN-13: 978-0-367-45176-9 (pbk)
ISBN-13: 978-0-8247-7105-8 (hbk)

Visit the Taylor & Francis Web site at
http://www.taylorandfrancis.com

and the CRC Press Web site at
http://www.crcpress.com

Library of Congress Cataloging in Publication Data
Main entry under title:

The Cosmetic industry.

    (Cosmetic science and technology series; v. 2)
    Includes index.
    1. Cosmetics--Law and legislation--United States.
2. Cosmetics industry--United States. I. Estrin,
Norman F. II. Series.
KF3896.C67 1984   344.73'0423   84-17030
ISBN 0-8247-7105-2   347.304423

ABOUT THE SERIES

The rapid growth of cosmetic science has made it virtually impossible for a single author or a single book to present a coherent review of the entire field. This series was conceived to permit discussion of the broad spectrum of current knowledge and theories of cosmetic science and technology. The series is made up of a number of books either written by single authors, or edited, with a number of contributors. Well-known authorities from industry, academia, medicine, and the government are participating in writing these books.

Our aim is to cover many facets of advances in the field of cosmetic science. Topics will be drawn from a wide spectrum of disciplines ranging from chemical, physical, analytical, and consumer evaluations to safety, efficacy, and regulatory questions. Organic, inorganic, physical and polymer chemistry, emulsion technology, microbiology, toxicology, dermatology and more—all play a role in cosmetic science. There is little commonality in the scientific methods, processes, or formulations required for the wide variety of cosmetics and toiletries manufactured. Products range from hair care, oral care and skin care products to lipsticks, nail polishes, deodorants, powders, aerosol products to over-the-counter products, such as antiperspirants, dandruff treatments, antibacterial soaps, acne creams, and suntan lotions. Thus cosmetics represent a highly diversified field with many subsections of science and art; for indeed, even today a lot of art and instinct are used and needed in the formulation and evaluation of cosmetics.

Since the early sixties, the regulatory climate has changed dramatically for the cosmetic industry. Under a modicum of control since the 1938 Federal Food, Drug, and Cosmetic Act, the passage of the 1962 Drug Amendments to this Act opened the door for a broad review of the

safety and efficacy of numerous cosmetics and toiletries, such as antiperspirants, baby creams, deodorant soaps, antidandruff products,
oral care preparations, etc. Regulatory actions by the F.D.A. and
other agencies soon followed, exemplified by the banning of tribromo-
salicylanilides, mercury compounds, bithionol, zirconium salts in
aerosol products, hexachlorophene, chloroform, and chlorofluorocarbons as propellants for aerosols. These concerns have resulted in increased attention by industry to regulatory matters. Large departments have been staffed to deal with these problems; millions of dollars
are spent annually on testing products to assure safety and efficacy
and compliance to a host of regulations. The second book of this series
entitled "The Cosmetic Industry: Scientific and Regulatory Foundations," edited by Norman F. Estrin, contains contributions by over
fifty experts presenting the different points of view of government,
industry, and trade associations regarding these important advances
in the field of cosmetic science.

Finally, I want to thank all contributors and editors who are participating in the development of this series, the editoral staff at
Marcel Dekker, Inc., and above all, my wife Eva, without whose editorial help and constant support I would never have undertaken this
project.

Eric Jungermann, Ph.D.
Series Editor

PREFACE

This text is designed as a practical guide for individuals with scientific and regulatory responsibilities or interests in the cosmetic industry. The term "scientific" is used in this text particularly to note the relationship of science to regulatory decisionmaking. The vast body of knowledge contained under the heading "cosmetic science" is therefore not included in this text.

For ease of reference, the text is organized into three major sections. The first section, "The Regulatory Environment," summarizes the authority of various regulatory agencies and regulatory programs as they pertain to the cosmetic industry. Chapters are included for areas that have a direct and indirect impact on the cosmetic industry.

The second section, "Functioning in the Regulatory Environment," which comprises the major portion of this text, offers practical advice on how to operate within the regulatory environment outlined in the first section. This section is itself divided into two parts. The first part summarizes industry-wide activities that have been undertaken in response to existing regulations or to promote self-regulation. The second part contains practical advice on how to cope with the maze of regulations to which the cosmetic industry is subject.

The final section, entitled "Challenges for Tommorow," introduces scientific and regulatory issues that now, or in the future, are likely to have an impact on cosmetic manufacturers in this decade.

Authors of chapters appearing in this text were invited to submit chapters based on their knowledge or contributions in a specific area. Readers are encouraged to review source material cited in most chapters. Armed with the information contained in this volume, the reader should be in a better position to phrase the right questions in soliciting expert advice and consulting primary sources. We hope that this text will find use both in the United States and abroad, not only by individuals with scientific and regulatory responsibilities within the cosmetic industry, but also by those in government, those studying at academic institutions, and those preparing for a career in the cosmetic industry.

Norman F. Estrin, Ph.D.

CONTENTS

CONTRIBUTORS

James M. Akerson   Regulatory Affairs, Clairol Research Laboratories, Stamford, Connecticut

Eve E. Bachrach   Legal Department, The Cosmetic, Toiletry and Fragrance Association, Inc., Washington, D.C.

Frank W. Baker   Research and Development, The Proctor and Gamble Company, Cincinnati, Ohio

John L. Beck*   Technical Services, The J. B. Williams Company, Inc., Cranford, New Jersey

Murray Berdick   Consultant, Branford, Connecticut

Jonathon T. Busch   The Cosmetic Ingredient Review, Washington, D.C.

Patricia A. Crosley   Information Resources, The Cosmetic, Toiletry and Fragrance Association, Inc., Washington, D.C.

Anita S. Curry   Science Department, The Cosmetic, Toiletry and Fragrance Association, Inc., Washington, D.C.

Betty J. Diener †   School of Business Administration, Old Dominion University, Norfolk, Virginia

---

Present affiliations:
*Quality Affairs Department, Berlex Laboratories, Inc., Wayne, New Jersey
†Secretary of Commerce and Resources, Commonwealth of Virginia, Richmond, Virginia

Thomas J. Donegan, Jr.   Hyman, Phelps & McNamara, P.C.,
Washington, D.C.

Robert L. Elder   The Cosmetic Ingredient Review, Washington, D.C.

Norman F. Estrin   The Cosmetic, Toiletry and Fragrance Association,
Inc., Washington, D.C.

William E. Gilbertson*   Division of OTC Drug Evaluation, Office of
Drugs, National Center for Drugs and Biologics, Food and Drug Administration, Rockville, Maryland

Robert P. Giovacchini   Corporate Product Integrity, The Gillette
Company, Boston, Massachusetts

Alan M. Goldberg   Department of Environmental Health Sciences, The
Johns Hopkins University, School of Hygiene and Public Health, Baltimore, Maryland

Martin Greift   TSCA Interagency Testing Committee, Environmental
Protection Agency, Washington, D.C.

Jerome H. Heckman   Keller and Heckman, Washington, D.C.

Victor R. Hirsh   Consultant, Chevy Chase, Maryland

James E. Huff   National Toxicology Program, National Institute of
Environmental Health Sciences, Research Triangle Park, North Carolina

Peter Barton Hutt   Covington & Burling, Washington, D.C.

Edward M. Jackson   Research Services Department, Noxell Corporation, Baltimore, Maryland

Joseph Kilsheimer*   Max Factor and Company, Hollywood, California

Monroe Lanzet   Research and Development, Max Factor and Company,
Los Angeles, California

Walter Larsen   Portland Dermatology Clinic, Portland, Oregon

James M. Long   Ohio Ethics Commission, Columbus, Ohio

Jerome K. Malbin   Kolmar Laboratories, Inc., Port Jervis, New York

Linda R. Marshall   Elysee Scientific Cosmetics, Inc., Madison, Wisconsin

Gerald N. McEwen, Jr.   Science Department, The Cosmetic, Toiletry
and Fragrance Association, Inc., Washington, D.C.

Present affiliation:
*Quality Assurance Department, Max Factor and Company, Oxford
†Deceased

Stephen H. McNamara   Hyman, Phelps & McNamara, P.C., Washington, D.C.

Philip C. Merker   Pharmacology and Toxicology, Vick Research Center, Shelton, Connecticut

Sanford A. Miller   Center for Food Safety and Applied Nutrition, The Food and Drug Administration, U.S. Department of Health and Human Services, Washington, D.C.

John A. Moore*   National Toxicology Program, National Institute of Environmental Health Sciences, Research Triangle Park, North Carolina

Emalee G. Murphy   Legal Department, The Cosmetic, Toiletry and Fragrance Association Inc., Washington, D.C.

Donald L. Opdyke†   Research Institute for Fragrance Materials, Inc., Englewood Cliffs, New Jersy

George Pollack   Cosmair, Inc., Clark, New Jersey

David P. Rall   National Toxicology Program, National Institute of Environmental Health Sciences, Research Triangle Park, North Carolina

Allan T. Reeves   The Procter and Gamble Co., Cincinnati, Ohio

Joseph V. Rodricks   Environ Corporation, Washington, D.C.

Ira Rosenberg   Research and Development, Clairol, Inc., Stamford, Connecticut

Andrew N. Rowan   School of Veterinary Medicine, Tufts University, Boston, Massachusetts

William H. Schmitt   Research Laboratories, Chesebrough-Pond's, Inc., Trumbull, Connecticut

H. Joseph Sekerke   Science Department, The Cosmetic, Toiletry and Fragrance Association, Inc., Washington, D.C.

John L. Smith   Microbiological Research, Chesebrough-Pond's Inc. Research Laboratories, Trumbull, Connecticut

Patricia Thompson   Center for Food Safety and Applied Nutrition, The Food and Drug Administration, U.S. Department of Health and Human Services, Washington, D.C.

Present affiliations:
*Environmental Protection Agency, Washington, D.C.
†Retired

William C. Waggoner*   Johnson & Johnson, New Brunswick, New Jersey

John A. Wenninger   Division of Cosmetics Technology, Bureau of Foods, Food and Drug Administration, Washington, D.C.

Theodore Wernick   Compliance Coordination, Gillette Medical Evaluation Laboratories, Rockville, Maryland

_____

Present affiliation:
*Thompson Medical Company, Inc., New York, N.Y.

THE COSMETIC INDUSTRY

Part I

THE REGULATORY ENVIRONMENT

1

REGULATION OF COSMETICS IN THE UNITED STATES—AN OVERVIEW

STEPHEN H. McNAMARA

Hyman, Phelps and McNamara, P.C., Washington, D.C.

I. BASIC PRINCIPLES

A. The Federal Food, Drug, and Cosmetic Act; the Fair Packaging and Labeling Act; and the Food and Drug Administration

In the United States, the composition and labeling of cosmetic products are governed primarily by the Federal Food, Drug, and Cosmetic Act (FDC Act)[1] and by the Fair Packaging and Labeling Act (FPLA).[2] These acts are enforced by the U.S. Food and Drug Administration (FDA), an agency of the Department of Health and Human Services (formerly the Department of Health, Education and Welfare).[3]

B. Definition of "Cosmetic"

The term "cosmetic" is defined by section 201(i) of the FDC Act as follows:

> (1) articles intended to be rubbed, poured, sprinkled, or sprayed on, or introduced into, or otherwise applied to the human body or any part thereof for cleansing, beautifying, promoting attractiveness, or altering the appearance, and (2) articles intended for use as a component of any such articles; except that such term shall not include soap.[4]

This definition is incorporated by reference in the FPLA.[5] If an article comes within this definition, it is subject to regulation as a "cosmetic" under the FDC Act and the FPLA.

II. COMPOSITION OF COSMETIC PRODUCTS

A. Prohibition of "Poisonous or Deleterious Substances"

Except for color additives (discussed in Sec. II.D), there is no official listing of ingredients "approved" for cosmetic use. Instead, section 601(a) of the FDC Act provides generally that a cosmetic shall be deemed to be "adulterated":

If it bears or contains any poisonous or deleterious substance which may render it injurious to users under the conditions of use prescribed in the labeling thereof, or, under such conditions of use as are customary or usual....[6]

Accordingly, manufacturers are under a general duty to avoid use of any cosmetic ingredient that may render the finished product injurious to users under expected conditions of use.

B. Regulations Restricting Particular Ingredients

The FDA has published regulations explicitly prohibiting or restricting use of the following ingredients in cosmetic products: bithionol,[7] mercury compounds,[8] vinyl chloride,[9] halogenated salicylanilides,[10] zirconium in aerosol products,[11] chloroform,[12] chlorofluorocarbon propellants,[13] and hexachlorophene.[14] The FDA would consider violative products to be adulterated within the meaning of section 601(a) of the FDC Act.

These regulations prohibiting or restricting use of certain ingredients do *not*, however, purport to be a complete listing of all ingredients that should not be used in cosmetics. The FDA may initiate regulatory action (as described in Sec. IV.A) whenever it concludes that a particular ingredient used in a cosmetic product violates the standard of section 601(a) of the FDC Act.

C. Regulation Requiring Safety Substantiation

The FDA has published a regulation stating that manufacturers have a general duty to substantiate the safety of each ingredient and finished product prior to marketing:

Each ingredient used in a cosmetic product and each finished cosmetic product shall be adequately substantiated for safety prior to marketing. Any such ingredient or product whose safety is not adequately substantiated prior to marketing is misbranded unless it contains the following conspicuous statement on the principal display panel:

Warning – The safety of this product has not been determined.[15]

D. Color Additives

Pursuant to the Color Additive Amendments of 1960 to the FDC Act, color additives are regulated in a different manner than other cosmetic ingredients.[16] A color additive may not be used in a cosmetic product unless such use first has been approved by an FDA regulation.

The FDA has published numerous regulations listing particular colors authorized for use in cosmetics.[17] For certain of these colors, FDA approval ("certification") of each production batch of the color also is required.

E. Coal-Tar Hair Dyes

The general prohibition against use of a "poisonous or deleterious substance," established by section 601(a) of the FDC Act (discussed in Sec. II.A), does not apply to coal-tar hair dyes. Instead, section 601(a) requires that a specific caution appear on the label of a coal-tar hair dye product:

> Caution – This product contains ingredients which may cause skin irritation on certain individuals and a preliminary test according to accompanying directions should first be made. This product must not be used for dyeing the eyelashes or eyebrows; to do so may cause blindness.[18]

Coal-tar hair dyes also are exempt from the general requirement for FDA approval of color additives used in cosmetics.[19]

Assuming that a coal-tar hair dye product bears the foregoing cautionary statement, insofar as section 601(a) and the Color Additive Amendments are concerned, a manufacturer may use any coal-tar ingredient(s) in the product.

III. LABELING

A. Labeling Requirements – Generally

Pursuant to the FDC Act, the FPLA, and FDA regulations, cosmetic products generally are required to provide the following information on their labels:

1. A statement of the identity of the product
2. A statement of the net quantity of contents
3. A statement of the name and place of business of the manufacturer, packer, or distributor
4. A list of the ingredients included in the product; and for certain products
5. Cautionary or warning language

Before reviewing these various labeling requirements, it is important to note a distinction between requirements imposed pursuant to the FDC Act and requirements imposed pursuant to the FPLA: Generally, labeling requirements premised on the FDC Act must appear *both* on the immediate container of the cosmetic product *and* on the outside container or wrapper of the retail package.[20] Labeling requirements premised on the FPLA must appear *only* on the *outside* container or wrapper of the retail package.[21] Also, certain label statements are required by FDA regulations to appear on the "principal display panel,"[22] while other mandatory labeling information may appear elsewhere on the label.

A cosmetic product that fails to bear required labeling information is deemed to be "misbranded" and is subject to regulatory action by the FDA. (See Sec. IV.A concerning FDA enforcement.)

B. Statement of Identity

Pursuant to FDA regulations premised on the FPLA, the principal display panel of the outside container or wrapper of the retail package of a cosmetic product must bear a statement of the identity of the product.[23]

This statement must be in terms of the "common or usual name" of the product, or "an appropriately descriptive name," or, "when the nature of the cosmetic is obvious, a fanciful name understood by the public to identify such a cosmetic," or "an appropriate illustration or vignette representing the intended cosmetic use."[24]

C. Net Quantity of Contents

Pursuant to the FDC Act and FDA regulations premised on both the FDC Act and the FPLA, a statement of the net quantity of contents must appear both on the immediate container of the cosmetic product and on the principal display panel of the outside container or wrapper of the retail package.[25]

The quantity may be stated in terms of weight, measure, or numerical count.[26]

Customary English units of measure must be used (avoirdupois ounces, fluid ounces, etc.).[27] Metric measure may be declared in addition, but the net quantity of contents declaration may not be solely in terms of metric units.[28]

The regulations provide numerous specific requirements with respect to the format, type size, and placement of information about net quantity of contents.[29]

D. Name and Place of Business of Manufacturer, Packer, or Distributor

Pursuant to the FDC Act and FDA regulations premised on both the FDC Act and the FPLA, the name and place of business of the manufacturer, packer, *or* distributor must be declared both on the immediate container and on the outside container or wrapper of the retail package.[30] The information is not required to be on the principal display panel of the package.

Note the "or": The actual manufacturer (e.g., a contract manufacturer) need not be identified if the packer or distributor is identified.

E. Listing of Ingredients

Pursuant to FDA regulations premised on the FPLA, the outside container or wrapper of the retail package of a cosmetic product must declare the ingredients contained in the product.[31] The information is not required to be on the principal display panel.

Generally, ingredients must be declared in descending order of predominance, although ingredients present at concentrations of not more than 1 percent and color additives may be grouped separately and declared at the end of the ingredient listing without regard to order of predominance. The CTFA (Cosmetic, Toiletry and Fragrance Association, Inc.) *Cosmetic Ingredient Dictionary*, explicitly incorporated into the FDA regulations, is the most comprehensive source of approved nomenclature for cosmetic ingredient labeling.[32]

Ingredients that have been accepted by the FDA as having trade-secret status need not be identified by name, but are required to be designated in the ingredient listing by the phrase "and other ingredient(s)."[33]

F. Warnings or Caution Statements

Pursuant to the FDC Act or FDA regulations premised on that act, warnings or caution statements are required both on the immediate container and on the outside container or wrapper of certain cosmetic products. The coal-tar hair dye warning established by section 601(a) of the FDC Act has been discussed in Sec. II.E. The warning required by FDA regulations for products whose safety has not been substantiated has been discussed in Sec. II.C. In addition, the FDA has issued regulations requiring warnings for cosmetics in self-pressurized containers[34] and for feminine deodorant sprays.[35] A regulation requiring a warning for bubble bath products has also been published, but this requirement has been stayed pending reconsideration. [36]

G. Exemptions

The FPLA was passed by Congress to facilitate value comparisons by consumers when making purchasing decisions. Accordingly, cosmetic products intended for use solely as free samples, or otherwise not intended to be sold to consumers (e.g., products intended to be used in beauty salons or as theatrical makeup) are exempt from FPLA requirements.[37] This exemption is important particularly in the case of ingredient labeling, which is required only pursuant to the FPLA. Thus, for example, free samples or products intended for use solely in beauty salons are not required to bear cosmetic ingredient labeling.

H. Prohibition of False or Misleading Labeling

All cosmetics are subject to a general prohibition against false or misleading labeling. Section 602(a) of the FDC Act provides that a cosmetic shall be deemed to be misbranded:

If its labeling is false or misleading in any particular.[38]

IV. FDA ACTIVITIES

A. FDA Enforcement

If the FDA believes that a cosmetic product is adulterated or misbranded because of violation of any of the requirements discussed earlier, the agency has several enforcement remedies.

The agency may send a *notice of adverse findings* or *regulatory letter* to the manufacturer or other responsible person.[39] These are formal warning correspondence. A notice of adverse findings usually states that the agency believes there has been a violation of law and requests a reply within 30 days that identifies "each corrective step taken or intended to be taken, including measures to prevent recurrence of the violation."[40] A regulatory letter is a more urgent document: It signifies that the "FDA is committed to initiate...administrative or legal action immediately if correction is not promptly achieved," and usually provides less time for response (often 10 days).[41]

In addition, if the FDA determines that a cosmetic is adulterated or misbranded, the agency may request the manufacturer to *recall* the product.[42] The FDA states that:

A request by the Food and Drug Administration that a firm recall a product is reserved for urgent situations and is to be directed to the firm that has primary responsibility for the manufacture and marketing of the product that is to be recalled.[43]

When the FDA believes that a serious violation of law has occurred, the agency may request a U.S. attorney to institute a *civil seizure action*,[44] *injunction action*,[45] or *criminal prosecution in a U.S. district court*.[46] (Violations of FPLA requirements are not subject to criminal prosecution.)[47]

Because cosmetics generally present fewer risks than other products the FDA regulates, a relatively small portion of the agency's resources is applied to cosmetics regulation. For example, in fiscal year 1982, the FDA spent on cosmetics approximately $2.5 million, that is, less than 1 percent of the agency's total operating budget of approximately $321 million.[48] This allowed the FDA, among other activities, to conduct approximately 375 inspections of cosmetic manufacturers.[49]

B. Imports and Exports

Special provisions apply in the case of imports and exports.

With respect to imports, the FDA may request the U.S. Customs Service, within the Department of the Treasury, to detain any cosmetic product offered for import that "appears" to be adulterated or misbranded.[50] FDA regulations provide for an informal hearing if an importer wishes to challenge an import detention.[51]

With respect to exports, the FDC Act provides that a cosmetic intended for export "shall not be deemed to be adulterated or misbranded" if it:

(A) accords to the specifications of the foreign purchaser,

(B) is not in conflict with the laws of the country to which it is intended for export,

(C) is labeled on the outside of the shipping package that it is intended for export, and

(D) is not sold or offered for sale in domestic commerce.[52]

Thus, a cosmetic intended for export may, for example, include ingredients not permitted in the United States, or omit labeling required in the United States, *if* it complies with the foregoing criteria.

C. Voluntary Reporting Programs

Cosmetics are subject to three voluntary reporting programs established by FDA regulations. These include:

1. Voluntary registration of manufacturing and packing establishments.[53] More than 950 companies have registered with the agency.[54]

2. Voluntary reporting of cosmetic raw materials and cosmetic product ingredients.[55] More than 3,600 raw materials and more than 19,500 finished product formulations have been registered with the agency.[56]

3. Voluntary reporting of cosmetic product experiences, including information about adverse reactions.[57] More than 70 companies currently are participating.[58]

The FDA has issued standard reporting forms, available from FDA district offices or from the FDA's Division of Cosmetics Technology at FDA headquarters in Washington, D.C., for use by companies that participate in the voluntary reporting program.

V. OTHER MATTERS

A. The Cosmetic Ingredient Review

The Cosmetic, Toiletry, and Fragrance Association (CTFA)[59] has established a comprehensive industry-funded program for reviewing the safety of cosmetic ingredients: the Cosmetic Ingredient Review (CIR).

Under the CIR program, an independent Expert Panel of scientists has been commissioned to evaluate the available literature, published and unpublished, on the safety of cosmetic ingredients. A permanent CIR staff is located in Washington, D.C. There is a liaison representative to the Expert Panel both from consumer organizations and from the cosmetic industry. In addition, the FDA has appointed one of its employees as a "contact person" to participate in Expert Panel proceedings.

Following peer review, panel reports assessing the safety of cosmetic ingredients are published by the CIR in a scientific journal.[60]

B. Cosmetic Drugs

Section 201(g) of the FDC Act defines a "drug" as including:

> articles intended for use in the...cure, mitigation, treatment, or prevention of disease in man...and...articles (other than food) intended to affect the structure or any function of the body of man....[61]

The "intended use" of an article may be determined by reference to the claims made for it in labeling or advertising.[62] Depending on the claims, a particular product may be subject to regulation as a cosmetic, a drug, or both.

Pursuant to section 201(g), certain cosmetic-type products have been subjected to regulation as drugs by the FDA. Examples include toothpastes represented to prevent tooth decay, antiperspirants, tanning products represented to prevent sunburn, and lip balm products represented to prevent chapping.

It is important to recognize that "drug" status entails significant consequences for a would-be cosmetic. Drugs are subject to different requirements with respect to manufacture, composition, and labeling, and to different FDA enforcement priorities. For example, drug manufacturers are required to register with the FDA their manufacturing establishments and each of their drug products.[63] The FDA has published regulations establishing minimum "current good manufacturing practice" (CGMP) requirements for drugs.[64] Labeling and composition may be subject to an applicable over-the-counter (OTC) drug monograph.[65] Active ingredients in drug products must be identified on the immediate container as well as on the outside container or wrapper.[66] Furthermore, drug establishments are likely to be inspected more often than are cosmetic establishments.[67]

For more information about cosmetic drugs, see Chap. 5.

C. Soap

The definition of cosmetic, quoted in Sec. I.B, explicitly excludes "soap."

The FDA has published a regulation interpreting the meaning of "soap" for the purpose of the cosmetic definition.[68] The FDA regulation takes the position that the "soap" exemption from the cosmetic definition applies only to articles that meet the following conditions:

(1) The bulk of the nonvolatile matter in the product consists of an alkali salt of fatty acids and the detergent properties of the article are due to the alkali-fatty acid compounds; and

(2) The product is labeled, sold, and represented only as soap.

If a product is a soap, it is exempt from the FDA's cosmetic regulations. Thus, for example, a soap may be manufactured with colors that are not approved for use in cosmetics, and a soap is not required to bear ingredient labeling.

Although exempt from cosmetic regulation under the FDC Act, soap products are subject to "consumer product" regulation by the Consumer Product Safety Commission (CPSC) and to labeling requirements established by the Federal Trade Commission (FTC). For more information about soap, see Chap. 4.

D. The Consumer Product Safety Commission

Cosmetics, as defined in the FDC Act, are explicitly excluded from regulation under the Consumer Product Safety Act[69] and the Federal Hazardous Substances Act,[70] the primary statutory authority of the Consumer Product Safety Commission (CPSC).

However, the CPSC has asserted authority over cosmetic-type products in a few respects: The CPSC has authority to regulate soap products, since "soap" does not come within the FDC Act definition of "cosmetic"; the CPSC asserts authority to regulate mechanical hazards presented by cosmetic product *containers*;[71] and the commission has authority under the Poison Prevention Packaging Act[72] to issue regulations establishing special packaging standards required to protect children from serious personal injury or illness.

E. Federal Trade Commission Regulation of Advertising

Advertising for cosmetic products (as distinguished from *labeling*) is regulated by the Federal Trade Commission (FTC), not the FDA. For information about the FTC, see Chap. 6.

F. Regulation by the States

In addition to federal regulation of cosmetics by the FDA, the individual states have the authority to regulate these products and to impose requirements that are not in conflict with federal provisions.

VI. CONCLUSION

In summary, the composition and labeling of cosmetic products offered for sale in the United States are regulated by the Federal Food, Drug, and Cosmetic Act, the Fair Packaging and Labeling Act, and regulations issued by the U.S. Food and Drug Administration. These requirements operate generally to prohibit the inclusion of harmful ingredients and to require informative labeling. Mandatory federal regulation is supplemented by a voluntary reporting program administered by the FDA, and by the industry-funded Cosmetic Ingredient Review evaluation of the safety of cosmetic ingredients. Products represented to affect the structure or function of the body, such as anticavity toothpastes or antiperspirants, are subject to regulation by the FDA as drugs. Soap is excluded from FDA cosmetic regulation and instead is subject to regulation by the Consumer Product Safety Commission and the Federal Trade Commission. Finally, the individual states have independent authority to establish regulatory requirements that are not in conflict with federal provisions.

NOTES

1   21 U.S. Code (U.S.C.) 301 et seq.
2   15 U.S.C. 1451 et seq.
3   The authority vested in the Secretary of Health and Human Services by the FDC Act and the FPLA has been delegated to the Commissioner of Food and Drugs, who directs the FDA. 21 Code of Federal Regulations (C.F.R.) 5.1(a).
4   21 U.S.C. 321(i).
5   15 U.S.C. 1454, 1456, 1459(a).
6   21 U.S.C. 361(a).
7   21 C.F.R. 700.11.
8   21 C.F.R. 700.13.
9   21 C.F.R. 700.14.
10 21 C.F.R. 700.15.
11 21 C.F.R. 700.16.
12 21 C.F.R. 700.18.
13 21 C.F.R. 700.23.
14 21 C.F.R. 250.250.
15 21 C.F.R. 740.10(a).
16 21 U.S.C. 321(t), 361(e), 362(e), 376.
17 21 C.F.R. 73, 74, 81, 82.

18   21 U.S.C. 361(a).

19   21 U.S.C. 361(e).

20   21 U.S.C. 321(k).

21   15 U.S.C. 1459(b).

22   21 C.F.R. 701.10.

23   15 U.S.C. 1453(a)(1); 21 C.F.R. 701.11.

24   21 C.F.R. 701.11(b).

25   21 U.S.C. 362(b)(2); 15 U.S.C. 1453(a)(2); 21 C.F.R. 701.13.

26   21 U.S.C. 362(b)(2); 15 U.S.C. 1453(a)(2); 21 C.F.R. 701.13.

27   15 U.S.C. 1453(a)(3); 21 C.F.R. 701.13(j), (k), (o), (p).

28   15 U.S.C. 1453(a)(3); 21 C.F.R. 701.13(j), (k), (o), (p), (r).

29   See generally 21 C.F.R. 701.13.

30   21 U.S.C. 362(b)(1); 15 U.S.C. 1453(a)(1); 21 C.F.R. 701.12.

31   15 U.S.C. 1454(c)(3); 21 C.F.R. 701.3.

32   21 C.F.R. 701.3(f); 21 C.F.R. 701.3(c)(2).

33   21 C.F.R. 701.3(a).

34   21 C.F.R. 740.11.

35   21 C.F.R. 740.12.

36   21 C.F.R. 740.17, published in *Fed. Reg.* *45*:55172 (August 19,
     1980). Notice of interim stay pending reconsideration published in
     Fed. Reg. *48*:7169, 7203-7204 (February 18, 1983).

37   15 U.S.C. 1451; FDA Inspection Operations Manual, section 694,
     Cosmetics, Exhibit 694.1-B, para. 1, Cosmetic Ingredient Labeling
     (TN 80-6), February 29, 1980.

38   21 U.S.C. 362(a).

39   FDA Regulatory Procedures Manual, part 8, Compliance, chaps.
     8-10, Notice of Adverse Findings and Regulatory Letters,
     September 1980.

40   FDA Regulatory Procedures Manual, part 8, Compliance, chap.
     8-10, Notice of Adverse Findings and Regulatory Letters,
     September 1980, p. 7.

41   FDA Regulatory Procedures Manual, part 8, Compliance, chap.
     8-10, Notice of Adverse Findings and Regulatory Letters,
     September 1980, pp. 7-8.

42   21 C.F.R. 7.40-7.59.

43   21 C.F.R. 7.40(b).

44   21 U.S.C. 334.

45   21 U.S.C. 332.

46   21 U.S.C. 331, 333.

47   15 U.S.C. 1456.

48   FDA Talk Paper T80-52, "FDA Budget for FY 1981," December
     18, 1980.

49   H. J. Eiermann, Cosmetic Regulatory Update – 1980, presented
     at Annual Educational Conference of the Food and Drug Law In-
     stitute, December 10, 1980, p. 13. (Mr. Eiermann is the Director
     of FDA's Division of Cosmetics Technology.) The number of in-
     spections fluctuates from year to year. At the time of final edit-

ing of this chapter, it appears that the FDA may conduct more than 900 cosmetic establishment inspections in 1983, although more than half of these are planned to be fairly cursory "abbreviated inspections."

50 21 U.S.C. 381.

51 21 C.F.R. 1.94.

52 21 U.S.C. 381(d)(1).

53 21 C.F.R. 710.

54 Statistics for the Voluntary Cosmetics Registration Program, Division of Cosmetics Technology (FDA), October 1, 1980.

55 21 C.F.R. 720.

56 See note 54.

57 21 C.F.R. 730.

58 See note 54.

59 The Cosmetic, Toiletry and Fragrance Association (CTFA) is the national trade association representing manufacturers and distributors of cosmetic, toiletry and fragrance products in the United States. The CTFA's offices are at 1110 Vermont Avenue, N.W., Suite 800, Washington, D.C. 20005. The CTFA includes more than 250 companies that manufacture or distribute approximately 90 percent of the finished cosmetic products marketed in the United States. In addition, the CTFA includes more than 230 associate member companies from related industries, such as manufacturers of cosmetic raw materials and packaging materials.

60 See *J. Environ. Pathol. Toxicol.* 4(4):2ff (October 1980).

61 21 U.S.C. 321(g)(1).

62 For example, *United States v. "Sudden Change,"* 409 F.2d 734, 739 (2d Cir. 1969). "The manufacturer of the article, through his representations in connection with its sale, can determine the use to which the article is to be put." S. Rep. No. 361, 74th Cong., 1st Sess. (1935), quoted in C. W. Dunn, *Federal Food, Drug, and Cosmetic Act*, Stechert & Co., N.Y.C., 1938, p. 240.

63 21 U.S.C. 360.

64 21 C.F.R. 210, 211.

65 21 C.F.R. 330.

66 21 U.S.C. 321(k), 352(e)(1).

67 21 U.S.C. 360(h).

68 21 C.F.R. 701.20.

69 15 U.S.C. 2051 et seq. Section 3(a) of the Consumer Product Safety Act provides that the term "consumer product" does not include "cosmetics...as...defined in [the FDC Act]."

70 15 U.S.C. 1261 et seq. Section 2(f) of the Federal Hazardous Substances Act provides that the term "hazardous substance" does not apply to "cosmetics subject to" the FDC Act.

71 CPSC Advisory Opinion No. 229, December 15, 1975.

72 15 U.S.C. 1471. Section 2(2) of the Poison Prevention Packaging Act specifically includes "cosmetic" within the act's definition of "household substance."

2

THE FDA's SCIENTIFIC AND REGULATORY PROGRAMS FOR COSMETICS

JOHN A. WENNINGER

Division of Cosmetics Technology, Bureau of Foods, Food and Drug Administration, Washington, D.C.

I. INTRODUCTION

The primary mission of the cosmetic program at the Food and Drug Administration (FDA) is to enforce the provisions of the Food, Drug, and Cosmetic Act (FDC Act) and the Fair Packaging and Labeling Act

(FPLA). These laws are intended to protect consumers from unsafe or deceptively labeled or packaged cosmetics and to provide consumers with adequate label information to enable them to make value comparisons. The laws apply to cosmetics that are shipped in interstate commerce, a term that includes imports as well as commerce between states.

The FDC Act defines cosmetics as articles or their ingredients applied to the human body for cleansing, beautifying, promoting attractiveness, or altering the appearance without affecting the body's structure or functions. Soap is exempt from the provisions of the FDC Act. Articles promoted as cosmetics but also intended to treat or prevent disease, or to affect the structure or functions of the human body, are drugs as well as cosmetics. Some of the products that fall into this category are hormone creams, sunscreen preparations, anti-caries toothpastes, antiperspirants, and antidandruff shampoos. These products must comply with both the drug and cosmetic regulatory requirements.

The FDC Act prohibits the introduction or receipt in interstate commerce of cosmetics that are adulterated or misbranded. A cosmetic is considered adulterated if it contains a poisonous or deleterious substance that may cause it to be injurious to users under customary conditions of use; if it contains a nonpermitted color additive or any filthy, putrid, or decomposed substance; or if it is manufactured or held under insanitary conditions. It is misbranded if its labeling is false or misleading, if it does not bear the required labeling information, or if it is deceptively packaged [1].

The FPLA is intended to ensure that packages and labels of consumer commodities provide accurate information on the identity of a product, the net quantity of contents, and the name of the manufacturer, packer, or distributor. The FPLA requires that the FDA publish regulations governing these requirements and authorizes the agency to promulgate other regulations necessary to prevent deception of consumers or to facilitate value comparisons [2]. In 1977, the agency, under the authority of the FPLA, required the listing of ingredients (other than those granted trade-secret status by the agency) in descending order of predominance (except flavor and fragrance ingredients) on cosmetic labels or packages [3]. It is important to note that under the FPLA the term "consumer commodity" is defined to include only those products customarily distributed for consumption by individuals in the home. Therefore products sold for use and consumption in professional establishments, such as beauty salons, are not required to bear a listing of cosmetic ingredients.

There is no statutory requirement that a cosmetic product be approved by the FDA before it is introduced into interstate commerce. A cosmetic manufacturer may, on his own responsibility, use essentially any ingredient or market any cosmetic until the FDA can demonstrate that it may be harmful under customary conditions of use [4]. Some ingredients have been prohibited or their use restricted in cosmetics by FDA regulation [5]. Color additives are the only cosmetic ingredients that require prior approval by the FDA before use. Cosmetic firms are not required by law to register their manufacturing establishments, product formulations, or file consumer reports of adverse reaction with the FDA; however, voluntary programs for these activities have been established as discussed below.

The FDA can remove a cosmetic product or ingredient from the market upon showing that it may be injurious to users under customary conditions of use. When such proof is available, the agency may take direct action against products containing the harmful ingredient or may publish a notice of rule making in the *Federal Register* proposing to prohibit or restrict the use of the ingredient. Comments on the proposal are solicited from the public and evaluated by the FDA, and a decision in then made as to whether or not a final regulation should be published in the *Federal Register*. If the regulation is adopted, it becomes part of the *Code of Federal Regulations*.

II. MAJOR COSMETIC PROGRAM ACTIVITIES

The FDA's major cosmetic program activities involve the following elements:

Field inspection and compliance program activities
Consumer complaint and injury evaluation
Cosmetic registration activities
Chemical analysis and method development
Health hazard research
Regulation development
Public assistance and education

Some of these activities are reviewed in detail below. Others, such as health hazard research and regulation development, are the subject of periodic reports by FDA personnel and are published in cosmetic trade journals [6,7].

A. Compliance Programs and Enforcement Policy

The FDC Act authorizes the FDA to inspect any establishment where cosmetics are manufactured or held before or after introduction into interstate commerce. FDA investigators may examine any equipment, finished product, raw material, container, or label pertaining to cosmetics. The inspection of cosmetic establishments gives the FDA an opportunity to uncover insanitary or other conditions that may cause adulteration of products and harm to users. During inspections FDA investigators frequently collect samples and product labeling for laboratory analysis and label review. The inspections are conducted by investigators from the FDA's district offices located throughout the United States.

To coordinate the inspectional activities of the district offices, appropriate headquarter units at the FDA each year issue a Compliance Program Guidance Manual for cosmetics. This manual, which is available to the public, provides background information for investigators on industry trends, products, and labeling issues that should be given emphasis during a given year. This information is frequently based on prevailing health hazard issues, pertinent scientific studies, and new regulations that may prohibit or restrict the use of an ingredient or require a warning statement on certain product labels. Often headquarters will designate some of the establishments to be inspected on the basis of past compliance problems, the types of product a firm is known to produce, or consumer reports of adverse reaction with specific products.

Some commodities under the FDC Act such as food, drugs, and medical devices are subject to established current good manufacturing practice regulations. No such regulation has as yet (1982) been established for cosmetics, although the authority to establish such regulations is incorporated in the FDC Act. Such regulations would benefit both the industry and the FDA by establishing the conditions under which a product would not be considered adulterated under section 601(c) of the FDC Act [1].

Enforcement actions may be taken against products that are found to be either adulterated or misbranded. The actions provided under the FDC Act include seizure of the product, injunction against the manufacturer or distributor to prevent further shipment of the product, or prosecution of an individual or the company responsible for the violation.

The most frequently used action to remove violative products from the market is product recall. Recall is a voluntary action on the part of a firm, but it is recognized that this action is an alternative to FDA-initiated court action against violative products. The FDA's policy regarding product recalls may be found in the *Code of Federal Regulations* [8]. The policy governs the practices, procedures, and guidelines for the voluntary retrieval or correction of violative products that have left the control of the responsible firm.

A cosmetic product being recalled or considered for recall because of a possible health hazard is evaluated by an ad hoc committee of FDA scientists who take into account a number of factors. Among them are the following assessments: (1) whether disease or injury has occurred from the use of a product; (2) whether any existing conditions could contribute to a clinical situation that might expose individuals to a health hazard; (3) the hazard for various segments of the population; (4) the severity of the health hazard to which the population at risk would be exposed; (5) the hazard's likelihood of occurrence; and (6) the immediate or long-term consequences of the hazard. Based on this determination the FDA assigns a recall classification for the product as follows:

Class I.   A situation in which there is a reasonable probability that the use of, or exposure to, a violative product will cause serious, adverse health consequences or death

Class II.   A situation in which the use of, or exposure to, a violative product may cause temporary or medically reversible adverse health consequences or where the probability of serious adverse health consequences is remote

Class III.   A situation in which the use of, or exposure to, a violative product is not likely to cause adverse health consequences

The depth of a recall action is determined by the degree of hazard a product may represent and the extent of the product's distribution as follows: (1) consumer or user level, including any intermediate wholesale or retail level; (2) retail level, including any intermediate wholesale level; or (3) wholesale level.

Recalls may also be requested for misbranding violations.

Each week the FDA publishes an Enforcement Report [9], which contains information on prosecutions, seizures, injunctions, and recalls undertaken by the agency. These terms are defined as follows:

Prosecution: A criminal action filed by the FDA against a company or individual charging violation of the law.

Seizure: An action taken to remove a product from commerce because it is in violation of the law. The FDA initiates a seizure by filing a complaint with the U.S. District Court where the goods are located. A U.S. marshal is then directed by the court to take possession of the goods until the matter is resolved.

Injunction: A civil action filed by the FDA against an individual or company seeking, in most cases, to stop a company from continuing to manufacture or distribute products that are in violation of the law.

Recall: Voluntary removal by a firm of a defective product from the market. Some recalls begin when the firm finds a problem; others are conducted at the FDA's request. Recalls may involve the physical removal of products from the market or correction of the problem where the product is located.

B. Evaluation of Consumer Adverse Reaction Reports

A major responsibility of the FDA's cosmetic program is to evaluate consumer reports of adverse reactions. Most such reports deal with acute toxic effects such as skin or eye irritation, contact sensitization reactions, skin or eye infections, and photocontact sensitization reactions. Normally, chronic toxic effects of products or their ingredients cannot be identified from data accumulated from consumer adverse reaction reports due to the length of time between the use of the product and the onset of the toxic effect and the inability of physicians and consumers to identify the causative agents for such reactions.

The FDA's consumer complaint review activities are designed to identify products, ingredients, and product categories that may be associated with significant adverse reactions. In cases of severe adverse reactions or when a single product is associated with several reports of injury the FDA conducts follow-up investigations. These may include interviewing the complainant and the treating physician, obtaining medical records, and collecting samples. Depending on the facts in a given case, the follow-up investigation may include a visit to the firm that manufactured or distributed the product.

There are three basic sources of consumer cosmetic adverse reaction information available to the FDA. They are reports received by the agency directly from consumers, physicians, or their representatives; reports from selected hospital emergency rooms under the National Electronic Injury Surveillance System (NEISS); and reports directly from industry under the voluntary product experience reporting program. Additional reports from poison control centers and special surveys conducted by the FDA are also valuable sources of information.

Reports Received Directly from Consumers

Since 1970 the FDA's Division of Cosmetics Technology has maintained a computer-based information system to track reports of cosmetic adverse reactions received directly from the public. The information entered into the system includes the name of the product, name of manufacturer or distributor, product category, type of alleged injury, body part affected, and the formulation number if the formulation was voluntarily filed with the FDA. Table 1 summarizes some of the information received for the 10-year period 1970 through 1980.

The number of adverse reaction reports the FDA recives directly from consumers is small compared to the number reported to cosmetic firms. For example, during 1975 the FDA received 583 reports directly from consumers. During the same year 124 cosmetic firms voluntarily filed product experience reports with the FDA reporting 7,229 alleged adverse reactions from consumers. The number of product units estimated to have been distributed by the 124 firms reporting this year was 3.2 billion.

TABLE 1   Summary of Cosmetic Adverse Reaction Reports Received by FDA Directly from Consumers (1970 - 1980)

Product category[a]	Number of adverse reaction reports
Shampoos (noncoloring)	321
Other personal cleanliness products	223
Deodorants (underarm)	214
Bubble baths	195
Moisturizing creams and lotions	181
Hair permanent waves	169
Suntan gels, creams, and liquids	167
Hair dyes and colors (all types requiring caution statements and patch tests)	161
Mascara	158
Other manicuring preparations	149
Hair sprays (aerosol fixatives)	145
Hair straighteners	145
Bath soaps and detergents	128
Skin care creams and lotions for face, body, and hand (excluding shaving preparations)	123
Cleansing creams, lotions, liquids, and pads	119
Eye shadow	113
Dentifrices (aerosol, liquid, pastes, and powders)	113
Other makeup preparations	111
Other eye makeup preparations	107
Depilatories	104
Lipstick	102
Hair conditioners	90
Other skin care preparations	88
Feminine deodorants	78

TABLE 1   (Continued)

Product category[a]	Number of adverse reaction reports
Nail extenders	77
Other hair coloring preparations	73
Nail polish and enamel	70
Tonics, dressing, and other hair grooming aids	68
Skin lighteners	60
Makeup foundations	59
Other hair preparations	55
Eyeliners	52
Makeup base	49
Hair bleaches	48
Colognes and toilet waters	47
Baby shampoos	46
Paste masks (mud packs)	42
Bath oils, tablets, and salts	39
Rinses (noncoloring)	36
Night creams and lotions	34
Eye makeup remover	31
Makeup blushers (all types)	28
Aftershave lotion	28
Baby lotions, oils, powders, and creams	26
Hair preparations, wave sets	23
Shaving creams	23
Perfumes	22
Powder fragrance preparations (excluding aftershave talc)	22
Skin fresheners	22
Cuticle softeners	21

TABLE 1   (Continued)

Product category[a]	Number of adverse reaction reports
Other fragrance preparations	19
Wrinkle smoothing preparations	19
Indoor tanning preparations	18
Face powders	17
Mouthwashes and breath fresheners (liquid and sprays)	16
Hormone creams and lotions	16
Other oral hygiene products	14
Other shaving preparations	13
Foot powders and sprays	12
Hair rinses (coloring)	11
Other bath preparations	10
Personal cleanliness douches	9
Other suntan preparations	9
Manicuring preparations, basecoats and undercoats	8
Bath capsules	7
Makeup preparations, rouges	7
Nail polish and enamel removers	7
Eyebrow pencil	6
Hair lighteners with color	6
All other categories	27
	4,856

[a]Product categories listed in accordance with *Code of Federal Regulations*, section 720.4(c) [10].

Although the total number of adverse reaction reports received by the FDA directly from consumers is relatively small, several significant problems associated with consumer use of cosmetics have been uncovered under the program. For example, in 1978 the FDA received about 50 complaints of hair breakage and scalp irritation associated with a hair straightener. An investigation disclosed that a compounding error had resulted in one batch of the product containing 60 percent more than the intended level of free caustic (sodium hydroxide). During the course of the FDA investigation the firm recalled the product.

In 1974 consumer complaints of fingernail injuries associated with the use of certain nail extenders led the FDA to investigate the problem. It was determined that the methyl methacrylate monomer used in these products was causing the injuries. The FDA obtained a court order to seize the offending product and an injunction against further distribution of the product [11]. Similar products were voluntarily recalled by other distributors.

During 1976 and 1977, consumer complaints that a nail hardener had caused serious allergic and irritant effects were received. An investigation and subsequent laboratory analyses demonstrated that the product contained formaldehyde at a potentially harmful concentration. The product was seized in August 1977 [12].

During 1976—1978 a significant number of consumer complaints concerning one brand of suntan product were received. The ensuing investigation disclosed that the firm also had received many complaints. Many of the adverse experiences appeared to be a form of photocontact dermatitis. Research sponsored by the firm and investigations by the FDA identified 6-methylcoumarin, a fragrance ingredient in the product, as a potent photocontact allergen [13].

National Electronic Injury Surveillance System

Each year since 1974 the FDA has prepared a yearly analysis of cosmetic-related injury data reported into the National Electronic Injury Surveillance System (NEISS) data base. The system is operated by the Consumer Product Safety Commission. The goal of the NEISS program is reduction of risk of injury to individuals from association with consumer products. To achieve this goal, it is necessary to gather information on the factors involved in such injuries. Literally every consumer product is included in the system, and cosmetics is only one of hundreds of product categories that come under the NEISS review. The data come from selected hospital emergency rooms located throughout the country. NEISS data for cosmetic products include injury diagnosis, frequency of injuries by product categories, part of body affected, age and sex of patient, severity of injury, and the disposition of the case.

For calendar year 1979, 739 cosmetic-related injury cases were reported to NEISS [14]. Based on this number, NEISS estimates that

30,965 cosmetic-related injuries (a low estimate of 22,468 to a high estimate of 39,462 with a 95 percent certainty) were treated in hospital emergency rooms throughout the United States and its territories. Of the 739 injuries, 266 cases were associated with misuse of cosmetic products such as ingestion or aspiration of foreign objects. Of the 473 noningestion cosmetic-related product injuries, dermatitis was diagnosed in 51 percent of the cases; contusions/abrasions 16 percent; foreign body injuries 11 percent; and chemical burns 11 percent. Nearly half of the injuries, 49.5 percent, involved injury to the eyes. Other frequently affected body parts include the face (16 percent) and head (10 percent). The data for 1979 were based on a data collection network that included 74 selected hospital emergency rooms.

Because NEISS data are obtained from hospital emergency rooms, they are not good indicators of adverse reactions associated with customary use of cosmetics. The data are heavily weighted to the type of adverse reactions that would prompt a person to seek emergency medical treatment. These include primarily accidental injuries most frequently associated with the eye, face, or head. The type of reactions most often reported include dermatitis, contusions/abrasions, foreign body (e.g., broken glass as an accidental contaminant in a product), and chemical burns.

Voluntary Product Experience Reporting Program

In 1974 the FDA promulgated a regulation that provided for the voluntary submission of product experience reports to the FDA by cosmetic firms [15]. The program is designed to provide information to the FDA on the type and frequency of cosmetic adverse reactions reported by consumers to cosmetic firms. The program is voluntary because under law the FDA does not have the authority to require firms to provide such information. All cosmetic firms are invited to participate and file semiannual reports. Except for summaries of the data, the information provided under the program is confidential and not available to the public. Among the data requested are the name of the product, an estimate of the number of product units distributed during a given reporting period, and the number of adverse reactions reported to the firm. This program is intended to identify products and product categories with rates of adverse reactions significantly higher than the determined norm. This information could serve as a basis for further in-depth review by the FDA to determine the need for toxicological testing or other appropriate action to improve consumer safety in the cosmetic area.

Poison Control Center Data

Data from poison control centers in the United States are also a valuable source of information for the FDA concerning the potential of cosmetic products to cause adverse effects through misuses of the products, that is, accidental ingestion or aspiration. Poison control

center cases are not necessarily poisonings. The majority, in fact,
are not. They usually are incidents of inappropriate exposure to
chemical substances that were presumed to pose at least a potential
threat to health at the time the control center was contacted. Whether
or not poisoning resulted from the exposure would depend on such
factors as the toxicity of the substance and the amount and route of
exposure. Thus, the total number of cases involving specific products is primarily a measure of the frequency of accidental misuse,
not poisoning. Cases that are reported with toxic signs or symptoms,
a hospital visit, or death are classified as "toxic." The FDA uses the
ratio of the "toxic" cases to the total number of cases reported to compare the relative harmfulness of various substances or products under
conditions of accidental misuse [16].

.Poison control center data involving cosmetics at the FDA include
the following information: total number of cases reported for all cosmetics and total cases broken down for 47 product categories. Within
each category the cases are further subdivided into the total number
and the number classified as "toxic." The data are provided for persons of all ages followed by totals for children under the age of 5
years. For example, for the years 1971 through 1978, 1.3 million
cases were reported to poison control centers for all products, of
which 83,000 were associated with cosmetics. Of these, 74,000 were
cases involving children under the age of 5, of which 6,510 were
"toxic" cases (approximately 6,000 cases had signs and symptoms,
500 resulted in a hospital visit, and none resulted in a fatality). The
product categories having the highest number of cases reported to
poison control centers were fragrance preparations, that is, perfumes,
colognes, and toilet waters (28,400); creams and lotions (15,722);
fingernail preparations (15,151); hair preparations (except shampoo)
(6,801); and shampoos (6,621) [16].

FDA Consumer Survey

In 1975 the FDA reported the results of a three-month study of
adverse reactions among a nationwide sample of 10,000 households
[17]. The study, a cooperative effort by the FDA and the American
Academy of Dermatology, was conducted in 1974 and involved 36,000
cosmetic users. This was the first attempt by the FDA to obtain
cosmetic-related injury statistics from a large group of consumers. In
the study, 703 consumer-perceived cosmetic reactions were reported,
of which 589 (84 percent) were judged by dermatologists as definitely
or probably product-related. Of these 589 cases, the vast majority,
505 (86 percent), were considered mild; 63 (11 percent) moderate; and
13 (2 percent) severe (in 8 cases the severity could not be determined)
[18].

A follow-up report on the further analysis of the data and information collected during the above survey was completed in 1977 [19].
This report contains a summary of the consumer-perceived adverse

reactions during the three-month survey broken down into 43 cosmetic product categories. For each product category the number of person-brand uses and number of product users are tabulated as well as the incidence rates of adverse reactions per 10,000 person-brand uses. The product categories having the highest number of adverse reactions were deodorants and antiperspirants, soap, hair spray lacquers, moisturizer lotions, bubble baths, shampoos, mascaras, and colognes. The 589 adverse reactions were associated with 1.11 million person-brand uses in all product categories. The average incidence rate of adverse reactions per 10,000 person-brand uses for all categories was 5.3.

C. Voluntary Cosmetic Registration Program

In 1972 and 1974, the FDA established programs under which cosmetic firms are provided the opportunity to voluntarily register their manu-facturing establishments, file product formulations, and report con-sumer adverse reaction information to the FDA. A detailed description of the regulations for the voluntary registration program may be found in the *Code of Federal Regulations* [10]. Firms interested in partici-pating in these voluntary programs may obtain information, instruc-tions, and the required forms by writing the Division of Cosmetics Technology, Food and Drug Administration, Washington, D.C. 20204.

The regulations were promulgated by the FDA as a result of petitions submitted to the agency by the Cosmetic, Toiletry and Fra-grance Association. The programs have enjoyed the strong support of the association and many cosmetic firms. Most of the major cosmetic firms in the United States participate in one or more parts of the pro-gram. Some foreign cosmetic firms that export cosmetic products to the United States also file information under the programs.

Although participation in these programs has been less than antic-ipated at the time the programs were implemented, the registered data provide valuable information to FDA. As of July 1982, cosmetics firms have voluntarily registered 1,002 manufacturing establishments with the FDA. Currently the Division of Cosmetics Technology's total in-ventory of cosmetic manufacturers and packers contains 2,080 location addresses.

Since the start of the program 32,666 product formulations have been filed by 1,003 firms. Also, 8,652 amendments to filed formula-tions have been received and 12,536 notices that formulations had been discontinued have been filed. Thus the number of current formula-tions on file as of July 1982 was 20,130. In addition to the product formulations, 4,091 Raw Material Composition Statements have also been filed by 164 cosmetic raw material suppliers.

Product experience reports are filed on a semiannual basis. The number of firms participating in this program during the first seven years (14 reporting periods) ranged from 94 to 140. Data tabulated

for the four reporting periods of 1975 and 1976 indicate that 14,240
alleged consumer adverse reactions were reported to the cosmetic
firms that participated in the program during this period (the number
of participating firms ranged from 112 to 128). The estimated number
of product units distributed by the reporting firms during this two-
year period was 6.9 billion. The FDA estimates that the firms filing
product experience reports represent 30 to 40 percent of U.S. cosmetic sales, although only 2 to 3 percent of the cosmetic firms participated in the program [20].

The information voluntarily provided by cosmetic firms on product
formulations is entered into a computer information data base. The
information in this system is not available from any other source and is
widely used by the FDA, industry, and other institutions as a basis
for establishing priorities for safety review programs and many other
purposes. For example, the monographs published by the Cosmetic
Ingredient Review Expert Panel, which is sponsored by the Cosmetic,
Toiletry and Fragrance Association, utilizes information voluntarily
submitted to the FDA by industry [21]. It is especially useful to the
FDA and others for gathering information concerning the usage of ingredients in cosmetics whose safety may be questioned on the basis of
new scientific information. The information can be used not only to
identify the individual ingredients in a brand-name product but also
permits the identification of ingredients used in specific product categories, their frequency of use, and use level. Some specific types of
ingredients, such as preservatives, colors, fragrances, and flavors,
are identified in the data base, and this information can be used to
compile special types of ingredient use information. For example, data
on the frequency of use of various preservatives in cosmetics have
been published [22,23].

D. Chemical Analysis of Cosmetics

The FDA's Division of Cosmetics Technology is responsible for conducting analyses of cosmetic products when such analyses are deemed
necessary for enforcement purposes. In addition, a major amount of
effort is expended conducting research in the field of analytical chemistry. This research is directed toward the application of new instru-
mental analytical techniques to the analysis of cosmetics, determining
the composition of complex cosmetic raw materials and fragrance ingredients, and, more recently, development of methods for the determination of potentially harmful trace contaminants of cosmetics, such
as nitrosamines [24].

Most cosmetic samples analyzed by the FDA are collected by the
agency either as the result of consumer adverse reaction reports or
for surveillance purposes during inspections of cosmetic manufacturing
establishments. The number of samples analyzed on an annual basis
varies widely and usually is decided on the basis of both the need and

the amount of resources allocated to the cosmetic program in a given year. In recent years, a substantial amount of sample analysis and analytical methods research has been devoted to supporting toxicological and microbiological research conducted either under contract for the FDA or in the FDA's own laboratories.

Most of the analytical methods developed in FDA laboratories for the analysis of cosmetic products are reported in scientific journals. The majority are published in the *Journal of the Association of Official Analytical Chemists* (AOAC). The AOAC is a scientific organization whose primary purpose is to serve the needs of government regulatory and research agencies for analytical methods. Its goal is to provide methods that will perform with the necessary accuracy and precision under usual laboratory conditions [25]. A compilation of official methods for the analysis of cosmetics may be found in Chapter 35 of *Official Methods of Analysis of the AOAC* [26]. Additional analytical methods for the analysis of cosmetics have been compiled in *Newburger's Manual of Cosmetic Analysis* [27]. This manual provides general as well as specific methods for the chemical analysis of cosmetic products.

E. Microbiological Examination of Cosmetics

The ability of microorganisms to grow in some types of cosmetic products has been known for many years [28–30]. Many cosmetic formulations provide a good medium for the growth of bacteria and fungi. Some of these organisms may be pathogenic and therefore products contaminated with such microorganisms may constitute a health hazard to the consumer. Eye area cosmetics are of special concern because of the serious consequences of contaminated products coming into contact with a scratched or damaged cornea. Several cases of corneal ulceration resulting from the use of contaminated mascaras have been reported in the literature [31].

There is no need for cosmetics to be sterile. However, microorganisms found should be low in number, must be nonpathogenic, and must not cause the product to decompose during the expected shelf life of the product. It is important that cosmetics contain preservative systems that prevent microbial contamination not only during manufacture but also under normal conditions of product use. A cosmetic that is contaminated with pathogenic microorganisms may be deemed to be adulterated under section 601(a) of the FDC Act [1]. A cosmetic that contains microorganisms commonly identified with filth may be deemed adulterated under section 601(b) of the FDC Act [1].

Microbial contamination of a cosmetic can result in separation of the emulsion, product discoloration, or formation of gas or odor. Such product changes are relatively easy to detect; however, in many cases a cosmetic may show no visible evidence of contamination and still contain unacceptable types or densities of microorganisms.

A variety of bacteria have been isolated from cosmetic products by FDA laboratories. Some of the various genera noted have been *Acinetobacter, Citrobacter, Clostridium, Enterobacter, Escherichia, Hafnia, Klebsiella, Morganella, Proteus, Providencia, Pseudomonas, Serratia,* and *Staphylococcus.* Of these, the ones encountered most frequently have been *Staphylococcus* (especially *S. aureus*), *Pseudomonas* (notably *P. aeruginosa*), and *Klebsiella* [32].

The FDA has not established any quantitative bacterial or fungal limit or standard for cosmetic products. The assessment of the health hazard associated with a given microbiologically contaminated cosmetic is a judgmental decision made by an ad hoc committee of physicians and scientists in the Bureau of Foods [8,33]. Each case is evaluated individually, taking into account such factors as the intended use of the product and the number, type, and pathogenicity of the microorganisms found in the cosmetic [32].

Because of reports of eye injuries associated with consumer use of eye area cosmetics, the FDA published a notice in the *Federal Register* in 1977 [34]. In this notice the FDA expressed the importance of having mascaras and other eye area cosmetics adequately preserved to reduce the risk of microbial contamination during use and possible eye injury. Mascaras can become contaminated with microorganisms when the consumer uses the product and reinserts the applicator wand into the container after application of the mascara to the eye lashes. When microorganisms are introduced into a mascara that is inadequately preserved, they may multiply inside the container. If a contaminated applicator wand comes into contact with a scratched or damaged cornea, the eye may become infected. The FDA has received reports of corneal ulceration associated with the use of mascara products containing pathogenic microorganisms [31]. The majority of incidents involved mascaras in which the microorganism *P. aeruginosa* has been found. *P. aeruginosa* infections, if not recognized and treated immediately, can cause corneal ulceration that leads to partial or total blindness in the injured eye. Thus, particular attention should be given to the microorganism *P. aeruginosa* in developing an adequate preservative system for all cosmetics that may come in contact with the eye during intended or customary conditions of use. The agency believes that the preservative systems used in mascara and other eye-contact products should be adequate not only to prevent the further growth of microorganisms but also to reduce significantly the number of microorganisms introduced during use [34].

Because of concerns in the area of microbial contamination of cosmetic products in general and eye area products in particular, the FDA funded research studies at Emory and Georgia State Universities in Atlanta. These studies resulted in a series of published scientific papers dealing with microbial contamination of eye area cosmetics and their association with ocular infections [31,35—40].

The methodology used in FDA laboratories for the microbiological examination of cosmetic products is contained in Chapter 23 of the *FDA Bacteriological Analytical Manual* [41]. This chapter is available as a separate publication and may be obtained by writing: Director, Division of Microbiology, Food and Drug Administration, Washington, D.C. 20204.

F. Safety Evaluation of Cosmetics

The FDC Act does not require that cosmetic manufacturers or distributors test their products for safety prior to marketing. However, if the safety of a cosmetic is not adequately substantiated, it may be considered misbranded unless the label bears the following statement: "Warning—The safety of this product has not been determined" [42]. In regard to this requirement, cosmetic firms frequently request information from the FDA on cosmetic safety substantiation. The agency has advised that the safety of cosmetics can be substantiated through reliance on already available toxicological test data on individual ingredients and on product formulations that are similar in composition to a particular cosmetic and by the performance of any additional toxicological and other tests that are appropriate in the light of such existing data and information. Although satisfactory toxicological information may exist for each ingredient in a cosmetic, the agency believes that it is necessary to conduct some toxicological testing on the complete formulation to adequately assure the safety of a finished cosmetic product [43]. The agency has cited two papers in the scientific literature that describe reasonable approaches to safety evaluation and that contain references to appropriate testing procedures [44,45].

When cosmetic firms have requested additional guidance on cosmetic safety testing, the agency has provided such assistance when certain information about the product and its ingredients is made available. The type of information usually requested includes but is not limited to the following: the quantitative formulation, intended use of the product, anticipated human exposure, proposed labeling, information on safety data available on the individual ingredients, composition of the raw materials used to formulate the product, manufacturing and processing information, product stability data, and preservative efficacy testing data.

III. FDA ORGANIZATIONAL UNITS AND FUNCTIONS

A. Headquarters

The FDA is a component of the Public Health Service within the Department of Health and Human Services (formerly the Department of Health, Education and Welfare). The Commissioner of Food and Drugs reports to the Secretary of Health and Human Services through the

Assistant Secretary for Health and Surgeon General of the Public
Health Service. The primary organizational units of the FDA are the
bureaus, each of which has responsibility for one or more of the commodities regulated by the FDA. The six bureaus of the agency are
Foods, Drugs, Veterinary Medicine, Radiological Health, Biologics,
and Medical Devices.

The responsibility for cosmetics at the FDA is assigned to the
Bureau of Foods. Within the Bureau of Foods the Division of Cosmetics Technology serves as the focal point for the FDA's cosmetic
program. The division has two branches. The Product Composition
Branch is charged with the responsibility of conducting chemical
analyses of products, developing analytical methods for cosmetic ingredients and product contaminants, and supporting research conducted by other units of the FDA. The other branch, the Registration and Product Experience Branch, administers the voluntary
cosmetic registration activities, monitors adverse reaction reports
from consumers, and assists FDA field offices in areas such as inspections and by providing technical and program advice on cosmetics in
general.

Other units of the Bureau of Foods that also provide scientific
support to the cosmetic program are the Division of Microbiology and
the Division of Toxicology. These units also conduct research studies
and conduct testing in their respective areas. The Division of Regulatory Guidance is responsible for developing policy and implementing
regulatory actions against violative cosmetic and food products.

B. Regional Offices

The FDA has District Offices in most of the major cities of the United
States. There are 10 major Regional Offices, each of which has one or
more District Offices. Most of the District Offices have laboratory
units and all have a staff of investigators who conduct inspections of
firms that manufacture, pack, or distribute commodities subject to the
FDA's jurisdiction. In addition, many of the District Offices maintain
offices known as Resident Posts in various cities in their districts.

IV. COSMETIC INFORMATION RESOURCES AT THE FDA

A. Publications

A variety of publications are available from the Bureau of Foods to
assist the cosmetic, toiletry, and fragrance industries in complying
with laws and regulations administered by the FDA. Information of
general interest to consumers is also available to the public on request. The FDA maintains a comprehensive listing of available
publications entitled "FDA's Catalog of Information Materials for the
Food and Cosmetic Industries," [the catalog may be obtained by

writing: Food and Drug Administration, Industry Programs Branch
(HFF-326), 200 C Street, S.W., Washington, D.C. 20204]. The catalog also provides addresses and phone numbers of some FDA headquarters units and Regional and District Offices as well as other U.S.
governmental agencies that have informational materials which may be
of interest to consumers and the cosmetics industry. Order forms,
stock numbers, and costs (if any) of publications are also provided in
the catalog. Most single copies of publications are available free of
charge. Information is provided in the catalog for such official publications as the *Federal Register,* the *Code of Federal Regulations,* and
the *FDA Consumer* magazine. These publications are available from
the Superintendent of Documents, U.S. Government Printing Office,
Washington, D.C. 20402 at a nominal cost.

The publications most frequently requested in the cosmetic area
are the Food, Drug and Cosmetic Act, Fair Packaging and Labeling
Act, Cosmetic Regulations—Reprint from Title 21, *Code of Federal
Regulations,* and Requirements of Laws and Regulations Enforced by
the U.S. Food and Drug Administration.

B. Corresponding with the FDA

Upon request, the FDA will be glad to reply to questions from firms
concerning cosmetic laws and regulations. If comments on a specific
product are needed, the inquirer should supply full information as to
the identity and quantitative amount of each ingredient in the product,
a copy of all labeling, and the dimensions of the container. Labeling
may be submitted in draft form and need not be printed.

Questions concerning compliance issues should be sent to the
Director, Division of Regulatory Guidance. Requests for scientific
information in the fields of toxicology or microbiology may be sent to
the Director, Division of Toxicology, or Director, Division of Microbiology. Questions relating to the use of color additives in cosmetics
should be sent to the Director, Division of Food and Color Additives.
Information requests in the field of chemistry or cosmetic science and
technology or other matters related to the cosmetic program at the
FDA may be directed to the Director, Division of Cosmetics Technology. The inquiries should be addressed to the appropriate division
at: Food and Drug Administration, 200 C Street, S.W., Washington,
D.C. 20204.

Questions relating to cosmetics that are drugs or inquiries
relating to the classification of a product as such should be addressed
to: Director, Division of Drug Labeling Compliance, Bureau of Drugs,
Food and Drug Administration, 5600 Fishers Lane, Rockville, Md.
20857.

Requests for records from FDA files under the Freedom of Information Act (FOI Act) should be addressed to: Public Records and

Document Center, HFI-35, Food and Drug Administration, 5600
Fishers Lane, Rockville, Md. 20857. Any request for records must
include a reasonable description of the record being sought so that it
can be identified and located. The policy and procedures established
for release of documents under the FOI Act may be found in the *Code
of Federal Regulations [46].*

V. ACKNOWLEDGMENT

The author wishes to thank H.J. Eiermann of the FDA for his helpful
comments on the content of this chapter.

VI. REFERENCES

1.   Federal Food, Drug, and Cosmetic Act, 21 U.S. Code (U.S.C.)
     321 et seq.
2.   Fair Packaging and Labeling Act, 15 U.S.C. 1451 et seq.
3.   21 Code of Federal Regulations (C.F.R.) 701.3.
4.   *Requirements of Laws and Regulations Enforced by the U.S.
     Food and Drug Administration,* Superintendent of Documents,
     U.S. Government Printing Office, Washington, D.C., 1980.
5.   21 C.F.R. 700.11, 700.13, 700.14, 700.15, 700.16, 700.18, and
     700.23.
6.   M. Greif, J. A. Wenninger, and N. Yess, *Cosmet. Technol.*
     2(4):40—46 (1980).
7.   H. J. Eiermann, *Cosmet. Technol.* 2(7):25—28 (1980).
8.   21 C.F.R. 7.40.
9.   FDA Enforcement Report, Press Office, Food and Drug Admin-
     istration, Rockville, Md.
10. 21 C.F.R. 710, 720, and 730.
11. *United States v C.E.B. Products Inc.,* 380 F. Supp 664, N.D.
     Ill., 1974.
12. FDA Enforcement Report, Press Office, Food and Drug Admini-
     stration, Rockville, Md., August 24, 1977.
13. K. H. Kaidbey and A. M. Kligman, *Contact Dermatitis* 4:277—282
     (1978).
14. *Cosmetic-Related Injuries, Study of NEISS Data, January
     1, 1979—December 31, 1979,* Food and Drug Administration,
     Bureau of Medical Devices, Silver Spring, Md.
15. 21 C.F.R. 730.
16. Mark I. Fow, Division of Poison Control, Bureau of Drugs, Food
     and Drug Administration, Rockville, Md., private communication,
     April 6, 1981.
17. Final Report, An Investigation of Consumers' Perceptions of
     Adverse Reactions to Cosmetic Products, available from National

Technical Information Service, Springfield, Va., Reference number PB 242479, June 1975.

18. News Release 75-25, Department of Health, Education and Welfare, Washington, D.C., June 29, 1975.

19. Supplementary Report, An Investigation of Consumers' Perceptions of Adverse Reactions to Cosmetic Products—Supplementary Data Base Preparation and Analysis Project, FDA Contract No. 223-76-8099, Food and Drug Administration, Rockville, Md., January 1977.

20. M. Novitch, *CTFA Cosmet. J. 12*(4):7 (1980).

21. R. L. Elder, *J. Environ. Pathol. Toxicol.* 4(4):1—170 (1980).

22. E. Richardson, *Cosmet. Toiletries 92*(3):85 (1977).

23. E. Richardson, *Cosmet. Toiletries 96* (3):91—92 (1981).

24. J. L. Ho, H. H. Wisneski, and R. L. Yates, *J. Assoc. Offic. Anal. Chem. 64*:800—804 (1981).

25. *Handbook of the AOAC*, 4th ed., Association of Official Analytical Chemists, Arlington, Va., 1977.

26. *Official Methods of Analysis of the AOAC*, 13th ed., Association of Official Analytical Chemists, Arlington, Va., 1980.

27. A. J. Senzel, *Newburger's Manual of Cosmetic Analysis*, 2nd ed., Association of Official Analytical Chemists, Arlington, VA., 1977.

28. M. G. DeNavarre, *The Chemistry and Manufacture of Cosmetics*, D. Van Nostrand Co., New York, 1941, p. 145.

29. A. P. Dunnigan, *Drug Cosmet. Ind. 102*:43—45, 152—158 (1968).

30. H. J. Eiermann, *Drug Cosmet. Ind. 119*:43—45, 114—115 (1976).

31. L. A. Wilson and D. J. Ahearn, *Am. J. Ophthalmol. 84*:112—119 (1977).

32. J. M. Madden, *Dev. Ind. Microbiol. 21*:149—156 (1980).

33. J. M. Madden and G. J. Jackson, *Cosmet. Toiletries 96*:75—77 (1981).

34. *Fed. Reg. 40*:54837—54838 (October 11, 1977).

35. D. G. Ahearn, J. Sanghvi, and G. J. Haller, *Soc. Cosmet. Chem. 29*:127—131 (1978).

36. L. A. Wilson, A. J. Jullian, and D. G. Ahearn, *Am. J. Ophthalmol. 79*:596—601 (1975).

37. D. G. Ahearn, L. A. Wilson, A. J. Jullian, D. J. Reinhardt, and J. Ajello, *Dev. Ind. Microbiol. 15*:211—216 (1974).

38. L. A. Wilson, J. W. Kuehne, S. W. Hall, and D. G. Ahearn. *Am. J. Ophthalmol. 71*:1298—1302 (1971).

39. D.G. Ahearn and L.A. Wilson, *Dev. Ind. Microbiol. 17*:23—28 (1976).

40. R. Bhadauria and D. G. Ahearn, *Appl. Environ. Microbiol. 39*: 005—667 (1980).

41. *FDA Bacteriological Analytical Manual*, Division of Microbiology, Food and Drug Administration, Washington, D.C., chapter 23.

42. 21 C.F.R. 740.10.

43. *Fed. Reg. 40*:8916 (March 3, 1975).

44. R. P. Giovacchini, *Toxicol. Appl. Pharmacol.*, Suppl. 3:13—18 (1969).

45. R. P. Giovacchini, *CRC Crit. Rev. Toxicol. 1*:361—378 (1972).

46. 21 C.F.R. 20.40.

3

THE TOXIC SUBSTANCES CONTROL ACT AND ITS POTENTIAL IMPACT ON THE COSMETIC INDUSTRY

MARTIN GREIF†

TSCA Interagency Testing Committee, Environmental Protection Agency, Washington, D.C.

I. INTRODUCTION

The Toxic Substances Control Act [1] (TSCA) is a far-reaching statute. It empowers the Environmental Protection Agency (EPA) to regulate most synthetic and natural chemicals to reduce the risk of injury to human health or the environment. Under the statute

---

†Deceased

the EPA may promulgate regulations to govern the manufacture, distribution in commerce, processing, use, or disposal of chemicals if the agency finds that such activities present an unreasonable risk of injury to human health or the environment. Use of the term "manufacture" under TSCA includes manufacturing, production, or importing of chemicals.

Congress adopted a policy, embodied in TSCA, that adequate data regarding safety of chemicals in commerce should be acquired and that the acquisition of such data should be the responsibility of those who manufacture or process the chemicals. It was also the intent of Congress to establish authority for the EPA to take quick action to prohibit or restrict the manufacture or use of chemicals that are imminent hazards to health or the environment. The act cautions that regulation of chemicals under its purview should not be carried out in a manner that will impede technological innovation. Furthermore, the EPA is directed to carry out the act in a reasonable and prudent manner that will consider the environmental, economic, and social impact of any actions taken under authority of TSCA.

TSCA authorizes the EPA to gather information on existing chemicals, require testing of those that impose risks of injury, screen new chemicals to determine if they present a risk of injury to health or the environment, and control chemicals proven to present an unreasonable risk.

The Toxic Substances Control Act specifically excludes jurisdiction over pesticides, tobacco or tobacco products, nuclear materials, firearms and ammunition, and articles regulated under the federal Food, Drug, and Cosmetic Act. It is logical, therefore, for one to ask: "Why include a chapter on TSCA in a reference book devoted, among other things, to the regulatory foundations of the cosmetic industry?"

There are several reasons why the cosmetic scientists, regulatory affairs specialist, or purchasing agent should be aware of actions proposed or taken under TSCA authority with respect to chemicals used in cosmetics. First and foremost, if the EPA finds that exposure to a chemical poses a risk of injury to humans, there is a distinct possibility that the intimate bodily exposure often encountered in cosmetic applications of the chemical may also be hazardous. If the EPA should prohibit or severely restrict the manufacture or use of a chemical, there is a possibility that the Food and Drug Administration (FDA) would find that use of the chemical in a cosmetic would cause the product to be adulterated under section 601 of the federal Food, Drug, and Cosmetic Act. The chemical would certainly undergo scrutiny by the FDA, and its use in cosmetics might ultimately be prohibited. Reformulation of products to remove or replace an ingredient is usually disruptive of ongoing production, and may often by very costly and time-consuming.

Even if the FDA decided not to take regulatory action against an ingredient found by the EPA to be injurious to human health, there are other problems the cosmetic producer must face. For example, the chemical may be produced primarily for use in TSCA-regulated applications, with purchases by the cosmetic industry amounting to a small fraction of the market. If use of the chemical is prohibited under TSCA, the producers may find it uneconomical to manufacture the relatively small amounts required by the cosmetic industry and choose to terminate all production, thus eliminating the source of the chemical.

Another consideration regarding a chemical given adverse publicity as a result of action taken by EPA under TSCA is the disclosure of that ingredient on the cosmetic labeling. This would undoubtedly dissuade many astute consumers from purchasing the cosmetic. It also might motivate public interest groups to discourage use of selected cosmetic products by means of adverse publicity. The negative impact of such publicity is often more expensive than the most costly reformulation or recall.

II. MAJOR PROVISIONS OF THE ACT

The Toxic Substances Control Act became effective January 1, 1977. It contains 31 sections, which deal with many major and minor provisions. This chapter will address itself primarily to eight sections of the act that should be of interest to the cosmetic industry. They are as follows:

Section 4: Testing of existing chemical
Section 5: Manufacturing notices
New chemicals
Significant new uses of existing chemicals
Section 6: Regulation of hazardous chemicals
Section 7: Imminent hazards
Section 8: Records and reports
Inventory
Exposure
Adverse reactions
Substantial risks
Section 9: Relationship to other laws
Section 14: Disclosure of data
Section 21: Citizens' petitions
Public information
Confidential information

TSCA divides the universe of chemicals in domestic commerce into two classes: "existing" (old) chemicals and "new" chemicals. This characterization is somewhat analogous to the FDA's definition of "old" drugs and "new" drugs. Under TSCA, an existing chemical is one that was manufactured for commercial purposes in the United States between January 1, 1975, and December 31, 1979. All other chemicals are considered to be new. A chemical may have been manufactured for commercial purposes in the past but been out of use for some time prior to 1975. That chemical would be classified as "new" if it were manufactured again after December 31, 1979.

A. Section 4: Testing of Existing Chemicals

Section 4(a) of TSCA authorizes the EPA to require testing of an existing chemical if the agency finds that its manufacture, distribution in commerce, processing, use, or disposal may precent an unreasonable risk of injury to human health or the environment. A finding by the agency that a chemical poses one of the aforementioned risks may be based on suspicion of adverse biological, biochemical, or physicochemical effects, taking into account the available test data and experience with the chemical. This may be coupled with reasonable expectations of the chemical's likely effects, based on correlation with the known effects of chemicals of similar structure (e.g., structure-activity relationships). Furthermore, the agency must establish that the information needed to resolve the issue is likely to be obtained through testing.

The EPA may also require testing of a chemical if it determines that it is or will be produced in substantial quantities and there is or will be substantial human or environmental exposure to the chemical. In such a case the agency need not find that a reasonable expectation of injury exists. All that is necessary is substantial production, substantial exposure, and the absence of sufficient data and experience on which to characterize the effects of the chemical on human health or the environment.

TSCA stipulates that the EPA shall promulgate rules governing the testing of chemicals meeting at least one of the criteria described in the two preceding paragraphs. The agency proposed its first testing rules under section 4 in July 1980 [2] and has continued to propose testing rules periodically since then. In some instances the EPA has accepted voluntary testing agreements in lieu of issuing testing rules. The agency also published an Advance Notice of Proposed Rulemaking governing phenylenediamines, a large group of chemicals, some of which are used extensively in the cosmetic industry [3].

Any testing of chemicals performed pursuant to an EPA-promulgated testing rule or negotiated testing agreement must be performed in accordance with good laboratory practices established by the agency. The results of these tests and all supporting test data must be submitted to the EPA and are considered to be public information. The act permits chemical firms to cooperate among themselves in conducting the required testing or designate an outside party, such as a trade association or independent testing firm, to conduct the testing. The act requires EPA to publish promptly in the *Federal Register* a notice of the receipt of test data submitted in accordance with the requirements of this section. The data are available for inspection by the public, and copies may be obtained from the EPA upon request, subject to the provisions of section 14 of TSCA.

Costs of testing may be distributed among the firms involved. The prorated sharing of costs would be based on proportionate shares of the total market held by the respective firms. In the event of a dispute over distribution of testing costs, the EPA administrator is authorized to determine what constitutes equitable sharing of costs by the firms manufacturing or distribut'ng the chemical.

More than 55,000 chemicals were reported to be in domestic commerce during the period between January 1, 1975, and December 31, 1977, the base period for distinguishing between existing and new chemicals. Section 4(e) of TSCA established a committee to make recommendations to the EPA administrator respecting chemicals that should be given priority consideration by the agency for the promulgation of testing rules for existing chemicals. This committee is known as the Toxic Substances Control Act Interagency Testing Committee (ITC). The statutory mandate and operating procedures of the ITC are discussed in detail in Sec. III of this chapter.

Section 4(f) of the act requires the EPA to take appropriate action to control or regulate a chemical if it receives test data or other information leading to a conclusion that the chemical presents a significant risk of serious or widespread harm to humans from cancer, gene mutations, or birth defects. If such a conclusion is reached, the agency must take appropriate action under sections 5, 6, or 7 of the act. The available options are described later in this chapter.

B. Section 5: Manufacturing Notices

Section 5 of TSCA requires that a firm notify the EPA prior to starting manufacture or processing of any chemical that does not meet the definition of an "existing chemical." Section 5 also authorizes the EPA to promulgate rules governing significant new uses of existing chemicals and notification procedures. These two notifications are referred to as Premanufacturing Notices (PMNs) and Significant New Use Reports (SNURs), respectively.

Although manufacturers are not required to perform any premarket testing of a new chemical or an existing chemical being considered for a significant new use, they are required to submit to the EPA any test data or information in their possession regarding hazards. The EPA must publish in the *Federal Register*, within five working days, a notice of receipt of a PMN or SNUR. The notice must include a summary of pertinent data received. TSCA permits firms to assert claims of confidentiality regarding information submitted to the agency in PMNs and SNURs.

The act allows the EPA 90 days to evaluate tho information received and notify a firm if it finds cause to delay, prohibit, limit, or restrict the manufacture of a chemical subject to either a PMN or SNUR. The agency may extend the evaulation period for an additional 90 days if it finds good cause to do so. Notice of an extension of the evaluation period and the reasons for it are sent to the firm and published in the *Federal Register*.

The EPA may require additional information or test data before it
will permit the manufacture of a chemical subject to the rules of section 5 to proceed. If so, the firm must be notified of the requirements
within the 90- or 180-day period mentioned in the preceding paragraph. If the firm is not notified within the period, it may proceed to
manufacture the chemical. A Notice of Commencement of Manufacture
must be sent to the agency at that time. Such notice may be sent by
letter. No special form is required.

If, in its review of information provided under this section of the
act, the EPA determines that the manufacture of a new chemical may
harm human health or the environment, it may prohibit or regulate use
of the chemical. The key provisions of TSCA concerning regulation
of hazardous chemicals will be discussed later. When manufacture of
a new chemical begins, it enters the universe of "existing chemicals."

A list of chemicals in commerce (e.g., existing chemicals), together with information on quantities produced, manufacturing sites,
and other pertinent information is maintained by the EPA. This data
base is known as the Toxic Substances Control Act Chemical Substance
Inventory. The reader is directed to Sec. II.E of this chapter for a
discussion of the Inventory and the statutory provisions governing it.

C. Section 6: Regulation of Hazardous Chemicals

The EPA may impose regulatory controls on a chemical if it concludes
that the manufacture, distribution, use, or disposal of the chemical
poses an unreasonable risk of injury to human health or the environment. Section 6 of TSCA contains the provisions governing the
regulation of such chemicals. The agency is authorized to promulgate
regulations imposing one or more of the following constraints on
commericial use of a chemical if the evidence supports such regulation:

> Prohibit entirely the manufacture or use
> Prohibit the manufacture for specified uses
> Impose a limit on the overall quantity produced
> Impose a limit on the quantity produced for specified uses
> Require that warnings and instructions for safe use accompany the
>   chemical
> Require that notice of unreasonable risk be given by manufacturers
>   to their distributors, processors, or to the general public, de-
>   pending on the nature and severity of the risk
> Regulate the methods used to dispose of a chemical

TSCA does not contain a section authorizing the promulgation of
rules governing current good manufacturing practices (GMPs) but it
does present an interesting alternative. The alternative, which appears
in section 6(b) of the act, authorizes the EPA to require a manufacturer
or processor of chemicals subject to TSCA to submit a description of
relevant quality control procedures used. This requirement may be

imposed if the EPA has a reasonable basis to conclude that the methods and controls used in producing a chemical cause the chemical to pose an unreasonable risk of injury to health or the environment.

If the agency determines that existing quality control procedures are not adequate, the EPA may order the firm to revise its procedures to the extent necessary to remedy the situation. By not stipulating quality control procedures in a GMP rule, the EPA and the regulated industry are in the enviable positions of not being locked into quality control standards that may become obsolete. Under this section of TSCA, the agency and industry may readily adapt to technological advances in manufacturing practices and their control.

A manufacturer may be required to notify distributors or processors of its chemicals if the deficiencies in quality control practices cause a chemical to pose an unreasonable risk of injury. Depending on the circumstances, including ultimate use of the chemical, a manufacturer may also be required to give notice of the risk to the general public. Finally, the agency may require the manufacturer of a chemical produced under inadequate quality control procedures to repurchase or replace chemicals in commerce that present a risk of injury that could have been reduced if adequate procedures were being used at the time of manufacture and distribution.

The EPA is required to consider several factors in supporting a regulation to control a hazardous chemical. The factors are prescribed in section 6(c) of TSCA, and the evidence must be included in a *Federal Register* Notice of Proposed Rulemaking. The first factor that must be addressed by the EPA is the effects of the chemical on health and the extent of human exposure to the chemical. Second, the agency is required to consider the effects of the chemical on the environment and the extent of environmental exposure to the chemical. The concept of benefit versus risk is included in a third factor, whereby the act requires that the agency consider and publish a statement regarding the benefits of the chemical for various uses and the availability of substitutes.

Finally, the statute requires the agency to include in its rulemaking record an analysis of the economic consequences of the proposed regulation. The agency must make every reasonable effort to predict these consequences after considering the impact on the national economy, small business, the environment, and public health. This section reiterates one of the policy statements of Congress written into the act: that the EPA shall consider the impact of proposed regulations on technical innovation. It also brings into focus the need to consider cost/benefit relationships, in addition to the benefit/risk assessment mentioned earlier. These should reduce the need for litigating such issues, as has been the case wih a number of rules promulgated under the federal Food, Drug, and Cosmetic Act, which does not address those issues.

D. Section 7: Imminent Hazards

An imminently hazardous chemical is one that presents an imminent (highly threatening) and unreasonable risk of serious or widespread injury to health or the environment. Such a risk is considered imminent under section 7(f) of the act if the EPA determines that its manufacture, use, or disposal in commerce is likely to result in injury to health or the environment before a Final Rule to regulate the chemical can be made effective under normal rule-making procedures.

If a chemical is found by the EPA to present an imminent hazard, the agency may propose a rule under section 6 of the act to prohibit or restrict its manufacture, use, or disposal and declare that rule to be effective immediately upon publication on the *Federal Register*. Regardless of whether or not a rule has been proposed under section 6, or testing initiated in accordance with section 4 of the TSCA, the EPA may proceed under section 7 to reduce or eliminate the risk of injury from an imminently hazardous chemical. The act authorizes EPA to seize a chemical or any article containing the chemical upon receipt of a court order authorizing the seizure. Publication of a *Federal Register* notice is not required. Consequently, action in the public interest can be taken more swiftly in situations where hazards are truly "imminent."

Whether or not a seizure order is received, the agency may seek relief in an appropriate district court to require recall of the chemical in question or articles containing the chemical. Other forms of relief available to the EPA include issuance of a mandatory order requiring the manufacturer of a chemical to notify purchasers of the risk, issue public notice of the risk, or replace or repurchase the chemical or articles containing it.

E. Section 8: Records and Reports

Section 8 of TSCA requires firms that manufacture chemicals for commercial purposes to maintain certain records and submit reports to the EPA in accordance with rules and procedures established by regulation.

Section 8(a) requires that manufacturers report to the EPA the following kinds of information, to the extent known, on selected chemicals:

Common name, chemical name, and molecular structure
Categories of use
Total amount of the chemical manufactured and the amount produced for each category of use
Description of the by-products resulting from the manufacture, use, or disposal
All existing data concerning health and environmental effects
Number of persons exposed to the chemical and duration of exposure in occupational areas
The manner of disposal

In June 1982 the EPA Issued its first Final Rule for section 8(a) reporting [4]. This rule covered about 250 chemicals that had either been recommended by the ITC for priority consideration, reported to EPA as presenting a substantial risk of injury to human health or the environment, or identified by the agency for other reasons. At the same time the agency proposed that 50 more chemicals recommended by the ITC be added to the list of those requiring section 8(a) reports [5]. This proposal includes a provision that all chemicals designated by the ITC in the future be automatically subject to reporting under section 8(a).

The information collected by the EPA under section 8 reporting rules is used, together with other information, in deciding which chemicals should be subject to testing requirements, regulatory controls, or other actions authorized by TSCA.

One of the highest-priority and most comprehensive undertakings required by TSCA was the compilation of a list of chemicals in commerce, the quantities produced during the initial reporting year, and the plant sites where those chemicals were manufactured. Section 8(b) of the act required the EPA to publish the initial list, known as the Toxic Substances Control Act Chemical Substance Inventory (the Inventory), and keep it current as new chemicals clear the Premanufacturing Notice process and are introduced into commerce. The Initial Inventory listed approximately 55,000 chemicals. By January 1984 the Inventory had grown to more than 60,000. All information in the Inventory is available to the public except for that which is claimed to be confidential business information.

Section 8(c) of TSCA authorizes the EPA to promulgate rules requiring manufacturers and distributors of chemicals to maintain records of significant adverse reactions to human health or the environment reported to have been caused by a chemical. The records referred to in this subsection include consumer allegations of personal injury or harm to health, reports of occupational disease or injury, and reports of adverse impact on the environment received from any source, whether or not such information qualifies for reporting in a Notice of Substantial Risk. These records are subject to inspection and copying by EPA investigators.

Section 8(d) requires manufacturers and processors of chemicals to submit to the EPA information or data concerning any health or safety studies known to them on chemicals designated by the agency. Most of the chemicals covered by the EPA rules are those recommended by the ITC and included in its Priority List. Information obtained under these rules will be used by the EPA to help make sound decisions on the need for testing rules governing the chemicals recommended by the ITC.

Finally, section 8(e) of TSCA requires that manufacturers or distributors of chemicals subject to TSCA report immediately to the agency any information supporting a conclusion that a chemical presents a substantial risk of injury to human health or the environment. The kinds of information to be submitted include previously unreported test results indicating that a chemical is carcinogenic, mutagenic, or tera-togenic. Data that show significant adverse acute or chronic health or environmental effects, and epidemiology or environmental monitoring data, must also be reported.

The EPA reviews promptly all Notices of Substantial Risk and judges whether the conclusions by the submitter appear to be valid. In many instances the agency requires additional information, and it communicates directly with the submitter. All Notices of Substantial Risk and the agency's evaluations are a matter of public record, ex-cept that a firm may request that portions of reports which are con-fidential business information be withheld from disclosure to the public. Information received in Notices of Substantial Risk may be used by EPA to support testing-rule or regulatory-control proposals.

F. Section 9: Relationship to Other Laws

Section 9 of TSCA should be of particular interest to the cosmetic industry. Under this section, if the EPA concludes that the manufac-ture, distribution, or use of a chemical is likely to present an unrea-sonable risk of injury to human health or the environment, and this risk may be controlled by another federal agency, the EPA is required to report the fact to the appropriate agency. The agency receiving such a report is required to evaluate the data, assess the risk, and report back to the EPA administrator what action, if any, it intends to take to reduce the risk. Both the EPA report to the other agency and the receiving agency's response must be fully documented in notices published in the *Federal Register*.

G. Section 14: Disclosure of Data

TSCA permits firms subject to the act to designate information dis-closed to the EPA as confidential business information (CBI), provided that such information meets the appropriate standards of the Freedom of Information Act. TSCA does not require that CBI be subject to presubmission review. Under section 14 of TSCA, a firm may claim and be given confidential status for information on quantities of a chemical produced, manufacturing sites, name of the firm producing the chemical, and even the name or structure of the chemical if it believes the information is entitled to confidential treatment. If the specific identity of a chemical is claimed to be confidential, it is listed under a generic name in an appendix to the TSCA Inventory.

If the agency proposes to disclose to the public, or to a person not authorized to receive it, information that a submitter has claimed to be CBI, the EPA must notify the submitter by certified mail or other verifiable means of its denial of the claim and its intent to release the information. The information may be released on the thirty-first day after the submitter is notified unless the EPA has first been notified that the submitter has begun an action in federal court to obtain judicial review and to prevent disclosure. Confidential information sought by any authorized Congressional committee must be released by the EPA upon written request after affected firms are provided 10 days' notice.

The EPA has developed strict security procedures to safeguard CBI. These procedures are described in the *TSCA Confidential Business Information Security Manual* [6]. TSCA imposes criminal penalties for wrongful disclosure of CBI.

H. Section 21: Citizens' Petitions

Section 21 of the act provides that any "person" (private citizen, firm, trade association, public interest group, etc.) may petition the EPA to issue, amend, or repeal a rule or order affecting the testing, regulation, recordkeeping, or reporting of information concerning chemicals subject to TSCA. The petition must be filed in the principal office of the administrator and must set forth the facts that are claimed by the petitioner to support the petition. The agency is authorized to hold a public hearing or conduct an investigation to determine if the petition should be granted.

The act requires the agency to either grant or deny the petition within 90 days after receipt. If the petition is granted, the EPA must initiate promptly proceedings to conform with the petitioner's request. If the petition is denied, the agency must publish in the *Federal Register* information about the petition and the reasons for denial. Denial of a Citizen's Petition is subject to review in federal court. The petitioner may challenge the denial in a U.S. district court within 60 days after the EPA's denial of the petition. The petitioner may also initiate a civil action to compel the agency to make a decision if it fails to act on the petition within 90 days after filing.

If the petitioner satisfies the court that the agency should be compelled to take the action sought in the petition, the court will order the agency to initiate the requested action. The court may award reimbursement of costs to the petitioner if it determines that an award is appropriate. The award may include costs of the suit and reasonable fees for attorneys and expert witnesses.

III. PRIORITIZING CHEMICALS FOR TESTING: THE TSCA INTER-AGENCY TESTING COMMITTEE

A. Statutory Mandate

Section 4(e) of TSCA requires that a committee be established to make recommendations to the EPA administrator respecting chemicals to which the EPA should give priority attention for the promulgation

of the testing rules discussed previously. The committee was established and held its first meeting within five weeks after the effective
date of the act. It adopted the name Toxic Substance Control Act
Interagency Testing Committee, and is usually referred to as the
TSCA/ITC or simply the ITC.

The ITC is composed of members from eight federal agencies, as
prescribed in the act. The agencies are:

> Council on Environmental Quality
> Department of Commerce
> Environmental Protection Agency
> National Cancer Institute
> National Institute of Environmental Health Sciences
> National Institute for Occupational Safety and Health
> National Science Foundation
> Occupational Safety and Health Administration

The act stipulates that members be appointed by the heads of the
respective agencies, and no person may serve as a member for more
than four years. Strict provisions, regarding conflict of interest by
appointed members and alternates, are set forth in section 4(e)(2)(C).

The EPA is required to provide administrative support services to
the ITC as necessary to enable the committee to carry out its function.
In this regard, the agency provides the services of a full-time executive secretary, clerical and legal support as required, access to available data processing programs and facilities, and funding for the
services of a Technical Support Contractor. Technical support to the
EPA members on the committee is also provided by the agency's Office
of Toxic Substances. The other member agencies provide administrative support to their respective representatives.

Soon after it was established, the ITC became aware that several
government agencies, not designated in TSCA to provide official
members on the committee, possess expertise and active programs
dealing with potentially toxic chemicals and related testing activities.
Five such agencies and one national program accepted inviations from
the ITC to provide nonvoting liaison representatives who regularly
attend the ITC meetings and make substantial contributions to the
committee's activities. The liaison representatives are from:

> Consumer Product Safety Commission
> Department of Agriculture
> Department of Defense
> Department of the Interior
> Food and Drug Administration
> National Toxicology Program

The statute requires the committee to issue, within nine months
after the effective date of the act, an initial list of chemicals recommended to be given priority consideration for the promulgation of
testing rules and to revise the list as it deems appropriate at least

every six months. The list, known as the Section 4(e) Priority List
(Priority List), may contain individual chemicals or groups (categories)
of chemicals. Categories may be established on the basis of similarities
in chemical structure; in physical, chemical, or biological properties;
in methods of use; or any other reasonable basis.

The committee is required to present its reasons for inclusion of
each chemical or category on the Priority List. In practice the reasons
have been presented in the form of scientific rationales, supported by
data and information screened from the open literature and those provided voluntarily by industry, academia, various government agencies,
and other sources. TSCA requires that the ITC recommendations
and their supporting rationales be published in the *Federal Register*,
affording interested persons opportunity to file written comments on
them. While there is no limit to the number of chemicals that may be
recommended on the Priority List, not more than 50 on the list at any
given time may be designated for mandatory response by the EPA administrator within 12 months of their being added to the list. In response to an ITC recommendation, the EPA must either initiate a rulemaking proceeding to require the industry to perform the tests recommended by the ITC or present reasons for not initiating the proceeding.

Contrary to misconceptions that have arisen from time to time, the
ITC is not part of the EPA. It is an independent committee, comprised
of members from eight different agencies. The EPA is only one of
those members, and it carries only one vote. The ITC recommends
chemicals to be given priority consideration for promulgation of testing
rules. The final decision on whether or not to require testing rests in
EPA.

B. Criteria for Designating Chemicals to the Priority List

The ITC is required by law to consider all relevant factors in determining which chemicals to add to the Priority List. The committee is
given broad discretion in selecting its criteria; however, section 4(e)
(1)(A) of the TSCA specifies eight factors about chemicals that must
be included. They are as follows:

Quantity manufactured
Amount that enters the environment
Extent of human occupational exposure
Extent of general population exposure
Similarity to chemicals known to have adverse health or environ-
    mental effects
Existence of data concerning effects on health or the environment
Extent to which testing may provide data adequate to predict or
    determine the effects of a chemical on health or the environment
Reasonably forseeable availability of facilities and personnel for
    performing the recommended testing

The act states that the committee must give special attention in establishing the Priority List to those chemicals that are known or suspected to be carcinogenic, mutagenic, or teratogenic. Other factors considered by the committee in its review include data on mammalian species with respect to acute toxicity (i.e., oral, dermal, ocular, and inhalation), reproductive effects other than teratogenicity, other subchronic or chronic effects (i.e., cardiovascular, behavioral, respiratory, and target organ effects), bioaccumulation potential in environmental species (especially aquatic organisms in the food chain), and general ecological effects (i.e., chemical fate, toxicology in aquatic, terrestrial, and avian species), and abiotic effects on the environment.

C. How Chemicals Are Screened and Scored

The TSCA Chemical Substance Inventory, currently containing more than 60,000 chemicals in commerce, provides the most comprehensive source list for chemicals to be screened by the ITC for selecting those that warrant in-depth review. The ITC makes its selections in a scoring procedure that is depicted schematically in Figure 1. The objective of the ITC scoring is to conduct a multistage screening of a large numbuer of chemicals to select a small number of candidates for detailed review.

In a typical scoring exercise, an Initial Listing of 2,000 to 3,000 chemicals from the TSCA Inventory is produced by applying various selection criteria in a computerized screening of the Inventory. The most widely used criterion is production volume. Other criteria may involve exclusion of "site-limited" chemicals (e.g., those that are manufactured and consumed as chemical intermediates in a single on-site process), high-molecular-weight polymers, chemicals of unknown or variable composition, complex reaction products, or biological materials.

The list of chemicals obtained from the Inventory is sometimes merged by computer with other lists of chemicals, including those maintained by other government agencies, to produce the ITC's Initial Listing. This is then screened manually by a panel to remove chemicals that have been scored previously, are already regulated under TSCA, are obviously well understood in terms of hazards (i.e., water, oxygen, many inorganic chemicals, etc.), and those that are so poorly characterized that they cannot be dealt with readily.

The committee had completed five scoring exercises by September 1983. In future exercises it intends to reexamine chemicals that were previously scored (some of them more than five years earlier) in the light of possible new information. The ITC intends also to begin to evaluate the poorly characterized chemicals mentioned previously.

The preliminary screening usually removes at least 75 percent of the chemicals from the Initial Listing. The remaining chemicals (usually 400 to 500) are then scored for human and environmental exposure by a panel of scientists with training and expertise in environmental chem-

SCREENING APPROACH:

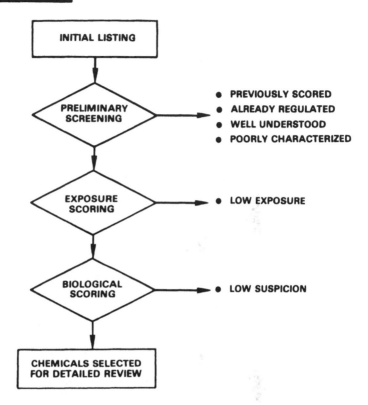

FIGURE 1 Flow diagram depicting the ITC screening and scoring
approach leading up to detailed review by the full committee.

istry, chemical engineering, industrial hygiene, ecology, and computer
science.

Exposure scoring is designed to rank the chemicals in terms of
human and environmental exposure by computing three indices, an
occupational exposure index, a general population exposure index, and
an environmental exposure index. Ten factors are currently used
in computing the indices. While the scope of this chapter does not
permit a detailed description of these factors and the algorithms used
in computing the indices, the factors are listed below:

Annual production
Fraction released in the plant
Number of workers exposed

Fraction released to the environment
Number of people exposed in the general population
Frequency of exposure
Intensity of exposure
Persistence in the environment
Penetrability in the host organism
Bioaccumlation potential

Numerical values for the factors are estimated by the exposure-scoring panel based on information obtained from existing data bases, reference books on industrial chemistry, chemical economics, chemical engineering and environmental chemistry, census data, and trade literature. Where data are lacking, the panel uses expert judgment to estimate the values for individual factors.

Upon completion of exposure scoring, the chemicals are ranked by computer in order of decreasing values of each of the three exposure indices and the ITC selects 200 to 250 chemicals with the highest exposure indices to be scored for likely biological effects. The objective of scoring for biological effects is to identify chemicals with high exposure that have the highest potential for causing harmful biological effects on human beings or the environment and for which adequate test data are lacking.

The general approach involves judgmental scoring and ranking of chemicals by experts using readily available toxicological data and extrapolating structure-activity relationships, where data are lacking. It must be appreciated that scoring is a priority-setting exercise. Its main purpose is to help improve the efficiency of the full evaluation by ITC by selecting the most likely candidates for attention first.

A consultant to the ITC once said: "A good exercise in priority setting involves some kind of compromise between completeness and speed. So almost be definition, the priority-setting exercise involves the use of incomplete data. Under ideal circumstances, expert judgment by the consultants would compensate for the paucity of data. But no situation is ideal. That's why selected chemicals are studied in depth by the full committee" [7].

In practice, a limited literature search is conducted by information specialists for each of the 200 to 250 chemicals being scored for biological effects. The information is summarized in a brief Information Profile. Abstracts of papers found in computerized data are appended to the profiles. Chemical and physical properties and structural formulas are obtained from standard handbooks and reference works.

A panel of 9 to 12 consultants with recognized expertise in various scientific disciplines is next engaged by the ITC to score the chemicals for eight biological effects. Those effects are

Carcinogenicity
Mutagenicity
Teratogenicity

Acute toxicity
Reproductive effects (other than teratogenicity)
Other subchronic and chronic effects
Bioaccumulation potential
Ecological effects

Normally, two or three consultants score all chemicals for those factors in which they have expertise. For example, mutagenicity and carcinogenicity would be scored by one team of consultants. A separate team of two or three consultants would score for teratogenicity and other reproductive effects, and so on.

Chemicals are scored on an algebraic scale of +3 to -3. Those that have been adequately tested and found to be negative for a particular biological effect are assigned a score of zero for that effect. Chemicals that have been tested and presented evidence for a particular biological effect are scored on a scale of +1 to +3, depending on the completeness of the studies and how convincing the data are. The more positive the results of testing are, the higher the numerical value assigned to a chemical and thus the less likely it is to need additional testing to determine its risk of injury related to that effect.

Chemicals that have not been tested for a particular effect, or have been tested with equivocal results presented, are scored on a scale of -1 to -3. Thus, a chemical that has not been tested, but presents no reason for suspicion, would be assigned a score of -1. One that has been tested, presented inconclusive or equivocal results, and is suspected of having the biological effect in question would be assigned a score of -2. Finally, a chemical that has not been tested and is strongly suspect because of a close structural or functional relationship to one or more compounds known to have the biological effect in question would be assigned a score of -3. The more negative the score, the more the chemical is judged to need testing.

At the completion of the biological-effects scoring exercise, the list of chemicals is rank-ordered by computer for each of the eight biological effects scored. The ITC then selects the most highly suspect chemicals, based on the scores, for in-depth study. Generally, those chemicals with scores between -2 and -3 for one or more effects are selected.

D. In-Depth Review of Data on Chemical Hazards

Following a typical scoring exercise, between 75 and 100 chemicals are selected by the ITC for in-depth review. The Technicals Support Contractor to the committee is charged with conducting for each chemical selected an exhaustive search of the open literature, information in the files of various government agencies, and other information sources. The ITC also publishes a notice in the *Federal Register* announcing the names of CAS registry numbers of all chemicals selected for study and requesting from the public any unpublished information on exposure,

health and environmental effects, current trends in production, and
any other factual information that might have a bearing on the committee's determination of the need for testing. In addition, the committee
makes direct inquiries by letter or telephone to manufacturers, soliciting
information on the chemicals. While the industry is under no legal
obligation to provide the information requested, voluntary cooperation
has been commendable.

The collected information is summarized by the Technical Support
Contractor in a document referred to as an Information Review (or
dossier), which includes a complete bibliography of references cited.
The committee studies the information on a chemical together with hard
copy of selected references and, guided by a lead reviewer, frames an
initial recommendation regarding the need for further testing.

The initial recommendation and its supporting rationale are reviewed at a meeting of the full committee. Data gaps, uncertainties,
and suspicions of adverse effects are identified and additional information is sought to complete the dossier on a chemical. Ultimately a
final recommendation and supporting rationale are approved by the
committee to either recommend the chemical (or a category of chemicals)
to the EPA for priority consideration or defer it from further consideration. A chemical is never dropped from the ITC roster, since at
some future time additional information might surface, which could re-
new the committee's interest in a chemical and cause it to be reevaluated
with regard to the need for additional testing.

The ITC meets approximately 24 times per year. A typical meeting
lasts five to six hours, during which 10 to 15 chemicals are reviewed.
A chemical will be reviewed at two or more meetings, depending on the
amount, clarity, and credibility of information found. The more ambi-
guity disclosed in the chemical review process, the longer it takes the
ITC to reach a firm conclusion on whether or not to recommend a
chemical or group of chemicals to the EPA for priority consideration.

During its first five years of existence the ITC scored about 2,000
chemicals for exposure and/or biological effects, studied more than 500
chemicals in depth, and recommended 53 chemicals and categories to
the EPA administrator for priority consideration in accordance with its
mandate under section 4(e) of TSCA.

E. Reports by the Committee and EPA Responses

The determinations of the ITC are summarized in reports issued to the
EPA Administrator every six months. Thirteen semiannual reports
were issued between October 1977 and November 1983. These were
published in the *Federal Register* [8—20] as required by the act. The
Initial Report contained recommendations for priority consideration of
four chemicals and six categories of chemicals. The nine subsequent
reports contained recommendations for 49 more chemicals and 14 additional categories, for a total of 73 entries on the Priority List. As of

January 1984 the EPA had responded to the recommendations covering 63 chemicals and categories.

One category, alkyltin compounds, was removed by the ITC from the Priority List in its Ninth Report [16] because information received in response to the *Federal Register* notice containing the recommendation [14] indicated that the category is too broad to be considered as a single category from the standpoint of chemistry, exposure, or effects. In the Eleventh Report [18] the committee designated 7 individual alkyltin compounds for priority consideration.

Under the provision of section 4(e)(1)(B) of TSCA, the EPA is required to give priority consideration to ITC testing recommendations and either initiate a rule-making proceeding to require the recommended testing or publish in the *Federal Register* the reason for not initiating the proceeding.

To date, the agency has actually proposed relatively few testing rules in response to ITC recommendations. In a few instances the EPA has determined from CBI that the hazards suspected by the ITC were not sufficient to warrant testing rules. In others the industry has initiated testing programs voluntarily. Some of the voluntary testing initiatives have been taken by individual firms, while other are being taken by consortia of firms. In some instances trade associations have designed testing programs in conjunction with member firms that manufacture chemicals recommended by the ITC. Several proposed testing programs have been negotiated with EPA prior to startup, and ultimately accepted by the agency.

In each instance, whether negotiated testing agreements, proposed testing rules, or advance notices of proposed rule making, the EPA has published its response to the ITC in the *Federal Register*. Some consumer and public interest groups have objected to the practice of negotiated testing agreements as EPA's means for complying with the statutory requirement of section 4(e)(1)(B); however, the practice has not been challenged formally. It would appear that the EPA has fulfilled its legal obligation to either propose rules to require the testing recommended by the ITC or publish its reason for not doing so. In these cases, the reason has been that acceptable voluntary testing programs have been undertaken by the regulated industry. Some day the courts may be asked to decide whether or not the agency has satisfied the intent as well as the letter of the law.

To fortify its position in permitting the industry to perform voluntary testing, in lieu of mandated testing under strict regulatory requirements, the EPA points out that the voluntary approach can be implemented more quickly, since the rule making procedure may take two years or more. Negotiated voluntary testing is often implemented within one year of an ITC recommendation. Furthermore, the EPA reserves the right to monitor the progress of voluntary testing programs and initiate a rule-making proceeding if it is not satisfied with the progress of a voluntary program.

IV. CONCLUSION

A brief overview of portions of the Toxic Substances Control Act has
been presented, with discussion of those sections that should be of
interest to the cosmetic industry. Although cosmetics are exempt from
regulation under TSCA, this chapter presents reasons why it be-
hooves cosmetic scientists, regulatory affairs specialists, and industry
purchasing agents to keep abreast of rules issued under authority of
the act and recommendations of the TSCA Interagency Testing Committee regarding testing of chemicals.

The ITC is organized and operates similarly to the Cosmetic Ingredient Review. Both organizations rely on the ability of their respective
expert panels to review available information on the toxic effects, if
any, of chemicals in commerce. Each has a staff of information specialists and scientists to search, collect, and summarize the available
information to facilitate the task of its expert panel.

The CIR is charged with judging the safety of chemicals used in
cosmetics and reporting its findings in reports that are available to the
public. Where data gaps exist, indicating the need for additional testing, they are pointed out in CIR Final Reports. If a chemical is found
to be safe for use in cosmetics, the CIR is required to report the
finding and the reasons supporting it.

The specific mandate of the ITC is to recommend to the EPA administrator chemicals that should be given priority consideration for
the promulgation of testing rules. The recommendations are based on
findings of substantial exposure to the chemicals and a reasonable
scientific basis to suspect that a chemical may be injurious. While the
ITC does not include in its formal reports chemicals that are clearly
safe or unsafe, indications of this nature are made by the committee
during its scoring and review procedures.

Many chemicals that have been studied by the ITC, or are expected
to be studied in the future, are used as cosmetic ingredients. Toxicology and other data in the ITC files are available to the public on request. The ITC and CIR have communicated in the past on chemicals
of mutual interest and will undoubtedly continue this practice as the
need arises in the future.

REFERENCES

1.   15 U.S.C. 2601—2629
2.   *Fed. Reg.* 45:48524—48564 (July 18, 1980).
3.   *Fed. Reg.* 47:973—983 (January 8, 1982).
4.   *Fed. Reg.* 47:26992—27008 (June 22, 1982).
5.   *Fed. Reg.* 47:27009-27018 (June 22, 1982).
6.   *TSCA Confidential Business Information Security Manual*, U.S.
     Environmental Protection Agency, Washington, D.C., October 1981.

7. Ian C. T. Nisbet, Ranking Chemicals for Testing: A Priority-Setting Exercise under the Toxic Substances Control Act, Report of the TSCA-ITC Workshop, Enviro Control, Rockville, Md., February 1979, pp. B-37—B-50.

8. Initial Report to the Administrator, Environmental Protection Agency, TSCA/ITC, *Fed. Reg.* 42:55026—55080 (October 12, 1977).

9. Second Report of the Interagency Testing Committee to the Administrator, Environmental Protection Agency, TSCA/ITC, *Fed. Reg.* 43:16684—16688 (April 19, 1978).

10. Third Report of the Interagency Testing Committee to the Administrator, Environmental Protection Agency, TSCA/ITC, *Fed. Reg.* 43:50630—50635 (October 10, 1978).

11. Fourth Reports of the Interagency Testing Committee to the Administrator, Environmental Protection Agency, TSCA/ITC, *Fed. Reg.* 44:31866—31889 (June 1, 1979).

12. Fifth Report of the Interagency Testing Committee to the Administrator, Environmental Protection Agency, TSCA/ITC, *Fed. Reg.* 44:70664—70674 (December 7, 1979).

13. Sixth Report of the Interagency Testing Committee to the Administrator, Environmental Protection Agency, TSCA/ITC, *Fed. Reg.* 45:35897—35910 (May 28, 1980).

14. Seventh Report of the Interagency Testing Committee to the Administrator, Environmental Protection Agency, TSCA/ITC, *Fed. Reg.* 45:78432—78446 (November 25, 1980).

15. Eighth Report of the Interagency Testing Committee to the Administrator, Environmental Protection Agency, TSCA/ITC, *Fed. Reg.* 46:28138-28144 (May 22, 1981).

16. Ninth Report of the Interagency Testing Committee to the Administrator, Environmental Protection Agency, TSCA/ITC, *Fed. Reg.* 47:5456—5463 (February 5, 1982).

17. Tenth Report of the Interagency Testing Committee to the Administrator, Environmental Protection Agency, TSCA/ITC, *Fed. Reg.* 47:22585—22596 (May 25, 1982).

18. Eleventh Report of the Interagency Testing Committee to the Administrator, Environmental Protection Agency, TSCA/ITC, *Fed. Reg.* 47:54626—54643 (December 3, 1982).

19. Twelfth Report of the Interagency Testing Committee to the Administrator, Environmental Protection Agency, TSCA/ITC, *Fed. Reg.* 48:24443—24452 (June 1, 1983).

20. Thirteenth Report of the Interagency Testing Committee to the Administrator, Environmental Protection Agency, TSCA/ITC, *Fed. Reg.* 48:55674—55684 (December 14, 1983).

4

FEDERAL REGULATION OF SOAP PRODUCTS

EVE E. BACHRACH

Legal Department, The Cosmetic, Toiletry and Fragrance Association, Inc., Washington, D.C.

I. INTRODUCTION

"Soap" is specifically excluded from the definition of "cosmetic" in the federal Food, Drug, and Cosmetic Act (FDC Act),[1] but the term "soap" is not defined elsewhere in the act.

True "soap" is *not* subject to FDC Act cosmetic requirements or any Food and Drug Administration (FDA) cosmetic regulations, including

61

the requirements for use of approved color additives and regulations requiring ingredient labeling. (Soap *is*, however, subject to regulation by other agencies, as discussed below.) However, the basic question is "What is soap?"

II. WHAT IS "SOAP"?

Although the FDC Act exempts "soap" from the definition of "cosmetic," it does not define the term "soap." However, the FDA has issued a regulation in which it interprets the term "soap" very narrowly.[2] The FDA views the term "soap" to apply only to articles that meet the following conditions:

> (1) The bulk of the nonvolatile matter in the product consists of an alkali salt of fatty acids and the detergent properties of the article are due to the alkali-fatty acid compounds; and
> (2) The product is labeled, sold, and represented only as soap.

The agency considers personal cleansing products that do *not* meet these soap conditions to be "cosmetics" and subject to all FDC Act cosmetic requirements.

There has been one court case[3] discussing the soap exemption that sheds light on the two requirements listed above. The case involved the question of whether bottles of shampoo labeled "Beacon Castile Shampoo with Lanolin" were "cosmetics" or "soap." The government had brought a seizure action against the shampoo, alleging that it was an "adulterated" cosmetic,[4] and the manufacturer had responded that the product was a "soap" and therefore not subject to FDA regulation.

The court divided the requirements into three parts and addressed each one, as follows:

> First: The bulk of the nonvolatile matter must consist of an alkali salt of fatty acids.

Nonvolatile matter constituted 15 percent of the shampoo. (Eighty-five percent of the product was water—volatile matter, i.e., capable of evaporating.) Of this 15 percent, between 11 and 14 percent—the "bulk"—was potassium oleate, an "alkali salt of fatty acids." (The remainder was a surfactant, which was not an alkali salt of fatty acids.) Thus, the shampoo met the first test.

> Second: The detergent properties of the article must be due to the alkali salt of fatty acids.

It was agreed that the potassium oleate—the alkali salt of fatty acids—gave the shampoo "detergent" or cleansing properties. However, the surfactant, which was *not* an alkali salt of fatty acids, *also* acted as an *independent* detergent or cleansing agent in the shampoo, albeit not a very effective one. The court considered the intent of Congress in exempting "soap" from the FDC Act: It concluded that the exemption should be read very narrowly to exclude only the "ordinary household and toilet soap" whose detergent properties are due *exclusively* to the alkali salt of fatty acids. Since the detergent properties of the shampoo were *not* due *solely* to the alkali salt of fatty acids (the potassium oleate) but were due *also* to the surfactant, the shampoo did *not* meet the second test.

> Third: The product must be labeled, sold, and represented only as soap.

The shampoo was labeled, sold, and represented as "Beacon Castile Shampoo." The term "soap" was not used. However, the court reasoned that even though "soap" was not used in the name of the product, the word "castile" was a "sufficient representative" of the word "soap." The evidence showed that "castile" means "soap." The court concluded that "by the use of the word castile, the [manufacturer] intended to label, sell, and represent the product as a soap (shampoo)."[5] In addition, the court concluded that a product called a "shampoo" is "not by virtue of that descriptive word precluded from receiving exempted status as a soap."[6]

Reviewing the three requirements as interpreted by the court, the following guides emerge:

> First: The bulk of the nonvolatile matter must consist of an alkali salt of fatty acids.
>
> Second: The detergent, or cleansing, properties of the article must be due *exclusively* to the alkali salt of fatty acids.
>
> Third: The product must be labeled, sold, and represented only as soap, but the word "soap" need not be used, so long as another term sufficiently represents the term soap. "Castile" and "shampoo" are examples of other acceptable terms.

The court ultimately found the Beacon Castile product subject to regulation as a cosmetic, since it did not meet all of the criteria set forth in FDA's soap definition: It did not meet the second requirement because some of the detergent properties were provided by a synthetic detergent.

III. BURDEN OF PROOF

If the FDA asserts that a product is a cosmetic, the agency is not required to show that the product does *not* fall within the soap exemption. In the *Beacon Castile Shampoo* case, the court followed the well-established rule that a party asserting that it qualifies for an exception to a statute has the burden to prove this assertion.

In reaching this conclusion, the court emphasized that the FDC Act is "remedial" legislation, which has the purpose of protecting the public. Accordingly, the statute should be construed in the manner that most effectively protects the public. This purpose is achieved here by requiring the party who claims that its product meets the soap exemptions, and therefore, not subject to the FDC Act, to prove its claim.

IV. ASSESSING THE SOAP REGULATION AND *BEACON CASTILE SHAMPOO*

In assessing the soap regulation and the *Beacon Castile Shampoo* case, the reader should bear in mind that the FDA's soap regulation was issued as an "informal statement of general policy or interpretation"[7] and was *not* the subject of notice and comment rule making under the Administrative Procedure Act (APA).[8] Of course, any government agency's pronouncement about a statute it enforces—such as the FDA statement on the scope of the soap exemption—is entitled to weight. However, an "informal statement of general policy or interpretation" generally is given *less* weight than is given to a regulation promulgated through notice and comment rule making.

The soap regulation defines soap in terms of *composition*—what the product is made of —rather than in terms of *function*—that is, whether the article cleanses as one expects from a product marketed as a soap. However, there is some credible evidence that Congress's intention in enacting the soap exemption was to exclude from regulation articles having the cleansing *function* of soap, regardless of their composition.[9] Unfortunately, in *Beacon Castile Shampoo*, the manufacturer of the shampoo did not vigorously challenge the FDA's soap regulation on this basis, nor on the basis that the regulation was only an informal statement of general policy and should not be applied strictly.

Instead, the manufacturer attempted, unsuccessfully, to show that its product *met* the three requirements of the FDA's soap regulation and therefore was exempt from the FDC Act. The manufacturer *might have* succeeded in this attempt except that the court narrowly read the already strict soap regulation to require that the detergent properties of a complying "soap" must be due *exclusively* to the alkali salts of fatty acids.

The Beacon Castile manufacturer challenged the soap regulation *itself only* insofar as the court might interpret it as having this "exclusivity" requirement. The manufacturer argued that such an interpretation was contrary to the intent of Congress, citing the history of the FDC Act and the soap exemption, and the practices of the soap industry in 1938, when the act with its soap exemption was passed. The manufacturer presented a document revealing that the additive rosin was added to some soaps of the time and that rosin had a detergent effect *independent* of the alkali salts of fatty acids. However, the court believed its "exclusivity" requirement was justified. The court dismissed the defendant's argument and evidence, and found that the detergent properties of soaps in 1938 were "commonly understood to be attributable to alkali-fatty acid compounds."[10] The court suspected that rosin could be used as a "point of entry" by which other detergent substances might be added to soap without subjecting the composite article to regulation as a cosmetic, and that it was not reasonable to attribute such a result to the intent of Congress.

No official definition of soap existed until the FDA issued its "informal statement of general policy or interpretation" in 1958, and that statement has yet to receive a head-on judicial challenge. *The Beacon Castile Shampoo* case is the only judicial treatment of the issue at all in the years since the regulation was issued.

Thus, while the FDA's soap regulation is the only extant official definition of "soap" and is entitled to weight as the statement of the agency charged with enforcing the applicable statute, the FDA analysis still could be challenged. A company that chooses to act conservatively and avoid a clash with FDA will adhere closely to the three guides in Sec. II. However, a company that is willing to challenge the FDA definition may prefer to interpret "soap" more broadly in formulating its soap product. If a company chooses to challenge the soap regulation head-on, the company should keep in mind that the long, unchallenged history of the soap regulation has cloaked it with an aura of legitimacy and given it an inertia that will be difficult to overcome. Also, despite any other shortcomings of the court's opinion in *Beacon Castile Shampoo*, one thing is clear: The burden of proving that a product meets the soap exemption to the FDC Act will be placed on the company.

V. COSMETIC OR DRUG CLAIMS WILL CAUSE A PRODUCT TO BE SUBJECT TO THE FDC ACT

A company must be cautious about any cosmetic-type claims it makes for a soap product: The FDA is likely to maintain that a "soap" product advertised or labeled as providing "moisturizing" or "beautifying" properties is a cosmetic, and thus subject to the FDC Act and all of its requirements for cosmetics—regardless of whether the bulk of the article comprises alkali salts of fatty acids and its cleansing properties are attributable exclusively to them.

In particular, a company should be careful about *fragrance* claims made for a product it wants to market only as a soap. The FDA would not challenge soap status for a product represented as "Lavender Scented Soap," for example. However, if the claims should go *further* and represent that Lavender Scented Soap will impart the lavender fragrance to the *user*, FDA might assert that this claim would bring the product within the FDC Act cosmetic definition because the product is being represented to "beautify" and "promote the attractiveness" of the user.[11]

Similarly, a soap for which *drug*-type claims are made may be subject to the FDC Act requirements for drugs. For example, the FDA has taken the position that products represented as "antimicrobial soap" or "acne soap" come within the FDC Act's definition for "drug"[12] and are subject to regulation as drugs.

VI. INCLUSION OF "SOAPS" IN FDA'S COSMETIC PRODUCT CATE-GORIES

The terms "bath soaps" and "shaving soap" appear in the FDA's voluntary program for filing of *cosmetic* product ingredient and *cosmetic* raw material composition statements.[13] The terms appear in the cosmetic product categories of the regulation under the headings of "personal cleanliness" and "shaving preparations," respectively.

However, the inclusion of these soap items among the FDA's "cosmetic" product categories does not mean that the items are cosmetics: *If* cosmetic claims are made for bath soaps or shaving soaps that otherwise qualify for "soap" status, the FDA may assert that they are "cosmetics" as defined by the FDC Act and are subject to the act's cosmetic requirements. It is likely that the "bath soaps" and "shaving soap" referred to mean only those "bath soaps" and "shaving soap" for which such cosmetic claims are made, and do not refer to "bath soaps" and "shaving soap" that do *not* make such claims and that otherwise qualify for soap status.

Further, these terms appear within a section that sets forth a *voluntary* program: The product categories are listed, in part, as a convenience for companies that wish to participate in the voluntary reporting program. The categories, one may conclude, were intended only to assist the participating company in this voluntary endeavor; they were not intended to be used in an adversarial proceeding where the FDA might challenge the soap status of a company's products.

VII. LIQUID SOAP

Recently, there has been some controversy over the marketing of so-called liquid soap products that do not come within the defintion of "soap" in the FDA regulation. These products are labeled as "liquid soap" or "cream soap"; they are marketed as personal cleansing products and are labeled in accordance with all of the FDA's requirements for labeling of cosmetics, including ingredient labeling.

In a series of letters, the FDA has advised several cosmetic companies that it recognizes the term "soap" to have two meanings:

First, "soap" for "jurisdictional purposes" (i.e., for the purpose of deciding whether a product is exempt from the cosmetic definition in the FDC Act and therefore exempt from FDA jurisdiction) is defined in the FDA regulation.

Second, "soap" is a term that may be used in a *generic* sense to refer to a variety of personal cleansing products, including products that are not technically "soaps" as defined in the FDA's regulation and thus are subject to regulation as cosmetics.

The FDA says that it would not object to this "generic" use of the term "soap" on the label of a synthetic detergent product (i.e., one that is not a soap within the meaning of the FDA regulation) *if* that term is "properly qualified" and the product otherwise is labeled in accordance with the FDA's cosmetic regulations, including ingredient labeling.

The FDA concludes that it would "take exception if the term was improperly qualified so as to be misleading."[14] However, the FDA does not explain what the phrase "properly qualified" means. And finally, the letters caution that although the FDA's resources and activities in the area of misbranded cosmetics currently are "extremely limited," the agency may pursue misbranding violations when resources become available.

VIII. FEDERAL TRADE COMMISSION LABELING REGULATIONS FOR SOAP

As stated earlier, if a product is a true soap, as defined in the FDA's regulation, it is not subject to the FDC Act or to any FDA cosmetic regulations, including those that require ingredient labeling. (The ingredient labeling requirements were issued by the FDA pursuant to the Fair Packaging and Labeling Act (FPLA).[15] The FPLA gives the FDA authority to issue regulations only "with respect to any consumer commodity which is a food, drug, device, or *cosmetic* as each term is defined by . . . *the Federal Food, Drug, and Cosmetic Act.*"[16] A cosmetic as defined by the FDA Act, of course, does *not* include *soap*. Thus, the FDA's cosmetic ingredient labeling regulations do *not* apply to soap.

However, there *are* labeling regulations issued by the *Federal Trade Commission* (FTC) that apply to soap.[17] (These do *not* require ingredient labeling for soap, although the FTC has the power to issue such regulations.)

The FTC regulations also were issued pursuant to the FPLA: The FPLA gives the FTC authority to issue regulations as to "any other consumer commodity"[18] (i.e., commodities other than foods, drugs, devices, or cosmetics). A true soap, since it is exempt from cosmetic status, *is* subject to the FTC regulations.

The FTC labeling regulations for soap require that the label of the *outer container* contain the following information (which parallels that which FDA requires for cosmetics):

1. *Identity of the product:* This labeling requirement[19] is similar to the FDA requirement for cosmetics. The commodity must be

identified by the name specified by any applicable federal law;
or in the absence of that, by the *generic name* or an *appropriate descriptive term* such as one that says what the commodity is used for. The name may not be misleading and may
not mention an ingredient that is not present in the commodity
in a *substantial or significantly effective amount*. The name
of the commodity should be a *principal feature of the principal
display panel*, easily read by consumers, and in lines positioned
generally parallel to the base of the package as designed to be
displayed.

2. *Name and place of business of the manufacturer, packer, or
distributor*: This labeling requirement[20] is similar to the FDA
labeling requirement for cosmetics. The statement must be
conspicuous. It must use the *actual corporate name*, which is
qualified by a phrase such as "Manufactured for _____" or
"Distributed by _____", or any other wording that expresses
the facts. It must include street address, city, state and zip
code. (The street address may be omitted if it appears in a
current city or telephone directory.)

3. *Net quantity of contents*: These requirements are similar to
FDA's requirements for cosmetics. An *accurate*[22] statement of
the net quantity of contents must appear as a *distinct item* on
the *principal display panel*[23] in conspicuous *and easily legible
type*.[24] It should be stated in terms of the *U.S. gallon* and
subdivisions such as the fluid ounce if in fluid measure.[25] It
should be stated in terms of weight if the product is solid,
semisolid, viscous, or a mixture of solid and liquid. It may be
stated in terms of *weight, measure, numerical count, or a
combination*.[26] The *metric equivalent* also may be stated on
the principal display panel or other panels.[27] The term "*net
weight*" must be used when stating the net quantity of contents in terms of weight;[28] for fluid measure, the terms "net"
or "net contents" are optional.[29]

   The statement should be *accurate* ("reasonable variations"
from the stated measure are allowed).[30] Common and decimal
fractions may be used, but decimal fractions should not be
carried out to more than two places.[31]

There are specific FTC regulations for labeling the net quantity of
contents on "multiunit packages," "variety packages," and "combination
packages."

A "multiunit package" is defined as a package intended for retail
sale that contains two or more units of an identical commodity.[32] If
the units are individually wrapped and labeled according to all the requirements discussed above, the outer container (which houses the
individual units) must declare the number of individual units, the
weight of each individual unit, and the total of all the units. For
example:

Soap bars: 6 Bars, Net Wt. 3 oz., Total Net Wt. 18 oz.

If the six soap bars are *not* individually wrapped and labeled, the outer container may instead state:

Soap bars: 6 Bars, Total Net Wt. 18 oz.

A "variety package" is defined as a package intended for retail sale containing two or more units of *similar but not identical* commodities.[33]

The outer container must declare the weight of each nonidentical unit, and then the total weight. For example:

2 Soap Bars Net Wt. 3 oz. each
1 Soap Bar Net Wt. 5 oz.

Total: 3 Soap Bars Net Wt. 11 oz.

A "combination package" is defined as a package intended for retail sale containing two or more units of *dissimilar commodities*.[34] Each group of dissimilar commodities in the package should be described separately on the outer container in terms of quantity as appropriate. For example:

Sponges and soap bars: 2 sponges each 4 in. × 6 in. × 1 in.; 6 soap bars, Total Net Wt. 18 oz.

IX. SOAP REGULATION BY OTHER AGENCIES

Finally, soap products are subject to regulation by other agencies in addition to the FTC. For example, soap is subject to regulation by the Consumer Product Safety Commission under the Consumer Product Safety Act[35] and the Federal Hazardous Substances Act.[36] Thus, exemption from the FDC Act by no means frees soap from thorough federal regulation.

X. NOTES

1 The definition of "cosmetic" is as follows:

(i) The term "cosmetic" means (1) articles intended to be rubbed, poured, sprinkled, or sprayed on, introduced into, or otherwise applied to the human body or any part thereof, for cleansing, beautifying, promoting attractiveness, or altering the appearance, and (2) articles intended for use as a component of any such articles; *except that such term shall not include soap.*

21 U.S. Code (U.S.C.) 201(i) (emphasis added).
2 21 Code of Federal Regulations (C.F.R.) 701.20.
3 *United States v. An article of Cosmetic . . . Beacon Castile Shampoo . . .* , Federal Food, Drug, and Cosmetic Act 1969—1974, Kleinfield, Kaplan and Weitzman, N. D. Ohio, November 6, 1973, p. 149.

4   "A cosmetic shall be deemed to be adulterated—(a) If it bears or
     contains any poisonous or deleterious substance which may render
     it injurious to users. . . under such conditions of use as are
     customary or usual." 21 U.S.C. 361.
5   *Beacon Castile Shampoo*, p. 156.
6   *Beacon Castile Shampoo*, p. 158.
7   See 21 C.F.R. part 3, subpart B, 3.652 (1970 ed.).
8   5 U.S.C. 551 et seq.
9   Association of American Soap and Glycerine Producers, Inc., Re-
     garding Statement of General Policy or Interpretation Issued by
     the Food and Drug Administration on September 26, 1958, memo to
     FDA, (January 9, 1959).
10 *Beacon Castile Shampoo*, p. 157.
11 21 U.S.C. 321(i).
12 21 U.S.C. 201(g).
13 21 C.F.R. 720.4(c) (10) and (c) (11).
14 See, e.g., Letter from Taylor Quinn, Associate Director for Com-
     pliance, to Bjorn Ahlgren, Consumer Goods International, Inc.,
     November 14, 1980. (On file at FDA Dockets Management Branch.)
15 15 U.S.C. 1451 et seq.
16 15 U.S.C. 1454.
17 See CTFA Labeling Manual, 4th ed., 1982, for a review of labeling
     requirements for cosmetics.
18 15 U.S.C. 1454.
19 16 C.F.R. 500.4.
20 16 C.F.R. 500.5.
21 16 C.F.R. 500.6—500.26.
22 16 C.F.R. 500.6.
23 16 C.F.R. 500.6.
24 16 C.F.R. 500.17.
25 16 C.F.R. 500.8.
26 16 C.F.R. 500.7.
27 16 C.F.R. 500.21.
28 16 C.F.R. 500.9.
29 16 C.F.R. 500.10.
30 16 C.F.R. 500.22.
31 16 C.F.R. 500.16.
32 16 C.F.R. 500.24.
33 16 C.F.R. 500.25.
34 16 C.F.R. 500.26.
35 15 U.S.C. 2051 et seq.
36 15 U.S.C. 1261 et seq.

5

THE IMPACT OF THE FDA's OVER-THE-COUNTER DRUG REVIEW PROGRAM ON THE REGULATION OF COSMETICS

WILLIAM E. GILBERTSON

Division of OTC Drug Evaluation, Office of Drugs, National Center for Drugs and Biologics, Food and Drug Administration, Rockville, Maryland

I. INTRODUCTION

The Over-the-Counter (OTC) Drug Review is one of the most innovative efforts of any federal regulatory agency in terms of the number of products reviewed and consumers to be affected. The program was instituted to carry out the Food and Drug Administration's (FDA's) statutory mandate to assure that all OTC drug products are safe and effective and properly labeled for their intended uses. The approach involves the development of drug "monographs" that define the conditions for which these products are generally recognized as safe and effective and not misbranded. Unfortunately, the review is a lengthy process due to its public nature and its inherent complexities. It will be several years before the recommendations of the various advisory panels are fully evaluated and promulgated as final regulations.

The approach in methods being used to conduct the review are of special importance to the cosmetic industry, not only for the direct effects on marketed cosmetic products but because this approach is likely to characterize more of the FDA's regulatory authority in the years ahead. The subject of this chapter is the impact of the FDA's Over-the-Counter Drug Review Program on the regulation of cosmetics. It is worth noting that cosmetic ingredients and labeling were truly not intended to be part of the OTC drug review. In fact, the procedures established to conduct the review fail to provide an evaluation and classification scheme for "cosmetic conditions." However, many consumer products are labeled with multiple claims or uses that include not only drug terms but terms traditionally viewed as cosmetic. The OTC Drug Review has thus found itself considering many ingredients and products that have historically been viewed as "cosmetic." Hence, one issue that has repeatedly surfaced is the drug-versus-cosmetic status of product categories. The agency cannot ignore the overlap that exists for these products when final regulations are ultimately established in the not-too-distant future.

II. THE OTC DRUG REVIEW

Before directly exploring the cosmetic-drug issue, it would appear appropriate to discuss first the overall OTC drug review program. The current authority and regulatory posture of the program is in essence an outcome of the evolution of the FDA. Most consumers do not question the safety and many do not question the effectiveness of OTC drugs. Before the establishment of the review, there had never been an evaulation of the available scientific evidence to determine both safety and effectiveness for most OTC drugs. In fact, prior to 1962 manufacturers were not legally obligated to demonstrate effectiveness for label claims.

Safety, effectiveness, and proper labeling have not always been characteristic of medications used in the United States. The history and evolution of our federal laws is a fascinating and complex story in itself, and is far beyond the scope of this chapter. It suffices to note that the first major legislation regulating drugs was the Pure Food and Drug Act passed in 1906. "Unsafe" and "nonefficacious" drug products were not dealt with directly. Rather, drugs were required only to meet standards of strength and purity claimed by manufacturers. Drug safety was not required by law until the passage of the 1938 Federal Food, Drug, and Cosmetic Act. The Act required that all drugs entering the marketplace after that date be shown to be safe for human use before they could be marketed.

The 1962 Drug Amendments to the Act required that all drugs also be shown to be effective for their intended uses. However, a review in the mid-1960s of drug products that had been approved for safety only since 1938 included 512 OTC drugs, only 25 percent of which were shown to be effective for their intended uses. Clearly it was time for the FDA to take a further look at the OTC marketplace.

It is estimated that more than 300,000 individual OTC drug products are marketed, but fortunately only some 700 active ingredients are claimed in the labeling of these products. In determining the logistics of the review, the FDA decided that a product-by-product review would not be feasible. Practicality dictated a review that focused on the ingredients used in these products, divided by therapeutic category as summarized in Table 1. Thus, instead of examining individual antacid products, of which there are estimated to be more than 8,000, it was deemed more practical to evaluate their active ingredients, such as aluminum hydroxide and magnesium carbonate. Clearly, the FDA's review of OTC drugs is different from its evaluation of prescription drugs, which is a review of finished dosage forms. For most OTCs there need not be an affirmative demonstration that specific formulations of active and inactive ingredients are safe and effective. In some

TABLE 1   OTC Drugs and Claims Originally Announced to Be Considered by Advisory Panels

Acne	Antispasmodics
Alcohol	Aphrodisiacs (oral)
Allergy products	Astringents (oral)
Antacids	Astringents (styptic pencil)
Antiasthmatic products	Athlete's foot (antifungal) products
Antibacterials	Baby creams (diaper rash, rash, prickly heat)
Antibiotics (topical)	
Antidiarrheals	Back and medicated plasters
Antiemetics	Bleaching (skin) preparations
Antimicrobials	Blemish remedies (skin)
Antiperspirants	Boil ointments
Antipyretics	Bronchodilator products
Antirheumatics (internal)	Burn products
Antirheumatics (topical)	Canker sore remedies
Antirheumatics (oral)	Chafing and chapping remedies

TABLE 1 (Continued)

Cold products	Insect bites
Cold sore remedies	Insect repellents
Contraceptives (vaginal)	Internal analgesics
Corn pads, plasters, and remedies	Laryngitis preparations
	Laxatives
Cough products	Liquid bandages (sprays)— protective skin preparations
Cradle cap remedies	
Dandruff products	Lozenges
Dental care agents	Medicated bandages
Dentifrices	Medicated bath preparations
Denture adhesives	Menstrual products
Denture irritation remedies	Mercurials
Denture pads	Minerals (excluding salt tablets)
Denture reliners	Minor skin irritation remedies
Detergents	Mouth fresheners
Digestive aids	Mouthwashes
Disclosing products	Nail biting deterrents
Diuretics	Nasal ointments, jellies, etc.
Dry skin remedies	Nausea remedies
Emetics	Ophthalmics
Eye drops	Oral cavity preparations
Feminine deodorant sprays	Otic products
Fever blister remedies	Otitis externa (swimmer's ear) products
Foot balms, baths, creams, etc.	
Hair growers	Parasiticides
Halogenated salicylanilides	Pharyngitis preparations
Hangover remedies	Poison ivy and oak remedies
Hemorrhoidal drug products	Premature ejaculation remedies
Hormone creams	Psoriasis remedies
Ingrown toenail remedies	Salt substitutes

TABLE 1   (Continued)

Salt tablets	Thumbsucking deterrents
Seborrhea remedies	Tonsilities preparations
Sebum hair loss remedies	Topical analgesics
Sedatives	Topical anesthetics
Skin cleaners	Topical fluoride preparations
Skin healing preparations	Toothache remedies
Skin protectants	Troches
Sleep aids	Universal antidotes
Smoking deterrents	Vaginal drug products
Sore throat products	Vaginitis remedies
Stimulants	Vitamins
Stomach acidifiers	Wart removers
Sunburn prevention products	Weight control products
Sunburn treatment products	Wet dressings
Sunscreen agents	Worm remedies
Sweeteners	Wrinkle removers
Teething lotions	

cases, however, a specific vehicle for the active ingredient may be identified.

A. The Advisory Panel Review

The first phase of the OTC drug review was the establishment of 17 advisory review panels (as described in Table 2) to evaluate collectively the numerous marketed products.

Each panel consisted of seven voting members, including a pharmacist, a pharmacologist or toxicologist, physicians, and other qualified scientists, as well as nonvoting technical liaison members representing consumer and drug industry interests. The panels were charged with reviewing the ingredients in OTC products to determine whether these ingredients could be generally recognized as safe and effective for use in self-treatment. They were also charged with reviewing claims and recommending appropriate labeling, including therapeutic indications, dosage instructions, and warnings against side effects and misuse.

TABLE 2   OTC Drug Advisory Review Panels

Antacids

Antimicrobial I—products used daily

Antimicrobial II—products used infrequently

Antipersipirants

Cold, Cough, Allergy, Bronchodilator, Antiasthmatic Products

Contraceptive and Other Vaginal Drugs

Dentrifrices and Dental Care Agents

Hemorrhoidals

Internal Analgesic, and Antirheumatic Products

Laxative, Antidiarrheal, Antiemetic, Emetic Products

Miscellaneous External—products used topically

Miscellaneous Internal—products used orally

Ophthalmics

Oral Cavity—Mouthwashes, Oral Antiseptics

Sedative, Tranquilizer, Sleep Aid, Stimulant Products

Topical Analgesics, Including Antirheumatic, Otic, Burn and Sunburn
    Treatment and Prevention Drugs

Vitamins, Minerals, and Hematinics

According to the terms of the review, ingredients were classified in three categories:

Category I: generally recognized as safe and effective for the
    claimed therapeutic indication
Category II: not generally recognized as safe and effective or
    making unacceptable claims
Category III: insufficient data available to permit final class-
    ification

Based on these determinations, the FDA promulgates monographs for each therapeutic class of drugs, and the drugs are reviewed in accordance with procedural rules that were established in 1972. Promulgation of OTC drug monographs is a three-step procedure involving (1) an advance notice of proposed rule making reflecting the recommendations of expert advisory panels; (2) a tentative final monograph (proposed rule) responding to comments and stating FDA's position for further comments and offering an opportunity for a legislative-type public hearing before the commissioner; and (3) a final monograph (final rule). Originally the regulations provided a period of time after publication of the final monograph for the testing of Category III ingredients. During this period, products containing such ingredients would be allowed to remain on the market. However, the concept of Category III was legally challenged. In *Cutler v. Kennedy* [1], the court held that the OTC drug review regulations were unlawful to the extent that they authorized the marketing of Category III drugs after a final monograph. Accordingly, the agency has revised the regulations to require that any necessary testing be done and the data resulting from such testing be submitted before the final monograph is published [2]. A manufacturer now has 12 months following publication of the tentative final monograph to complete any necessary studies and submit the data to the agency. During this time the agency will assist the manufacturer in developing testing guidelines and provide interim feedback on data submitted.

The panel phase of the OTC drug review extended over a period of almost 10 years, with more than 300 individuals participating in this unprecedented project. The first panel met in February 1972 to review OTC anatacid ingredients. The last panel meeting was held in October 1981 to review OTC menstrual products. In between, an initial determination was made of the safety and effectiveness of more than 700 ingredients for therapeutic claims ranging from antiflatulents to antimicrobials, from hair restorers to pinworm remedies. These findings were based on a review of more than 14,000 volumes of data submitted largely by manufacturers, but also by concerned consumers, pharmacists, and other interested parties. The panels' judgments were based on their own clinical experience and expertise, on marketing experience of ingredients, and both controlled and uncontrolled clinical trials. The panels also relied on the published literature; but isolated case reports, random experience, testimonials, and reports lacking sufficient details to permit scientific evaluation were not considered.

Overall, FDA received nearly 60 panel reports, which are described in Table 3. These reports and advanced notices of proposed rule making summarizing the panels' recommendations to the Commissioner of Food and Drugs were published in the *Federal Register*. With each publication, public comment was invited.

B. The FDA's Review and Proposal

The second phase of the review process amounts to the agency's review of the ingredients in each class of drugs, based on the panel's findings, on public comment, and new data that may have become available. The

TABLE 3   OTC Drug Product Categories Included in Panel Reports

Acne products

Alcohol (topical)
  first aid products

Ammonia inhalants

Antacid products

Anorectal products

Anthelmintic products

Antibiotic (topical) first aid
  products

Anticaries products

Antidiarrheal products

Antidotes (acute toxic ingestion)

Antiemetic products

Antimicrobial first aid products

Antimicrobials (hospital use)

Antimicrobials (soaps and skin
  cleansers)

Antiflatulent products

Antifungal products

Antipersipirant products

Aphrodisiac products

Astringent products

Boil relief products

Cholecystokinetic products

Corn and callus removers

Cough/cold products
  Anticholergics
  Antihistamines
  Antitussives
  Bronchodilators
  Expectorants
  Nasal decongestants

Dandruff relief products

Daytime sedatives

Deodorants (internal use)

Diaper rash products

Digestive aids

Diuretic products

Emetic products

Exocrine pancreatic insufficiency
  products

External analgesic products

Fever blister/cold sore remedies

Hair grower products

Hair loss prevention products

Hormone products (topical)

Hyperphosphatemia products

Hypophosphatemia products

Ingrown toenail relief products

Insect repellents (internal use)

Insect sting and bite products)

Internal analgesic products

Laxatives

Male genital desensitizers

Menstrual products

Mercurial first aid products
  (topical)

Nailbiting/thumbsucking
  deterrents

Nighttime sleep aids

Ophthalmic products

Oral health care products

Oral discomfort relief products

Oral mucosal injury products

TABLE 3   (Continued)

Otic products (topical)	Smoking deterrent products
Overindulgence remedies	Stimulant products
Pediculicides	Stomach acidifier products
Psoriasis relief products	Sunscreen products
Poison ivy and poison oak prevention products	Vaginal contraceptives
	Vaginal drug products
Seborrheic dermatitis relief products	Vitamin and mineral products
	Wart removers
Skin bleaching products	Weight control products
Skin protectants	

agency, in turn, publishes its conclusions in the form of a tentative final monograph. This is actually the FDA's proposal and offers the first clear signal as to the agency's ultimate intentions. A period of time is allotted for consideration of objections or requests for a hearing before the agency. New data may also be submitted. After this lengthy but necessary public process, involving both scientific and legal resources, a final monograph is established.

This second phase of the review is awesome, not only in terms of size, but in terms of complexity. During this phase, we must decide whether to accept a panel's recommendations or revise them on the basis of comments received. Many questions must be resolved, including:

1. When there is a need, are the policies in the agency's proposal consistent with related policies for prescription drugs?
2. Are the decisions legally and scientifically supportable?
3. What precedents will be established in one rulemaking that will affect other rulemakings?
4. What is the quality of the scientific evidence supporting a particular recommendation?
5. How will products with cosmetic claims be regulated when these products are also subject to OTC drug monographs?

C. Establishment of OTC Drug Monographs

The third and last phase of the OTC review process is publication of final regulations in the form of drug monographs. The monographs provide the regulatory standards for the continued marketing of OTC drug products not covered by new drug applications. Preclearance by

FDA is not required if these standards are followed. Few final monographs have been established thus far, and it may well be that further
changes in structure and content will occur, but final monographs
generally will contain the following components:

1. *Subpart A, General Provisions*, details the scope of the monograph, i.e., the therapeutic category or categories to which it applies
and defines specific terms as they are used in the monograph.

2. *Subpart B, Active Ingredients*, will identify the specific ingredients and the amounts that can be used in the product. Any stipula-
tions on the number of combined active ingredients or other such
criteria will also be identified including details as to the age, dosage,
and frequency of administration recommended for proper, safe, and
effective use. Additionally, active ingredients that may be combined
with other active ingredients, not only from that particular monograph
but from other monographs, will be identified; for example, the combination of an antacid with a nonantacid active ingredient such as an
analgesic. In such instances, the specific ingredients will be identified
under the pertinent subsections of the final monograph.

3. *Subpart C, Labeling*, will contain the indications for the
product; warnings for proper use, which may also include drug interaction precautions; directions for use, including the time interval or
time period (frequency) and amount for use; and any special labeling.
This subpart may also contain professional labeling, which is labeling
for the product that is provided to health professionals but not to the
general public.

The feature of the OTC drug review most heavily criticized has
been the policy that the exact labeling terms listed in a monograph are
the only terms that may be used. Since the inception of the OTC drug
review, the agency has maintained that a monograph describing the
conditions under which an OTC drug will be generally recognized as
safe and effective and not misbranded must include both specific active
ingredients and specific labeling. (This policy has become known as
the "exclusivity rule.") The agency's position has been that it is
necessary to limit the acceptable labeling language to that developed
and approved through the OTC drug review process in order to ensure
the proper and safe use of OTC drugs. The agency has never contended, however, that any list of terms developed during the course
of the review literally exhausts all the possibilities of terms that appropriately can be used in OTC drug labeling. Suggestions for additional terms or for other labeling changes may be submitted as comments
to proposed or tentative final monographs within the specified time
periods or through petitions to amend monographs [3].

During the course of the review, the FDA's position on the "exclusivity rule" has been questioned many times in comments and objec-

tions filed in response to particular proceedings and in correspondence with the agency. At the time this chapter was submitted, the FDA announced plans to conduct an open public forum on September 29, 1982, where all interested parties could present their views. The forum, a legislative-type administrative hearing [4] was being held in response to a request for a hearing on the tentative final monograph for nighttime sleep-aids [5].

It should be pointed out that FDA does not regulate or have authority over OTC drug advertising, e.g., TV commercials or store displays. That authority rests with the Federal Trade Commission (FTC), which, until recently, had proposed to allow in OTC drug advertising only those indications allowed in the FDA final monographs. The FTC, however, has rejected this across-the-board approach and will instead use a case-by-case review, giving support to the FDA's findings on the safety and effectiveness of OTC drugs in weighing advertising claims for them.

4. *Subpart D, Testing Procedures*, will identify any testing necessary before marketing of the product. For example, the antacid monograph requires each antacid ingredient to be included in the product at a level that contributes at least 25 percent of the total acid-neutralizing capacity of the product and, in addition, the finished product must contain a specific minimum acid-neutralizing capacity at the end of a specified test period. Other proposed monographs contain tests such as preservative testing of antimicrobials and testing to determine the sun protection factor (SPF) of sunscreen products.

III. IMPACT OF THE REVIEW ON MARKETING PRACTICES

Although the review is still probably several years away from completion in terms of final regulations, it has already had an impact on the public's attitude toward self-medication. Publicity given the review by the news media has resulted in a heightened public awareness of OTC drugs and their usefulness in health care.

The review has also generated substantial scientific research that has produced impressive amounts of new data. Approximately one-third of the ingredients reviewed by the panels were shown to be safe and effective for their intended uses. Additional data are being developed on many of the ingredients in the balance in an effort to demonstrate their safety or effectiveness. However, some ingredients have already been dropped from many formulations and will probably eventually disappear completely from the marketplace.

A few ingredients were found to be so unsafe that they were removed before completion of the full rulemaking process. In these special cases, the panels' recommendations were published as the ag-

ency's proposals to expedite market removal. For example, the antimicrobial ingredient hexachlorophene was removed from the OTC drug
market in 1972 because of potential neurologic toxicity. The drug is
still available, but by prescription only. In 1975, other antimicrobials,
including tribromsalan and similar halogenated salicylanilides that were
found to be photosensitizing ingredients, were removed from the market;
and in 1977 zirconium, widely used in antiperspirants, was removed
from "aerosolized" drug and cosmetic products because of the potential
for particulate zirconium to cause granuloma formation in the lungs.

It becomes apparent that, since all drugs that have been marketed
OTC cannot meet the requirement of demonstrated safety and effectiveness for their labeled use, the future holds an OTC drug market
with fewer ingredients. However, the number of OTC drug ingredients
is being augmented in one not altogether expected way. Several panels
have recommended changing the marketing status of particular ingredients from prescription use to OTC drug availability.

Prior to passage of the 1951 Durham-Humphrey Amendment to the
U.S. Food, Drug, and Cosmetic Act, there was no clear-cut distinction
between prescription and OTC drugs. This amendment requires that
drugs that cannot be used safely without medical supervision be dispensed only on the prescription of a practitioner licensed by law to
administer them. Such drugs include drugs that are habit-forming,
drugs that are toxic or have potential for harmful effects, and drugs
limited by approved new drug applications to use under the supervision of a practitioner licensed by law to prescribe them. In effect,
all other drugs in the marketplace are OTC. Panels that have recommended changing the marketing status of ingredients from prescription
to OTC thereby judged that these ingredients could be safely used by
consumers in self-treatment without professional supervision.

The FDA's policy with respect to such recommendations has been
to permit the marketing change to take place upon publication of the
panel report unless it found compelling reason to dissent. It is important to remember that these reports merely represent recommendations
to the agency. Unfortunately, it takes many months or even years to
establish final regulations. Where the consumer can clearly benefit,
the agency has allowed prescription-to-OTC drug switches at this
earlier stage.

Even when the agency does not dissent from the OTC marketing of
prescription ingredients during the rulemaking process, it should be
recognized that it does not reach a final decision until publication of
the final monograph. Should new data become available indicating that
OTC use of these drugs is not appropriate, their marketing status
could revert to prescription only. The reverse situation may also exist in which the agency may initially disagree with a panel recommendation but ultimately conclude that the ingredient may be marketed
over-the-counter. In any event, only the final regulation will establish
each drug's marketing status.

As of May 1982, panel reports published in the *Federal Register* recommended that the active ingredients shown in Table 4 be switched from prescription to over-the-counter drug status.

IV. COSMETIC OR DRUG?

The drug or cosmetic status of an OTC product has been one of the most perplexing and complex issues of the review. The Federal Food, Drug, and Cosmetic Act defines the meaning of a drug and of a cosmetic in sections 201(g) and 201(i), respectively. Drugs and cosmetics seem to be clearly separate entities. According to statutory defintion, a "drug" is something intended to cure, treat, or prevent a disease, or, without reference to disease, something intended to affect the structure or function of the body. On the other hand, a "cosmetic" is something intended to be applied to the body for cleansing, beautifying, promoting attractiveness, or altering the appearance. The intended use of a product, therefore, determines whether the product is a "drug," a "cosmetic," or both. This intended use may be inferred from the product's labeling, promotional material, advertising, and any other relevant source [6].

In issuing the procedural regulations governing the OTC drug review, the agency stated that "any product for which only cosmetic claims are made and which is therefore not a drug will not be reviewed." When the advisory panels were asked to review drug products with both drug and cosmetic claims, the agency stated that the panels were charged with reviewing the safety and effectiveness of the active ingredients in drug products and not with reviewing the safety and effectiveness of these same ingredients used in products for cosmetic purposes. Upon review of the products submitted to the advisory panels, many were found to have labeling that also contained cosmetic claims. The 1962 amendments to the act [7] added the provision that a cosmetic product could be subject to the new drug provisions if the product was also a drug. Therefore, the conclusions of a panel with respect to an ingredient's drug use could be utilized to determine whether the ingredient should continue to be used in cosmetic products. For example, as mentioned above, zirconium in aerosolized containers was declared unsafe for use in OTC drug products because of the potential for aerosolized zirconium to cause lung disease. This ban was extended to cosmetics as well.

The regulation of drugs is much stricter than the standards applied to cosmetics. Unlike cosmetics, drugs must be shown to be effective for their labeled uses. Under existing law, there are no equivalent premarketing requirements for cosmetics. Although the law prohibits adulteration or misbranding of cosmetics, they are not required to be produced in compliance with good manufacturing practice regulations, such as those established for drugs to help assure safety, potency, and purity. Cosmetics manufacturers are not required to register themselves and their products with FDA as are drug manufacturers.

TABLE 4   Prescription Drug to OTC Drug Switches Recommended by Panels in Published Reports

Panel report	Active ingredient	Indication
Anorectal (published 5/27/80)	Epinephrine hyrochloride, 100–200 µg aqueous solution	Vasoconstrictor
	Phenylephrine hydrochloride, 0.5 mg aqueous solution	Vasoconstrictor
	Ephedrine sulfate, 2–25% aqueous solution	Vasoconstrictor
Anthelmintic (published 10/3/80)	Pyrantel pamoate (oral), 11 mg/kg	Anthelmintic (for pinworms)
Anticaries (published 3/28/80)	Sodium fluoride rinse (0.05%)	Anticaries
	Stannous fluroide rinse (0.1%)	Anticaries
	Stannous fluoride gel (0.4%)	Anticaries
	Acidulated phosphate fluoride rinse (0.02% fluoride)	Anticaries
Antifungal (published 3/23/82)	Haloprogin (1%)	Antifungal, anticandidal[a]
	Miconazole nitrate (2%)	Antifungal, anticandidal[a]
	Nystatin[a] 100,000 µ/g	Anticandidal[a]
	Hydrocortisone[a] or hydrocortisone acetate[a] (0.5–1%) combined with an antifungal	Anti-inflammatory-antifungal combination
Cold, cough, allergy, bronchodilator and antiasthmatic (published 9/9/76)	Brompheniramine maleate (oral) (4 mg/4–6 hrs)	Antihistamine
	Chlorpheniramine maleate (oral) (4 mg/4–6 hrs)	Antihistamine
	Diphenhydramine hydrochloride[a] (oral) (25–50 mg/4–6 hrs)	Antihistamine
	Doxylamine succinate[a] (oral) (7.5–12.5 mg/4–6 hrs)	Antihistamine
	Promethazine hydrochloride[a] (oral) (6.25–12.5 mg/8–12 hrs)	Antihistamine
	Diphenhydramine hydrochloride[a] (oral) (25 mg/r hrs)	Antitussive
	Methoxyphenamine hydrochloride (oral) (100 mg/4–6 hrs)	Bronchodilator

	Theophylline preparations[a] (oral)	Bronchodilator
	Aminophylline	
	Theophylline calcium salicylate	
	Theophylline sodium glycinate (theophylline equivalent 100—200 mg/6 hrs)	
	Oxymetazonline hydrochloride (topical), 0.5% aqueous solution	Nasal decongestant
	Pseudoephedrine hydrochloride (oral), 60 mg/4 hrs	Nasal decongestant
	Pseudoephedrine sulfate (oral), 60 mg/4 hrs	Nasal decongestant
	Xylometazoline hydrochloride (topical), 0.1% aqueous solution	Nasal decongestant
External anlgesic (published 12/4/79)	Hydrocortisone (topical) 0.25—0.5%	Anti-inflammatory
	Hydrocortisone acetate (topical), 0.25—0.5%	Anti-inflammatory
Nighttime sleep aid (published 12/8/75)	Diphenhydramine hydrocloride[b] (oral), 50 mg	Nighttime sleep aid
	Diphenhydramine monocitrate[b] (oral), 76 mg	Nighttime sleep aid
	Doxylamine succinate (oral), 25—50 mg	Nighttime sleep aid
	Phenyltoloxamine[a] dihydrogen citrate[a] (oral), 100—200 mg	Nighttime sleep aid
Oral cavity (published 5/25/82)	Dyclonine hydrochloride, 0.05—0.10% concentration in the form of a rinse mouthwash, gargle, or in spray not more than 3 to 4 times daily; 0.05—0.10% concentration in the form of a lozenge (equivalent to 1.0—3.0 mg/lozenge) every 2 hrs if necessary	Anesthetic/analgesic

[a] The agency dissented from OTC marketing at the time of publication.
[b] The agency originally dissented, but recommended OTC use on 4/23/82.

Nor are cosmetics manufacturers required to undergo inspection as
frequently as drug manufacturers. Cosmetics are not limited to specific ingredients as OTC drugs will be (with respect to active ingredients) when the results of the FDA's OTC drug review are fully
implemented in the marketplace.

In reviewing panel recommendations, it is worth noting that the
Antimicrobial I Panel was concerned that soaps containing antimicrobial
ingredients for which only deodorant claims are made may be legally
classified as cosmetics rather than drugs and therefore would not be
subjected to the same safety and effectiveness requirements that would
be imposed if these same products made drug claims. To label these
products as cosmetics was not the concern of the panel, provided that
adequate safety and effectiveness were demonstrated. If this were not
the case, the panel recommended that the agency initiate the necessary
actions to reclassify their legal status from cosmetics to drugs. At the
Tentative Final Monograph (TFM) stage in the agency's proposal, the
FDA indicated that safety testing would not be required for cosmetic
products containing antimicrobial ingredients that the panel had placed
in Category III for safety. The agency had taken the position that the
claim determined the drug or cosmetic status of the product, not the
ingredient. However, the agency further stated that if the antimicrobial ingredient was finally placed in Category II for safety as a drug,
separate action would be taken to ban its use in cosmetics. It should
be further noted that the TFM provided additional labeling statements
for antimicrobial soaps. All products would be required to contain a
statement of the indications for use under the heading "Indications"
that were limited to one or more of the following phrases: "Antimicrobial soap," "antibacterial soap," "antibacterial." In addition, the
labeling could also contain the phrases "reduces odor" and "deodorant
soap," provided that such phrases were neither placed in direct conjunction with information required to appear in the labeling nor occupy
labeling space with greater prominence or conspicuousness than the
required information. In view of recent developments, it is quite
likely that the amended TFM, which will be published in the near
future, will include only drug claims for the active ingredients.

The Antiperspirant Panel accepted that deodorants are legally
classified as cosmetics, whereas antiperspirants are drugs. Deodorants
use one odor to mask another, but antiperspirants actually affect a
function of the body. The panel recognized that antiperspirant ingredients may also exert a deodorant effect because these ingredients
have antimicrobial activity that could inhibit the bacteria that contribute to body odor. However, the panel made no specific recommendations regarding regulating deodorant claims on antiperspirant products,
nor was it concerned that a product containing an antiperspirant ingredient might be promoted only as a deodorant. In other words, that

panel adhered strictly to its charge and made recommendations only on the drug aspects of the products reviewed.

The panel reviewing sunscreens concluded that, regardless of the claims made, products containing sunscreen ingredients should be regarded as drugs. It was the panel's view that the ingredient determined the drug or cosmetic status of the product and not the claim. The panel recommended that claims such as "promotes suntan," "increases your ability to achieve a tan," "rapid tanning, " and so on, be placed in Category II even though these claims have traditionally been considered cosmetic claims. The panel did not place in Category I such claims as "permits tanning and reduces chance of sunburning," and "prevents sunburn and limits tanning." Whether a suntan lotion containing an active sunscreen ingredient is a drug or cosmetic has yet to be fully resolved. As far back as 1940 a trade correspondence addressed the issue, and the conclusion at that time was that products which refer to sunburn or any other condition of disease or injury are drugs, but products promoted exclusively for the production of an even tan are cosmetics. This issue came to the surface again in 1976 when a law firm questioned the legal status of a suntan product containing a sunscreen ingredient (cinoxate) but that was promoted solely as a suntan aid. This problem was deliberated for some time, resulting in a letter from the agency to the law firm explaining that, even though the 1940 trade correspondence was the latest formal agency position, the agency has now taken the position that products containing sunscreen ingredients would be considered drugs even though only cosmetic claims are made. At this point the determination of the drug or cosmetic status of a product has begun to shift from the claim to the ingredient. The agency, as yet, has not published the tentative final sunscreen monograph. However, there has been no change in policy formally announced, and the enforcement posture outlined in the 1940 trade correspondence remains.

In a final example, the Oral Cavity Panel's review of antimicrobial mouthwash product ingredients, it was recognized that two different groups of mouthwash products currently exist, one promoted primarily for cosmetic and cleansing purposes and another promoted for relief of symptoms of sore mouth and sore throat. Early in the panel's evaluation of these ingredients, the agency advised the panel that any claim for cleansing would be a cosmetic claim and any claim for killing germs in the mouth would generally be considered a drug claim, as such a claim implies a therapeutic effect. The exception, however, would be the claim "kills germs that cause mouth odor," since the end result is a cosmetic one. The panel considered these views and proceeded to conclude that claims for suppression or elimination of mouth odor using a pharmacologically active ingredient, that is, an antimicrobial ingredient, are Category II drug claims. The panel further recommended that antimicrobial ingredients should not be used to achieve a cosmetic effect. It should be pointed out that the panel was not opposed to products containing strictly cosmetic ingredients to be used for mouth freshening or suppression of mouth odor, provided that there was no implication that such a product is to be used for some therapeutic effect.

V. A FINAL WORD

So far in the course of the review, the advisory panels have dealt with the drug-cosmetic issue on an ad hoc basis with respect to the ingredients under their review. The agency is in the process of reviewing the panel's recommendations, and the balance of the review will undoubtedly provide the opportunity to resolve the drug-cosmetic issue. Any predictions at this time would be of little value. Several issues still need to be resolved within the agency before formal opinions are expressed in projected *Federal Register* documents for further comment.

The threshold question that must be answered in finalizing monographs is a determination of the criteria to be used. Will the *ingredient*, the *claim*, or *both* be determining factors? Products containing basically inert ingredients for which questions of safety and effectiveness would not necessarily arise may be classified primarily on the claims made. For example, products containing talc, cocoa butter, petrolatum, and so on, would be considered cosmetics unless some therapeutic claim is made. For products containing pharmacologically active ingredients, the *ingredient* and its concentration in the product may be the major determining factor as to whether the product is a drug or a cosmetic. Using this approach, all products containing antimicrobials (in higher than preservative-level concentrations), sunscreens, antiperspirants, skin bleaching ingredients, hormones, and so on, would be considered drugs regardless of the claims made for the product and thereby be subject to monograph standards. This would not preclude the use of cosmetic labeling on these products as long as the labeling is not included in any of the labeling required by the monograph. However, for a product to contain such ingredients at pharmacologically active concentrations, making only cosmetic claims for the ingredients could be construed as misleading and could even be considered fraudulent.

It is obvious that the time is appropriate for resolving these issues. We hope the next few years will provide the final word.

REFERENCES

1. *Cutler v. Kennedy*, 475 F. Supp. 838 (D.D.C. 1979).
2. *Fed. Reg. 46*:47730 (September 29, 1981).

3. 21 Code of Federal Regulations (CFR) 330.10(a)(12).
4. Under 21 CFR 15.
5. *Fed. Reg.*, *43*:25544 (June 13, 1978).
6. See, e.g., *National Nutritional Foods Association v. Mathews*, 557 F. 2d 325, 334 (2d Cir. 1977).
7. 21 U.S. Code (U.S.C.) 359, section 509.

6

REGULATION OF ADVERTISING BY THE FEDERAL TRADE COMMISSION

THOMAS J. DONEGAN, JR.

Hyman, Phelps & McNamara, P.C., Washington, D.C.

I. INTRODUCTION

For companies engaged in the marketing of foods, drugs, cosmetics, or medical devices, the Food and Drug Administration (FDA) is clearly the most critical agency on the regulatory map in terms of premarketing approval of the product for safety and efficacy where necessary and scrutiny of the claims that will be made in labeling for the product. The powers of the FDA and the criteria it applies in regulating cosmetic products are fully described in Chap. 1.

However, the FDA is only the beginning of the inquiry. Another federal agency that can have a critical impact on the marketing of a food, drug, cosmetic, or medical device is the Federal Trade Commission (FTC). At times in its history since it was established in 1914, the FTC has been extraordinarily active in monitoring and regulating the advertising and other marketing practices utilized in the sale of consumer products. Because of safety and efficacy concerns that inevitably accompany the use of such products, the FTC has always paid particular concern to foods, drugs, cosmetics, and medical devices. In this chapter, we will review the authority of the FTC to regulate the marketing of cosmetics and drugs and medical devices with cosmetic uses, the procedures under which the FTC takes action, the division of authority over these products between the FDA and FTC, the standards that the FTC has established for the advertising of those products, the changing nature of FTC activity in this area due to political pressures and changes, and the standards that have been adopted by private regulatory and self-regulatory bodies and the courts that also have a significant impact on the advertising of these products.

II. THE FEDERAL TRADE COMMISSION—GENERAL SCOPE OF AUTHORITY

The Federal Trade Commission is an independent federal agency composed of five commissioners and a staff of approximately 1,400 employees located at its Washington headquarters and 10 regional offices around the country. The commissioners are appointed by the President for seven-year terms, subject to Senate approval. No more than three of the commissioners may be members of the same political party. However, the chairmanship of the FTC can be changed for purely political reasons and usually shifts with a change of administration. The commission acts by majority vote.

The commission's authority over advertising and other marketing practices is quite broad. Section 5 of the Federal Trade Commission Act (FTC Act) prohibits "unfair methods of competition in or affecting commerce, and unfair or deceptive acts or practices in or affecting commerce."[1] In the past, it has been left to the commission to further define those acts that will be considered deceptive or unfair through

individual cases or rulemaking proceedings. Current efforts to legislatively define "unfairness" and "deception" and to restrict the ability
of the commission to initiate cases or rulemaking proceedings may
have a significant effect in reducing the ability of the FTC to question the appropriateness of certain advertising or marketing practices unless it is able to challenge them under a traditional definition of deception. [2]

In addition, section 12 of the FTC Act prohibits false advertising
for foods, drugs, devices, and cosmetics. [3] Section 15 of the act
defines "false advertisement" to mean an advertisement, other than
labeling, that is misleading in a material respect. In determining what
is misleading, the commission must look not only at what is said but
also what is not said but should be stated in the advertisement. An
advertisement is considered false if it does not reveal

(1) facts which are material in light of representations in the
advertisement, or
(2) facts which are material regarding possible consequences
resulting from customary and usual use of the product or under
conditions prescribed in the advertisement. [4]

The failure to disclose a fact that is material to a buyer's purchase
decision is also a clear violation of the more general provisions of section 5 of the FTC Act, even if the advertisement does not affirmatively
state any false facts. For example, in *Simeon Management Corp. v.
FTC*, [5] the advertiser of a weight-control program was required to disclose that a prescription drug used as part of the program had not
been approved by the Food and Drug Administration for safety and
efficacy. The court stated:

The Commission found that the advertisement could reasonably lead
consumers to believe that the claim of safety and effectiveness are
(sic) based on a determination by the appropriate administrative
agency. In view of the currently pervasive level of governmental
regulation, particularly in the medical field, we cannot say that
this determination is unreasonable, arbitrary, capricious, or an
abuse of discretion. [6]

Similarly, an advertisement for a cosmetic that makes no affirmatively false statement could nevertheless be considered a false advertisement in violation of sections 5 and 12 if the cosmetic produced harmful side effects in significant numbers of people and the advertisement
did not disclose that fact.

One of the FTC's most powerful tools in preventing deceptive advertising is its ability to require that an advertiser have adequate
substantiation for a claim at the time that claim is first made to consumers. A program to ensure that advertisers have such substantiation
has been pursued since this doctrine was established by the commission

in the *Pfizer* case.[7] The type, quantity, and quality of substantiation required varies depending on the nature of the claim, the product, and the advertiser.[8]

The commission's advertising substantiation requirement led to a number of advertising substantiation orders directed to specified claims made by certain companies in an industry targeted for investigation. The compulsory orders included demands for substantiating data for advertising claims for shampoos, deodorants and antiperspirants, toothpaste, denture adhesives and cleansers, and acne products.[9] Information submitted was reviewed by the FTC staff and expert consultants and placed on the public record for consumers to review.[10] Some cases resulted from these broad inquiries, but the process was often slow and the cases stale by the time they were brought. Substantiation orders aimed at entire industries became less frequent with the advent of industrywide rulemaking proceedings on a large scale at the FTC, but still remain a possible investigative tool.[11] However, in recent years the commission and its staff have more frequently chosen to enforce the requirement that advertisers possess substantiation for claims by bringing individual cases.

The FTC's legal jurisdiction over advertising also extends to labels and labeling, items also within the jurisdiction of the Food and Drug Administration. The two agencies have executed a liaison agreement to establish the procedure for apportioning overlapping jurisdiction. In essence, the agreement provides that the FTC will have primary jurisdiction over advertising of food, drugs (except prescription drugs), cosmetics, and medical devices, and that the FDA has primary jurisdiction over prescription drug advertising and labeling for food, drugs, cosmetics, and medical devices. The agreement provides, among other things, for discussions between the agencies when the same or similar claims are found in both labeling and advertising.[12] In one recent rulemaking proceeding involving protein supplement products, the agencies decided it was more efficient for one agency— the FTC—to promulgate regulations covering both advertising and labeling for these products.[13]

III. TYPES OF FTC ACTION

A. Case-by-Case Law Enforcement

One way in which the FTC can proceed when confronted with false or misleading advertising advertisements or advertising campaigns by individual companies is to initiate an investigation that will usually focus on the potential liability of both the advertiser and its advertising agency. Generally, an informal investigation involving a maximum of 100 staff hours can be conducted with a minimum of higher-level approval and with no involvement of the commission itself. The usual procedure is for the staff to request submission of all material in the advertiser and advertising agency files bearing on substantiation for the questioned claims. At this stage of the investigation, they must depend on the voluntary cooperation of those to whom the requests are made. This is a stage of the inquiry at which misunderstandings by the FTC staff can be clarified or minor problems in the advertisement can be corrected and formal government action avoided. However, it is also a juncture where great care must be exercised by companies dealing with the FTC. The materials voluntarily provided to the FTC staff at this time can be used as evidence against the companies in a later proceeding.

If a company refuses a request for voluntary submission of sub-stantiation, or, for other reasons, the staff determines that it is necessary to force the companies under investigation to submit sub-stantiation for the advertising claims or other materials, the staff must ask the commission to approve the investigation and a Civil Investiga-tive Demand must be approved by a commissioner. This is, in effect, a subpoena for documents or testimony and, subject to intermediate rulings by the commission on motions to quash or limit, can be enforced in federal court if the recipient does not comply. [14]

If, after a review of the materials collected during the investiga-tion, the staff determines that a significant violation of law has oc-curred and that the commission can provide a cost-effective remedy for consumers by intervening, they may recommend to the commission that a complaint be issued. however, prior to the issuance of a com-plaint, the matter may be resolved by an agreement between the parties called a consent order. The consent order must be tentatively accepted by the commission and placed in the public record for comment before it is finally accepted by the commission and takes effect. Although it is not an admission by the advertiser that the law has been violated, a consent order has the same effect as a final order entered by the com-mission following administrative litigation. In both instances, violation of an order may result in court-imposed fines of up to $10,000 per violation. A consent order will usually prohibit certain practices that have been questioned in the FTC investigation, may prohibit other prospective law violations of a similar type, and may include any addi-tional relief on which the parties can agree—including restitution, corrective advertising, other forms of affirmative disclosures, or other remedies.

If an agreement between the advertiser, the advertising agency, and the commission cannot be reached, the commission may issue a complaint and the issues will be litigated before an administrative law judge. (Of course, with the judge's consent, the case can always be settled at any time during litigation by submission of a proposed order to the commission and approval of that order by the commission.) Once a complaint is issued, and for the entire time the matter is in litigation, the FTC staff and other parties may communicate with the commission and the administrative law judge only on the public record. If, after litigation, the judge finds that a violation has occurred, he or she may issue an order prohibiting certain practices or imposing other remedies. The judge may not, however, impose penalties such as monetary fines but may only take the action necessary to restore the marketplace to its previolation condition and to prevent future violations. There is a wide latitude for selecting a remedy, as long as the remedy chosen has a reasonable relation to the unlawful practices found to exist.[15] The judge's decision is subject to review by the commission. If it decides against the advertiser, the case may be appealed to the U.S. Court of Appeals and, possibly, the U.S. Supreme Court.

Remedies available to the administrative law judge and the commission in litigated cases range from the traditional cease-and-desist order to more far-reaching remedies such as corrective advertising. A cease-and-desist order generally prohibits the claims found to violate the law and similar claims for the category of products involved in the litigation. On occasion, the commission has extended an order to apply to all products of a particular company. Extending the scope of an order beyond the exact practice and product involved in a case is known as "fencing in." This practice is based on the theory that once a company has shown to have violated the law, some protection for consumers is necessary to ensure that the advertiser does not engage in similarly deceptive practices in related product categories.[16] However, the courts can be expected to scrutinize carefully the extension of any order beyond the product and practice involved in the case, and orders extending to all products manufactured by a company are likely to be rare.[17]

A cease-and-desist order may prohibit a claim completely, or it may prohibit the claim unless the advertiser has adequate substantiation for the claim in its possession at the time it is first made. In the food and drug area, the commission has often spelled out the type of substantiation necessary to justify a certain claim. For example, *Karr Preventative Medical Products*,[18] a consent order against a manufacturer of acne remedies, specified that two well-controlled clinical studies must be conducted to support claims that the product will treat acne effectively or will be more effective in the treatment of acne than other products. Similarly, in a case involving advertising of over-the-counter analgesics, a commission determination that a company must have two well-controlled clinical studies to support any claim of superiority in efficacy or freedom from side effects was upheld.[19] That order sets out the standards for those tests with some specificity. A cease-and-desist order also may go so far as to prevent a product from being marketed under a particular trade name if that name is deceptive and no other remedy will cure the deception.[20] However, the commission and the courts will go to considerable lengths to find another acceptable remedy before they will ban the use of a particular trade name.[21]

The commission may also order affirmative disclosures if certain claims made in advertising would be deceptive without those disclosures. Furthermore, if it is necessary to dispel lingering deception from an advertising campaign, the commission may go one step further and order corrective advertising. Corrective advertising differs from a traditional affirmative disclosure remedy in that the commission may direct that corrective advertising be disseminated regardless of whether the offending claims continue to be made. In contrast, a traditional affirmative disclosure remedy is generally triggered by a particular claim and must be included in the advertisement only when that claim is made to cure the potential deception.

The FTC's authority to require corrective advertising was affirmed by the U.S. Court of Appeals for the D.C. Circuit in *Warner-Lambert v. Federal Trade Commission.*[22] In that case, after finding the claim that Listerene was an effective cold remedy to be false, the commission ordered Warner-Lambert to stop advertising for Listerene unless each advertisement clearly and conspicuously disclosed that "contrary to prior advertising, Listerene will not help or prevent colds or sore throats or lessen their severity." The corrective advertising requirement applied to the next $10 million of advertising for Listerene. With the exception of the preamble to the disclosure, "contrary to prior advertising. . . ," which the Court of Appeals found unnecessary and unduly harsh, the court upheld the corrective advertising requirement in the face of a challenge to FTC authority and charges that the corrective advertising requirement violated the First Amendment. In a later case, an administrative law judge ordered American Home Products Corporation to run corrective advertising in the amount of the average annual advertising budget for Anacin from April 1968 to April 1973 (approximately $24 million) to inform consumers that "Anacin is not a tension reliever."[23] However, this portion of the order was eliminated by the commission on appeal because it determined that the claims had not been made for a long period of time and therefore corrective advertising was not necessary.[24]

Under section 19 of the FTC Act, the commission may seek direct consumer redress in U.S. District Court after a final cease-and-desist order has been issued in a commission case.[25] Such action can result in advertisers being required to return money to consumers and may be ordered by the court if a deceptive advertisement is one that a reasonable person would have known under the circumstances was dishonest or fraudulent. This is a much more difficult standard than the one that must be met to prove deceptive or unfair advertising in violation of section 5 of the FTC Act, where it must only be demonstrated that the challenged advertising had the capacity and tendency to deceive consumers. In the *Karr* consent order, the respondents, including singer Pat Boone, who endorsed the product, set up a restitution fund totaling $250,000 for the benefit of consumers who paid approximately $10 per bottle for the acne remedy involved in that case.[26] The respondents placed advertisements in various publications to alert consumers to their refund rights under the order.

Finally, it is important to note that the commission may authorize its staff to seek an injunction against false or misleading advertising in U.S. District Court under section 13 of the FTC Act.[27] The commission is particularly likely to consider such action in any case it believes to present a health or safety issue, thus making it easier to convince the court that there is a probability of immediate and irreparable harm if the advertising is permitted to continue. If an injunction is granted by the court, the questioned advertising would be prohibited or restricted pending trial of the issues before an administrative law judge.

B. Rulemaking

If a commission believes that particular advertising problems are prevalent in a given industry, it has the option of proposing a trade regulation rule to deal with those problems. This avoids the necessity of bringing several cases against different companies involving the same issues and has the advantage of putting the industry on notice for the future as to what practices will be considered a violation of the law. This is a quasi-legislative process, and the rules define conduct that is considered to be a violation of the FTC Act. Once a rule is in place, advertisers who violate the terms of that regulation can be fined up to $10,000 per violation, just as if they had violated an order in a litigated case. There are a variety of procedural requirements and safeguards for trade-regulation rulemaking proceedings. The process involves numerous opportunities for interested parties to make their views on the proposed regulations known to the commission and its staff. As a result, rulemaking proceedings tend to be very time-consuming, and additional safeguards contained in the FTC Improvements Act of 1980 are likely to make the rule-making process even more complex and lengthy in the future.[28] Simplified, the stages in the rule-making proceedings are as follows:

1. The commission publishes an advanced notice of proposed rule-making setting forth its objectives and possible regulatory alternatives under consideration, and it invites responses and suggestions from interested parties. This advanced notice of proposed rulemaking must be forwarded to the Senate Committee on Commerce, Science, and Transportation and the House Committee on Interstate and Foreign Commerce.
2. Thirty days before publication of a notice of proposed rule-making, the commission forwards that notice to the same two Congressional committees.

3. The commission issues a notice of proposed rulemaking, which is followed by a period of written comment, hearings, written or oral rebuttal, a staff report, a presiding officer's report, additional public comment, final staff recommendations to the commission, public meetings at which outside parties may make oral and written presentations to the commission and at which the staff publicly discusses its recommendations to the commission, and, finally, a commission decision. In the past this process has consumed a period of several years in many rule-making proceedings.
4. Within 60 days of promulgation a rule may be challenged by an interested person in the U.S. Court of Appeals.

As can be seen, the FTC must make a very substantial commitment of staff resources and time to undertake a rulemaking proceeding. In addition, because of Congressional review and the substantial lobbying efforts that frequently accompany a rule aimed at an entire industry, the commission incurs substantial potential political liability in undertaking a rulemaking proceeding. Although in recent years the commission has undertaken rulemaking on a large scale, it is likely in the future to use this process much more selectively and to use case-by-case litigation much more often as its primary enforcement tool.

C. Effect of Decisions in Litigated Cases on Nonparty Advertisers

In the past, advertisers have followed FTC decisions in cases involving other companies and have generally applied relevant principles from those cases to their own future conduct. However, those cases had no binding effect on nonparties. Even though the exact same practice was involved, it would be necessary for the commission to bring and prove an entirely new case against a company engaging in conduct that had previously been found to be a violation of the law. Only then could the second company be placed under an order where monetary penalties could be imposed for any future actions that violated the order.

As part of the Magnuson-Moss Warranty—FTC Improvements Act of 1975, Congress amended the FTC Act to give orders in litigated cases a broader application. Section 5(m)(1)(B) of the FTC Act— more popularly known as section 205 of the Magnuson-Moss Act— provides that any person or company with actual knowledge that an act is a violation of any litigated order, even though not directly subject to that order, can be held liable for civil penalties in the amount of up to $10,000 per violation of the order in an action in U.S. District Court.[29]

Although there has been a substantial amount of speculation regarding the constitutionality of section 205 on due process grounds,

a U.S. District Court has upheld the provision in the face of such a
challenge in *U.S.* v. *Braswell*.[30] However, the court made this
finding on the basis of an explicit understanding that a defendant
in such a proceeding has a right to a de novo proceeding. The government had conceded that the defendants could raise lack of actual knowledge, changed circumstances since entry of the previous order, and,
most significantly, the issue of whether the FTC's prior determinations
are legally sustainable interpretations of the FTC Act.[31] It remains to
be seen whether this interpretation of one U.S. District Court will be
generally accepted. It also remains unclear whether section 205 will
become a significant enforcement tool if the U.S. District Court hearing
will, in fact, be such a broad-ranging inquiry.

IV. CASE SELECTION CRITERIA IN FTC ADVERTISING ACTIONS

The policy planning criteria for deceptive and unsubstantiated advertising claims are set forth in a document that was originally developed
by the FTC staff in 1975 and has been followed to varying degrees by
subsequent administrations at the FTC. This document was highlighted
as containing the criteria currently in use by Chairman Miller in a
speech to the American Advertising Federation in December 1981.[32]
The document sets out several general areas of inquiry that will be
examined before a case will be brought. These are set out below with
a discussion of their relative nature and importance.

A. Consumer Interpretations of the Claim

Consumer interpretation of the claim is perhaps the single most critical
factor in an advertising matter at the FTC, since so many advertising
cases turn on the issue of implied claims. Often the only contested
issue in an investigation or case is whether a particular implied claim
is contained in the advertisement. Historically the courts have upheld
the concept that the commissioners have expertise to make the judgment
as to whether an advertisement contains a particular implied claim without reliance on consumer research studies or consumer testimony.[33]
This procedure has often been criticized, since the commissioners are
usually lawyers and occasionally economists and seldom have special
training or experience in interpreting advertisements. This criticism
has intensified recently as the benefit derived from FTC's advertising
cases is carefully examined. Although there may not be a rigid insistence on consumer surveys to support advertising cases involving
implied claims in the future, there is likely to be a marked increase in
the use of empirical data to interpret advertisements in these cases.
Most disputes over the existence of implied claims and advertisements
will most likely be resolved in the future on the basis of a review of
consumer surveys.

B. Scale of the Deception or Lack of Substantiation

The FTC is most likely to invest a portion of its budget in an advertising case if the advertisement reaches a large number of consumers
and is likely to affect those consumers' purchasing decisions in a significant way. Thus, national advertising campaigns on television or
in highly visible print media will attract FTC attention before more
limited campaigns. In addition, the FTC will have a greater interest
in objectively verifiable claims about significant product characteristics
or uses than in subjective claims.

C. Materiality

The materiality criterion addresses the real affect of the alleged deception in the marketplace. Under the law, the staff need only prove that
an advertisement has the tendency and capacity to deceive and need
not show an actual deception or effect on purchase decisions. However, the internal policy analysis of cases at the commission in the
foreseeable future is likely to focus heavily on how many consumers
are likely to have purchased the product as a result of the deceptive
claim and on the nature and extent of the injury they incur as a result of that purchase decision. For example, if a consumer, deceived
into purchasing a household product that costs only $3.00, can deter-
mine from actual use of the product that the claim was deceptive, and
runs no risk of physical harm as a result of being deceived, the commission is unlikely to bring a case to correct that deception, even
though there is some economic harm. Theoretically, if the consumer
is dissatisfied, the market will be self-correcting. However, if the
deception results in the purchase of an automobile costing several
thousand dollars, in the purchase of a drug that is not effective as
represented in the advertisement, or in the purchase of a cosmetic that
may cause physical harm to the consumer, the commission is much more
likely to take action.

D. Adequacy of Corrective Market Forces

The adequacy of corrective market forces ties into the consumer injury
analysis above and is a critical factor. If at relatively low cost the
consumer can discover that the advertising is deceptive, the commission
is not likely to take action. This is based partially on the assumption
that if consumers do not get what they bargained for, they will purchase a different product next time. The commission will also look to
see whether there are other souces of information that will correct the
effect of a deceptive adverstising. Some possible "self-correcting"
sources of information include labels or labeling information for the product, consumer education programs, or a competitor's advertising,
which may correct a deceptive claim.

E. Effect on Flow of Truthful Information

Under the criterion termed effect on flow of truthful information, the commission will look at whether the action being taken, if successful, will reduce the flow of truthful information by making it impossible to make certain types of claims. If a small number of consumers are deceived but a large number of consumers receive useful information from a challenged claim, the commission may be more willing to overlook the deception in favor of the benefits of the claim. For the same reasons, the commission will be reluctant to issue an order that is broader than necessary to stop the deceptive practices out of concern for restricting the flow of truthful and useful information.

F. Deterrence

There are a large number of important questions that fall within the criterion of deterrence. The commission will ask itself questions such as whether the violation is so evident that its failure to act will reduce the agency's credibility and deterrent effect among other industry members. It will look at the significance and breadth of possible relief obtainable in that particular case and the significance of that particular product and advertising campaign in the marketplace as a whole. It will also look at whether a particular case has the right characteristics to establish a rule of law that will have application to other members of a particular industry. One important factor here might be the usefulness of a litigated order as an industrywide standard that could be enforced under section 5(m)(1)(B) of the FTC Act.

G. Law Enforcement Efficiency

The law enforcement efficiency criterion considers whether FTC action is the most efficient way of dealing with a problem or whether it would be more appropriate and efficient for a different federal agency or a state or local agency to handle the problem. In addition, the commission will rely heavily on industry self-regulatory mechanisms to solve many advertising problems and is even likely to consider the possibility of private lawsuits between competitors as a way in which a particular problem might be appropriately resolved.

H. Additional Considerations

Additional considerations include some of the social considerations that in past administrations have been more prominent criteria. For example, the commission will look at whether the claim directly affects a vulnerable group such as children or whether it will ". . . exploit legitimate concerns of a substantial segment of the population. . . ."

In applying these criteria, the commission is currently under heavy pressure to justify its actions in very specific terms. This is a signifi-

cant burden given that advertisers themselves often have difficulty
measuring the impact of a particular advertisement or advertising
campaign on their sales. The commission will be called upon to satisfy
both itself and Congress that the burdens it imposes on an individual
company or an entire industry are justified by benefits to consumers.
The FTC is unlikely to take action in any matter unless there is a clear
demonstration of significant consumer injury and of the ability of the
commission to remedy the problem in an efficient and cost-effective way.

V. FTC ACTIONS AFFECTING THE COSMETICS INDUSTRY

An assessment of the impact of the FTC on the cosmetics industry in
the past and the likely impact of the agency on that industry in the
future depends in large part on how broadly you define cosmetic advertising. To the extent that the term is defined to include only
beauty and personal appearance claims, cosmetic advertising has not
been a particularly active area for the FTC. However, the courts have
clearly supported the FTC's legal authority to take action against such
claims if it desires. Inaction has been the result of policy and case-
selection decisions rather that any uncertainty over legal authority to
take such action.

In a landmark decision, *Charles of the Ritz Dist. Corp. v. FTC*, 34,
the U.S. Court of Appeals for the Second Circuit upheld the commission finding advertising for Rejuvenescence Cream to represent falsely
that the product would restore youth or the appearance of youth to the
skin regardless of condition. The court stated:

> And, while the wise and worldly may well realize the falsity of any
> representations that the present product can roll back the years,
> there remains "that vast multitude" of others who, like Ponce de
> Leon, still seek the fountain of youth. As the commission's expert
> further testified, the average woman, conditioned by talk or mag-
> azines and over the radio of "vitamins, hormones, and God knows
> what," might take "rejuvenescence" to mean that this "is one of
> the modern miracles" and is "something which would actually cause
> her youth to be restored." It is for this reason that the commission
> may "insist upon the most literal truthfulness" in advertisements
> [citation omitted] and should have the discretion, undisturbed by
> the courts, to insist if it chooses, upon a form of advertising clear
> enough so that, in the words of the prophet Isaiah "wayfaring men,
> though fools, shall not err therein."[35]

Although that is still the law, such an attitude is not likely to be
prevalent in the FTC of the 1980s. The commission is much more likely
to view beauty and personal appearance claims in much the same way
that they view taste claims for food products. In essence, such claims
are viewed as very subjective and to fall into an area where the con-

sumer is perfectly capable of assessing the performance of the product and making a determination for him-/or herself as to whether the product works as advertised. If consumers are not satisfied with their appearance after using certain types of cosmetics, they are not likely to repurchase the product. Therefore, the commission and its staff are not likely to foresee sufficient consumer injury in an overzealous beauty claim to warrant the expense of bringing a case against the advertiser.

However, if cosmetic claims are viewed broadly to include those products for which therapeutic claims are made (usually resulting in their classification as drugs or medical devices), then there has been significant commission activity in this area and, under the policy planning criteria discussed above, there is substantial reason to expect commission activity in the future. In addition, many of the drug and medical device cases brought by the commission in the past 10 years are based on theories and have imposed relief that could be equally applicable to products that have cosmetic as well as therapeutic uses. Some of the more recent cases have involved electric razors represented as effective in preventing razor bumps, a skin affliction primarily affecting black persons,[36] an oral irrigating device that was claimed to prevent gum disease,[37] several acne remedies,[38] hair implants and other hair care products, [39] denture cushions, [40] and contraceptives.[41]

Clearly the most prominent signal for FTC action in this area is the threat of significant risk of physical injury when the product is used normally by the consumer. If the threat to safety is perceived to be a real one, commission action seeking disclosure of that risk in advertising is possible. Advertising that does not contain such a disclosure could become a target for an immediate injunction. Furthermore, it is not a sufficient defense to the threat of such action to argue that health risk has not been proven. The fact that there is a lively controversy over the existence of a risk and that some responsible experts believe the risk is significant may be enough to cause the commission to move for a disclosure. If the risk has been fully proven, the FDA most likely will have already acted to take the product off the market or require warnings or the manufacture will have voluntarily taken such action.

In addition there have been three proposed trade regulation rules recently under consideration by the commission that could have some impact on the cosmetic and related industries.

The proposed trade regulation rule for OTC drug advertising was rejected by the commission in 1981 after full hearings and commission consideration.[42] That rule would have prohibited claims in advertising that would have been prohibited in labeling by a final order issued as part of the Food and Drug Administration's OTC drug review. One proposal under consideration would have required advertising claims to be made only in the specific language approved for labeling by the

FDA. The commission finally rejected a rulemaking approach in this area and indicated that it would deal with problems through case-by-case enforcement. Part of the reasoning in rejecting this approach was based on the fact that the FDA's OTC drug review is so far behind its originally projected schedule that the issuance of a rule by the commission at that time would have had little meaning. However, subsequently, there have also been some indications from the commission that the FTC may be less willing in the future to issue regulations that put it in lock-step with the Food and Drug Administration.

A second rule, still under commission consideration, is the antacid warning rule, which may require certain warnings required in labeling for OTC antacid products by the FDA also to be disclosed in advertising.[43] Also under consideration is a general "see the label" type warning in all antacid advertising. Although this proceeding does not have direct impact on cosmetic products, it does emphasize the importance that the commission places on disclosure of side effects or risks. In addition, the antacid rulemaking was originally donceived as the flagship proceeding for a series or rulemaking proceedings dealing with warnings for categories of drug products after final FDA orders establish requirements for warnings on the label for a specific drug category. Such a series of rulemaking proceedings, which might have affected cosmetic products directly, now seems unlikely.

Finally, the commission still has under consideration a comprehensive rulemaking initiated in 1975 involving the marketing and advertising of hearing aids.[44] The primary remedy proposed in this proceeding is a requirement that each hearing aid seller give the consumer a 30-day period in which to cancel the purchase of a hearing aid. The original proposal also contains many specific requirements governing advertising and direct sales practices of hearing aid devices, but most of these provisions seem unlikely to survive in a final regulation.

VI. THE POLITICAL CLIMATE AFFECTING THE FTC

Normally, when discussing the enforcement powers of a federal agency, it is not necessary to become involved in a detailed discussion of the political forces affecting that agency. Generally, as administrations change, subtle changes in enforcement policy occur, but the basic, mission of the agency remains the same. However, the FTC has been the focus of controversy for many years. Once severely criticized as a sleepy and relatively ineffective agency,[45] in the 1970s the FTC became aggressive in its enforcement of its consumer protection mandate and in its efforts to expand the scope of its authority. Much of this activity focused on advertising issues, and often the products involved were foods, drugs, cosmetics, or medical devices. In the first half of that decade, these efforts generally met with favorable responses in

Congress. In 1975, Congress passed the Magnuson-Moss FTC Improvements Act, which firmly established the FTC's rulemaking authority and gave the commission other forms of expanded authority.

However, the latter half of the 1970s proved a different story. There was a negative reaction to the agency's activities by many members of Congress, fueled by intense industry lobbying efforts to curtail several actions by the FTC during this time period. Perhaps most notably, many persons took issue with the commission's initiation in 1978 of a rulemaking proceeding regarding children's advertising. One of the proposed remedies was a complete ban on advertising to young children, an action that was perceived as having unfair and catastrophic consequences for certain industries. It was also seen by many as the inappropriate insertion of government into family decisions regarding what children should be able to see on television. At the same time, the commission had several other proposed trade regulation rules under consideration. These proposals, if passed, would have resulted in comprehensive regulations for advertising and other marketing practices for several product categories. In addition to the proposals discussed previously, these rules covered food, used cars, funerals, mobile homes, credit practices, and many other product categories. Many in Congress felt that the FTC was becoming too intrusive and restrictive. Because it was so visibly active at the time, the FTC became a lightning rod for much of the deregulatory movement that gained strength during the latter part of the Carter administration. In 1980, as a result of all these factors, Congress passed the FTC Improvements Act of 1980, which placed restrictions on the agency's authority.[46]

Deregulation has been pursued with vigor by the Reagan administration. The FTC has been directly affected by the appointment of James C. Miller, III, an economist, and one of President Reagan's strongest advocates of deregulation, as chairman. The chairman has a substantial ability to influence the direction of the commission, which goes well beyond his one vote in formal commission matters. He controls the appointment of major FTC officials such as the directors of the Bureaus of Competition, Consumer Protection, and Economics, who, in turn, have a substantial influence on the case-selection process and the programs that will ultimately be sent to the full commission for consideration. In addition, he is the chief spokesman for the commission with the Congress on both budgetary and substantive issues and in explaining FTC policies to business and consumer constituencies.

Chairman Miller has strongly urged additional legislative action to limit permanently both the unfairness and deception authority of the commission. He has also enforced strict economic analysis of the costs and benefits of all proposed actions within the commission before those actions are formally considered.[47] In addition, he has expressed doubts about the benefits of advertising substantiation, a program that

has gained broad support and compliance from industry.[48] FTC law
enforcement and regulatory activity in the field of advertising is one
of the activities that Dr. Miller and some Congressional critics have
brought into question, and little new action has been taken in this
area in recent years.

Therefore, in analyzing FTC standards and potential liability of an
advertising program to FTC action, it is important to review the latest
legislative changes and internal policy developments. This chapter is
written to reflect the law as it currently exists. However, several
legislative proposals have been considered that could significantly
change the legal standards for commission action.[49] Those amendments to the law, if passed, will result in the incorporation into
the FTC Act itself of certain policy planning criteria currently in effect. Generally, the changes will impose additional burdens of proof
on the government in bringing an advertising case. However, the
volatile nature of the debate surrounding the FTC necessitates a constantly updated analysis of the agency's policy stance and the status
of the current Congressional debate over the FTC in determining
whether certain types of advertising claims are likely to provoke action.

VII. OTHER TYPES OF ADVERTISING REGULATION

The deregulatory atmosphere throughout government in the early 1980s
and the particular controversies surrounding the FTC that seem at
least temporarily to have brought its enforcement efforts to a near
standstill have resulted in a shift in many disputes over allegedly deceptive advertising from the government to private forums. At this
writing, advertisements, particularly comparative advertisements, are
much more likely to be challenged by a private party than by the
government. There are three major forums of national importance
where such activity can occur.

The first involves a hurdle that all television advertisements must
clear. All such advertisements are subject to review by the commercial
clearance authorities of the television networks. Each major commercial
network has a staff that reviews all advertisements submitted for
dissemination on that particular network. These organizations review advertising not only for proper substantiation and for the absence of false or misleading claims but also for taste. Current indications are that the network clearance authorities give advertisements a
rigorous review and that advertisers are often called upon to make
changes or submit additional substantiation before the advertisements
are approved to be shown on network programming.

A second source of advertising review is the National Advertising
Division of the Council of Better Business Bureaus, Inc. (NAD).
This organization evaluates challenges to the truth and accuracy of

national advertising. The challenges are self-initiated or brought by others, usually competitors. If a decision of the NAD is disputed, the National Advertising Review Board (NARB) acts as an appellate body and a five-member panel drawn from advertisers, advertising agencies, and public members decide the appeal.

The NAD is more likely than the FTC to challenge more subjective claims for cosmetic products. For example, the NAD has recently reviewed substantiation for the following claims for fingernail products:

> "Which side is wet nail color, which side is three days old? With Cover Girl Nailslicks, it is almost impossible to tell the wet side from the side that was polished three days ago. Both sides look shiny-wet and new. NailSlicks gives nails lasting strength, lasting protection and lasting shine."[50]

> "Use it twice daily (even over your own nail polish!) and you should see remarkable results within 3 to 4 weeks. As your nails stop peeling, splitting, breaking. And begin to grow stronger, longer and lovelier."[51]

and the following claim for a mascara product:

> "Walking, working, laughing, loafing, dancing, romancing, snuggling, sleeping, 24 hours later . . . your mascara looks fresh."[52]

Under the FTC case-selection criteria, such claims would most likely be deemed unworthy of serious inquiry since, with minimal economic risk, a consumer could decide whether these products lived up to their promise in the advertising claims. The judgment whether the product performs as promised is, in addition, a subjective one that, under FTC guidelines, would best be left to each individual consumer who buys the product.

The NAD and the NARB have had extraordinary success in gaining compliance from advertisers when they have called for changes in particular advertisements or advertising campaigns. Of course, in the past, much of this organization's leverage has come from the ultimate threat of turning an unresolved matter over to the FTC. It remains to be seen if the NAD's effectiveness can be maintained if the FTC does not purse an active role in regulating advertising.

The networks and the NAD are, of course, forums where competitors can level challenges at each other's advertising, and the recent proliferation of comparative advertising claims explicitly naming competitors or competing products has accelerated this trend. However, a third private remedy available to companies who feel that they are the victims of false and misleading advertising by their competitors is to bring a private lawsuit against the competitor in a federal district court under section 43(a) of the Lanham Trademark Act.[53] Such a suit is available against both explicit and implied claims.[54] The suit may be brought by a competitor or a party who has suffered commercial harm as a result of the claim.[55] Only a defendant's false or misleading claim about its own product can be the subject of such a suit.[56] The remedy most frequently sought is a preliminary injunction, which can, in some cases, be obtained in a matter of weeks. To obtain a preliminary injunction under the Lanham Act, it is necessary to show that the advertisement is false or misleading and that there is a likelihood of irreparable harm to the plaintiff if the advertising is allowed to continue.[57] Monetary damages are also available in appropriate cases, but are usually not pursued because of a much more difficult burden of proof in showing that actual losses have resulted from a false or misleading advertisement.[58] In most cases the only real goal is to bring an end to the offending advertising campaign.

If the current trend of government inactivity continues, the importance of each of these remedies in the private sector is likely to increase. Even if the FTC does become more active in the future in its law enforcement efforts against false and misleading advertising, efficiency and resource constraints are likely to result in its reliance on the networks and the NAD to police advertising claims and resolve most problems. Similarly, there are certain situations where companies are likely to perceive potential competitive harm to be so significant that they will be unwilling to wait for the rather slow FTC processes to work. In those cases, they may take matters into their own hands by bringing a suit under section 43(a) of the Lanham Act. Therefore, in constructing an advertising review program within a company, it is wise not only to ensure compliance with FTC standards but also to minimize the risks that the company's advertising claims can be successfully challenged in other forums.

NOTES

1   15 U.S. Code (U.S.C.) §45.
2   Chairman Miller and others have proposed amendments of the FTC Act to define "deceptive" and "unfair" practices.
3   15 U.S.C. §52.
4   15 U.S.C. §55.
5   579 F.2d 1137 (9th Cir. 1978).
6   *Id.* at 1146.
7   81 Federal Trade Commission (F.T.C.) 23 (1972).
8   National Dynamics Corp., 82 F.T.C. 488 (1973), aff'd. and remanded on other grounds 492 F.2d 1333 (2d Cir. 1973), cert. denied 419 U.S. 993 (1974), reissued 85 F.T.C. 391 (1976).
9   See 2 Trade Reg. Rep. (CCH), par. 7573, for a list of advertising substantiation orders from 1971 through early 1974.
10 It was originally believed that consumers would make use of the substantiation materials that were publicly available in making

purchase decisions. However, few consumers actually looked at
the materials.

11 Debate over the effectiveness of industry-wide advertising substantiation orders has centered on their efficiency. They impose
a substantial burden on the companies affected, and the commission
staff is often ill equipped to analyze expeditiously the large quantities of material submitted. More narrowly targeted investigations
of suspected deceptive advertising would avoid large and unnecessary costs to an entire industry.

12 See Updated FTC-FDA Liaison Agreement—Advertising of Over-
the-Counter Drugs, *36 Fed. Reg.*, 18539 (September 16, 1971),
reprinted in 3 Trade Reg. Rep. (CCH), par. 9851.

13 *40 Fed. Reg.*, 41,444 (1975).

14 See 15 U.S.C. §57b-1 for statutory requirements for issuance of
civil investigative demands.

15 *Jacob Siegel Co. v. FTC*, 327 U.S. 608 (1946).

16 *FTC v. Colgate-Palmolive Co.*, 380 U.S. 374, 390 (1965); *FTC v.
National Lead Co.*, 352 U.S. 419, 431 (1957); *FTC v. Ruberoid Co.*,
343 U.S. 470, 473 (1952).

17 See *FTC v. Colgate-Palmolive Co.*, *Supra.* at n. 16.; *Litton Industries, Inc. v. FTC*, 676 F.2d 364 (9th Cir. 1982).

18 94 F.T.C. 1080 (1979).

19 *American Home Products Corp. v. FTC*, 695 F2d 681 (3d Cir.
1982). This order was later modified to make it consistent with
Commission decisions in other cases involving over-the-counter
analgesics. The modified order still contains the two-clinical test
standard but leaves open a possibility that something less may be
adequate substantiation, *3 Trade Reg. Rep.* (CCH), par. 22,110.

20 *U.S. v. Algoma Lumber Co.*, 291 U.S. 67 (1934).

21 *Jacob Siegel Co. v. FTC* 327 U.S. 608 (1946); *FTC v. Royal
Milling Co.*, 288 U.S. 212 (1933).

22 *Warner-Lambert Co. v. FTC.*, 562 F.2d 749 (D.C. Ar. 1977),
cert. denied 435 U.S 950 (1978).

23 *American Home Products Corp.*, Docket No. 8918, order of administrative law judge, September 1, 1978, Trade Reg. Rep. (CCH),
1976—1979 Transfer Binder, par. 21,465.

24 *American Home Products Corp.*, final order to cease and desist,
September 9, 1981, 3 Trade Reg. Rep. (CCH), par. 21,874.

25 15 U.S.C. §57b.

26 See *Karr Preventative Medical Products, Inc.*, 94 F.T.C 1080
(1979); *National Media Group*, 94 F.T.C 1096 (1979); *Cooga Mooga,
Inc.*, 92 F.T.C. 310 (1978), modified October 15, 1981, 3 Trade
Reg. Rep. (CCH), par. 21,880.

27 15 U.S.C. §53.

28 The legislative veto for FTC rules, a part of this statute, has
been held to be unconstitutional. *Consumers Union of U.S. Inc.*

v. FTC, 691 F.2d 575 (D.C. Cir. 1982), aff'd *United States Senate v. FTC*, 103 S.Ct. 3556 (1983); see also *INS v. Chadha*, 103 S.Ct. 2764 (1983).

29 15 U.S.C. 45(m)(1)(B), Public Law 93-637, sec. 205(a).

30 *U.S. v. Braswell* (Interlocutory Order, N.D. Ga., Civil Action No. C 81-558 A, September 28, 1981, aff'd. No. 81-9666 (11th Cir. March 3, 1982.

31 *U.S. v. Braswell*, slip opinion, pp. 4—5.

32 See *Advertising Age*, December 14, 1981, p. 6.

33 *FTC v. Colgate Palmolive Co., Supra* at n. 16; *Resort Rental Car Systems, Inc. v. FTC*, 518 F.2d 962 (9th Cir. 1965).

34 143 F.2d 676 (2d. Cir. 1944).

35 Id. at 680.

36 *Sperry Corp.* and *DKG Advertising, Inc.*, consent orders finally accepted July 17, 1981, 3 Trade Reg. Rep. (CCH), par. 21,817, and *North American Phillips Corp.*, consent order announced September 23, 1981, 3 Trade Reg. Rep. (CCH), par. 21,869. These cases both involved claims that use of a particular electric razor would prevent razor bumps.

37 *Teledyne Inc.*, 97 F.T.C. 320 (1981). The central allegation of this complaint was that claims that use of the Water Pik significantly contributes to the prevention of gum disease.

38 *Hayoun Cosmetique, Inc.*, 95 F.T.C. 794 (1980); *AHC Pharmacal Inc.*, 95 F.T.C. 528 (1980); *San-Mar Laboratories, Inc.*, 95 F.T.C. 236 (1980); *Harvey Glass, M.D.*, 95 F.T.C. 236 (1980); and *Karr Preventative Medical Products, Inc.*, 94 F.T.C. 1080 (1979). These cases all involved false and unsubstantiated efficacy claims for the cure or prevention of acne.

39 *Hair Extension of Beverly Hills, Inc.*, 95 F.T.C. 361 (1980). The order in this case prohibits selling, advertising, or performing hair implants for treatment of balding or hair loss unless the respondent complies with requirements to inform consumers that hair implants are unsafe.

40 *The Mentholatum Co.*, 96 F.T.C. 757 (1980). The complaint alleged that denture cushions were represented to be appropriate for long-term use in contradiction of label warnings that the product should only be used temporarily until a dentist is seen.

41 *American Home Products Corp.*, 95 F.T.C. 884 (1980); *Jordan-Simmer, Inc.*, 95 F.T.C. 871 (1980), *Morton-Norwich Products, Inc.*, 95 F.T.C. 899 (1980). The complaints challenged efficacy and comparative efficacy claims for vaginal contraceptive suppositories.

42 Initial Notice, *40 Fed. Reg.*, 52,631 (1975); Termination of Rule-making, *46 Fed. Reg.*, 24,584 (1981).

43 Initial Notice, *41 Fed. Reg.*, 14,534 (1976).

44 Initial Notice, *40 Fed. Reg.*, 26,646 (1975).

45 Report of the ABA Committee to Study the FTC, 1969, reprinted
    in 427 Antitrust & Trade Reg. Rep. (BNA, Special Suppl., Sep-
    tember 16, 1969); E. Cox, *The Consumer and the Federal Trade
    Commission; a Critique of the Consumer Protection Board of the
    FTC*, 1969.
46 Public Law 96-252.
47 One action that will have significant impact is the strengthening
    of the Bureau of Economics and the increasing emphasis on eco-
    nomic analysis at every stage of the decision-making process.
48 Transcript of Press Conference of Chairman Miller, October 26,
    1981; speech of Chairman Miller to Association of National Adver-
    tisers, November 10, 1981; speech of Chairman Miller to the
    American Advertising Federation, December 8, 1981 (See *Adver-
    tising Age*, November 2, 1981, pp. 1, 102; November 16, 1981,
    pp. 1, 107; December 14, 1981, p 6.
49 Bills to amend the FTC's authority have been considered by com-
    mittees in both houses of Congress. They differ in material re-
    spects, but several proposals contain a statutory definition of
    unfair practices.
50 Noxell Corporation (Cover Girl NailSlicks), NAD Case Report,
    August 16, 1982, p. 26.
51 The Wilkes Group, Inc. (Barielle Nail Strengthener Cream), NAD
    Case Report, May 17, 1982, p. 16.
52 Schering-Plough Corp./Maybelline Co. (Fresh Lash Mascara), NAD
    Case Report, April 15, 1982, p. 11.
53 15 U.S.C. 1125(a).
54 *American Home Products Corp. v. Johnson & Johnson*, 577 F.2d
    160 (2d Cir. 1978).
55 *Dallas Cowboys Cheerleaders v. Pussycat Cinema, Ltd.*, 467 F.
    Supp. 366 (S.D.N.Y. 1979), aff'd., 604 F.2d 200 (2d Cir. 1979).
56 *Bernard Foods Industries, Inc. v. Dietene Co.*, 415 F.2d 1279
    (7th Cir. 1969). *cert. denied* 397 U.S. 912 (1970).
57 *Dallas Cowboy Cheerleaders v. Pussycat Cinema, Ltd.*, Supra at
    n. 57.
58 *Toro Co. v. Textron, Inc*, 499 F. Supp. 241 (D. Del. 1980).

7

THE IMPACT OF OTHER REGULATORY AGENCIES ON THE COSMETIC INDUSTRY

PETER BARTON HUTT

Covington & Burling, Washington, D.C.

I.   INTRODUCTION

Virtually every aspect of American business is regulated by one or
more federal agencies under a variety of regulatory statutes enacted
by Congress during the last 80 years. Some of these statutes are relatively old and familiar, such as the antitrust laws. Others are
quite new and bewilderingly unfamiliar, such as the recent environmental laws.

Chapters 1 and 6 describe the impact of the Food and Drug
Administration (FDA) and the Federal Trade Commission (FTC) on
the cosmetic industry. This chapter surveys the impact of all other
regulatory agencies. Because of its breadth, it necessarily provides
only a brief overview of some of the more important regulatory statutes.
It is neither comprehensive nor detailed. It does serve, however,
to illustrate the remarkable scope of current regulatory law affecting
the cosmetic industry and to identify areas of potential concern to the
industry.

The chapter is organized according to the type of activity or area
regulated, not according to the regulatory agency or statute involved.
Thus, statutory requirements are considered in the context in which
the issues arise.

II. RESEARCH

A. Compliance with FDA Regulations for Nonregulated Cosmetic Industry Research

The FDA regulates some aspects of cosmetic research but not others.
Essentially, the FDA requirements (such as the good laboratory practices requirements[1] and the protection of human subjects[2] are

imposed for research that is to be submitted to the FDA for regulatory
consideration (such as in a color additive petition). Even for nonregulatory research, however, it is advisable for the cosmetic industry
to comply in all pertinent respects with these requirements, not only
for sound ethical and business reasons but also because it is never
entirely certain when research initially conducted for nonregulatory
purposes may become important in a regulatory context.

B. The Animal Welfare Act

In 1966, Congress enacted the Animal Welfare Act[3] to protect research animals. This legislation authorizes the U.S. Department of
Agriculture (USDA) to promulgate standards regarding the humane
handling of animals used for research. It covers both regulatory and
nonregulatory research, and thus is applicable to all cosmetic industry
research on test animals.

The USDA has promulgated extensive regulations pursuant to this
statute[4]. As a result of the increased activity of animal rights advocates, moreover, enforcement of these provisions has become more
stringent.

C. The Economic Recovery Tax Act of 1981

The Internal Revenue Code is not often considered a regulatory statute.
Some of its provisions, however, have important regulatory aspects.
Sections 221—223 of the Economic Recovery Tax Act of 1981[5], for
example, contain significant tax incentives for cosmetic industry
research and development.

Section 221 provides a nonrefundable 25 percent tax credit,
effective 1981—1985, for qualified research expenses in excess of past
research expenses averaged over a base period. Qualified research
expenses include both company research and contract research, and
corporate grants to charitable institutions for basic research.

Section 222 allows an increased charitable deduction for qualified
corporate contributions of newly manufactured scientific equipment
to charitable institutions for research or experimentation. Section
223 provides favorable allocation of research and experimental
expenditures under the Internal Revenue Code.

Thus, the impact of tax legislation on cosmetic industry research
and development must always be considered.

III. THE WORKPLACE

A number of federal statutes apply generally to the area of labor
relations, and are beyond the scope of this chapter. Two statutes
in particular, however, have become of increasing importance in regulating workplace practices. One prohibits discrimination and the
other protects the health and safety of workers.

A. Equal Employment Opportunity

Title VII of the Civil Rights Act of 1964 [6] and other antidiscrimination statutes broadly prohibit non-job-related discrimination in employment. These statutes are enforced by the Equal Employment Opportunity Commission (EEOC).

It has been widely understood for many years that Title VII prohibits job discrimination based on race, sex, or religion. In recent years, however, sexual harassment has become a significant equal employment issue. The EEOC has issued guidelines stating that sexual harassment violates Title VII[7]. Under the guidelines, an employer is liable for the acts of its supervisors or agents regardless of whether the conduct was authorized or even specifically forbidden, and is also liable for the acts of nonsupervisory personnel and even customers if it knows or should have known of the conduct involved.

Since enactment of the Age Discrimination in Employment Act of 1967[8], age discrimination has also become an increasingly important matter. The act prohibits employers with 20 or more employees from discriminating because of age against any employee or applicant for employment between the ages of 40 and 70. New guidelines issued by EEOC in 1981 [9] interpret this law and apply it to a wide variety of situations. Bona fide occupational qualifications (BFOQs) are permitted, but are scrutinized very closely to assure that they are not simply a means of disguising age discrimination.

B. The Occupational Safety and Health Act

Congress passed the Occupational Safety and Health Act (the OSH Act)[10] in 1970 to protect the health and safety of employees. The OSH Act imposes both general and specific duties on employers to provide employees with safe and healthful places in which to work. It authorizes inspection of all workplaces, including cosmetic workplaces, by the Occupational Safety and Health Administration (OSHA), an agency in the Department of Labor.

It is important to understand that all aspects of cosmetic work are subject to the OSH Act. The fact that cosmetics are also subject to FDA regulation and that, in particular, color additives are subject to premarket approval by the FDA, in no way diminishes the legal authority of OSHA to control those products and substances. For example, OSHA has the authority to prohibit or limit the use of a color additive previously approved as safe by the FDA.

The "general duty clause" in the OSH Act required each employer to:

> . . . furnish to each of his employees employment and a place of employment which are free from recognized hazards that are causing or are likely to cause death or serious physical harm to his employees[11].

In addition to enforcing this general duty, OSHA is authorized to promulgate specific occupational safety and health standards, with which the regulated industry must comply. In accordance with this specific authority, OSHA has promulgated a number of safety standards that apply to all aspects of the working environment.

To enforce the general and specific requirements of the OSH Act, OSHA is authorized to enter any workplace at reasonable times, and to inspect within reasonable limits and in a reasonable manner[12]. Just as it is important for every cosmetic company to have a general procedure governing FDA inspections, it is equally important to have a general procedure governing OSHA inspections. Recommended guidelines for such inspections have been published in the CTFA Cosmetic Journal[13]. In general, company representatives should always be cooperative and polite in dealing with OSHA inspectors, but should be aware of the limits of OSHA authority and the rights of the company being inspected.

The OSH Act establishes penalties for two types of violations: serious and nonserious. Serious violations, which are likely to result in death, serious bodily injury, or serious illness, are subject to a mandatory penalty of up to $1,000 per violation. Nonserious violations are subject to penalties of up to $1,000 per violation at the discretion of OSHA officials[14].

IV. THE ENVIRONMENT

No area has been the subject of greater public concern or more comprehensive statutory control during the past 10 years than the environment. Indeed, it is likely that more federal statutes and amendments have been enacted in this area during the past decade than in all other areas of health and safety combined. Public opinion polls continue to show deep public commitment to protecting the environment, both to assure its esthetic quality and to prevent a deterioration of health and safety.

It is particularly difficult to summarize current regulatory requirements in this area, because they are so detailed and complex and change so rapidly. Environmental requirements today represent the largest single burden of federal regulation on American business. Even with efforts at regulatory relief and regulatory reform, their impact has been and will remain enormous. This section therefore merely sketches an outline of current requirements.

A. The Clean Air Act

The Clean Air Act (CAA)[15] is comprised of a basic statute enacted by Congress in 1963 and a number of major amendments since then that have gradually imposed more stringent air quality requirements

on industry. All of the requirements of the CAA apply to manufacturers of cosmetics.

The EPA is authorized to issue primary and secondary national
ambient air quality standards (NAAQSs) for air pollutants that may
reasonably be anticipated to endanger public health or welfare. Primary standards are to protect public health with an adequate margin
of safety, and secondary standards are to protect public welfare from
any known or anticipated adverse effects.

Pursuant to this authority, the EPA has established primary and
secondary ambient air quality standards for seven categories of chemicals specified by Congress: sulfur dioxide, particulate matter,
nitrogen oxide, carbon monoxide, photochemical oxidents, hydrocar-
bons, and lead[16]. The CAA also requires the EPA to consider
establishing standards for four other chemical categories: radioactive
pollutants, cadmium, arsenic, and polycyclic organic matter. These
pollutants were selected because of congressional concern about their
potential risk to health.

The NAAQSs are to be achieved through legally enforceable state
implementation plans (SIPs). These SIPs must provide for attainment
of the primary NAAQS by specified deadlines (1982—1987, depending
on the pollutant and other conditions) and for attainment of the secondary NAAQSs within a reasonable time. SIPs must contain emission
limitations, compliance schedules, and such other measures as may
be necessary to ensure attainment and maintenance of NAAQSs.

Separate criteria for SIPs are established for "nonattainment areas,"
where ambient standards are not being met, and for "attainment areas,"
where the standards have been achieved. Regulation is required in
attainment areas for the purpose of the prevention of significant
deterioration (PSD) of air quality.

A number of stringent provisions control emissions by new and
existing sources in nonattainment areas. State plans must include
requirements for technology-based emission reductions from existing
sources, a comprehensive and current emissions inventory, and
"growth allowances" that identify and quantify the allowable emissions
from major new or modified stationary sources in that area over time.
Major new or modified sources must obtain a permit and certification
before construction may begin.

Sources subject to PSD review are major new sources or major
modifications of existing sources in 28 enumerated industry categories
with potential to emit more than 100 tons per year of a pollutant, or
any other source with potential to emit more than 250 tons per year.
These facilities are subject to a detailed preconstruction review and
permit procedure.

EPA is also authorized to issue new source performance standards
(NSPSs) for categories of stationary air pollution sources that cause
or contribute significantly to air pollution and that may reasonably

be anticipated to endanger public health or welfare. Such NSPSs are
to be based on the best adequately demonstrated technology system
of continuous emission reduction, taking into account costs, energy,
and non-air quality, health, and environmental impacts.

NSPSs are to apply to those categories of stationary sources that
have a significant impact on air pollution. Thus far, regulations for
20 categories of stationary sources have been issued[17]. There is no
regulation for the manufacture of cosmetics. An owner or operator of
a facility subject to an NSPS must provide the EPA with extensive
information on the magnitude of emissions and the performance of
emission control equipment.

The EPA may require the states to adopt performance standards
for existing sources that apply to "designated pollutants" that are
subject to a NSPS but have not been listed as "hazardous" and thus
are not subject to air quality criteria. A number of chemicals and
industries have been regulated by the EPA under this authority[18].

Finally, the EPA is authorized to issue national emissions standards
for hazardous air pollutants (NESHAPs), which apply to both new and
existing sources. A "hazardous" air pollutant is one for which no
NAAQS is applicable but that may reasonably be anticipated to result
in increase in mortality or in serious irreversible illness. The NESHAPs
must provide an ample margin of safety to protect the public health.
To date, such standards have been established for four chemicals:
beryllium, asbestos, mercury, and vinyl chloride[19]. These regulatory authorities do not presently have a major impact on the cosmetic
industry. As the EPA continues to consider the hazard of industrial
chemicals, however, the cosmetic industry could come under scrutiny
at a later time.

B. The Clean Water Act

The Clean Water Act (CWA)[20] originated with a Congressional statute
in 1948 that has since been amended a number of times to provide pro-
gressively more stringent requirements to protect the quality of the
nation's water supply.

For industrial plants that discharge directly to surface waters,
the CWA authorizes the EPA to issue industry-by-industry effluent
limitation guidelines. These guidelines place limitations on the types
and amounts of pollutants that different categories of source can discharge. In developing effluent limitation guidelines, the EPA must
achieve control of industrial discharge in two basic steps. By 1977,
all discharges were to achieve limitations based on the "best practicable
technology currently available" (BPT). By 1984—1987, all discharges
are to achieve more stringent control levels by various deadlines depending on the types of pollutants being discharged.

For the 65 "toxic" pollutants incorporated by reference in the
CWA, dischargers must achieve "best available technology economically

achievable" (BAT) by 1984. The EPA may designate other pollutants as "toxic," and dischargers of these pollutants must achieve BAT within three years after the limitations are established. EPA may also establish limitations for certain "conventional" pollutants, which are usually defined in terms of pollution parameters such as pH levels or suspended solid and fecal coliform content. Dischargers of these conventional pollutants must achieve "best conventional pollutant control technology" (BCT) by 1984. For all other pollutants (i.e., those not on a toxic or conventional list), dischargers must achieve BAT no later than 1987.

In developing effluent limitation guidelines, the EPA must also establish "new source performance standards" (NSPSs) that embody the "best available demonstrated control technology" (BACT). The CWA specifies a minimum of 27 named industries for which such standards are to be developed. New dischargers must achieve any applicable standards by the time they commence operations, regardless of the type of pollutant to be discharged. The EPA has to date issued effluent limitation guidelines for some 42 industries, encompassing most of the nation's major point sources, but not including cosmetic manufacture. These guidelines are framed to cover not only specific pollutants but also pollutant categories and often pollutant parameters. Underlying these guidelines is the EPA determination of the pollutants that are commonly discharged in significant quantities by the particular industry to which the guidelines apply.

For industries covered by the EPA's effluent limitation guidelines, existing pollutants generated during the manufacture or use of new chemicals may be subject to discharge restrictions. Dischargers of new chemicals themselves may also be controlled if those discharges fall within the pollutant classes or pollutant parameters covered by the EPA's guidelines.

Because many smaller industrial plants do not discharge directly into the nation's surface waters, Congress authorized the EPA to set "pretreatment" standards, designed to assure that pollutants discharged into publically owned treatment works (POTWs) by "indirect" discharges neither interfere with nor pass through POTWs.

The EPA's pretreatment limitations, like its limitations on direct discharges, have been developed on an industry-by-industry basis. They apply to specific pollutants, pollutant classes, and pollution parameters that the EPA has determined are of particular concern for the industry in question.

Pursuant to these technology-based limitations and pretreatment standards, the EPA has exercised broad control over continuous "through-the-pipe" discharges to surface waters and to POTWs. The CWA also contains additional "best management practices" authority for significant release of toxic pollutants or hazardous substances to receiving waters. Under this authority, the EPA can, when setting

BAT and NSPS limitations for any industrial category, prescribe regulations to control all significant discharges of toxic and hazardous pollutants with respect to plant site runoff, spillage or leads, sludge or waste disposal, and drainage from raw material storage.

The EPA is authorized under the CWA to impose site-specific limitations even more stringent than BAT and BCT whenever, in the EPA's judgment, achievement of such limitations would still be insufficient to achieve a high level of water quality (generally referred to as "fishable-swimmable") in any particular water body. In such an event, the EPA may impose not only effluent limitations, but also effluent control strategies. The EPA is required to assess the relationship between the costs and benefits of any such limitations.

The CWA requires that states (and the EPA, in the absence of adequate state action) issue "water quality standards" and revise them at least every three years. Such standards must protect the public health or welfare and enhance the quality of water. These standards are to be set taking into consideration their use and value for public water supplies, propagation of fish and wildlife, recreational purposes, and other beneficial uses.

Water quality standards issued under the CWA have set permissible levels for particular pollutants. They have also defined overall parameters for acceptable water quality. In addition to the EPA's effluent limitation guidelines and other standards, water quality standards have been used during the permit-issuing process to control the release practices of particular dischargers.

In addition to the authority to set industry-by-industry technology-based (BAT) limitations for toxic pollutants, the EPA has the authority under the CWA to set pollutant-by-pollutant health/environmental effects-based discharge standards (including prohibitions) for toxic pollutants. Such a standard must provide an ample margin of safety. The EPA has established toxic effluent standards for five major pollutants, including polychlorinated biphenyls (PCBs).

Every discharger to the surface waters must have a permit under the national pollutant discharge elimination system (NPDES). This permit includes site-specific limitations derived from the effluent guidelines, water quality standards, new source performance standards, toxic standards, and other limitations described above. Permits must be reissued at least every five years to take into account changes in the applicable substantive requirements. If the EPA has not yet issued effluent limitations guidelines or other standards for a particular industry category, such as the cosmetic industry, a plant in such a category may still be required to meet BAT, BCT, and other limitations on a case-by-case basis.

Because this permit-issuing process incorporates the discharge limitations and water quality standards developed by the EPA and the states, permits will often place controls not merely on specific

pollutants but also on pollutant categories and pollution parameters.
Thus, not only will discharges of existing pollutants generated during
the manufacture or use of new chemicals be controlled during the permitting process, but discharges of new chemicals may fall within
restrictions as well.

Under current EPA regulations, permits will almost invariably
require the discharger to monitor discharges, to keep records of such
monitoring, and to report periodically on discharge activities. It will
often be necessary to inform the EPA of changes in the composition
of the discharger's effluent, and such changes may often necessitate
the modification of permit provisions. This system assures that discharge practices related to the manufacture or use of chemicals come
to the attention of the EPA or state authorities during the permitting
process or after the permit is issued.

In addition to the "best management practices" authority, the CWA
contains significant additional authority for regulating releases that
are not of the "through-the-pipe" variety. The discharge of a reportable quantity of any hazardous substance by spilling, leaking, pumping, pouring, or dumping is prohibited. The EPA has designated 297
substances as hazardous[21]. Where releases of these substances
occur in reportable quantities, the responsible parties must immediately
notify the appropriate federal agency.

C. The Solid Waste Disposal Act

The Solid Waste Disposal Act (SWDA)[22] resulted from the consolida-
tion and revision of many Congressional environmental laws in the
Resource Conservation and Recovery Act of 1976 (RCRA). It authorizes a wide variety of federal controls and programs respecting disposal of solid waste. It is quite accurately described as a "cradle-
to-grave" management system for the generation, transportation,
storage, treatment, and disposal of hazardous wastes.

All hazardous wastes are subject to the cradle-to-grave management
system. A "waste" includes solid, liquid, semisolid, or contained
gaseous material that is discharged. The statute defines a hazardous
waste as one that may, in addition to other factors, pose a substantial
present or potential hazard to human health or the environment when
improperly treated, stored, transported, or disposed of, or otherwise
managed.

Under this authority, the EPA has developed regulations defining
hazardous wastes in a comprehensive fashion. First, the EPA has
developed three different lists that specifically identify certain types
of wastes as hazardous. These lists are for types of chemicals from
nonspecific and specific sources and specifically discarded commercial
chemical products and residues. The EPA has thus far included more
than 400 substances on such lists.

Second, even for wastes that are not specifically identified on such a list, EPA regulations require that wastes be tested against four geneneral characteristics, and that a waste exhibiting any of such characteristics will be deemed hazardous. The four characteristics are toxicity, ignitability, corrosivity, and reactivity.

Once a waste is determined to be hazardous, either through a specific listing or because it has one of these four characteristics, all aspects of its life as a waste are strictly regulated by the EPA. Regulations generally apply to those who handle hazardous wastes in three steps: generators, transporters, and storers/treaters/disposers. Under the RCRA, the requirements applicable to each must be sufficient to protect human health and the environment. Ultimately, the program is designed to assure that specific performance standards will apply to all storage, treatment, and disposal facilities and that these performance standards will be enforced through permits. Specific requirements have been imposed by regulation on generators [23], transporters[24], and storers/treaters/disposers[25]. All of these requirements apply directly to the cosmetic industry.

D. The Comprehensive Environmental Response, Compensation, and Liability Act

The Comprehensive Environmental Response, Compensation, and Liability Act[26], commonly called the "Superfund" Act, was passed in 1980 to provide substantial authority for the EPA to deal with releases of toxic substances to the environment. A release includes any spilling, leaking, pumping, pouring, emitting, emptying, discharging, injecting, escaping, leeching, dumping, or disposing. The environment includes surface waters, ground water, land surface or subsurface, or ambient air.

Under the Superfund Act, any company must report release of a hazardous substance in excess of the reportable quantity. Until the EPA issues regulations, that quantity is defined to be either the quantity established pursuant to the Clean Water Act, or one pound for substances not covered by the Clean Water Act quantity designations.

Consistent with a "national contingency plan," the EPA may take appropriate removal or other remedial action whenever there is a release of a hazardous substance or any other pollutant or contaminant that may present an imminent and substantial danger to health or welfare. The Superfund Act broadly defines hazardous substances to include, among others, the 297 substances listed by the EPA as hazardous in 40 C.F.R. Part 261, the 65 toxic pollutants incorporated by reference in the Clean Water Act, and any waste considered hazardous under the RCRA. The EPA is authorized to add a substance to this list if it presents substantial danger to the public health or welfare or the environment.

E. The Noise Control Act

The Noise Control Act (NCA)[27] was enacted in 1972 to provide the EPA with authority to establish federal noise emission standards and to take other action to reduce noise. This statute has had particular applicability in the manufacture of cans and other products that generate substantial noise, but has not been important to the cosmetic industry and is unlikely to have any significant applicability to cosmetics in the future.

V. COSMETIC INGREDIENTS AND PRODUCT FORMULATION

A number of specialized regulatory statutes govern particular ingredients or types of hazard. In most instances, they have little applicability to the cosmetic industry. In some specific cases, however, they have become quite important in the development of particular cosmetic products.

A. The Endangered Species Act

The Endangered Species Act[28], enacted in 1973, makes it illegal to export or to sell in interstate or foreign commerce any endangered species or any part or product of an endangered species. It is enforced by the National Marine Fishery Service (NMFS) of the National Oceanic and Atmospheric Administration (NOAA) in the Department of Commerce.

Sperm whale oil has been placed on the list of endangered species pursuant to this statute. Both sperm whale oil and spermacetti obtained from the sperm whale have been used as cosmetic ingredients. Under this statute, however, such use is unlawful. The statute provides for civil fines of up to $10,000 for each violation and criminal penalties of up to one year imprisonment and a fine of up to $20,000 for those who knowingly violate the act. The products themselves are subject to seizure and forfeiture.

B. The Federal Alcohol Administration Act

Following the repeal of Prohibition, Congress enacted the Federal Alcohol Administration Act (FAA Act) [29] authorizing the Secretary of the Treasury to exercise comprehensive regulation of alcohol in interstate commerce. That authority is now exercised by the Bureau of Alcohol, Tobacco, and Firearms (BATF) in the Department of the Treasury.

Use of alcohol in cosmetic products must comply with BATF regulations. The regulations of particular importance to cosmetic manufacturers govern the use of specially denatured alcohol[30].

As a general rule, specially denatured alcohol may be used in a cosmetic without being taxed if there is full compliance with the BATF regulations. The regulations prescribe how a manufacturer obtains the specially denatured alcohol, what records must be kept, and what reports must be filed with the BATF. In particular, the BATF regulations prescribe formulas for specially denatured alcohol and authorize particular uses for each formula.

C. The Federal Caustic Poison Act

The Federal Caustic Poison Act[31] provides for labeling of dangerous caustic or corrosive substances. Although this statute technically is applicable to cosmetics, the nature and use of cosmetic products make it unlikely that any "caustic" ingredient would ever be used.

D. The Poison Prevention Packaging Act

The Poison Prevention Packaging Act (PPPA)[32] was passed in 1970 to authorize a requirement of special packaging to protect children from dangerous household substances. Although cosmetics are technically subject to the PPPA, and many drugs are subject to special packaging requirements, it is unlikely that any cosmetic product would be sufficiently dangerous to require special packaging under this statute.

E. The Flammable Fabrics Act

The Flammable Fabrics Act[33], enacted in 1953, authorizes regulation of wearing apparel and fabrics that are flammable. It does not specifically exempt "cosmetic device" products, and could possibly be applicable to such products as wigs. The act was formerly enforced by the FTC and is now enforced by the Consumer Product Safety Commission, through the promulgation of regulations[34].

F. The Hazardous Materials Transportation Act

The Hazardous Materials Transportation Act (HMTA)[35], enacted in 1975 to consolidate a number of earlier laws governing the transportation of hazardous materials, is administered by the Department of Transportation. It governs regulation of the transportation of hazardous materials by water, air, railroad, and highway vehicles.

A hazardous material is defined as a substance that has been determined by the Secretary of Transportation to be capable of posing an unreasonable risk to health, safety, and property when transported in commerce. Pursuant to the HMTA, the Department of Transportation (DOT) has designated and classified extensive lists of hazardous materials[36]. Transportation of any such listed material is lawful only if it complies with the DOT regulations governing shipping papers, markings, labeling, and placarding. Because the hazards to which the HMTA applies includes such things as flammable materials, a number of cosmetic ingredients are subject to these provisions.

VI. ADVERTISING AND PROMOTION

In addition to the regulation of advertising by the FTC discussed in
Chap. 6, other federal regulatory statutes have a direct impact on
cosmetic product promotion.

A. The Postal Fraud Act

The Postal Fraud Act[37], which dates back to 1872, prohibits any
fraud involving use of the mails. It has been used over the years to
combat false promotional materials for consumer products, including
cosmetics. Although the FDA has primary jurisdiction over cosmetic
labeling that is false or misleading, as described in Chap 4, the Postal
Fraud Act has been and continues to be used for exactly the same
purposes. It is often reserved for clearly fraudulent get-rich-quick
schemes.

B. The Tariff Act

Under the Tariff Act of 1930[38], the Department of the Treasury has
promulgated regulations governing the marking of foreign articles imported into the United States with the country of origin. These requirements are administered by the Customs Service of the Department
of the Treasury. They supplement the FTC requirements for marking
the country of origin discussed in Chap. 1. The CTFA Labeling
Manual contains detailed information on these requirements[39].

C. The Fair Packaging and Labeling Act

The Fair Packaging and Labeling Act (FPLA)[40] was enacted in 1967
to provide greater authority to the FDA and the FTC to regulate consumer packaging and labeling. The FPLA authorizes these agencies
to promulgate regulations limiting the number of package sizes for any
consumer commodity, regulating "cents off" claims, and preventing
nonfunctional slack fill of packages. The FDA has established regulations for package size savings and "cents off" claims for cosmetics[41],
but has not established regulations limiting package sizes or preventing
slack fill.

D. Patent, Trademark, and Copyright Law

The laws relating to patents, trademarks, and copyright are not
"regulatory" in the limited sense of that term. Broadly speaking, however, they are part of the law of unfair competition and thus must be
considered in formulating and promoting cosmetic products.

E. The Lanham Act

The Lanham Act[42] was enacted in 1946 as a comprehensive revision
of the U.S. law of trademarks. Section 43(a) of the Lanham Act[43]

provides a private civil cause of action for damages and an injunction for any false representation made in connection with any goods or services, including cosmetic products. Although this statute is invoked by private lawsuits directly in the courts, and not by any government agency, its frequent use by companies to assure fair competition from competitors has made it increasingly important as a source of private "regulation."

VII. LAWS FROM WHICH COSMETICS ARE EXEMPT

The provisions of four important regulatory statutes specifically exempt cosmetics from coverage. In each instance, however, action taken by regulatory agencies pursuant to these statutes can have a major impact on the availability and consumer acceptability of cosmetic ingredients and thus cosmetic products formulation and promotion. Accordingly, action taken pursuant to these statutes may well be of major importance to the cosmetic industry in spite of the technical exemption of cosmetic products from this coverage.

A. The Consumer Product Safety Act

Under the Consumer Product Safety Act (CPSA)[44], enacted in 1972, the Consumer Product Safety Commission (CPSC) is authorized to prevent unreasonable risks of injury from consumer products. Products regulated by the FDA are exempt from the CPSA. To the extent that the CPSC regulates cosmetic ingredients that are used in other non-exempt products (e.g., formaldehyde), however, the impact of the CPSC on the cosmetic industry can be substantial. The FDA and the CPSC have also agreed that the CPSC may exert jurisdiction over cosmetic containers that present a physical hazard rather than a hazard of toxicity.

B. The Federal Hazardous Substances Act

The Federal Hazardous Substances Act (FHSA)[45], enacted in 1960 and amended to expand its coverage several times prior to enactment of the CPSA, is now administered by the CPSC. Any consumer product that is toxic, corrosive, an irritant, a strong sensitizer, flammable or combustible, or generates pressure is required to be labeled with appropriate cautions and warnings if it may cause substantial personal injury or illness during customary or reasonably foreseeable handling or use. Although cosmetics are exempt from this statute, the FDA has indicated that a cosmetic product that would otherwise be required to be labeled under the FHSA might be regarded as misbranded under the FDC Act unless it bears an appropriate caution or warning. As a result, some cosmetic products do contain cautions or warnings similar to the FHSA labeling requirements under appropriate circumstances.

C. The Federal Insecticide, Fungicide, and Rodenticide Act

The Federal Insecticide, Fungicide, and Rodenticide Act (FIFRA)[46], enacted in 1947 and modernized by a series of amendments in the 1970s, authorizes the EPA to register and otherwise regulate all pesticides, including all preservatives. Although the statute does not explicitly exempt cosmetics, EPA regulations have exempted preservatives used in cosmetics[47]. Any EPA action taken against a preservative on safety grounds, however, would have a clear impact on the use of that preservative in cosmetics.

D. The Toxic Substances Control Act

The Toxic Substances Control Act (TSCA)[48], enacted in 1976, establishes a comprehensive mechanism for regulating unreasonable risks of injury to health or the environment from new and existing chemicals. Although chemicals used in cosmetic products are specifically exempt from the TSCA, any action taken by the EPA to limit or ban such chemicals for nonexempt uses would undoubtedly have a major impact on their cosmetic uses as well, as discussed more fully in Chap. 3.

VIII. CONCLUSION

This chapter has presented a sampling of the more important regulatory statutes, in addition to the FDC Act and the FTC Act, which are discussed more fully in Chaps. 1 and 6 that must be taken into consideration in marketing cosmetic products. It illustrates the regulatory complexity and uncertainty facing cosmetic manufacturers today, and particularly small businesses. Even if efforts at regulatory relief and regulatory reform are successful, this complexity and uncertainty are unlikely to disappear.

REFERENCES

1.   21 Code of Federal Regulations (C.F.R.) Part 58.
2.   21 C.F.R. Parts 50 and 56.
3.   7 U.S. Code (U.S.C.) 2131 et seq.
4.   9 C.F.R. subchap. A.
5.   95 Stat. 172 (1981).
6.   42 U.S.C. 2000e et seq.
7.   29 C.F.R. 1604.11.
8.   29 U.S.C. 621 et seq.
9.   29 C.F.R. Part 1625.
10. 29 U.S.C. 651 et seq.
11. 29 U.S.C. 654(a)(1).
12. 29 U.S.C. 657(a).

13. Frank and Rupp, Recommended Guidelines for Cosmetic Manufacturers Subject to Occupational Safety and Health Administration Inspections, *CTFA Cosmet. J.*, October/November/December 1981, p. 28.

14. 29 U.S.C. 666.

15. 42 U.S.C. 7401 et seq.

16. 40 C.F.R. Part 50.

17. 40 C.F.R. Part 60.

18. 40 C.F.R. Part 61.

19. 40 C.F.R. Part 61.

20. 33 U.S.C. 1251 et seq.

21. 40 C.F.R. Parts 116, 117.

22. 42 U.S.C. 6901 et seq.

23. 40 C.F.R. Part 262.

24. 40 C.F.R. Part 263.

25. 40 C.F.R. Parts 264, 265, 267.

26. 42 U.S.C. 9601 et seq.

27. 42 U.S.C. 4901 et seq.

28. 16 U.S.C. 1531 et seq.

29. 27 U.S.C. 201 et seq.

30. 27 C.F.R. Parts 211, 212.

31. 15 U.S.C. 401 et seq.

32. 15 U.S.C. 1471 et seq.

33. 15 U.S.C. 1191 et seq.

34. 16 C.F.R. Parts 1700—1704.

35. 49 U.S.C. 1801 et seq.

36. 49 C.F.R. Part 172.

37. 18 U.S.C. 1341—1343 and 39 U.S.C. 3005.

38. 19 U.S.C. 1304.

39. CTFA Labeling Manual, 4th ed., app. C., 1982.

40. 15 U.S.C. 1451 et seq.

41. 21 C.F.R. 1.31, 1.35.

42   60 Stat. 427 (1946).

43. 15 U.S.C. 1125(a).

44. 15 U.S.C. 2051 et seq.

45. 15 U.S.C. 1261 et seq.

46. 7 U.S.C. 135 et seq.

47   40 C.F.R. 162.3(ee) and (ff)(2)(ii).

48. 15 U.S.C. 2601 et seq.

8

AN OVERVIEW OF WORLDWIDE REGULATORY PROGRAMS

WILLIAM H. SCHMITT

Research Laboratories, Chesebrough-Pond's, Inc., Trumbull, Connecticut

EMALEE G. MURPHY

Legal Department, The Cosmetic, Toiletry and Fragrance Association, Inc., Washington, D.C.

I. INTRODUCTION

This chapter discusses the development of cosmetic regulations and their impact both in the country where enacted and in other countries affected by foreign regulatory programs. As communications improve, regulations in one part of the world invariably have an effect in other countries. As more and more companies from all countries attempt to extend their business beyond national boundaries, a general knowledge of worldwide regulations becomes increasingly important. This chapter is concerned chiefly with the impact of overseas regulations on companies located in the United States.

We have attempted to highlight regulations and regulatory approaches in various countries. However, regulations are in a constant state of change; new regulations are adopted, existing regulations are challenged, and it is impossible to provide a completely up-to-date review. Therefore, although we will present the current status of the regulations, the most important message in this chapter is for companies to seek local advice about regulations and trade customs and practices.

There are many available resources that can improve your own knowledge and make execution of business programs outside of home countries earier. It is vitally important, however, to recognize that translations of regulations or even an ability to understand the regulations in the native language may still result in misinformed opinions and practices and possibly disastrous results.

Individuals responsible for compliance with regulatory programs outside their national boundaries should have adequate references, make use of contacts through trade associations, and have local staff or consultants who are expert on local regulations and their implementation. It is extremely important that local staff or consultants have knowledge of and access to people in the government who can help clarify regulatory requirements. The U.S.-based executive should have a good overview of national regulations but should rely on local experts for their interpretation and execution. These local experts should be thoroughly familiar with your business practices.

II. INFORMATION SOURCES

Fortunately, there are many good sources of information on the regulation of cosmetic products in international markets. The Cosmetic, Toiletry and Fragrance Association* (CTFA) has established a relationship with other associations throughout the world and through these can often provide documents for member companies. In addition, the

---

*The complete name and address of all organizations or publications cited herein appear in the appendix at the end of this chapter.

CTFA's International Committee has prepared the *CTFA International Resource Manual*, a compilation of information sources for 47 countries, including names, addresses, and telephone numbers for key government agencies, trade and professional organizations, and listings for relevant government and private publications.

The Department of Commerce of the U.S. government is interested in aiding U.S. companies in establishing overseas businesses. General information on developing export markets for international businesses can be obtained from the Office of Export Marketing Assistance in the Bureau of Export Development. Specific information on exporting to or doing business in a country may be obtained from the Office of Country Marketing in the Bureau of Export Development, which employs specialists knowledgeable in the customs and laws of specific countries. If you are attempting to do business in a country and encounter a problem that you consider an unfair trade practice, you can contact the Trade Advisory Center in the Office of Trade Policy, U.S. Department of Commerce. The Trade Advisory Center deals directly with foreign countries to adjudicate unfair or restrictive trade practices.

The U.S. government can also be helpful in the country in which you wish to do business. In larger countries, the U.S. Embassy or Consulate will have a commercial officer who is a member of the Foreign Commercial Service under the U.S. Department of Commerce. In smaller countries, there will usually be an economic officer whose duties will include the development of trade. Very often these individuals are knowledgeable about regulations and practices within the country. In addition, they will probably have contacts within the local government who may be helpful. They should also be able to inform you as to whether or not there are associations of U.S. businessmen within the country who may be useful sources of information. There may be a local chapter of the Chamber of Commerce or perhaps an American club where contacts can be made.

Another valuable source of information within the U.S. government is the National Bureau of Standards. Its Standards Information Center obtains standards and regulations on a routine basis from most countries in the world. If standards or regulations are not on hand, the Standards Information Center may be able to procure the relevant documents from the country in question.

A most valuable source of information is local trade associations. The short summaries of national regulations that follow in this chapter provide the names of local trade associations whose full addresses are listed in the appendix. Your local staff or consultants should be encouraged to join and support the local trade association, as this is a most valuable source of information, as well as a vehicle for expressing opinions and influencing regulations.

There are excellent publications that provide coverage of international issues of importance to our industry. These are invaluable and are gratefully acknowledged as a resource used in assembling the highlights on regulations presented in this chapter.

The International Information Center of Cosmetic Industries (IICCI) is an organization that consists of 15 national trade associations. These trade associations regularly contribute national news for dissemination to members and subscribers through the *IICCI Bulletin*. The *IICCI Bulletin* is published monthly and covers legislation, regulations, and abstracts of scientific and trade-related matters. In addition, the IICCI sponsors a semiannual conference on worldwide regulatory matters.

The *International Drug and Device Regulatory Monitor* is a legal, medical, and scientific information publication for the pharmaceutical, device, and cosmetic industries. It is published by Monitor Publications on a monthly basis and occasionally covers cosmetic regulations.

There are other publications that, although not specifically oriented toward legislation and regulations, are nevertheless valuable references. They are *Exporters Encyclopedia* and *World Marketing*, published by Dun & Bradstreet International Publications; *Business International, Business Asia, Business Latin America, Investing Licensing and Trading Conditions Abroad*, and *Doing Business in Eastern Europe*, publications of the Business International Corporation; and *Common Market Reports*, which includes *Doing Business in Europe* and *Euro Market News*, all published by the Commerce Clearing House. In addition, the United Nations Environmental Program publishes a bulletin entitled *The International Register of Potentially Toxic Chemicals*. The above listing is by no means inclusive, but represents references that are extremely helpful.

III. THE EUROPEAN ECONOMIC COMMUNITY

The European Economic Community (EEC) is an organization made up of 10 member states. They are Belgium, Denmark, France, Germany, Great Britain, Greece, Ireland, Italy, Luxembourg, and the Netherlands. Spain will be admitted to the EEC in the future, probably in 1984 or 1985. The population of the EEC member states approaches 300 million, and taken as a whole, the EEC is roughly the same economic size as the United States.

Member states function independently and freely within the EEC; however, the EEC issues rules and regulations designed to harmonize laws affecting trade. The organization of the EEC is complex. The European Commission functions as an executive branch to ensure that EEC rules and principles of the common market are executed. It also makes proposals to the EEC Council of Ministers that advance EEC policies. The Council of Ministers consists of foreign ministers from each member state and is the legislative body responsible for enacting EEC directives.

The European Parliament consists of members elected directly from EEC member states. The European Parliament supervises and gives counsel to the European Commission and the Council of Ministers. The EEC Economic and Social Committee consists of interest groups representing consumers, trade unions, industry, and other interested groups. The Economic and Social Committee reviews proposals promulgated by the European Commission and directives under consideration by the Council of Ministers.

The most important EEC action to affect the cosmetic industry is the Council of Ministers enactment of a cosmetic directive (76/768/EEC) on July 27, 1976. The Council Directive outlines the need for EEC-wide cosmetic legislation:

> *Whereas* the provisions laid down by law, regulation or administrative action in force in the Member States define the composition characteristics to which cosmetic products must conform and prescribe rules for their labeling and for their packaging;
> *Whereas* these provisions differ from one Member State to another;
> *Whereas* the differences between these laws oblige Community cosmetic producers to vary their production according to the member state for which the products are intended;
> *Whereas*, consequently, they hinder trade in these products and, as a result, have a direct effect on the establishment and functioning of the common market;—
> *Whereas*, however, this objective must be attained by means which also take account of economic and technological requirements;
> *Whereas* it is necessary to determine at Community level the regulations which must be observed as regards the composition, labeling and packaging of cosmetic products; . . . .

The directive consists of 14 articles and 4 original annexes. Annex I is a listing by category of cosmetic products. Annex II is a listing of prohibited substances consisting of 361 materials. Annex III, Part 1, is a list of 29 restricted substances. Annex III, Part 2, lists the coloring agents permitted in products intended to come into contact with mucous membranes, including restriction levels. Annex IV, Part 1, is a list of 33 substances that are provisionally allowed in cosmetic products. Annex IV, Part 2, is a list of provisionally allowed coloring agents intended for contact with mucous membranes; and Annex IV, Part 3, lists provisionally allowed coloring agents for products that do not come into contact with mucous membranes. Annex V is a list of substances excluded from the scope of the directive. In addition to the specific information on substances contained in Annexes I through V, the directive outlines labeling requirements (including expiry dates for products with a stability of less than three years), provisions for establishing microbiological and chemical purity criteria, and for developing methods to check compliance.

The original Council Directive on cosmetics was amended on July 24, 1979 (79/661/EEC), to extend the deadline for compliance, and a second amendment (82/368/EEC) was signed May 18, 1982, to regulate trace materials also included in the prohibited list (Annex II), to amend Annexes III and IV, and to add Annex VI. Annex VI, Part 1, contains 11 preservatives definitely permitted for use; Part 2 contains 56 preservative materials that are provisionally permitted. A seventh annex listing sunscreens that are either permitted or provisionally permitted has been published as part of the third amendment to the 1976 directive (83/574/EEC). The third amendment would permit the use of certain color lakes in cosmetic products and regulate composition of antiperspirants. Four Commission directives have also been issued to amend the cosmetic directive and to adapt certain of its provisions to technical change, (82/147/EEC, 83/194/EEC, 83/341/EEC, and 83/496/EEC).

The EEC Council of Ministers has adopted other trade-related directives that affect the cosmetic and other industries. One of these is the Prescribed Quantities Directive (80/232/EEC) adopted on January 15, 1980. This directive, along with prior directives 76/211/EEC, 75/106/EEC with amendments 79/1005/EEC, and 78/891/EEC, outlines the nominal quantities and nominal capacities permitted and labeling required for certain prepacked products. It specifies designation of declaration by product type. The purpose of these directives is to promote harmonization of uniform metric product sizes. Directive 80/232/EEC is an optional directive that is, however, being adopted widely throughout the EEC. Member states are free under this directive to allow other prescribed quantities within their national boundaries, but they cannot prevent sale of imported products complying with the directive.

The general purpose of EEC directives is to harmonize legislation within the European Economic Community and thereby eliminate trade barriers created by national laws. Member countries are allowed a fixed time in which to enact national legislation that is harmony with directives. What actually happens is that the national legislation may be somewhat different than the EEC directive. There are many examples of local legislation that goes further than requirements specified in the directives.

The Court of Justice is the judicial branch of the European Economic Community. In 1979, the Court of Justice made a series of landmark rulings. The court decided that any product that was legally manufactured and marketed in one member state must in theory be admitted to the market of any other member state. The only exceptions would be those regulations in the originating treaty of the EEC that cover consumer protection and public health and safety. This decision implies that national commercial or technical rules may not restrict free movement of goods among member states.

A. COLIPA

The Comite de Liaison des Associations Europeenes de l'Industrie de la Parfumerie, des Produits Cosmetiques, et de Toilette (COLIPA) is the federation of cosmetic trade associations within the EEC. COLIPA represents 10 national associations from European countries whose members include more than 1,000 manufacturers and importers of cosmetics, toiletries, and perfumes. COLIPA interacts with the EEC Directorate General and with the working groups of the European Commission and provides a forum within which industry and association groups can arrive at a consensus of viewpoints and defend these viewpoints before the EEC and other international bodies.

COLIPA's Board of Directors is responsible to a General Assembly that consists of national delegations. There is a permanent, professional staff that comprises a secretary general, a scientific officer, and other administrative personnel who handle the day-to-day functioning of COLIPA. COLIPA responds to technical and regulatory issues through its committees and working parties of experts from various disciplines. There are technical committees in each of the areas covered or proposed by EEC regulations, that is, for cosmetic colorants, hair dyes, and positive ingredient lists, as well as committees for microbiology, analytical methods, traces, and nitrosamines. A toxicology submission group examines any data before it goes to the EEC to ensure that it meets adequate scientific standards. There are also socioeconomic committees covering consumer relations, packaging, alcohol, distribution, and interests of small companies. COLIPA appoints expert groups and can assign specific tasks to various national trade associations within its membership.

For instance, COLIPA works with its member associations assembling toxicological and usage data to support currently used ingredients under review by the EEC Scientific Committee for Cosmetology. To support this program, the CTFA has forwarded appropriate data available in the United States for submission to the COLIPA data bank. Information from the U.S. Food and Drug Administration's Over-the-Counter Drug Review program and U.S. industry's Cosmetic Ingredient Review program has been most helpful. Because regulatory actions in one part of the world directly affect or initiate regulatory actions in other parts of the world, it is imperative that cooperation between national associations continue at the highest level.

B. Belgium

Belgium has adopted the EEC Cosmetic Directive by Arrete Royal Relatif aux Produits Cosmetiques du 10 Mai 1978. This regulation was published in the *Moniteur Belge* on September 1, 1978, and officially took effect April 1, 1979. The Belgian directive is quite similar to the original EEC directive. It does, however, clarify the status of Annex

V substances that are excluded from the scope of the directive. Some, such as lead acetate, are prohibited, and others, such as strontium and zirconium salts, are permitted for specific uses. The trade association in Belgium, DETIC/BELLUCO, also represents the industry in Luxembourg and is headquartered in Brussels.

C. Denmark

The EEC Cosmetic Directive has been implemented in Denmark by Bekendtgorelse om Kosmetiske Produkter No. 38 of January 2, 1979. The regulation was effective April 1, 1979, and is very similar to the EEC directive. It did, however, go beyond the scope of the directive and banned some hair dyes. The trade association in Denmark is Saebe Parfumeri Toilet & Kemisktekniske Artikler (SPT).

D. France

France has adopted a cosmetic directive that is more stringent than the EEC directive. This is the Loi Relative aux Cosmetiques du 10 Julliet 1975, which was published in the *Journal Officiel* on July 11, 1975. The cosmetic law took effect on July 12, 1976, and covers, along with subsequent regulations, registration of manufacturers, packers, and importers. Subsequent regulations were issued on July 12, 1979, requiring manufacturers to prepare information dossiers for cosmetic products. Regulations on poisonous substances were published in Arret du 22 Mars 1977, Fixant la Liste des Substances Veneneuses Pouvant etre Incluses dans les Produits Cosmetiques, effective October 16, 1977. Further regulations covering information required for poison control centers were published in Arret du 27 Janvier 1978 and Arrete du 15 Fevrier 1978.

French law requires that cosmetic packages must indicate the quality and quantity of all ingredients featured in the product's labeling or advertising. If a label warning is required, the full text must appear on both primary and secondary containers. In 1980, the Ministry of Health issued a rule regarding moisturization claims in labeling. The rule indicated that claims for moisturization may be made in labeling or advertising only if moisturization is superficial and if the test of labeling or advertising explains how the water is retained. Terms such as "supermoisturizing" or "remoisturizing" may not be used. Products exerting action beyond the stratum corneum must obtain a special permit from the ministry. The French cosmetic trade association is the Federation Francaise de l'Industrie des Produits de Parfumerie, de Beaute et de Toilette, located in Paris.

E. Germany

Germany has adopted the EEC Cosmetic Directive with a harmonization decree, the Verordnung über Kosmetische Mittel von 16 Dezember 1977,

published in *Bundesgesetzblatt* of December 21, 1977, and effective
January 21, 1978. The German law is similar, but not identical, to the
EEC Cosmetic Directive. Germany adopted the use of zinc pyrithion
in a first amendment to its cosmetic regulation in December 1978. A
second amendment extended the time allowed for use of materials from
the EEC provisional list. In October 1979 a third amendment was issued
prohibiting two hair dyes and permitting use of FD&C No. 5, D&C Red
No. 6 barium lake in nail polish, and strontium chloride in toothpaste.
As of 1984, the Kosmetic Verordnung had been amended nine times.
The trade association in Germany is Industrieverband Körperpflege und
Waschmittel e.V. (IKW), located in Frankfurt.

F. Great Britain

British cosmetic legislation compliments the EEC directive and is contained in Cosmetic Products Regulations 1978, Statutory Instruments
No. 1354, Consumer Protection, taking effect October 15, 1978. The
EEC list of colors for non-mucous membrane application has not been
included in the British legislation. Great Britain adopted a weights
and measures directive, effective January 1, 1978, which, in harmony
with EEC standards, calls for product label declaration of solids in
grams and liquids in milliliters. Britain has not, however, adopted
the EEC prescribed quantities directive. The trade association in
Great Britain is the Cosmetic, Toiletry and Perfumery Association Ltd.
(CTPA), located in London.

G. Greece

Greece joined the EEC on January 1, 1978, and has a five-year transitional period within which to align itself with EEC rules. Draft cosmetic legislation has been proposed by the Greek Ministry of Social
Services, based on the EEC directive. In the past, all products sold
in Greece required registration with KEEF, the government laboratories
for the control of drugs. The new law is expected to continue the requirement for product registration. The trade association in Greece
is Union Panhellenique des Industriels et Agents de Produits
Cosmetiques et de Parfumerie, located in Athens.

H. Ireland

Ireland implemented the EEC Cosmetic Directive on December 10, 1979,
by regulation entitled Statutory Instruments No. 402 of 1979. This
regulation enacted as national law the EEC directive 76/768/EEC of
July 27, 1976, as amended by directive 79/661/EEC extending the
implementation timetable. The Federation of Irish Chemical Industries
(FICI), located in Dublin, represents the cosmetic industry.

I. Italy

The EEC Cosmetic Directive has been only partially implemented in
Italy. However, recent Ministry of Health ordinances and decrees go
beyond the Cosmetic Directive. For instance, the ordinance of March
7, 1979, prohibits use of certain hair dyes, and a ministerial ordinance
of February 15, 1980, limits the use of boric acid in talcum powders
and oral hygiene products and prohibits its use in powders sold for
babies. A ministerial ordinance published January 27, 1979, limits the
use of fluoride compounds and indicates labeling requirements. Two
important trade associations in Italy are Associazione Nazionale
dell'Industria Chimica (ASCHIMICI) and Unione Della Profumeria e Della
Cosmesi (UNIPRO), both located in Milan.

J. Luxembourg

Regulations implementing the EEC Cosmetic Directive were adopted
December 4, 1978, by the Grand Duchy of Luxembourg. They provide
for marketing of cosmetic products containing provisionally allowed
substances from Annex IV, Part 1, until August 31, 1979. Article VI
of the Luxembourg regulation prohibits marketing of cosmetics containing the hormonal materials estradiol and progesterone. The trade
association representing cosmetic companies in Luxembourg is the same
as that representing companies in Belgium: DETIC/BELLUCO, headquartered in Brussels.

K. The Netherlands

The Netherlands implemented the EEC Cosmetic Directive in a decree
that took effect September 1, 1980. The decree is very specific in its
definition of cosmetics and requires mandatory warning statements for
awide variety of materials. For example, inner soles of shoes containing methenamine are required to bear a warning label. In addition,
products containing ethisteron must indicate that they are to be used
exclusively for aging skin. The restrictions covering color additives
cover 12 separate product categories, and there are also regulations
governing aerosol formulations, packaging, and labeling, including a
strong warning for products containing chlorofluorocarbons. The
trade association in the Netherlands is Nederlandse Cosmetica
Vereniging in Utrecht.

L. Spain

Spain plans to join the EEC, and although no official date has been set,
entry is tentatively scheduled for 1983 or 1984. Spain currently regulates cosmetics under chap. 38 of the 1975 Code on Foodstuffs, the
Codigo Alimentario Espanol, which covers cosmetics and toiletries in
addition to foods. The code lists substances not permitted for use in

cosmetics, indicates general requirements for labeling and packaging
that can also apply to cosmetics (chap. 4), establishes good manufacturing practices (chaps. 3 and 4), and contains special requirements
for soaps.

A government decree of December 16, 1968 (No. 3339/1968, chap.
38), authorizes the development of standards for registration of cosmetic laboratories and products. Since then, the Scientifie Research
Council of Spain has worked to establish standards through the Institute for Rationalization and Normalization. Standards have been issued
for some raw materials used in the chemical industry. Spain is expected to adopt the EEC Cosmetic Directive upon its formal admission,
and a new cosmetic law (July 1, 1981) has been drafted, but is not yet
ratified. The trade association, located in Madrid, is S.T.A.N.P.A.

IV. OTHER WESTERN EUROPEAN COUNTRIES

European countries not included in the EEC are Austria, Finland,
Norway, Portugal, Sweden, and Switzerland. Because of their proximity to the EEC countries, some regulations are similar to the EEC
directive. Each, however, has its own distinctive approach to
cosmetic law.

A. Austria

In Austria, cosmetic products are governed by the Austrian Food Law,
which includes specific references to cosmetic products, Bundesgesetz
Nr. 86 vom 23 January 1975, über den Verkehr mit Lebensmitteln,
Verzehrprodukten, Kosmetischen Mitteln und Gebrauchsgegenstanden.
The law is referred to as the Lebensmittelgesetz 1975. It defines cosmetic products and specifies labeling requirements. A proposed
amendment to the law would restrict the use of pharmacologically active
substances in cosmetic products. On October 19, 1979, Austria adopted
a new law covering additional labeling requirements for cosmetic products, such as labeling of net weight, storage indications, featured
ingredients, and expiration dating. A new law, effective September
1984, specifies permitted ingredients and claims for pharmacologically
active materials in cosmetic products.

Fill weight regulations in Austria require that the minimum production fill is the amount to be declared on the label. There are also
specific regulations in Austria that require a minimum quantity of
active ingredient in bubble baths and shampoos. The trade association
is Fachverband der Chemischen Industrie Österreichs, located in
Vienna.

B. Finland

Finland issued a decree in 1977 entitled Statute Concerning Cosmetic
Preparations. Authority for this legislation was the Finnish Foodstuffs

Act of 1941. The decree covers requirements for labeling, including ingredient listing for products coming into direct contact with mucous membranes (other than the lips) for childrens' products, and for prohibited materials; a proposed revision was published in 1983. An extensive standards list for aerosol products was issued as a Decree of State Council, February 1, 1973. Although detailed, the standards are generally reasonable. Finland's Consumer Protection Act, in force since September 1978, established an independent authority, the consumer ombudsman, who assists individual consumers and consumer groups. Recommendations made by the ombudsman are reviewed by the Market Court, which has authority to prohibit some marketing practices. The trade association is Teknokemian Yhdistys r.y. in Helsinki.

C. Norway

The cosmetic regulations in Norway are contained in the Foods Act. This general legislation provides that cosmetics should be safe and should not contain poisonous substances. Enforcement of the act as it applies to cosmetic products is carried out by the Board of Health and Welfare. Consumer complaints are arbitrated by a consumer ombudsman and the Board of Consumer Policies. Most actions concerning cosmetics have been directed toward advertising. As a result, Norway has adopted a regulation prohibiting sexual discrimination in advertising. There have also been proposals in Norway to strictly control suspected carcinogens. It is hoped that any action will be consistent with the EEC lists of prohibited substances and trace materials. Norway prohibited manufacture and import of aerosol products containing chlorofluorocarbons by regulation promulgated by the Ministry of Environment on June 1, 1979. The act specifies that on or before July 1, 1981, chlorofluorocarbon-propelled aerosols must be withdrawn from the market. The trade association in Norway is Kosmetikkleverandorenes Forening in Oslo.

D. Portugal

Portugal issued a cosmetic regulation to govern cosmetic products published June 8, 1973, in *Diario do Governo*. The regulation includes provisions for good manufacturing practices and labeling standards, such as warnings for aerosol and hair dye products. The four annexes to the regulation contain a list of prohibited substances (Annex I), substances prohibited for use above certain concentrations (Annexes II and III), and colorants that may be used in cosmetics that come into contact with mucous membranes (Annex IV).

E. Sweden

At the present time, Sweden has no legislation dealing specifically with cosmetics, but bills calling for control of cosmetics have been brought before Parliament. The Board of Health and Welfare in the Ministry of Health and Social Affairs is currently reviewing the need for greater control of cosmetics. There is, however, an act giving authorities the right to control substances hazardous to health or the environment. Chlorofluorocarbons were prohibited by a statute issued December 15, 1977, which banned the use of these propellants from July 1, 1979. Methylene chloride and dichloromethane have also been prohibited.

   The Swedish Medical Board has also issued a negative list of materials that cannot be used in cosmetic products. For instance, materials such as methyl alcohol are banned, certain hair dyes have usage restrictions, and hexachlorophene can be used only in soaps or as a preservative at a prescribed level under specified conditions. Restrictions on the use of boric acid have also been issued. Sweden has a consumer ombudsman to handle consumer complaints and a Market Court for trade disputes. The trade association in Sweden is Kemisk-Tekniska Leverantorforbundet (KTF).

F. Switzerland

Switzerland's Federal Department published a decree regulating cosmetics February 12, 1970, including a list of substances with pharmacological action that may be used in cosmetics at specified maximum concentrations. The concentrations allowed vary depending on the kinds of product in which the substances are used. Group A includes ingredients for oral hygiene. Group B lists ingredients that remain on the skin. Group C includes substances that are either removed entirely or remain in traces on hair or nails. In addition, if a listed substance is mentioned in the product labeling, it must be present at a minimal concentration of between one-fifth and one-tenth that of the maximum concentration permitted. The same decree established guidelines for permissible claims for cosmetic products, again divided into the same three groups. There is also a list of substances prohibited in cosmetics. Cosmetic manufacturing firms must be registered, and products containing either vitamins or placental extract must be registered with the Swiss Vitamin Institute. In early 1982, the Swiss Federal Health Department proposed new rules for premarket testing and some ingredient labeling for products containing vitamins, "hypoallergenics," and baby products. There is also a labeling requirement in Switzerland for net weight and pricing declaration. The trade associations are ASCOPA, headquartered in Geneva, and Verband der Kosmetik Industrie, in Zurich.

V. PACIFIC AREA COUNTRIES

A. Australia

The cosmetic industry in Australia is largely self-regulated. The Therapeutic Goods and Cosmetic Act of 1972 enacted in New South Wales is the primary statute dealing directly with the cosmetic industry. In addition, the trade association in Australia, the Cosmetic, Toiletry and Fragrance Association of Australia in Sydney, has issued Codes of Voluntary Standards and Procedures, including the Code for Registration of Ingredients in Cosmetics and Toiletries, the Code of Good Practice to Avoid Excess Free Space and Deception in Packaging of Cosmetics, and the Code of Good Manufacturing Practice. This self-regulation has been very effective.

Registration of ingredients has been on a voluntary basis. Each company keeps a register of the ingredients used in its products and authorizes at least one person within the company to supply ingredient information to members of the public or the medical profession. The ingredients are grouped according to type and function. For instance, various vegetable waxes would be listed simply as vegetable wax.

The code of deceptive packaging is very specific and effectively curtails deceptive packaging.

Sunscreen products receive special attention in Australia because of the level of usage and the incidence of actinic damage in the population. Products on the market have been tested by the Department of Science and Consumer Affairs, University of Queensland. Results have been released to the public. Sunscreens have been classified into six groups according to the degree of protection afforded. The State of Victoria has passed regulations requiring registration of sunscreens, and the Standards Association of Australia has published a sunscreen testing and evaluation model.

B. Japan

Cosmetic regulations in Japan are extensive and complex. All cosmetic products in Japan, whether manufactured locally or imported, are subject to premarket clearance by the Ministry of Health and Welfare (MHW). MHW, through its Pharmaceutical Council, has established a standard for raw materials most frequently used in cosmetic formulations. One hundred fourteen ingredients were listed in August 1967. In 1970, one ingredient was deleted and 91 ingredients were added; in 1973, another 227 ingredients were added, for a total of 431 approved ingredients. This listing is published in the *Japanese Standards of Cosmetic Ingredients*. The government plans to add about 300 ingredients to the official standards in the near future.

In addition, the Japan Cosmetic Industry Association (JCIA) is working with MHW to publish on an unofficial basis the several thousand ingredients that have been approved by MHW when submitted in specific formulations by individual companies. CTFA has compiled an alphabetical list of generally known and permitted cosmetic ingredients. New cosmetic ingredients not listed in the official standards, nor approved for a specific type of formulation, could require a wide battery of tests, including acute, subacute, and chronic toxicity, mutagenicity, carcinogenicity, topical irritancy (such as skin and mucous membrane), allergenicity (such as skin and photo allergenicity), absorption, distribution, metabolism, and excretion testing.

A new amendment to the Pharmaceutical Affairs Law, effective September 30, 1980, requires ingredient labeling for 97 cosmetic raw materials, 83 coal tar colors, and 33 materials used in "quasi-drugs," the Japanese term for cosmetic drugs. The purpose of the amendment is to inform consumers of those cosmetics that contain ingredients deemed to have potential for sensitization or irritation. There are also voluntary standards for cosmetic label warning statements that vary according to product type. For instance, all cosmetics applied to the skin should be labeled "Attention: This product may cause untoward reaction to the skin on rare occasion. If so, please discontinue use." Nail cosmetics should indicate: "Do not use when there is any abnormality in your nail."

There are also specific regulations covering the cosmetic ingredients that may be used in certain products. For example, hair dyes may contain hydrogen peroxide as part of the oxidation reduction system, while home permanents may not. Formaldehyde was banned in the Harmful Substance Control Law 112, October 12, 1973, and MHW Ordinance 34, enforced on September 26, 1975. This essentially prohibits formaldehyde or formaldehyde donors from cosmetics. The Ministry of Health and Welfare has also listed 135 colors that are approved for specific uses.

There are also regulations covering cosmetic products with pharmacological action, that is, quasi-drugs. The list of substance considered quasi-drugs is contained in Appendix II MHW Ordinances 14 and 378 (1961). Specific restrictions covering concentration of active substances and products in which these substances can occur are outlined. Additional regulations have been passed covering manufacturing establishment registration. Specific requirements depend on the type of product manufactured in the plant.

Aerosols are regulated by the High Pressure Gas Control Law. Under this law, propellents such as liquefied petroleum gas and dimethyl ether, which are highly flammable, have been banned in cosmetic products since 1966 by ministerial regulation. Vinyl chloride was banned because of toxicity problems in 1974. The Japanese government reviewed consumption of chlorofluorocarbons following the Molina ozone depletion theory, and has achieved a reduction in the use of chlorofluorocarbons of approximately 25 percent. This was made easier because household products and insecticides were permitted to use

liquified petroleum gas as propellents, and because some cosmetic
manufacturers voluntarily changed aerosols to pump sprays. Future
government policies may further reduce the consumption of chlorofluorocarbons.

The extent and complexity of cosmetic regulations in Japan man-
dates obtaining excellent local advice. The trade association is the
Japan Cosmetic Industry Association, located in Tokyo.

C. The Philippines

The Philippines Food, Drug and Cosmetic Act No. 3720 was enacted
June 22, 1963. The act establishes statutory authority to regulate
cosmetic products and defines cosmetic misbranding. Specific administrative orders under the act ban various raw materials such as
chloroform; they restrict the use of other raw materials such as boric
acid; and they require registration of cosmetic products. Other
orders include regulations for the enforcement of the Food, Drug and
Cosmetic Act, requirements for cosmetic laboratories, product labeling
requirements, and lists of approved color additives. These administrative orders are detailed and in general follow U.S. cosmetic regulation. The trade association is the Chamber of Cosmetic Industry of
the Philippines, headquartered in Manila.

D. Sri Lanka

Sri Lanka has adopted a very interesting Drug, Cosmetic and Device
Law, Act No. 27, September 19, 1980. The law was developed as a
model by the World Health Organization for less developed countries.
It establishes good manufacturing practices and product licensing by
the Sri Lankan cosmetic, device, and drug authority, and requires
that cosmetic products be safe and adequately labeled. Food and drug
inspectors are authorized to inspect cosmetic products and establishments for violations in order to ensure compliance.

E. Other Pacific Countries

Other Pacific countries present different problems. For instance,
cosmetic regulations in India are general in nature and not restrictive.
The primary difficulty in India is the requirement for local ownership
or "Indianization," which is beyond the scope of this chapter. The
trade association is the Indian Soap and Toiletry Makers Association,
located in Calcutta.

Indonesia is a country where it is important to know the customs
and procedures. It is therefore necessary to have knowledgeable local
representation familiar with the various regulatory agencies. At the
time of this writing, free importation of finished goods is not possible.
The most appropriate trade association is PERKOSMI in Jakarta.

In Korea most of the problems are trade-related, such as how to repatriate profits. Specific cosmetic regulations do not generally present problems. The trade association is the Korea Cosmetic Industry Association, located in Seoul.

Malaysia and Singapore, which are governed separately, both allow relatively free trade and importation of cosmetic products. There are no specific government regulations in Malaysia covering cosmetic products. The most appropriate trade group in Malaysia is the Malaysian Pharmaceutical Trade and Manufacturers Association.

New Zealand severly restricts importation of finished products and raw materials, including packaging. However, government regulations for local cosmetic manufacture are general. The major problems relate to trade. The trade association in New Zealand is the New Zealand Cosmetic and Toiletry Manufacturers Federation in Wellington.

Taiwan passed a cosmetic regulation in December 1972. It includes general definitions, requires licensing of manufactured and imported goods, and provides for establishment inspection and good manufacturing practices regulations. A list of approved colorants as well as a list of medical or poisonous materials are contained in the cosmetic regulation. Featured ingredients must be listed on the package. The primary problems in Taiwan are trade-related, such as restriction on the import of finished goods and repatriation of profits.

Thailand's cosmetic act was issued in 1974, and on October 12, 1976, the Ministry of Public Health published three notifications covering a prohibited materials list (Notification 3), a quality standard for heavy metal traces with an annex covering permitted color additives and pigments (Notification4), and a list of products referred to as controlled substances with an annex of restricted materials contained in these controlled cosmetics (Notification 5), as well as usage limitations and labeling requirements for the products. The limitations are detailed but not restrictive beyond normal cosmetic regulation. The trade group in Thailand is the Cosmetic Manufacturers Association in Bangkok.

VI. AFRICAN COUNTRIES

There are many countries in Africa, most of which do not have specific cosmetic regulations. Free trade, importation of goods, and repatriation of profits are difficult in some of these countries. The major countries with regulatory programs are discussed below.

A. Kenya

Kenya does not have a cosmetic regulation, but there is a Pharmacy and Poisons Act, which was revised in 1972. Part II, chapter 23, covers cosmetics and cosmetic drugs. The administration of this act is by the Pharmacies and Poisons Board in Nairobi. Kenya also requires import licenses for all equipment and raw materials, and places limitations on ex-patriots conducting business in the country.

B. Nigeria

Nigeria passed a Food and Drugs Decree in 1974. This decree covers drugs, devices, and cosmetics and is administered by the Drug Control Division of the Food and Drug Administration. Federal Ministry of Health and Social Welfare. The decree of 1974 became effective November 2, 1978, and requires documentation for products already on the market, including product brand name, generic name, composition, labeling, data from experimental and clinical trials, anme and address of the manufacturer, country of origin, name and address of the importer, name and address of the Nigerian marketer, and date of introduction onto the Nigerian market. New products require the same types of data together with samples, product specifications, and methods of analysis. There is also a Weights and Measures Decree of 1974 that became effective on April 1, 1978. Schedule 8, section 28, of the decree requires prepackaged goods to use standard metric quantities.

C. South Africa

South Africa's Foodstuffs, Cosmetics and Disinfectants Act was published June 2, 1972, in the *Government Gazette*. The act controls the sales, manufacturing, and importation of cosmetics, including control of packaging, labeling, and advertising. It also authorizes Inspectors from the Ministry of the Department of Health to enforce the regulation and includes a published list of materials not permitted in cosmetics.

In 1978, South Africa issued proposal legislation for comment that would require ingredient labeling, aerosol warnings, and indicated that some of the EEC Cosmetic Directive annexes could be adopted by South Africa. Industry made comments that the proposed legislation should be amended. The legislation has not been enacted at this date.

D. Other African Countries

Zambia has established a Standards Institute in its Bureau of Drugs, which is currently developing detailed standards of acceptability for cosmetic raw materials such as talc, mineral oil, and petroleum jelly. The standards conform in general to standards previously adopted in the U.S. Pharmacopoeia or British Pharmacopoeia, with some differences due to local preferences for certain analytical test methods or specifications.

Zimbabwe does not register cosmetics but has a full set of regulations for cosmetics drugs, which does include registration. Major problems in Zimbabwe involve restrictions on imports and repatriation of profits.

VII. LATIN AMERICAN COUNTRIES

Although the Latin American countries share a common culture and history, there are wide differences from country to country with respect to cosmetic regulation. In general, technical regulations are not as important as product registration. In almost all Latin American countries, products must be registered with the government before they can be sold. Product registration generally precludes any further regulatory problems. However, registration requirements tend to be extremely detailed and require the submission of formulas, analytical data, specifications for raw materials, and examples of labeling and claims, including advertising.

A. Argentina

Regulations in Argentina are contained in articles issued by the Health Department under the authority of Decree No. 141 (1953). These articles cover extensive documentation requirements for cosmetics and their ingredients. Article 890 lists color additives and their specifications and usage. Article 900 covers requirements for shaving creams and soaps. Article 901 covers requirements for talcum powder, including limitations on boric acid content. Article 902 goes into detail on combinations of talcum powders, corn starches, zinc oxides, and so on. Article 903 covers hair dyes and simply states that official certication is required. Article 904 covers warning labels for hair dyes. Article 905 indicates warnings for flammable materials. Article 906 covers labeling for irritating substances. Articles 907 covers tanning products, while Article 908 limits antimicrobial substances.

In addition, in 1981 the government issued Resolution 360 containing the General Technical Rules for Formulations, Elaboration and Marketing of Cosmetic Products, including revised positive lists for cosmetic ingredients and a provision for yearly review of the rules. This resolution was modified by Ministerial Resolution 1390, issued July 2, 1982.

Cosmetic aerosols are regulated by Regulation 710, covering definitions and labeling, and Regulation IRAM 3793, December 1978, covering detailed constraints on flame extension. In addition, the regulation does not permit exclusive use of hydrocarbon propellents in cosmetics, but allows free use of chlorofluorocarbons. Uncoated glass containers may not have pressures greater than 1 atmosphere. The trade association is Camara Argentina de la Industria de Productos de Higiene y Tocador in Buenos Aires.

B. Bolivia

Bolivia's cosmetic regulation is Resolucion Ministerial 0393, issued by
the Minister of Public Health on June 11, 1975. It is a general decree
that does, however, require company registration. Individual products
are not registered unless claims for pharmacological activity are made,
in which case products must be registered in compliance with drug
regulations.

C. Brazil

Brazil has issued extensive cosmetic regulations contained in the Sanitary Vigilance Law, No. 6360, published in September 1976, and in
Decree No. 79,094 (January 5, 1977). Brazil required products to be
registered before they can be marketed. Such registration is valid
for only five years, but it can be extended. There are also detailed
lists of "safe" substances published by the Minister of Health, and if a
product contains substances not on these lists, the petitioner must convince the ministry of the safety of the product. The positive lists are
included in Normativ Resolucions for colorants #1/78, dye-stuff intermediates #2/78, preservatives #3/78, antioxidants #4/78, chelating
agents #5/78, thickening agents #1/79, additives such as silk powder
#3/79, surfactants #4/79, a separate decree for emollients #5/79, and
astringents #6/79. In addition, a new list of ultraviolet absorbers was
published in 1980. These lists are extensive but do not limit a manufacturer to the sole use of the materials on the list if safety data for
the product containing a new raw material is submitted in the registration and accepted. There are several trade associations in Brazil.
Three of the major organizations are Associacio Brasileria da Industria
de Productos de Limpeza e Afins do Sao Paulo, Sindicado da Industria
de Perfumaria e Artigos de Toucador do Estado de Sao Paulo, and
Sindicado da Industria de Perfumaria e Artigos de Toucador do Estado
de Trio de Janeiro.

D. Chile

Chile requires detailed information for registration of cosmetic products.
Importation of finished goods is permitted; however, imported products
must be registered and must be accompanied by a Certificate of Free
Sale and a health certificate from the country of origin. Chilean regulations specify which colorants may be used in cosmetic products and
in general follow U.S. Food and Drug Administration color regulations.
The trade association is Camara de Industria Cosmetica in Santiago.

E. Colombia

Colombia also requires detailed registration of cosmetic products and
of manufacturing facilities. A certificate of trademark ownership must

be submitted to the Ministry of Health with the registration data and product samples. Guidelines for good manufacturing practices have been issued by the Director General of Surveillance in the Division of Surveillance, Ministry of Health. The trade association in Columbia is ACOPER, located in Bogota.

F. Ecuador

Ecuador also has complex registration requirements. Documentation needed for imported materials is more extensive than for products made locally, and includes technical information as well as health certificates and Certificates of Free Trade approved by the Ecuadorian Consulate in the exporting country.

G. Guyana

Guyana published a full cosmetic regulation in its official gazette on August 6, 1977, that details standards for cosmetics and provides instructions for cosmetic labeling and advertising. The law establishes government authority to inspect cosmetic manufacturing facilities and to assure compliance with the regulation.

H. Mexico

Cosmetic requirements in Mexico are contained in a very detailed regulation published by the Ministry of Health and Welfare in its *Diario Official* of August 16, 1960. The regulation consists of 85 separate articles covering every facet of the cosmetic industry, including, for example, a requirement specifying the number of toilets per person in manufacturing facilities.

This regulation also lists substances that may be sold as raw materials from perfumeries and cosmetic shops. Full details of the requirements for product registration are contained in articles 46 through 65. Details about advertising are contained in articles 66 through 79, again in great detail.

Augmeting the cosmetic regulation is the Sanitary Code of Mexico, published in the *Diario Official* on March 13, 1973, which includes articles outling specific requirements for products and packaging. The trade association in Mexico is Camara Nacional de la Industria de Perfumeria y Cosmetica.

I. Panama

Panama traditionally has enjoyed free and open trade for many years. While there are various government health regulations, there are no specific cosmetic regulations. Free importation of finished goods and raw materials is allowed.

J. Peru

Peru's cosmetic regulation is contained in the Codigo Sanitario 17505, dated March 18, 1969, and Decree Law 21199, dated June 26, 1975. These cover general requirements for good manufacturing practices and outline the government's authority in regard to cosmetics. For instance, products must be registered with the Pharmacy Division of the Ministry of Health. Such registration is valid for three years and can be renewed. The product registration must indicate that the manufacturing facility has also been registered and that the company is authorized to sell the product being registered. The trade association in Peru is Associacion de Fabricantes de Articulos de Tocador in Lima.

K. Uruguay

Product registration is required in Uruguay. In addition, it is difficult to import raw materials and packaging. The trade association in Uruguay in CUPCAT in Montevideo.

L. Venezuela

The government of Venezuela actively regulates cosmetic products. Initial cosmetic legislation was passed in Venezuela on August 26, 1975 (Decree No. 1118), and includes general requirements for cosmetics, such as product registration, good manufacturing practices, and labeling and advertising. Registration is required for certain cosmetic products, including hair dyes; deodorants and antiperspirants; other hair products, including hair dressings, shampoos, and rinses; as well as tanning products. Regulations to require registration of other cosmetic products are being developed.

A Consultant Committee for cosmetic products in the Ministry of Health issues bulletins for discussion before final regulations are adopted. The "Unique" Bulletin of cosmetic regulations, issued December 1981, incorporates and supersedes all previous bulletins. It covers materials not permitted in cosmetics, such as salicylic acid, resorcinol, and benzoyl peroxide. There are also lists of prohibited or restricted colorants, concentration limits for materials such as camphor and vitamins and for trace substances in pigments, and for microorganisms in the finished product. In addition, labeling requirements and guidelines for adequate safety testing are included. The trade association is CAVEINCA, located in Caracas.

VIII. OTHER COUNTRIES

A. Canada

Canada regulates cosmetics under its Food and Drugs Act, sections 16 and 18, and has issued several implementing regulations such as those specifying proper labeling copy, and in 1972 a regulation restricting the use of hexachlorophene in cosmetics was issued. The Canadian Consumer Packaging and Labeling Act of March 1, 1974, enacted by the Minister of Consumer and Corporate Affairs, requires standardized packaging sizes for many cosmetic products.

An additional cosmetic regulation published in the *Canada Gazette* of October 12, 1977, pursuant to section 25 of the Food and Drug Act, requires manufacturers or importers to notify the government of all new cosmetic products within 10 days of first sale if manufactured in Canada or prior to first importation. Notification includes the name of the product, ingredients, concentrations of ingredients, form of the product, and name and address of the manufacturer or distributor, as well as warning requirements for various cosmetic products and detailed testing and labeling of aerosols.

Claims for advertising and labeling do not require preclearance, but all broadcast advertising, copy, and illustrations must be approved, and claim substantiation may be required. The Charter of the French Language of Quebec requires that all products marketed within the Province of Quebec be labeled in both the French and English languages after July 1978.

Canada's Weights and Measures Act was amended December 28, 1977, to cover metric designations. Canada has also passed regulations prohibiting the use of chlorofluorocarbons as propellents in hairsprays, deodorants, and antiperspirants, but allowing chlorofluorocarbons to be used in other cosmetic products such as fragrances. This regulation was published in the *Canada Gazette* March 24, 1979, effective May 1, 1980, as part of the Environmental Contaminant Act.

The trade association in Canada is the Canadian Cosmetic, Toiletry and Fragrance Association (CCTFA), which published a Code of Consumer Advertising Practices for Cosmetics, Toiletries and Fragrances in November 1976. This code is self-enforced by an arm of the Canadian Regulatory Advisory Board, and covers such items as conformity with appropriate legislation, claim substantiation, competitive advertising, and good taste. In addition, the CCTFA has published guidelines for good manufacturing practices (GMP) for cosmetic manufacturers, a voluntary code followed by many member companies.

B. Egypt

Egypt requires product registration. Its cosmetic law is general in nature but prohibits certain raw materials, such as hormones, all vitamins except A, D, B, and E in cosmetics, heavy metals above certain limits, hexachlorophene at a level greater than 0.1 percent, and any more than a trace of any drug listed in the official pharmacopoeia.

C. Iran

An Iranian Food, Drink, Cosmetic, and Hygiene Act was published in 1967. However, it is not known whether that particular act is still in force. The act was general in nature and included specific criminal penalties for violations. For example, the act authorized mandatory imprisonment for the person responsible for a purposely adulterated cosmetic that caused disfigurement, and if a purposely adulterated food or cosmetic caused death, the penalty could be as severe as death.

The establishment of a cosmetic factory required a permit from the Ministry of Health and from the Ministry of Economy. The act also included lists of permitted colors, and any materials not on the permitted list were prohibited. Import permits from the Ministry of Health for cosmetic products required full formula disclosure.

D. Israel

Israel's cosmetic regulations require individual product registration for both products manufactured locally and those imported into Israel. A general license must be obtained for the importer, manufacturer, or distributor. The registration must include details about the product, such as formula, packaging, labeling, and advertising. If materials used in a cosmetic product do not conform to the U.S. Pharmacopoeia, British Pharmacopoeia, another recognized pharmacopoeia, or the CTFA Cosmetic Ingredient Descriptions, then safety data must also be submitted with the registration. Color additives are permitted for use in Israel if they have been approved by the U.S. Food and Drug Administration or are included in the EEC Cosmetic Directive. Labeling requirements specify that some aspects of the label copy must appear in Hebrew.

IX. EASTERN EUROPEAN COUNTRIES

Most Eastern European countries have extensive cosmetic regulations and guidelines, which vary from country to country.

A. Bulgaria

Bulgaria published a regulation and hygienic standard for cosmetic products on January 20 and July 21, 1970. This general regulation includes a detailed list of raw materials authorized by state standard of by established industry norms. Some positive lists are similar to those used in the United States and the EEC. Importer products require full government registration, a free sales permit, and approval by health authorities.

B. Czechoslovakia

Czechoslovakia published a cosmetic regulation on January 1, 1971. The Czechoslovakian regulations are general in nature except for specific details regarding microbiological acceptability of cosmetic products. There is also a list of prohibited substances and a list of

materials subject to usage limitations. Imported cosmetic products
must be registered.

C. German Democratic Republic

The German Democratic Republic regulates cosmetics under its Food
and Commodities Law and has published positive lists of permitted
colorants. Cosmetics that have some drug action are covered under
the drug law. All products sold in East Germany must be registered.

D. Hungary

Hungary issued a cosmetic regulation on February 2, 1969, and a
regulation on allowable colors in 1963. All cosmetic products must be
registered with the government.

E. Poland

Poland does not have a published cosmetic law but does require product
registration. In addition, some materials are prohibited for use in cosmetics in Poland. A list of prohibited substances can be obtained from
the Public Health Department, as can the lists of colorants approved
for cosmetic usage.

F. Rumania

Rumania does not have a formal cosmetic law. Products must be registered in great detail with the government before they may be sold.

G. Union of Soviet Socialist Republics

The Soviet Union does not have a specific cosmetic law, but rather
publishes individual standards that cover all details required for each
product, for example, fragrances. The Goss Standard 7237 of 1971,
reissued in July 1974, concerning perfumes covers all aspects of product, package, manufacture, and analysis. There are additional Goss
Standards covering such items as the kinds of separatory funnels and
pipettes to be used in manufacture. These standards must, of course,
be cross-referenced in order to ensure compliance with the regulation.
    Permitted concentrations are specified by fragrance type. For
instance, a youth fragrance may contain 9 percent concentration of
essential oil, while camomile fragrance may contain only 4.1 percent
essential oil. Details such as whether shipments must be encased in
wood, depend on how they will be shipped. Regulations for the wire
to encase the wood are outlined in one Goss Standard, while steel packing bands are outlined in another. All product types are covered by
these detailed standards. However, in general, imported products are
exempt from these norms as long as all specifications for the product
are spelled out in great detail in the negotiated agreement to purchase.

H. Yugoslavia

Yugoslavia, although an Eastern European country, is not a member of the Warsaw Pact or of COMECON. Yugoslavia published a cosmetic regulation in the *Official Gazette* No. 13, March 14, 1974, that includes positive lists for cosmetic antioxidants, preservatives, colorants, and a list of restricted materials. Yugoslavia does not require product registration but does monitor compliance with the general regulation. The trade association in Yugoslavia is Savez Farmaceutiskih Drustave Jugoslavije Sekcija za Kozmetologiju, located in Ljubljana.

X. WORLD HEALTH ORGANIZATION

The World Health Organization (WHO) is the part of the United States involved with medical and environmental concerns. It sponsors a World Health Assembly, and at its May 1977 meeting, assembly members pointed out the need for a program to evaluate effects of all chemicals on health. During 1978 and 1979, an action plan was developed to initiate this program. The program established a central World Health Organization unit for planning and coordinating the work to be done by a network of international institutions assigned to specific tasks. The objectives set were to disseminate evaluations of the effects of chemicals on human health and the quality of the environment; to develop guidelines on exposure limits in air, water, food, and the working environment for all types of chemicals; to develop guidelines for appropriate methods for toxicity testing, epidemiologic and clinical surveys, and risk and hazard identification; to coordinate laboratory testing and epidemiological studies where an international approach is appropriate; to promote research on dose-response relations and the mechanisms of biological actions of chemicals; and to develop information for coping with chemical accidents. This is indeed an extremely ambitious program.

Coordination with many existing national regulatory bodies as well as multinational trade and scientific organizations has been initiated. For instance, the WHO program will evaluate chemicals according to a priority list, and the U.S. Food and Drug Administration has given WHO its own regulatory list of food additives, food contaminants, animal feed additives, cosmetic ingredients, and natural flavors. Cosmetic ingredients are fourth on the FDA priority list. The FDA has indicated to WHO that although cosmetic ingredients have not received a high-priority rating for work to be done through 1981, it feels that attention must be paid to these materials. The FDA has indicated a willingness to serve as a lead institution for evaluation of cosmetic raw materials, partially because of the efforts already underway in the United States to evaluate cosmetic ingredient safety through the Cosmetic Ingredient Review undertaken by the CTFA, as well as the FDA's own work.

The goal of WHO is to have these lead groups prepare annual documents on environmental chemicals for evaluation by the World Health Assembly Program. WHO also intends to prepare monographs in order to assess risks for chemicals for which substantial data exist. Data received from industry will be funneled through the International Center for Industry and the Environemnt, as association representing industries and various industrial associations. Continuing review of this program will have to be made in order to ensure that the latest, most appropriate data are considered rather than simply existing prior data that have been reviewed and found technically deficient. Extensive coordination will be the key.

WHO has also published guidelines for model drug and cosmetic regulations and drug good manufacturing practices (GMP) standards. These have been disseminated to U.N. members. As a result, Sri Lanka recently adopted a cosmetic law modeled on these guidelines, and Venezuela has published for comment a draft cosmetic GMP document that was modeled on the WHO drug GMP guidelines.

XI. SUMMARY AND CONCLUSION

We have attempted in this chapter to highlight cosmetic regulations and regulatory approaches. There are some conclusions that can be drawn regarding the regulatory approach taken by various countries. Generally, regulations in Europe and the English-speaking countries tend to require product labeling and to prohibit or restrict certain cosmetic ingredients; Japan requires premarket approval for each individual product, which varies from the practice throughout most of the world; Lation American and Easter European countries tend to rely on detailed product registration requirements to control cosmetic manufacture and importation.

Because national cosmetic regulations are constantly changing, the information in this chapter should be used only as a guide. In fact, there will undoubtedly be some changes before this chapter goes to print, and, of course, it is also possible to misinterpret rules and regulations.

Therefore, it is absolutely essential that you use as many references as possible to keep up with international cosmetic regulations and, more important, that you develop local sources of information who assess the regulatory climate within the country and keep you apprised. You will be served well only if you tell local resource persons about your company's needs and business practices, and then rely on them for advice and execution.

XII. APPENDIX

The Cosmetic, Toiletry and Fragrance Association, Inc., Suite 800, 1110 Vermont Avenue N.W., Washington, D.C. 20005

Office of Export Marketing Assistance, Bureau of Export Development, U.S. Department of Commerce, Washington, D.C. 20230

Office of Country Marketing, Bureau of Exporting Development, U.S. Department of Commerce, Washington, D.C. 20230

Trade Advisory Center, Office of Trade Policy, Room 3036, U.S. Department of Commerce, Washington, D.C. 20230

Standards Information Center, Room D162, Bldg. 225, National Bureau of Standards, U.S. Department of Commerce, Washington, D.C. 20234

IICCI Bulletin, Box 1542, Hollandargatan 23, S-111 85 Stockholm, Sweden

International Drug Regulatory Monitor, Monitor Publications, 1545 New York Avenue, N.E., Washington, D.C. 20002

International Federation of Society of Cosmetic Chemists, 56 Kingsway, London WC2B6DX, England

International Cosmetic Regulations, Allured Publishing Corporation, P.O. Box 318, Wheaton, Ill. 60187

Dun & Bradstreet International Ltd., 99 Church Street, New York, N.Y. 10007

The Business International Corporation, 1 Dag Hammarskjold Plaza, New York, N.Y. 10017

Commerce Clearing House Inc., 4025 W. Peterson Avenue, Chicago, Ill. 60646

International Register of Potentially Toxic Chemicals Bulletin, United Nations Environmental Program, World Health Organization, Avenue Appia, 1211 Geneva 27, Switzerland

COLIPA, Rue de la Loi, 223 (Bte 2), B-1040 Brussels, Belgium

DETIC/DELLUCO, Square Marie-Louise 49, B-1040 Brussels, Belgium

Saebe Parfumeri Toilet & Kemisktekniske Artikler (SPT), Ostergade 22, 1100 Copenhagen, Denmark

Federation Francaise de l'Industrie des Produits de Parfumerie, de Beaute et de Toilette, 8, Place du General Catroux, F 75017 Paris, France

Industrieverband Körperpflege und Waschmittel e.V. (IKW), Karlstrasse 21, D-6000 Frankfurt/M 1, Germany

Cosmetic, Toiletry and Perfumery Assn. Ltd. (CTPA), 35 Dover Street, London WIX 3RA, England

Union Panhellenique des Industriels et Agents de Produits Cosmetiques et de Parfumerie, 28, rue Academias, Athens, Greece

Associazione Nazionale dell'Industria Chimica (ASCHIMICI), Via Fatebenefratelli 10, I-20121 Milano, Italy

Unione Della Profumeria e Della Cosmesi (UNIPRO), Via Buonarroti 38-Milano, Italy

Nederlandse Cosmetica Vereniging, Gebouw Trinderborch, Catharij-nesingel 53, 3511 GC-Utrecht, Netherlands

S.T.A.N.P.A., San Bernardo, 23-2, Madrid 8, Spain

Fachverband der Chemischen Industrie Osterreichs, Gruppe Korper-pflegemittelindustrie, Schliessfach No. 69, A-1011 Wien, Austria

Teknokemian Yhdistrys r.yu., Fabianinjatu 7B, SF-00130 Helsinki 13, Finland

Kosmetikkleverandorenes Forening, Boks 6780 St. Olavs Pl., Oslo 1, Norway

Kemisk-tekniska Leverantorforbundet (KTF), Box 1542, S-111 85 Stockholm, Sweden

ASCOPA, Case postale 230, CH-1211 Geneva 3, Switzerland

Verband der Kosmetik-Industrie, Breitingerstrasse 35, 8002 Zurich, Switzerland

Cosmetic, Toiletry and Fragrance Assn. of Australia, 60 York Street, Sydney, New South Wales 2001, Australia

Japan Cosmetic Industry Assn., 4th Floor Hatsumei Bldg., 9-14, 2-chome Toranomon, Minato-Ku, Tokyo 105, Japan

Chamber of Cosmetics Industry of the Philippines, P.O. Box 4541, Manila, Philippines

Indian Soap and Toiletries Makers Assn., P-11, Mission Row Extension, Calcutta 1, India

PERKOSMI, C/O P. T. Kalbe Farma, J1 Jendral a Yani, Jakarta, Indonesia

Korea Cosmetic Industry Assn., 45-1, Pil-Dong, Jung-Ku, Seoul, Korea

Malaysian Pharmaceutical Trade and Manufacturers Assn., 3rd Floor, Jaya Supermarket, Jalan Semangat, Petaling Jaya, Malaysia

New Zealand Cosmetic and Toiletry Manufacturers Federation, P.O. Box 9130, Wellington, New Zealand

The Cosmetic Manufacturers Assn., 292/37 Lan luang Road, Siyek Mahamark, Khet Dusit, Bangkok, Thailand

The Proprietary Association of South Africa, P.O. Box 933, Pretoria 0001, South Africa

The Grocery Manufacturers Association of South Africa, P.O. Box 10435, Johannesburg 2000, South Africa

Camara Argentina de la Industria de Productos de Hygiene y Tocador, Paraguary 1857, Capital Federal, Buenos Aires (1121), Argentina

Associacao Brasileria da Industria de Produtos Limpezae Afins, 1570-8 Andar, San Paulo, Brazil

Sindicado da Industria de Produtos de Perfunarias e Artigos de Toucador, Av. Paulista, 1319-9 Andar, San Paulo, Brazil

Sindicado da Industria de Perfumaria e Artigos de Toucador, Avenida Calogeras, 15, 4° andar, CEP 20030, Rio de Janeiro RJ, Brazil

Camara de Industria Cosmetica, San Antonio 427, Santiago, Chile

ACOPER, Igancio Chiappe Lemos, Calle 37, No. 7-43, Bogata, Colombia

Camara Nacional de la Industria de Perfumeria y Cosmetica, Talsan No. 54 Despacho 6, Mexico 1, D.F., Mexico

Associacion de Fabricantes de Articulos de Tocador, Los Laureles 365, San Isidro, Lima, Peru

CUPCAT, Adva. Agraciada 1670 Piso 1, Montevideo, Uruguay

Camara Venezolana de la Industria de Cosmeticos y Afines (CAVEINCA), Edificio "IASA" Plaza la Castellana, Oficiana 106 Apartado 3577, Caracas, Venezuela

Canadian Cosmetic, Toiletry and Fragrance Assn. (CCTFA), 24 Merton Street, Toronto, Ontario M4S 1A1, Canada

Savez Farmaceutiskih Drustave Jugoslavije Sekcija Za Kozmetologiju, Askercerva 9, 6100 Ljubljana, Yugoslavia

9

COSMETICS: THE LEGISLATIVE CLIMATE

EVE E. BACHRACH

Legal Department, The Cosmetic, Toiletry and Fragrance Association, Inc., Washington, D.C.

I. INTRODUCTION

Cosmetics became the subject of federal legislation with the passage of the 1938 Federal Food, Drug, and Cosmetic Act (FDC Act). Jurisdiction to enforce the adulteration and misbranding provisions of the act was invested in the Food and Drug Administration (FDA).

163

The adulteration provision of the Act[1] deems a cosmetic to be adulterated if it contains any poisonous or deleterious substance that "may" render it injurious to users under customary conditions of use. The adulteration provision contains an exemption for coal-tar hair dyes bearing the prescribed warning label.

The misbranding provision of the Act[2] prohibits labeling that is "false or misleading in any particular" and also requires the conspicuous label designation of quantity of contents and name and address of the manufacturer, packer, or distributor.

To enforce these prohibitions against adulterated or misbranded cosmetics, the FDA has the power to inspect factories producing cosmetics, to institute seizure proceedings against violative products, to institute proceedings to enjoin their production, and to prosecute criminally those who produce them.[3]

II. THE COLOR ADDITIVE AMENDMENTS

The first major change in the FDC Act affecting cosmetics came in 1960 with the passage of the Color Additive Amendments.[4] The amendments provide that all color additives used in foods, drugs, and cosmetics must be proved safe to the satisfaction of the FDA before they can be marketed commercially. This premarket clearance imposed vast new responsibilities on the food, drug, and cosmetic industries to undertake safety research on the hundreds of color additives used in their products.[5]

III. THE DRUG AMENDMENTS

A second change in the FDC Act occurred in 1962 with passage of the Drug Amendments, which called for premarket clearance of "new" drugs.[6]

The Drug Amendments did not have a direct effect on cosmetics but served to focus attention on the FDC Act as a whole and were the likely impetus for H.R. 8418, the Cosmetic Amendments of 1963, introduced by Congressman Harris. The bill mirrored the Drug Amendments of 1962 and contained provisions calling for premarket clearance of "new" cosmetics for safety, including filing of "new cosmetic applications." The Harris bill also contained a Delaney anticancer clause and requirements for cautionary labeling. The bill was referred to the House Committee on Interstate and Foreign Commerce, but no hearings were held and the bill was not enacted.

IV. THE 1970s—THE RISE OF CONSUMERISM

The early 1970s saw a rise in the consumer protection movement in the United States that affected every segment of industry.

Private initiatives to improve safety of goods sold to the public were accompanied by increased government activity: The FDA increased its regulation of cosmetics within the existing provisions of the FDC Act, and pursuant to the (then) recently enacted Fair Packaging and Labeling Act (FPLA).[7] Ingredient labeling of cosmetic products was required by the FPLA,[8] and warning statements for aerosols and feminine deodorant sprays were required by the FDC Act.[9] The FDA also required products not substantiated for safety before marketing to bear a warning informing consumers of that fact.[10] Similarly, the FDA banned several substances from use in cosmetic products; bithionol,[11] zirconium aerosols,[12] vinyl chloride,[13] mercury,[14] halogenated salicylanilides,[15] and chloroform[16] were included on the list.

During the same period, the first real attempts to expand the underlying legislative authority of the FDA were undertaken. Early initiatives in the House of Representatives included bills sponsored by Congressmen Evans (H.R. 13417 and H.R. 3261), Sullivan (H.R. 1235), and Roybal (H.R. 1107). The Evans and Sullivan bills would have eliminated the hair dye and soap exemptions,[17] required FDA-approved premarket safety testing of cosmetics, directed submission of adverse reaction data to poison control centers, and mandated even more detailed ingredient labeling of cosmetics than that required by the FPLA. In addition, the Sullivan bill would have included a Delaney anticancer clause applicable to cosmetics. The Roybal bill contained provisions requiring specific safety tests to be undertaken before a cosmetic could be marketed, mandatory registration of cosmetic manufacturers and cosmetic formulas with FDA, and ingredient labeling. These provisions are typical of those found in all the proposed cosmetic legislation of the 1970s.[18]

The bills were referred to the House Committee on Interstate and Foreign Commerce, but no hearings were held. However, in anticipation of possible House activity on Congressman Evans's first bill, H.R. 13417, the cosmetic industry through its trade association, the Cosmetic, Toiletry and Fragrance Association (CTFA), prepared a statement setting forth its position on the provisions.

The CTFA statement emphasized the good safety record of the cosmetic industry. It opposed the requirements for premarket safety testing using prescribed tests and for FDA premarket approval, pointing out that allergic reactions of susceptible individuals to some cosmetic ingredients cannot be prevented by safety testing or premarket approval. Industry also objected that a rigid and burdensome testing requirement would cause industry research costs to skyrocket out of proportion to the minor and transient injuries suffered by allergic individuals. Furthermore, industry was already working of a partial solution to the allergic reaction problem through support of FPLA-based ingredient labeling. Industry also supported the forwarding of adverse reaction information to poison control centers voluntarily, obviating the need for the mandatory scheme contained in the bills.

Similarly, industry took the position that the existing voluntary program of cosmetic establishment registration was adequate to supply
the FDA with useful information, such that a mandatory registration
program was unwarranted.[19]

Throughout its statement CTFA stressed the benefits consumers
derive from cosmetics and the enviable safety record of the cosmetic
industry, all of which belied the need for further legislation. These
themes surfaced throughout industry's responses to legislative initiatives in the 1970s.

V. THE EAGLETON BILL

At the time the House bills described were languishing, Senator
Eagleton introduced S. 863 in 1973. This began the most important
legislative initiative of the 1970s relating to cosmetics.

The bill was referred to Senator Kennedy's Subcommittee on
Health of the Senate Committee on Labor and Public Welfare, where
two days of hearings were held on February 20 and 21, 1974.[20]

The major cosmetic issues were identified and discussed in these
hearings.

VI. THE MAJOR COSMETIC ISSUES

A. Premarket Clearance

Premarket *clearance* of cosmetics by the FDA versus premarket *testing*
by the industry without FDA oversight constituted a major issue.
Premarket *clearance* was the most restrictive, requiring the same oversight by the FDA as the agency devoted to drugs. Premarket *testing*
requirements contained in the Eagleton bill were extremely restrictive,
and designated specific product tests that industry would be required
to perform before marketing.

B. Soap and Hair Dye Exemptions

A second issue addressed by the hearings was proposed deletion of
both the soap[21] and hair dye[22] exemptions of the FDC Act. Industry
opposed the deletions.

With respect to hair dyes, industry testified that hair colorants have
a proven record of safety, as demonstrated by millions of safe uses of
these products. The patch test warning on hair colorant products is
designed to, and adequately does, protect users from possible ill effects.
Hair colorants bear ingredient labeling pursuant to the FPLA, and this
labeling provides an additional warrant of safety for hair dye users:
They may scan the list of ingredients to ascertain if the product contains an ingredient to which they know they are allergic. They can
avoid use of products containing these ingredients.

With respect to the soap exemption, industry argued that soap is *not* free from governmental regulation: Although soap *is* exempted from FDA scrutiny, soap is subject to regulation by the Consumer Product Safety Commission under two statutes—the Federal Hazardous Substances Act[23] and the Consumer Product Safety Act.[24] The Federal Trade Commission also has jurisdiction over soap for labeling purposes.[25]

C. Mandatory Cosmetic Registration

A third issue raised by the legislation was mandatory FDA registration of cosmetic products. The Eagleton bill called for registration of all cosmetic manufacturing establishments, submission of cosmetic formulas, and reports on adverse reactions. This program was essentially a mandatory version of the existing voluntary cosmetic registration program. Industry opposed the attempt to make the program mandatory, on the ground that the voluntary program allowed the industry needed flexibility that a mandatory filing program would not permit. The voluntary program provided the FDA with much valuable information about cosmetics, and 100 percent industry participation—which a mandatory program would provide—is not required to provide the FDA with necessary and useful information. (Industry later changed its position, stating that at that time it would *not* oppose making the voluntary program mandatory.)

D. Ingredient Labeling

Related to the formula disclosure provision of the registration program was a fourth issue: Even more detailed ingredient labeling requirements than those already in place. For example, Congresswoman Lenore Sullivan led a drive to require declaration of all flavors and fragrances, by name. The CTFA approved of ingredient labeling generally, and ingredient labeling was already in practice, pursuant to regulations promulgated under the FPLA. However, flavors and fragrances were required to be declared only by the generic terms "flavor" and "fragrance," rather than by chemical name, partly because many flavors and fragrances constitute trade secrets. Also, there typically may be dozens or hundreds of fragrance and flavor components in a single cosmetic so that it would be impossible to list them all in limited label space.

E. Complaint Files

A fifth provision of the Eagleton bill would have required manufacturers to compile and retain consumer complaints. A manufacturer would have no discretion to distinguish among various complaints or to make judgments about their veracity. Industry of course favored promoting consumer safety, but rejected the rigid complaint retention method as an inappropriate means to achieve that end. The CTFA pointed out that many consumer complaints are spurious and filed only to obtain cash settlements from companies. Permitting companies to use discretion and to conduct appropriate evaluation of complaints could produce a meaningful body of data for the FDA. An arbitrary program of compilation and retention of complaints with no screening would produce an exaggerated, nonmeaningful injury total.

F. Access to Records

A sixth issue involved FDA access to records and information at cosmetic companies.[26] The CTFA opposed expansion of existing powers because they would exceed the FDA's inspection powers over drugs, and such breadth was not justified by the relative innocuousness of cosmetics.

G. Poison Control Centers

A seventh issue involved the forwarding of adverse reaction data to poison control centers. The CTFA supported this provision.

VII. INDUSTRY INITIATIVES

A common theme running through the CTFA responses to many of these issues was the industry's voluntary self-regulatory programs.

The voluntary cosmetic registration program (which would have been made mandatory by the Eagleton bill) was the result of industry initiatives. A petition filed by the CTFA urged the FDA to adopt the program, now in place, whereby the cosmetic industry voluntarily submits information to the FDA on product establishments, formulas, and adverse reactions. Industry participation provides the FDA with extremely valuable information.

The CTFA has also been at the forefront of ingredient labeling. The ingredient labeling regulations would not be possible without the voluntary efforts of the cosmetic industry in compiling and publishing the *CTFA Cosmetic Ingredient Dictionary*, the indispensible volume for labeling cosmetic products with their ingredients. It is a compendium of uniform nomenclature for common cosmetic ingredients, which otherwise might be identified on cosmetic labels by different names, depending on the manufacturer of the product. Uniform nomenclature allows the consumer to compare cosmetic product labels to ascertain the presence of potentially allergenic ingredients and to avoid them.[27]

VIII. THE EAGLETON BILL PASSES THE SENATE

Following the 1974 hearings, the Senate failed to pass the Eagleton bill. However, on May 7, 1975, Senator Eagleton reintroduced his bill, designated S. 1681, the Cosmetic Safety Amendments of 1975.

The bill contained virtually the same provisions as the earlier S. 863, with an additional provision authorizing good manufacturing practices (GMPs) for cosmetics, similar to GMPs in existence for drugs.[28]

Hearings were held on S. 1681 on June 12, 1975,[29] with the CTFA reiterating its positions on the issues set forth above.

S. 1681 passed the full Senate on July 30, 1976, late in the session. No similar bill was considered in the House.

In 1977, Eagleton reintroduced his bill, designated S. 2365, and it was referred to the Committee on Human Resources. By this time, however, interest in cosmetic legislation was diminishing and the relevant committees in the House and Senate had more demanding, higher-priority items on their agendas, such as drug regulatory reform, food labeling, and saccharin. No hearings were held.

IX. FOCUS SHIFTS TO CANCER IN THE LATE 1970s

Beginning in the late 1970s, Congress's focus began to shift from plenary cosmetic legislation to concerns about specific cancer-causing agents. This concern with cancer resulted in oversight hearings on cosmetics in the House in 1978 and 1979.

X. THE 1978 OVERSIGHT HEARINGS

The 1978 oversight hearings focused on the safety of hair dyes. They were held January 23 and 26 and February 2 and 3, 1978,[30] in the Subcommittee on Oversight and Investigations of the House Committee on Interstate and Foreign Commerce, chaired by Congressman Moss.

The CTFA testified at the hearings and asserted that cosmetics are among the safest consumer products available, as evidenced by their history of safe use.

The CTFA explained that cosmetics are oriented toward the fashion industry and are, like fashion, subject to rapid change. Thus, rapid product changes are key to the well-being of the industry. However, cosmetic products use ingredients with historically good safety records such that these rapid changes do not adversely affect cosmetic safety.

The CTFA said that company records show that reactions to cosmetics are generally transient and reversible and rarely serious. The CTFA was aware of no case of cancer shown to have been caused by any cosmetic.

The CTFA cited the existing laws to which cosmetics are subject, and the actions the FDA can take to enforce those standards. The CTFA also pointed to industry self-regulation, citing ingredient labeling made possible by the *CTFA Cosmetic Ingredient Dictionary*, the numerous technical guidelines issued by the association, and the FDA's voluntary registration program adopted in response to a CTFA petition.

The hearings produced no recommendation for legislation.

XI. THE 1979 OVERSIGHT HEARINGS

On July 19, 1979, another set of hearings was held in the House of Representatives on the safety of hair dyes and cosmetic products.[31] Congressman Bob Eckhardt chaired the hearings held by the Subcommittee on Oversight and Investigations of the House Committee on Interstate and Foreign Commerce.

Congressman Eckhardt characterized the hearings as a continuation of the 1978 hearings, and he identified several goals: to update the problems uncovered in earlier hearings, to explore evidence of new hazards in cosmetics, and to examine whether more effective regulation of cosmetics was possible under the current statute or whether additional legislation was required.

Acting FDA Commissioner Sherwin Gardner testified on behalf of the agency. He said that findings about hazards in cosmetics should be viewed in the broader perspective of overall risks to consumers from various consumer goods. For example, he observed that the agency's risk model revealed a much lower risk to consumers from suspect cosmetic ingredients than the risk presented by saccharin.

He discussed four substances in cosmetics. 4-Methoxy-*m*-phenylenediamine (4-MMPD) is an ingredient that was used in some hair dyes and that had been shown to be an animal carcinogen. The FDA had proposed label warnings for products containing the ingredient.[32]

N-Nitrosodiethanolamine (NDELA) is a nitrosamine that is a contaminant of some cosmetic products formed during the manufacturing process. Gardner stated that the FDA had taken steps to warn manufacturers of the problem and to ask them to maintain surveillance of suspect products.

A substance known as 6-methylcoumarin (6MC) was being used as a fragrance ingredient, but was shown to be a sensitizer for some people. The FDA contacted manufacturers about 6MC to request voluntary recalls and discontinuance of its use.

Last, dioxane is an animal carcinogen that is a contaminant of cosmetics. Risk to consumers was unknown, but the FDA believed contamination could be avoided by precautions taken in manufacturing.

Because of scheduling difficulties, the CTFA did not testify at the hearings. However, the association did submit a detailed statement for the record that addressed charges of lack of safety in general, and discussed three of the items cited by Commissioner Gardner in particular.

In the general sections of its statement, the CTFA emphasized the good safety record of the cosmetic industry, and cited the ample authority of the FDA to regulate cosmetics: mandated color additive testing, detailed ingredient labeling, label warning statements, bans on certain ingredients, and others. This supplemented the CTFA's voluntary program of self-regulation, including the Cosmetic Ingredient Review (CIR),[33] and publication of the *CTFA Cosmetic Ingredient Dictionary*.

Other CTFA programs dealt specifically with problems addressed by Dr. Gardner, such as nitrosamines: Two years before FDA published its *Federal Register* notice about NDELA,[34] the CTFA had begun work through a Nitrosamine Task Force toward a comprehensive program of research to define nitrosamines, to initiate a screening program to detect their presence in cosmetics, and to plan a nitrosamine-inhibition program.

With respect to 4-MMPD in hair dyes, the CTFA disputed the validity of the feeding studies on which the FDA relied to assert that the ingredient is a carcinogen. The association also cited a risk assessment study[35] demonstrating that the greatest hypothetical lifetime cancer risk to users of hair dyes containing 4-MMPD was one in 16 million. By comparison, saccharin is far riskier.

The CTFA explained that dioxane is not a cosmetic ingredient, but like NDELA, it is a cosmetic contaminant in very minute amounts. The CTFA took steps to work with the FDA to minimize the presence of dioxane in cosmetics.

The hearings concluded and no proposed legislation followed.

XII. THE CURRENT CLIMATE

No hearings have been held since 1979, and it is unlikely that any will be held soon. Congress has moved away from its stance on cosmetics in the 1970s and is involved in more pressing concerns, such as the safety of the food supply in the United States.

For its part, the FDA has moved away from the heavier regulation of the 1970s and has recognized that the cosmetic industry truly does self-regulate effectively and has a good safety record.[36] Further, the FDA has moved to embrace the risk-assessment approach advanced by the CTFA, as evidenced by the consent order in the 4-MMPD court, case,[37] the agency ruling permanently listing lead acetate,[38] and the proposed trace constituents policy.[39]

NOTES

1 21 U.S. Code (U.S.C.) 361.
2 21 U.S.C. 362.
3 21 U.S.C. 374, 334, 332, and 333.
4 21 U.S.C. 376.
5 Of the more than 200 color additives that required testing in 1960 to gain permanent FDA approval, fewer than seven still are being tested by industry as of this writing (1982).
6 21 U.S.C. 355.
7 15 U.S.C. 1451 et seq.
8 15 U.S.C. 1454(c) (3); 21 Code of Federal Regulations (C.F.R.) 701.3.
9 21 U.S.C. 362; 21 C.F.R. 740.11, 740.12.

10 21 C.F.R. 740.10.
11 21 C.F.R. 700.11.
12 21 C.F.R. 700.16.
13 21 C.F.R. 700.14.
14 21. C.F.R. 700.13.
15 21 C.F.R. 700.15.
16 21 C.F.R. 700.18.
17 The FDC Act excludes "soap" from the definition of "cosmetic":
    "The term cosmetic means (1) articles intended to be rubbed,
    poured, spinkled, or sprayed on, introduced into, or otherwise
    applied to the human body or any part thereof for cleansing,
    beautifying, promoting attractiveness, or altering the appearance,
    and (2) articles intended for use as a component of any such
    articles; *except that such term shall not include soap*" [emphasis
    added]. 21 U.S.C. 321(i). For a review of the soap exemption,
    see Chap. 4.
18 Table 1 contains a list of the cosmetic bills and their provisions.

TABLE 1   Cosmetic Bills and Their Provisions

Bill number	H.R. 8418	H.R. 13417	H.R. 1235	H.R. 1107	H.R. 1527	H.R. 3261
Sponsor	Harris (D-Ark.)	Evans (D-Colo.)	Sullivan (D-Mo.)	Roybal (D-Calif.)	Helstoski (D-N.J.)	Evans (D-Colo.)
Date of introduction	9-12-63	2-28-72	1-3-73	1-3-73	1-9-73	1-30-73
Title of bill	Cosmetic Amendments of 1963	Cosmetics Act of 1972	Consumer Protection Amendments of 1973	Cosmetics Act of 1973		Cosmetics Act of 1973
Hearings held	No	No	No	No	No	No
Passed house						
Passed Senate						
Principal provisions Premarket clearance	●					
Cosmetic Delaney clause	●		●			
Warnings	●					
Safety pretesting		●	●	●		●
Delete soap exemption		●	●			●
Delete hair dye exemption		●	●			●
Make voluntary program mandatory		●		●		●
Ingredient labeling		●	●	●		●
Maintain complaint files for inspection		●		●		●
Increase fines		●				●
Send information to poison centers		●		●		●
Administrative subpoenas			●			
Expand inspections			●			
Administrative detentions						
GMPs						
Recall authority						
Mercury labeling					●	
Set ingredient tolerances						

[a]For Ford administration by request

19 For a review of the FDA's voluntary program, see Chap. 10.
20 Hearings on S. 863 and S. 3012 Before the Subcommittee on Health of the Senate Committee on Labor and Public Welfare, 93rd Cong., 2d Sess., February 20 and 21, 1974.
21 21. U.S.C. 321(i).
22 21 U.S.C. 361(a).
23 15 U.S.C. 1261 et seq.
24 15 U.S.C. 2051 et seq.
25 15 U.S.C. 1454; 16 C.F.R. Part 500.
26 This provision was an important section in companion bills, most notably the Rogers bill, H.R. 14009, introduced April 4, 1974, after Congressman Rogers had the opportunity to see the developments in the Senate hearings on the Eagleton bill.
27 For a review of the *CTFA Cosmetic Ingredient Dictionary*, see Chap. 24.
28 21 C.F.R. Part 210.
29 Hearings on S. 1681 Before the Subcommittee on Health of the

S. 863	S. 3012	H.R. 14009	H.R. 14805	S. 1681	H.R. 1993	S. 2365
Eagleton (D-Mo.)	Javits[a] (R-N.Y.)	Rogers (D.-Fla.)	Koch (D-N.Y.)	Eagleton (D-Mo.)	Roybal (D-Calif.)	Eagleton (D-Mo.)
2-15-73	2-18-74	4-4-74	5-15-74	5-7-75	1-17-77	12-15-77
Cosmetic Safety Act of 1974	Food, Drug, and Cosmetic Amendments of 1974	Cosmetic Amendments of 1974		Cosmetic Safety Amendments of 1975	Cosmetics Act of 1977	Cosmetic Safety Amendments of 1977
2-20/21-74	2-20/21-74	No	No	6-12-75	No	No
				7-30-76		
•						
•		•		•	•	•
•		•		•	•	
•		•			•	•
•		•		•	•	•
•		•		•	•	•
•		•		•	•	•
	•					
•		•		•	•	
	•					
	•	•		•		•
	•			•		
		•		•		•
			•			
						•

Senate Committee on Labor and Public Welfare, 94th Cong., 1st Sess., June 12, 1975.

30 Hearings on the Safety of Cosmetics and Hair Dyes Before the Subcommittee on Oversight and Investigations of the House Committee on Interstate and Foreign Commerce, 95th Cong., 2d Sess., January 23 and 26, February 2 and 3, 1978.

31 Hearings on the Safety of Hair Dyes and Cosmetic Products Before the Subcommittee on Oversight and Investigations of the House Committee on Interstate and Foreign Commerce, 96th Cong., 1st Sess., July 19, 1979.

32 FDA issued a final rule requiring label warnings on October 16, 1979 (*Fed. Reg. 44*: 59509), which seven major hair-coloring companies challenged in a federal court, obtaining a consent order staying the rule. *Carson Products Company et al. v. Department of Health and Human Services*, CV-480-71, CCH Food and Drug Cosmetic Law Reporter 38,334, September 18, 1980.

33 The Cosmetic Ingredient Review (CIR) is the only safety program of its kind in any industry. Sponsored by the cosmetic industry, it gathers worldwide safety data on commonly used cosmetic ingredients. An independent panel of scientists reviews these data to determine whether the ingredients, as used in cosmetics, are safe for the consumer. When an ingredient has been reviewed by the panel, the results are published in the open scientific literature. The CIR program provides continuing assurance that cosmetic ingredients are safe for use. For a review of the CIR program, see Chap. 12.

34 *Fed. Reg. 44*: 21365 (April 10, 1979). The notice stated that the FDA had evidence that some cosmetics were contaminated with nitrosamines, and urged the cosmetic industry to take steps to eliminate nitrosamine contamination to the extent possible.

35 For a review of risk assessment, see Chap. 36.

36 Statement of Dr. Mark Novitch, Deputy Commissioner, Food and Drug Administration, FDLI Annual Conference, Washington, D C., December 16, 1981.

37 *Carson Products et al. v. HHS*

38 *Fed. Reg. 45*: 72112 (October 31, 1980).

39 *Fed. Reg. 47*: 14464 (April 2, 1982). See *Scott v. Food and Drug Administration*, Nos. 82-3544/3759 (6th Cir.) February 23, 1984.

Part II

FUNCTIONING IN THE REGULATORY ENVIRONMENT

10

VOLUNTARY SELF-REGULATION PROGRAMS OF THE COSMETIC INDUSTRY

EMALEE G. MURPHY

Legal Department, The Cosmetic, Toiletry and Fragrance Association, Inc., Washington, D.C.

I. HISTORY OF THE VOLUNTARY REPORTING PROGRAMS

A. CTFA Petitions

The cosmetic industry participates in three voluntary programs designed to provide the Food and Drug Administration (FDA) information about cosmetic product establishments, ingredients, and product experiences. The programs were initiated in 1971 and 1972 by peti-

tions to the FDA filed by the Cosmetic, Toiletry and Fragrance Association (CTFA) on behalf of its members. The programs, which
have been incorporated into the Code of Federal Regulations (C.F.R.)
include: (1) voluntary registration of cosmetic manufacturing and
packing establishments;[1] (1) voluntary filing of cosmetic raw material
composition and cosmetic product ingredient statements;[2] and (3)
voluntary filing of cosmetic product experiences.[3]

At the time the association petitions proposing the voluntary regulations were submitted to the FDA, the CTFA characterized the programs as "a desirable and innovative example of industry cooperation
with government to better consumer protection."[4] The CTFA stated
that the programs were conceived and developed in the public interest
to support efficient enforcement of the Federal Food, Drug, and Cosmetic Act (FDC Act).[5] Several interested parties assisted the CTFA
in developing the proposed regulations, including the Commissioner
of Food and Drugs and his staff, the President's Special Assistant for
Consumer Affairs, and former Congressman Paul Rogers (D.-Fla.),
Chairman of the Subcommittee on Public Health and Welfare, Committee
on Interstate and Foreign Commerce.

The CTFA's petitions to establish voluntary reporting programs
for the cosmetic industry were filed at a time of heightened public
interest in cosmetic products. Virginia Knauer, Special Assistant to
the President for Consumer Affairs, in 1970 called on the cosmetic industry to adopt ingredient labeling and to support a voluntary reporting system.[6] In 1971 and 1972 the FDA moved to restrict use of mercury[7] and hexachlorophene[8] in cosmetic products and introduced its
first proposal for cosmetic ingredient labeling.[9] During the same
years, Congress began to consider legislation that would have
amended the FDC Act to authorize FDA premarket safety clearance of
cosmetic products.[10] In light of the prevailing regulatory climate,
the CTFA's recommendation for a system of voluntary regulation was
designed to demonstrate both industry's willingness to supply information to the FDA about cosmetic establishments, ingredients and experiences, and to discourage Congressional legislation.

B. FDA Proposals

In general, the final regulations reflect the recommendations and
language of the CTFA petitions. However, the FDA added three new
provisions to the *Federal Register* proposals for registration of cosmetic establishments and filing cosmetic ingredient statements, two of
which were retained in the final program [*Fed. Reg. 36*:16934 (August
26, 1971)]. The first provision would have identified certain categories of cosmetics as ones the FDA considered to be drugs as well as
cosmetics. The CTFA opposed this provision, stating that such a determination of regulatory status could be made only after consideration of each product's intended use and full labeling.[11] The FDA
subsequently deleted the paragraph from the final regulation.

Two other provisions proposed by the FDA were eventually included in the final regulations. These established the filing of cosmetic raw material composition statements[12] and extended the reporting program to foreign cosmetic establishments whose products are imported for sale to the United States.[13]

C. Screening Reportable Product Experiences

The program for voluntary filing of cosmetic product experiences was proposed in the *Federal Register* of November 2, 1972 (*Fed. Reg.* 37: 23344) and included the text suggested in the CTFA's third petition to the FDA. In the statement of grounds accompanying the petition, the CTFA stated its belief that the FDA should have reasonable access to cosmetic product experience information and that such information should be sufficient to establish reliable baseline data against which to assess products and ingredients. Moreover, the CTFA stated that any reporting system should be designed to notify the FDA promptly of experiences indicating a possible threat to the public health.[14]

The CTFA's petition emphasized that companies must be reasonably able to conclude that reactions or injuries reported to the FDA actually occurred. The association noted that despite the best efforts of manufacturers and distributors, many adverse reaction complaints remain unconfirmed. The CTFA recommended that a "reportable experience" refer only to reactions or injuries verified by the manufacturer, distributor, or physician to be the result of a cosmetic product or ingredient.[15]

Industry's insistence that it be permitted to screen nonlegitimate complaints was made part of the final regulation for reporting cosmetic product experiences, published October 17, 1973, in the *Federal Register* (*Fed. Reg.* 38:28914).[16] In a statement made shortly before the final regulations issued, the FDA's Dr. Robert M. Schaffner, now Associate Director for Physical Sciences, Center for Food Safety and Applied Nutrition, commented on the screening provision: "In principle, I believe we all can agree that an acceptable procedure should be one that would not 'screen out' any actual cases of adverse reactions. . . ."[17]

Thus, to avoid screening procedures that might eliminate valid experience reports, the final regulations suggest that each participating company file its screening mechanism with the FDA for public inspection.[18] In addition, the regulations permit the FDA to audit company screening procedures to determine that the procedure is consistently applied and is not eliminating reportable information. Companies participating in the voluntary reporting program must either use a screening procedure on file with the agency or report all complaints of alleged bodily injury.[19]

Following publication of the final regulations, the CTFA sought to clarify the scope of the FDA's screening procedure audits. In a December 1973 letter to FDA Commissioner Alexander Schmidt, the CTFA asserted its belief that the term "audit" as used in the regulation does not automatically render all company complaint files open to FDA inspection. In the CTFA's view, a company under audit is obligated only to provide data sufficient to assure consistent application of a valid screening procedure. For instance, this might include examples of reported and unreported consumer complaints, plus an explanation of the procedure used to evaluate the validity of each complaint.

D. The CTFA's Suggested Screening Procedure for Cosmetic Product Experiences

To assist companies participating in the program, the CTFA issued a five-step suggested screening procedure as a guide for industry members to use when reporting product experiences:[20]

1. If the initial consumer communication provides enough information about the alleged injury and the information does not appear to be unfounded or spurious, the experience should be considered "reportable."
2. If the initial consumer communication does not contain sufficient information to determine whether it is a reportable experience, the company should promptly contact the consumer and request additional information.
3. If, after reasonable efforts have been made to contact the consumer, further information is not available, the experience should be considered not reportable.
4. If the company receives information from the consumer that confirms the alleged injury described in the initial communication, the company may choose to consider the consumer response adequate to determine reportability, whether or not the consumer received medical treatment.
5. If the company decides to investigate an experience to determine if it is unfounded or spurious, a reasonable attempt should be made to obtain any necessary additional information, including a medical report when applicable. A final reporting decision is then made on the basis of all information received.

The CTFA screening guidelines also suggest useful questions for follow-up interviews with consumers and medical personnel, such as information about the affected consumer (name, address, age, and sex), information to identify the product alleged to have caused the injury (brand name, description, container, product or package codes, and place of purchase), and information about the specific complaint

(a description of the symptoms and whether they appeared shortly
or sometime after the product's application). Other questions might
explore whether the product had been used previously by the affected
consumer, whether there had been similar instances of symptoms,
either from the same product or from other brands of similar products,
and whether a physician was consulted about the symptoms.

Physicians who treat the alleged injuries might be asked to supply
(with appropriate patient permission) a description of the problem,
the pertinent patient history, any treatment prescribed, any self-
medication used by the patient, and whether the physician would be
willing to patch-test the patient to determine whether the cosmetic
actually caused the complaint.

II. CONFIDENTIALITY OF REPORTED DATA

In its petitions to the FDA, the cosmetic industry stated that information submitted under the voluntary reporting programs should be
treated as confidential. The CTFA emphasized that such information
is not ordinarily released by a company to competitors or to the public, and that such information, if publicly divulged, could substantially harm a reporting company.[21]

When in 1974 the FDA issued its comprehensive regulation to implement the Freedom of Information Act (FOI), *Fed. Reg.* 39:44601
(December 24, 1974), information submitted under the voluntary programs was not made subject to the same disclosure policy as were data
submitted to the FDA under mandatory reporting programs.[22] Accordingly, the Food and Drug Administration's FOI regulations do not
allow the agency to disclose consumer complaints and adverse reactions voluntarily submitted by a manufacturer, except as part of a
blind compilation that does not reveal the name of the manufacturer
or the brand name of the product involved.[23] Manufacturing processes and controls and product formulations are *per se* exempt from
disclosure, except to the extent that they have already been publicly
disclosed.[24] Quantitative and semiquantitative formulas are also held
in confidence by the FDA and not released to the public, unless they
have already been publicly disclosed by the company or are related to
a product or ingredient that has been abandoned and thus no longer
constitutes trade secret or confidential commercial information. Similarly, information about production, sales, and distribution is available for public disclosure only if it appears in a compilation of aggre-
gated data prepared in a way that does not reveal protected material.[25]

On the other hand, information such as lists of ingredients contained in a cosmetic, whether or not in descending order of predominance, will be made available to the public, unless a company can
show that a particular ingredient or group of ingredients falls within

the exemptions established in 21 C.F.R. 20.61 for trade secrets and confidential commercial information.[26]

The FDA's policy for voluntarily submitted reports was addressed in 1978 by Donald Kennedy, Commissioner of Food and Drugs, in a letter to the CTFA regarding reports that the FDA might be considering public release of product experience data.[27] The letter reaffirms the FDA Freedom of Information Act regulations prohibiting release of voluntarily submitted information except as part of a "blind" compilation. Dr. Kennedy stated, however, that information otherwise exempt from public disclosure may be revealed in administrative or court enforcement proceedings when the data are relevant. Such regulatory proceedings presumably would include a FDA recall request, should the agency decide that a cosmetic product should be removed from the market.

A company that wishes to claim trade-secret status for the identity of an ingredient may ask the FDA to make such a determination through a presubmission review of the data supporting its claim (21 C.F.R. 20.44). Pending a decision on the request, the records are kept confidential by the agency. All determinations of trade-secret claims are made in writing, and if the petition for trade-secret status is granted, the records are marked accordingly, and are exempt from public disclosure.

When such a request for confidentiality is denied, the person requesting the presubmission review may withdraw the records and all copies, summaries, related correspondence, and memoranda from the FDA's files. As an additional safeguard, 21 C.F.R. 720.8 provides that denial of trade-secret or confidential commercial information constitutes final agency action subject to judicial review under the Administrative Procedure Act. If a suit is brought within 10 days following the denial, the FDA will not disclose the records involved or require that the disputed ingredient be disclosed in labeling until the matter is finally determined in the courts.

Before implementation of the FDA's cosmetic ingredient labeling regulations in 1975, the ability to withdraw the record following denial of a petition appeared to provide complete trade secret protection. However, the mandatory ingredient disclosure requirement of the FDA's subsequent cosmetic ingredient labeling regulation, 21 C.F.R. 701.3, left the recipient of a FDA denial for trade-secret status under the voluntary filing program with only three rather unsatisfactory options: (1) to disclose the disputed ingredient on product labels; (2) to take the product off the market entirely; or (3) to continue to sell without disclosure, thereby risking civil penalties.

The FDA procedures originally established in 21 C.F.R. 20.44 for determination of trade-secret status proved to be inadequate in light of the increased risks presented by the new ingredient labeling regulations. When in 1976 the RDA denied trade-secret status to a product ingredient, the manufacturer challenged the procedures in the U.S. District Court for the District of Columbia, *Zotos International, Inc. v. Kennedy*, 460 F.Supp. 268 (D.D.C. 1978). The court in *Zotos* ruled that the procedures followed by the FDA in denying the company's trade-secret claim violated the Due Process Clause of the Fifth Amendment because it (1) failed to provide sufficient notice of the basis for the agency's determination that the ingredient was not a trade secret and (2) failed to provide a sufficient opportunity for "focused dialogue" on the relevant issues prior to a final agency ruling.

On February 7, 1979, the CTFA petitioned the FDA to amend its procedures for considering claims of trade-secret exemption ". . . to provide cosmetic manufacturers fair notice and opportunity for hearing when FDA believes an ingredient may not be eligible for the trade secret exemption from cosmetic ingredient labeling."[28] The FDA published its proposed trade-secret procedures on August 31, 1982 (*Fed. Reg.* 47:38353). The proposal would codify the policy adopted by the agency after the *Zotos* decision, that firms be granted 30 days to respond with additional data following a tentative determination by the FDA denying trade-secret status. The proposed regulation enumerates the criteria the FDA will consider in making its determination, and provides that the agency give the petitioning manufacturer the grounds for its tentative determination in writing. However, the proposal would not grant a closed regulatory hearing to firms requesting confidentiality for a cosmetic ingredient (as requested by the CTFA in its 1979 petition.)[29] It seems likely, however, that if a meeting is requested during the pending of a petition for recognition of trade-secret status, the FDA would agree to meet [21 C.F.R. 10.65(d)].

III. CURRENT REPORTING PROCEDURES

Almost a decade after initiation of the voluntary reporting programs, the FDA published proposed amendments [*Fed. Reg.* 45:73960 (November 7, 1980)] to reduce the reporting burden for participating companies. Final regulations implementing the proposed changes became effective on July 24, 1981 (*Fed. Reg.* 46:38073). The CTFA filed comments with the FDA supporting the changes and stating that it hoped a reduction in paperwork would encourage more companies to take part in the programs.[30] The following description of the reporting procedures includes a discussion of the original requirements and an explanation of the changes.

A. Voluntary Registration of Cosmetic Product Establishments
(21 C.F.R. 710)

Owners and operators of cosmetic product manufacturing and packaging establishments, including foreign establishments whose products are sold in the United States, should register within 30 days of beginning operations, whether or not the products enter interstate commerce. FDA Form FD-2511 (Registration of Cosmetic Product Establishment) request the establishment's name and address, all its business trading names, and whether the establishment manufactures or packages cosmetic products.

It is no longer necessary that the establishment designate its type of ownership—for example, partnership, corporation—and all reference to establishments that merely distribute cosmetics has been deleted.

The FDA provides each registrant with a validated copy of Form FD-2511 and assigns a permanent registration number to each cosmetic product establishment. Assignment of a registration number does not denote FDA approval of the firm or its products, and any representation in a company's labeling or advertising that could create an impression of official approval is considered by the FDA as misleading and will render the products misbranded.

Some classes of persons are asked *not* to register under the voluntary program. They include beauty shops, cosmetologists, retailers, pharmacies, physicians, hospitals, clinics, public health agencies, and persons who manufacture or process cosmetic products solely for use in research, pilot plant production, or chemical analysis.

B. Voluntary Filing of Cosmetic Product Ingredient and Cosmetic Raw Material Composition Statements (21 C.F.R. 720)

Either the manufacturer, packer, or distributor of a cosmetic product should file Form FD-2521 (Cosmetic Product Ingredient Statement, CPIS). The CPIS requests the name and address of the manufacturer, packer, or distributor listed on the product label or of the manufacturer or packer of the product if it differs from the name on the label. Form FD-2512 requests the brand name(s) of each cosmetic product in "commercial distribution," that is, those with over $1,000 in sales, and asks the reporting company to indicate the product's proper cosmetic category based on its intended use.

There are 13 major product categories by which to identify a product's intended use. They include baby products, preparations for the bath, eye makeup, fragrance, noncoloring hair preparations, hair coloring preparations, non-eye makeup, manicuring products, oral hygiene products, products for personal cleanliness, shaving preparations, skin care preparations (creams, lotions, powders, and sprays), and suntan and sunscreen preparations.

The CPIS also requests that product ingredients be listed in descending order of predominance and the amount of each ingredient in a product is specified by use levels, which are assigned letter designations, for example: A, over 50 percent; B, 25 to 50 percent; C, 10 to 25 percent; and so forth. Level H represents 0 percent and is used only to indicate that a particular color additive is absent in certain shades of a product. The use level information is confidential and not subject to public disclosure by the FDA under FOI regulations [21 C.F.R. 20.111(d) (4)].

All companies are urged to provide poison control centers with information, so that quick diagnostic and therapeutic measures can be taken in cases of ingestion or other accidental use. Product information should also be available to physicians when requested in connection with treatment of patients.

Information about cosmetic raw materials should be submitted using Form FD-2513 (Cosmetic Raw Material Composition Statement, CRMCS) within 60 days after beginning commercial distribution. Forms for cosmetic raw materials that are mixtures should identify each ingredient in descending order of predominance in the same manner described in the procedures for filing cosmetic product ingredient statements.

Companies should identify ingredients in products and raw materials by the common or usual name, if there is one, or by the chemical name (except for trade-secret ingredients), or the chemical description. The trade name of the ingredient and the name of the supplier or manufacturer may be listed when other information is not available. However, the regulations do not require the individual ingredients in fragrances and flavors to be identified on the form when such information is not available to the person filing the product ingredient or raw material composition statement. In such cases, the trade name of the fragrance or flavor and/or the registration number and name of the supplier should be listed.

The regulations also provide for the filing of amendments to previously submitted information and for notices that a company has discontinued commercial distribution of a cosmetic product (Form FD-2514).

The *changes* to the regulations delete the previous requirement in sections 720.4(d) (2) and 720.5(c) (1) that a reporting company also list standards, such as those in the National Formulary or U.S. Pharmacopoeia, with which an ingredient or raw material complies.

Upon receipt of a company's cosmetic product ingredient statement, the FDA assigns a Cosmetic Product Ingredient Statement Number (CPIS No.) to the ingredient or an FDA reference number in those cases where a permanent number cannot be assigned. Cosmetic raw materials are assigned a Cosmetic Raw Material Composition Statement Number (CRMCS No.), which may be used to identify the material on

container labels, trade literature, catalog citations, and other correspondence. However, use of these numbers does not denote FDA approval of the product, and the CRMCS number may be used on shipping container labels and trade literature only if the following dis-
claimer appears on the same page or on the principal display panel of
the label:

> The FDA Cosmetic Raw Material Composition Statement Number is
> assigned for raw material identification purposes only and does not
> in any way denote approval of the firm or the raw material by the
> Food and Drug Administration.

Any use of CPIS or CRMCS numbers assigned by the FDA that creates
the impression of official approval by the agency may be considered to
be misleading and the product therefore misbranded.

C. Voluntary Filing of Cosmetic Product Experiences (21 C.F.R. 730)

Manufacturers, packers, and distributors of cosmetic products are
asked to file cosmetic product experience reports voluntarily, regardless of whether they also participate in the program to register cosmetic product establishments or to file cosmetic product ingredient and
raw material composition statements. Until the FDA's recent changes
in the regulations, three forms were used to submit cosmetic product
experience data to the agency.

The basic Form FD-2704 (Cosmetic Product Experience Report),
still in effect, should be filed on a semiannual basis for the periods
January—June and July—December, not later than 60 days after the
close of the reporting period. The information requested includes the
name and address of the person designated on the label of each cosmetic product; the complete product name and CPIS number, if available; the product's appropriate "product category"; the total number
of reportable experiences received by the company for the product
during the reporting period; and an estimate of the number of product units distributed to consumers during the reporting period.

It is no longer necessary to report the cosmetic establishment
registration number, former 21 C.F.R. 7304(a) (8), or the estimated
rate of reportable experiences per million units distributed, former
21 C.F.R. 730.4(a) (7). Also discontinued is the provision in 21
C.F.R. 730.4(a) (5) formerly requiring reporting companies to identify reportable experiences by category as described in the FDA in-
struction booklet "Voluntary Filing of Cosmetic Product Experiences. [31] The provision for filing Form FD-2705 (Cosmetic Product
Unusual Experience Report), former 21 C.F.R. 730.4(b) has also been
eliminated and the definition of "unusual reportable experiences" has
been deleted from former 21 C.F.R. 700.3(r). [32]

Finally, companies still are requested to submit a summary report
of cosmetic product experiences by product categories using Form FD-

2706 within 60 days after the end of each six-month reporting period. The product categories on Form FD-2706 are those listed in 21 C.F.R. 720.4(c) to indicate the product's intended use when filing cosmetic product ingredient statements. [In 1984 CTFA asked that Form FD-2706 be required only annually. At publication time, FDA had agreed to annual filing of Forms FD-2706 and FD-2704.]

The regulations state that filing a product experience report does not denote FDA approval of the firm or the cosmetic product, nor does filing an experience report consitute an admission by the reporting company that the alleged experience was the result of an ingredient in the cosmetic product or of any other fact.[33]

V. PROGRAM PERFORMANCE AND FUTURE

Shortly after the establishment of the voluntary reporting program, former FDA Commissioner Alexander Schmidt stated: "I know of no industry which has a better record of voluntary accomplishment."[34] Only two years after the initial petitions were filed with the FDA, Dr. Robert M. Schaffner, Associate Director for Physical Sciences, FDA Bureau of Foods, reported that 754 cosmetic manufacturing establishments had registered under the program, a figure he estimated as representing 70 percent of all cosmetic manufacturing plants and 85 percent of all cosmetic production.[35] Moreover, Dr. Schaffner stated that 7,327 cosmetic product formulas had been submitted to the FDA as of October 1973. A summary of the product experience data for the first six reporting periods (1974—1976) reveals that an average of 125 cosmetic companies submitted experience reports during each reporting period, representing an estimated distribution of 9,754,380,000 units at a low rate of 2.03 experiences per million units distributed.[36]

More recent figures indicate that by July 1, 1980, 975 cosmetic establishments had registered with the agency.[37] In addition, between 1972 and June 30, 1981, a *net* total of between 19,000 and 20,000 product formulations (those in current use) were filed with the FDA,[38] a number estimated as approximately one-half of the formulations actually marketed.[39] Finally, the average number of firms that have participated in the product experience reporting program during all the previous reporting periods was reported to be 122,[40] a figure that represents approximately 30 to 40 percent of U.S. cosmetic sales.[41]

In a speech to cosmetic manufacturers in October 1980, Deputy FDA Commissioner Dr. Mark Novitch praised the industry for its support of the voluntary reporting programs, but stated that: "There is still an uncomfortably high degree of non-participation."[42] Referring to the small number of firms filing product experience reports, he stated:

I don't think you can be satisfied with those numbers and you can hardly expect us to be. The challenge is to both industry and FDA to find ways of improving on them through cooperation.

Overall, these and other essentially voluntary programs have
proven their value, but what of the future? Will they suffice to
meet the challenges of the 80's and beyond?[43]

In 1984, CTFA confirmed its support for all the voluntary reporting
programs through new active plans to encourage member company participation. As Dr. Novitch stated: "Responsible regulation of the
cosmetic industry can continue to depend substantially on a healthy
balance of efforts between the industry and the FDS and regulatory
measures. The record of the past testifies abundantly to that."[44]
The voluntary programs initiated and maintained by the cosmetic industry are a key element in that balance.

NOTES

1   21 Code of Federal Regulations (C.F.R.) 710.
2   21 C.F.R. 720.
3   21 C.F.R 730.
4   Letter from CTFA President James Merritt to Beryl McCullar,
    Hearing Clerk, U.S. Department of Health, Education and Welfare,
    September 24, 1971.
5   CTFA Petition for the Issuance of a Regulation Providing for the
    Voluntary Filing of Cosmetic Product Experience, June 8, 1972,
    Attachment B, p. 4.
6   Proposed Amendments to the Federal Food, Drug, and Cosmetic
    Act: Hearings on S. 863 and S. 3012 Before the Subcommittee on
    Health of the Committee on Labor and Public Welfare, 93rd Cong.,
    2d Sess. 104 (1974) (statement of Virginia H. Knauer).
7   21 C.F.R. 700.13.
8   21 C.F.R. 700.11.
9   *Fed. Reg.* 37:16208 (August 11, 1972).
10 S 863, 93rd Cong. 1st Sess. (1973).
11 See note 4.
12 21 C.F.R. 720.5.
13 21 C.F.R. 710.1, 720.1, and 730.1.
14 CTFA Petition for the Issuance of a Regulation Providing for the
    Voluntary Filing of Cosmetic Product Experience, June 8, 1972,
    Attachment B, p. 4.
15 CTFA Petition, Attachment B, p. 4.
16 21 C.F.R. 730.4(d) (2) and 21 C F.R. 700.3(p) and (q).
17 Speech by Dr. Robert Schaffner before The Foragers of America,
    September 26, 1973, reported in *Drug and Cosmetic Industry* 35
    (November 1973).
18 21 C.F.R. 700.3(p) (1).
19 21 C.F.R. 700.31(p) (1).
20 Memorandum to CTFA Members from CTFA President James Mer-
    ritt, February 20, 1974.

21 Preamble to Proposal for Cosmetic Product Experience Voluntary Filing, *Fed. Reg.* 37:23344 (November 2, 1972).

22 21 C.F.R. 20.111.

23 21 C.F.R. 20.111(c) (e) (ii) (c).

24 21 C.F.R. 20.111(d) (2).

25 21 C.F.R. 20.111(d) (3) and (4).

26 21 C.F.R. 20.111(c) (4).

27 Letter to CTFA President James Merritt from FDA Commissioner Donald Kennedy, Ph.D., December 13, 1978.

28 CTFA Citizen Petition filed February 7, 1979, with the FDA seeking amendment of the FDA's procedural regulations governing trade-secret exemption from cosmetic ingredient labeling regulations.

29 At the time this chapter was prepared, the final regulations had not yet been issued but were expected to be substantially identical to the FDS's proposals.

30 CTFA comments on FDA-proposed rule to reduce filing requirements of voluntary cosmetic reporting program, January 6, 1981.

31 Food and Drug Administration, Instructions and General Information for the Voluntary Filing of Cosmetic Product Experience in Accordance with Title 21, Code of Federal Relations, Part 730 (1974): "1) *Local irritation or allergic reaction*, such as reddening, swelling of eyelids, watering eyes, includes burning sensations and soreness in the mouth and throat areas, tenderness of teeth, and reddening or other irritation of the mucosa. This category does not include any alleged experience entailing damage or corrosion of skin, nails or teeth or any alleged infection. 2) *Skin or nail damage* includes destruction of skin tissue, split or broken nails. 3) *Infection*, and 4) *Other*, which includes categories of reactions such as choking, difficulty in breathing, blurring of vision, dizziness, damage to teeth, hair, eyes and other parts of the body other than skin and nail damage."

32 "Unusual" experiences were described as those that might be due to deliberate abuse of a cosmetic product (use of a hair dye to dye eyelashes, use of underarm deodorant as feminine spray, or deliberate inhalation of aerosols to achieve intoxication). Accidental ingestion of a cosmetic product that results in temporary indisposition or bodily injury was also considered an "unusual" experience. FDA, Instructions and General Information.

33 21 C.F.R. 730.8(a) and (b).

34 Speech by FDA Commissioner Alexander M. Schmidt, M.D., at the CTFA Annual Meeting, February 27, 1974.

35 Speech by Dr. Robert Schaffner at a metting of the New York Chapter of the Society of Cosmetic Chemists, reported in *Drug and Cosmetic Industry 132* (November 1973).

36 Statistics for the Voluntary Cosmetics Registration Program at the Food and Drug Administration as of September 30, 1977.

37 Statistics for the Voluntary Cosmetics Registration Program at the Food and Drug Administration as of June 3, 1979, and October 1, 1980.

38 Novitch, The FDA & Cosmetics: Looking to the 1980's, *CTFA Cosmet. J.* *12*(4):8 (1980).

39 Novitch, The FDA & Cosmetics, p. 8.

40 Novitch, The FDA & Cosmetics, p. 8.

41 Novitch, The FDA & Cosmetics, p. 8.

42 Novitch, The FDA & Cosmetics, p. 8.

43 Novitch, The FDA & Cosmetics, p. 9.

44 Novitch, The FDS & Cosmetics, p. 9.

11

ORGANIZATION FOR ACTION—DEVELOPMENT OF CTFA's SCIENTIFIC PROGRAMS

FRANK W. BAKER

Research and Development, The Procter and Gamble Company, Cincinnati, Ohio

NORMAN F. ESTRIN

The Cosmetic, Toiletry and Fragrance Association, Inc., Washington, D.C.

The Cosmetic, Toiletry and Fragrance Association (CTFA) scientific effort has developed in response to the needs of the industry. As time and external circumstances have changed those needs, the CTFA has responded by developing a committee and task force structure capable of meeting the new challenges. The current scientific organization did not develop de novo but came about as a natural evolution of the industry's concern for the quality and safety of its ingredients and products.

I. ORGANIZATIONAL DEVELOPMENT

Looking back to the period of the 1940s and 1950s, we can now see in hindsight how simple and quiet a time it was for the cosmetic industry. The major issues were availability of raw materials and the need for adequate raw material specifications to ensure consistency and quality in product formulations. The CTFA (then the Toilet Goods Association, TGA) provided the needed specifications and analytical methods in the form of TGA Standards, prepared by the Standards Committee. Thanks to the war effort and the subsequent inflation, raw materials were at a premium. When materials were available, they varied in quality and purity to such an extent that adequate compensation for these differences made cosmetic formaultion more of a nightmare than

an art. The TGA responded through its annual scientific conference
by providing a focal point for discussion on the latest information regarding ingredient quality and purity and how best to formulate with
what was available.

With the passage of the Color Additive Amendments to the federal
Food, Drug and Cosmetic Act in 1960, a new set of challenges faced
the industry. In fact, it is fair to say that now, some 20 years later,
the same basic issues remain to challenge us. The Color Additive
Amendments thrust upon the cosmetic industry extensive new testing
requirements for safety. The TGA responded by establishing a Color
Additive Committee charged with responsibility for development, execution, monitoring, and evaluating the testing program, which, it was
hoped, would lead to the permanent listing of colors that were of interest to the cosmetic industry. It can readily be seen that this initial effort at developing safety data supporting the cosmetic use of
colors is identical in thrust, if not in scope, to the much more sophisticated program undertaken in 1977. Relationships with the Food and
Drug Administration (FDA) became strained in 1965 when the TGA
filed suit against the agency in an attempt to block an effort by the
FDA to broaden its powers over cosmetic ingredients and products,
relying on what industry felt was an over-reading of the Color Additive Amendments. In fact, the FDA was attempting to establish its
rights to preclear finished cosmetic products through a very broad
interpretation of the powers it felt resided in the amendments of 1960.
In 1970 the association's position was upheld. The industry had
successfully responded to both a technical and a legal challenge from
the government, and a precedent was set establishing that this industry was willing to defend its rights on the basis of sound legal and
scientific arguments.

In the late 1960s, Ralph Nader and the *Corvair* ushered in a new
era of consumer awareness and concern. This concern centered on a
growing lack of trust in industry's commitment to product quality and
safety. The resulting consumerist movement generated an adversarial
relationship between some parts of American industry and its customers. As a result, both industry and the consumerists called upon
government to use its powers to protect each side from the other.
Consumerists wanted stricter controls over product quality, with
greater evidence of safety publicly available and more severe penalties
for "wrongdoers." Industry wanted protection of its trade secrets
and confidential information from the growing demand for full public
exposure. In some instances, industry desired federal preemption
from a growing list of local, often conflicting, legislation and regulation. The resulting pressures on the federal government gave birth
to a series of new regulatory agencies and enhanced powers for those
already in existence. Thus, in the 10-year period following the late
1960s, agencies such as the Consumer Product Safety Commission

(CPSC), the Occupational Safety and Health Administration (OSHA), and the Environmental Protection Agency (EPA) came into existence. The Federal Trade Commission (FTC) acquired broader powers and became aggressively pro-consumerist. The FDA gained additional powers through the Device Admendments to the Food, Drug and Cosmetic Act (FDC Act), and, perhaps most important of all, the passage of the Toxic Substances Act extended to the EPA preclearance powers over all chemicals not covered under the FDC Act or by the federal Insecticide, Fungicide and Rodenticide Act (FIFRA).

The winds of change ushered in by this new era did not leave the cosmetic industry untouched. The CTFA recognized early that it must organize itself for the new challenges that were becoming ever more apparent. In 1967 a group of highly motivated microbiologists from within the industry formed a committee whose purpose was to enhance the awareness of the cosmetic industry to the potential problems of microbiological contamination. Thus was born the association's Microbiology Committee, and with its inception came an ever-growing commitment to high-quality microbiological research and to an educational program to inform all members of the industry of ways in which products can be safely formulated, processed, and packaged to ensure the highest level of microbiological quality.

With the consumerist movement of the late 1960s and early 1970s came another influence that was ultimately to have a much greater effect on the cosmetic industry. For simplicity, we will just call this the influence of environmental concern. The keystone of the environmental movement was an intense distrust and suspicion of "man-made" chemicals. This "chemiphobia" was fed by a growing list of familiar chemicals recently found to have new and unexpected adverse biological activity and by a rapid growth in analytical capability and sophistication resulting in the ability to find chemicals potentially of concern in minute quantities all about us. Although the growing science of toxicology identified a number of biological responses of concern among the chemicals it was investigating, none had the impact of cancer. Two developments appear key in explaining why cancer came into the forefront of concern. One was the clear identification, based on epidemiological evidence, that cigarette smoking was a major factor in lung cancer. With this discovery came the realization that some cancer, at least, had a directly identifiable cause. The second development was the appointment, study, and report of the Mrak Commission. This commission, formed in 1969 by Robert Finch, then Secretary of the Department of Health, Education and Welfare, was the first to have sufficient political power to deal with the new environmental safety concerns. Its primary focus was on environmentally persistent pesticides and the resulting potential biological hazard to the various life forms thus exposed. Aside from the obvious concerns regarding the impact of such materials on wildlife, a central focus was the potential carcinogenic activity of some of these materials. The

growing concern around chemicals and cancer led to the acceleration of
efforts by the National Cancer Institute to identify chemicals in the
environment that might be potential causes of cancer in humans.

The stage was now set for the development of a philosophy that
was to have a major impact not only on the cosmetic industry but on
all chemical-based industrial concerns. Knowledgable experts were
claiming that up to 80 percent of all cancers were caused by environmental agents. The primary candidates for suspicion were synthetic
chemicals now being found via new and even more sensitive analytical
techniques in the water, in the air, in the soil, and in the food supply.
Given the state of ignorance with regard to the mechanism of cancer
production, it was natural that a philosophy of "zero risk" began to
develop. Enunciated legislatively in the Delaney Amendment of some
years before, this philosophy became the battle cry of those wishing
to ban any material that had a suspicion of oncogenic activity. Where
a suspicion existed, no risk was acceptable. A necessary corollary to
such a philosophical position was that of absolute safety, a concept
that could be held emotionally but that did not withstand any rational
examination.

During this period, the cosmetic industry found itself faced with
concerns about asbestos contamination of talc, toxicity of heavy metals
(e.g., mercury-containing preservatives), even more pressure on
color additives, and embryonic legislative activity aimed at giving the
FDA preclearance powers over cosmetic products. These challenges
resulted in the establishment of a number of additional committees within the Scientific Department of the CTFA. The Pharmacology/Toxico-
logy Committee was created to deal with the ever-growing list of issues centered on the potential biological activity of cosmetic ingredients. The Hair Color Technical Committee was created to further provide reassurance of the safety of materials used in hair dyes, bleach-
es, and colorants. The Quality Assurance Committee was formed to
help the industry ensure that ingredients are processed in such a way
as to produce the highest-quality products possible.

The CTFA understood that if it were to be responsive efficiently
to safety issues, it must develop an organized data base on cosmetic
ingredients. Through a survey of its members, the CTFA developed
a list of ingredients reported by chemical, trade, and common names.
The Cosmetic Ingredient Nomenclature Committee came into being with
the responsibility for development of a uniform nomenclature for cosmetic ingredients. The product of its effort was the *CTFA Cosmetic
Ingredient Dictionary*. This unprecedented work helped the CTFA
respond in a technically sound way to growing pressures for ingredient labeling.

The Scientific Conference Program Planning Committee dramatically changed the conference format to workshops emphasizing scientific
and regulatory developments rather than cosmetic chemistry. This
burgeoning scientific effort was coordinated by a growing CTFA staff

receiving guidance from the Scientific Advisory Committee, whose members were responsible for the oversight of the scientific and technical programs of the CTFA.

But even this structure proved inadequate to answer some of the new challenges of the 1970s. Concerns with regard to first, safety, then the environmental impact of fluorcarbons, the discovery of minute levels of nitrosamines in cosmetic products, the biological activity of dioxane present as a contaminant in some cosmetic materials, led to the formation of task forces devoted to developing an appropriate industry response to these issues. Added to this was the development within the FDA of the Over-the-Counter (OTC) Drug Review. Since some products of interest to the cosmetic industry are also drugs, task forces formed to work with and respond to the expert panels appointed by the FDA.

In order to provide for an independent evaluation and documentation of the safety of cosmetic ingredients, a unique and proactive program was initiated by the CTFA. This effort, called the Cosmetic Ingredient Review, brought together a panel of independent experts from academia who were charged with evaluating all data available either from the literature or from public or industry submissions regarding the safety of cosmetic ingredients. Their deliberations were to be held in public and the reports made available to all who had an interest. As the program grew in maturity, additional organizational structures were created within the Scientific Department of the CTFA to aid the industry in responding to the call for data and any issues of concern the panel might raise.

The growing animal rights movement provided the CTFA with its first major challenge of the 1980s. The CTFA's innovative response, developed by appropriate scientific committees and staff, was the establishment of the Johns Hopkins Center for Alternative Testing. The center was provided with a seed grant of $1 million and operates completely independently from the CTFA. Its objective is to fund research that could lead to replacement of certain animal tests or reduction of numbers of animals needed for testing.

It should be clear that the CTFA's ability to act and respond in a climate of escalating safety and environmental concerns was based on three factors. First and foremost was an industrywide commitment to assuring the public that its products and their ingredients were of the highest quality and safety. Second was the dedication of a group of well-qualified scientists and technical experts within the industry willing to take on the job of dealing with all the new issues that continued to arise. And finally, there was within the Scientific Department the organizational flexibility to respond appropriately as the need demanded.

This organizational flexibility is inherent in the structure of the Scientific Department. With the guidance and oversight of the Scien-

tific Advisory Committee, the activities of its standing committees,
and the ability to create special interest task forces, the CTFA is
responding to the challenges of the 1980s.

II. ORGANIZATIONAL STRUCTURE

The Scientific Advisory Committee (SAC) is the parent scientific committee of the CTFA. Membership on the Scientific Advisory Committee
is open to every CTFA active member company. In addition, the committee has been opened to accept 10 percent of its membership from
associate member companies. Members have the responsibilities within
their own companies for research management and regulatory affairs.
The major function of the Scientific Advisory Committee (SAC) is
oversight of the scientific and technical programs for the CTFA. In
this role, the SAC reviews the program priorities of the scientific
committees, as well as audits their progress in meeting these goals.
The Scientific Advisory Committee meetings provide a forum for discussion of technical issues. If the situation warrants, the Scientific
Advisory Committee will assign projects to existing committees or subcommittees or form special task forces to review problems and make
recommendations. In order to fulfill properly its function as chief
scientific advisor to the CTFA Board of Directors on technical issues,
the Scientific Advisory Committee must keep abreast of all scientific,
regulatory, and legislative developments that could affect the industry. This is done through committee and task force reports at SAC
meetings, and distribution of written materials. The collection and
dissemination of these materials is one of the important ways the CTFA
assists its member companies to individually meet regulatory and scientific challenges.

In its role as advisor to the CTFA Board of Directors, the Scientific Advisory Committee alerts the board to impending technical issues
of import and makes specific recommendations for appropriate action
through the Board subcommittee on research. The chairman of the Scientific Advisory Committee is an ex officio member of the CTFA Board of
Directors. As one of the principal committees of the CTFA board, the
Scientific Advisory Committee maintains liaison with the Government
Relations Committee, as well as the Legislative Planning Committee. The
Government Relations Committee advises the board on legal and regulatory developments, while the Legislative Planning Committee reviews proposed legislation and makes recommendations for appropriate responses.
Contact is maintained through liaison representatives appointed by
each committee and through communication among CTFA staff.

In order to fulfill its responsibilities, the Scientific Advisory Committee is organized into standing committees and task forces. The
standing committees of SAC are illustrated in Fig. 1. To augment the
standing committees, task forces have been established to deal with

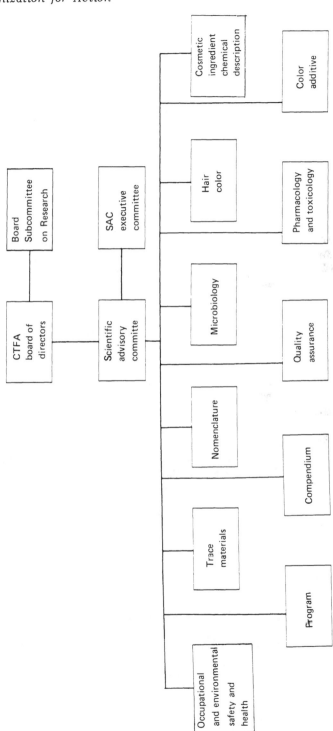

FIGURE 1 Standing committees of the Scientific Advisory Committee.

specific areas. For example, task forces were formed to respond to
the FDA's Over-the-Counter Drug Review Program and to assist industry in preparing responses to the Cosmetic Ingredient Review Program. In addition, task forces have been formed to respond to specific scientific and regulatory problems. To assist the SAC in its every-
day operations, a SAC Executive Committee has been established.
The SAC Executive Committee reports to the SAC and carries out
those functions assigned to it. It acts to facilitate consideration of
issues by the full committee.

One of the principal challenges for the cosmetic industry today is
the reaffirmation of the safety of its ingredients. To assist the industry in this task, a Pharmacology and Toxicology (P/T) Committee was
established. The Pharmacology and Toxicology Committee reports to
the SAC on matters concerning ingredient safety testing programs.
The P/T Committee consists of a major subcommittee, as well as a
variety of task forces. The major activity of the CIR Subcommittee is
to ensure quality to industry's response to the CIR Program. The
subcommittee provides expertise to individual task forces formed in
response to CIR, reviews industry submissions and CIR reports.
The P/T Committee has among its responsibilities the development and
updating of safety testing guidelines for the industry, as well as
studying modifications in protocols and alternate test systems designed to provide for minimum use of animals and maximum humane
treatment. The P/T Committee prepares protocols in response to requests by other CTFA scientific committees. For example, it prepared
protocols for testing certifiable colors in the area of the eye at the request of the Color Additive Committee.

The objective of the Color Additive Committee is to ensure that
the broadest possible pallet of colors consistent with quality and safety standards is available to the cosmetic industry. In order to accomplish this objective, the committee has initiated an extensive color
safety research program. The committee also investigates both new
uses for existing colors, as well as examines the potential to make
available new colors through permanent listing for use in cosmetics.
For example, the committee is attempting to obtain approval from the
FDA for the use of certain FD&C and D&C colors for eye-area cosmetics. A program has been initiated on a new color, Pigment Violet
19 (Quinacridone). While this material provides a variety of shades
of interest to cosmetic formulators, its approval can be obtained only
upon completion of an extensive chemistry and safety program.

The Hair Color Technical Committee has the responsibility for
confirming the safety of hair color ingredients. In order to accomplish this task, the committee has recommended, monitored, and evaluated research in independent laboratories designed to further establish the safety of hair color ingredients. Most recently, the Hair
Color Technical Committee has been engaged in developing the neces-

sary data to respond to proposed government regulations, CIR as well as consumerist and Congressional concerns.

In response to concerns with regard to the presence in cosmetic ingredients and formulations of such trace contaminants as nitrosamines and dioxane, task forces were organized to develop appropriate responses and research programs. These task forces maintain an active dialogue with the FDA and other interested trade associations, thus providing reassurance to the public and to the government of industry's responsible activity in this area. With the advent of increasingly sophisticated analytical technology with the attendant ability to detect increasingly smaller quantities of materials, it became clear that a Trace Materials Committee would have to be organized. Presently the Nitrosamine Task Force and Dioxane Task Force report to this committee. Its objectives are to monitor task force activities and to keep abreast of developments in the analytical and regulatory areas pertaining to trace materials.

The CTFA's Cosmetic Ingredient Nomenclature Committee has as its responsibility the development of uniform nomenclature for cosmetic ingredients in order to simplify cosmetic manufacturers' task for ingredient labeling all cosmetic products. The committee is responsible for the publication of the *CTFA Cosmetic Ingredient Dictionary*, which is officially recognized by the FDA as a controlling compendium for ingredient labeling. By providing uniform nomenclature, the cosmetic industry has avoided confusion among consumers, the medical profession, and industry alike in attempting to meet FDA ingredient labeling requirements. The third edition of the *Cosmetic Ingredient Dictionary* provides the most comprehensive compilation of information about cosmetic ingredients available to the industry.

Evidence of industry's commitment to maintaining the highest microbiological standards for its products is provided by the Microbiology Committee. This committee, working through five subcommittees, has developed an important series of guidelines and scientific papers useful to all cosmetic companies. The Information and Education Subcommittee prepares tapes and slide presentations and plans regional workshops. A broad range of issues of concern to cosmetic manufacturers and supplies are covered by the Microbial Content, Preservation, Raw Material, and Quality Assurance Subcommittees.

A Quality Assurance Committee was formed in order to assist industry members in developing practical programs to assure the quality of their products. Educational activities have included development of technical guidelines, tape/slide presentations, and special workshops in areas of quality assurance.

The Quality Assurance, Microbiology, and Pharmacology and Toxicology Committees have compiled their guidelines into the *CTFA Technical Guidelines*, an impressive educational tool, especially valuable to small and medium-sized companies interested in building or expanding in-house programs.

   The primary purpose of the Compendium Committee is to define
chemically the raw materials used by cosmetic and toiletry manufacturers. The committee has developed an outstanding compilation of
specifications, analytical methods, and spectra used by cosmetic manufacturers. The specifications are supported by round-robin testing
of cosmetic raw materials by committee members. In addition, the
committee compiles and publishes chemical descriptions of other raw
materials for which no specifications have yet been set. The compendium is of great value to the cosmetic industry in general with regard
to the items of commerce used in the formulation of its products and
to small businesses in particular, since the information contained in
the *CTFA Compendium of Cosmetic Ingredient Composition* represents
the collaborative work of many chemists. These volumes assist individual companies in establishing in-house specifications for the quality
of raw materials used in their products.
   In addition to the forum provided by the Scientific Advisory Committee itself, the CTFA, on an annual basis, holds a Scientific Conference. At these meetings, scientists, attorneys, regulators, and
representatives of the medical professions gather and participate in
workshops covering the important current issues confronting the industry. The conference provides a valuable place for formal and informal discussions of scientific and regulatory problems. The program for the conference is formulated by the Scientific Conference
Program Planning Committee.

III. CONCLUSION

As we enter the era of the 1980s, a new atmosphere is emerging.
There is growing public concern that perhaps government has overre-
sponded to health and environmental issues resulting in a situation of
overregulation. There is a recognition that absolute safety or its
corollary, zero risk, is not realistic. Risk assessment, despite its
inadequacies, is clearly coming into fashion. This is reflected in FDA
actions on lead acetate, color additives, and hair dyes, and in court
decisions such as those involving benzene.
   Animal testing, color safety, self-regulation (e.g., continued need
for the CIR Program), will all likely remain issues of great import well
into the mid-1980s. If these trends continue, the activities of the
Scientific Department will have an even greater effect on the public
image of the CTFA than in recent years. Regulatory pressures seem
to be waning, while pressure from special interests groups (e.g.,
animal rights advocates) is on the rise. Nevertheless, other agencies,
such as OSHA and the EPA, whose regulations also affect the cosmetic
industry, will continue to require careful monitoring and perhaps development of industrywide programs. Infact, in 1983, CTFA established the Occupational and Environmental Safety and Health Committee
for this purpose. Certainly the challenges of the 1980s will be as

vigorous and stretching as those of the 1970s have been. However, it is already clear that they will be sufficiently different in type and content to require the continued application of those strengths that have served the industry and the CTFA so well in the past—commitment to high quality, safe products, dedicated, competent scientists devoted to applying the best technical techniques, and flexibility to organize appropriately to meet whatever developments arise.

12

THE COSMETIC INGREDIENT REVIEW

ROBERT L. ELDER AND JONATHON T. BUSCH

The Cosmetic Ingredient Review, Washington, D.C.

I. INTRODUCTION

In 1976, the Cosmetic, Toiletry and Fragrance Association (CTFA) established the Cosmetic Ingredient Review (CIR) for the purpose of reviewing and evaluating the safety of ingredients used in cosmetics. The cosmetic industry has conducted safety testing on individual ingredients as well as cosmetic formulations for many years. Although some of this safety information has already been published, a significant portion is in industry files and not available for public or scientific review. The CIR program is designed to bring together for review and evaluation all of the worldwide data, both published and unpublished, on cosmetic ingredient safety. The review is conducted by a panel of distinguished scientists representing the various disciplines appropriate for safety assessment. This unique program of industrial self-regulation plays a key role in protecting the consumer, and in supporting the cosmetic industry's continuing commitment to the safety of its products*.

Although funded by the CTFA, the CIR is staffed and operated independently from the CTFA, except as the latter contributes directly and substantially to the unpublished industry information available for assessment. The total number of ingredients to be reviewed is large, and the spectrum of ingredients is expected to change over time. With the exception of proprietary material and trade-secret information, which remains confidential, the review is conducted by a process of open scientific deliberation.

II. OPEN PUBLIC REVIEW

In order for the safety review to be conducted in an unbiased manner, it was necessary to separate the review process from the CTFA and the cosmetic industry, and to function in a open public forum. This was accomplished in part by the establishment of a formalized set of written procedures under which the CIR program is required to operate; these procedures specifically call for a public review of all safety data. Meetings of the CIR are open to the public, during which time interested parties may present their own viewpoint, expertise, or information regarding ingredient safety. Public comments are also requested on all data and materials used in the review process, as well as on the reports issued by the CIR. The participation of consumer, industry, and government organizations as liaison members in the program has further assured the open nature of the review. In addition, all consultants and panel members used by the CIR in the review process meet the same conflict-of-interest standards applicable under federal law to special government employees.

III. ORGANIZATION

A Steering Committee was selected by the CTFA to provide policy and direction for the review process. The President of the CTFA chairs this committee, which includes a dermatologist representing the American Academy of Dermatology, a toxicologist from the Society of Toxicology, the chairman of the CTFA's Scientific Advisory Committee, and the CTFA Senior Vice President—Science.

---

*The CIR program received the 1979 Consumer Product Safety Award from the National Association of Professional Insurance Agents in recognition of the CIR's important contribution to consumer safety.

The Steering Committee selected an Expert Panel of seven distinguished scientists, who in turn were publicly nominated by government agencies, industry, consumer, and scientific and clinical societies. The Expert Panel members were selected from the scientific disciplines of dermatology, pharmacology, chemistry, and toxicology. The CIR Procedures give these experts independence to review and assess the safety data on cosmetic ingredients. All Expert Panel reports containing safety evaluations and recommendations are publicly announced without any private review or comment by the cosmetic industry. Nonvoting industry and consumer liaisons have been named to assist the panel; it is their responsibility to represent industry and consumer interests in all deliberations. Expenses of the consumer representative (appointed by the Consumer Federation of America) are all paid for by the CIR. The industry liaison person also serves as an intermediary through which unpublished industry data are requested. The Food and Drug Administration (FDA) has provided a nonvoting "contact person" to attend the Expert Panel meetings and assist the CIR in obtaining unpublished but publicly available data from the FDA. A scientific and clerical staff for the CIR were selected, and a Director and an Administrator appointed to support the Expert Panel and Steering Committee.

IV. REVIEW OF INGREDIENTS ON A PRIORITY BASIS

The Expert Panel evaluates the safety of each ingredient or group of chemically related ingredients according to a priority list developed by the Steering Committee and CIR staff. In accordance with CIR Procedures, the Expert Panel may revise the order of this list at any time by adding new cosmetic ingredients, deleting materials no longer used, or by giving a lower or higher ranking to ingredients. To establish an ingredient priority list for the Expert Panel, review is of critical importance. The Steering Committee was faced with a problem similar to that faced by many federal government agencies involved with prioritizing the vast number of chemicals for toxicological safety testing. (The CIR priority list does differ from the latter. All chemicals are not nominated for testing by the federal agencies; however, all cosmetic chemicals will be reviewed by CIR. Thus the priority list establishes a time sequence for ingredient review, rather than deleting an ingredient from consideration.) Any such attempt eventually resolves down to using an inadequate data base for many of the multiple factors that must be considered. After lengthy discussions with outside toxicological consultants, seven selected variables were selected for use in determining an ingredient priority. The selected variables were (1) frequency of use in cosmetic formulations; (2) concentration in cosmetic formulations; (3) area of use; (4) frequency of application; (5) use by sensitive population subgroups; (6) suggestion of biological activity; (7) consumer complaints about products containing the ingredient. In addition, a weighting coefficient for each variable was also established. The value of the variable and the weighting factor was assigned on the basis of the consultants' opinion of the significance of the variable *and* a judgment on the accuracy of the data base that was available to them. The equation expressing both terms is as follows:

$$R = C_1 t + C_2 u + C_3 v = C_4 w + C_5 x + C_6 y + C_7 z$$

where

C_1–C_7 are the weighting coefficients assigned to each variable t–z

and

R represents the score summation for each ingredient. Ordering of the R values for all ingredients from high to low establishes a numerical ranking of priority for review.

A summary of the values that were assigned to the weighting coefficient, and the possible values for the seven variables, is as follows:

Weighting coefficient (C)	Variable	Range of possible variable values
$C_1 = 0.20$	t = frequency of occurrence	1–10
$C_2 = 0.05$	u = use in high concentrations	0–1
$C_3 = 0.15$	v = area of normal use	1–8
$C_4 = 0.05$	w = frequency of application	0–1
$C_5 = 0.05$	x = use by particular groups	0–1
$C_6 = 0.40$	y = suggestion of biological activity	0–3
$C_7 = 0.10$	z = frequency of consumer complaints	0–1

By taking the maximum value assinged to each variable, one may order the relative contribution of each variable (t–z) in the final score (R). They are as follows:

Frequency of occurence [$C_1 t_1 = (0.2)(10) = 2.0$]
Suggestion of biological activity [$C_6 y = (0.4)(3) = 1.2$]
Area of normal use [$C_3 v = (0.15)(8) = 1.2$]
Frequency of consumer complaints [$C_7 z = (0.1)(1) = 0.1$]
Use by particular groups [$C_5 x = (0.05)(1) = 0.05$]
Use in high concentration [$C_2 u = (0.05)(1) = 0.05$]
Area of normal use [$C_3 v = (0.05)(1) = 0.05$]

A review of the preceding makes it obvious that frequency of occurrence, suggestion of biological activity, and area of normal use are the main variables affecting the priority score for each ingredient. The remaining four factors provide a basis for establishing priority among ingredients that have all received a maximum, and/or equal, total score.

The "frequency of occurrence" is based on the voluntary FDA program for receiving data from cosmetic manufacturers on ingredients used in cosmetic formulations [see Code of Federal Regulations (21 C.F.R. 720)]. This variable attempts to estimate the potential for public exposure and assumes that an ingredient that is used in a great number of formulations has the potential for exposing more people than one that is used less frequently. Although some estimates of production of raw cosmetic ingredients have been made, the use of this data base to estimate exposure presents serious problems. Estimates have not been made for production quantities of all cosmetic ingredients, and no reasonable allowance can be made for the noncosmetic product use of these chemicals. Therefore, raw material production values could seriously overestimate the potential of an ingredient for exposing the public from cosmetic products. Even though some manufacturers have not submitted data to the FDA on their product formulations, it was assumed that the data available did approximate a sampling of the large as well as small manufacturers. The assigned constant $C_1 = 0.2$ lowers the contribution, as compared to biological activity, where $C_6 = 0.4$ due to the questionable nature of the data base for frequency of use to effectively estimate population exposure. The factor 1–10 was assigned on the basis of a distribution curve for the frequency of use for all ingredients used in more than 25 formulations. Thus a mid-range value of 5 would be assigned to an ingredient with a frequency use of between 49 and 61. This indicates that for the 239 ingredients used in more than 25 formulations, one-half had a formula frequency below 61, and the remaining 50 percent had a frequency of formula use above 62.

The variable "biological activity" used in the equation extends beyond a indication of toxicity that might be documented in the published literature. It also includes any indication that the ingredient may effect a biochemical function. A maximum value of 3 was assigned to the variable where there was a indication in the literature that it might cause some detrimental effect. An assigned value of 0 was used when there was evidence that the ingredient was toxicologically inert. As previously noted, a value of 1 was assigned when it was known that an ingredient may affect the rate of a reaction but not inhibit it. A value of 2 was assigned when there was information available to indicate that the ingredient could alter a condition, or could accelerate a potentially adverse effect that by itself it could not do.

The highest weighted factor of 0.4 ($C_6 = 0.4$) was assigned to biological activity. This was considered appropriate due to the direct implication on the potential for toxicity, as well as the more reliable data base that was used to evaluate this parameter. The latter was accomplished by a computerized literature search of all ingredients used in more than 25 formulations. More than 17,500 abstracts were reviewed by the CIR staff for the 239 ingredients being prioritized. The abstracts for each ingredient were reviewed by two separate staff members. Where a difference in rating occurred between the two reviewers, a third staff member reevaluated the ingredient and made a final recommendation. It must be stressed that the review was a judgment process, and that the assigned value does not necessarily indicate the presence or absence of toxicity.

The variable entitled "area of normal use" relates directly to the type of cosmetic product and the normal area of the body to which it is usually applied. Products that may be inhaled or used near the mouth were given a maximum value of 8. Products associated with use near the eye or mucous membranes were rated at 6. These were followed by skin application, with a value of 2; and nails and hair, rated at 1. The values for the variable "area of use" primarily emphasized the potential for absorption into the body, whereas the assigned weighting factor $C = 0.15$ stressed the potential for effect on the immediate area of the body through which the ingredient was absorbed. This, in effect, was an additive consideration to that which may have already been included under the variable "biological activity."

The "frequency of consumer complaints" was based on the FDA records of consumer complaints on cosmetic formulations that contained the ingredient. The variable contribution is all or none (0 or 1), and assumes that the specific ingredient being prioritized was the one causing the complaint. The error in failing to report a complaint is offset by the assumption that the ingredient was the complaint-causing factor. Due to the high probability for error, both the assigned weighting factor ($C = 0.1$) and the variable factor do not contribute significantly to the overall priority value.

The "frequency of application" attempts to give recognition that repeated use, more than one time per day, may be of significance. Its overall contribution is low due to the fact that many cosmetics are designed to remain in contact with the body surface throughout the day without reapplication.

"Use by particular groups" recognizes that some products are designed for infants or older people; both groups may be more sensitive to cosmetic products that other population subgroups. In that most products are used by all groups, the contribution of this variable (0 or 1) is significant only in that it was assigned to those formulations specifically marketed for specific groups that may be more sensitive than the general population. Where equal scores occurred for general-use ingredients, the calculated value for this ingredient would raise its final score and require an earlier review.

The "concentration of use" variable was introduced to give an increase in priority scoring for ingredients used at concentrations above 25 percent. It was not intended to stress the relative significance of potential toxicity, but to indicate concern for ingredients used at high concentrations on the body surfaces. Thus, even though poorly absorbed, it may be of safety significance at the high concentration of use. The contribution to the score for this variable was all or none (0 or 1), and was given a low weighting factor of $C = 0.1$.

Establishing a priority list for safety evaluations where all ingredients can and will be reviewed is different from a priority list for toxicological testing, for the latter faces a time and cost problem that could be so great that the total list of chemicals could not be tested in a reasonable period of time and at a reasonable cost. It is anticipated that the CIR review will be completed in less than the time required for two cycles of a laboratory carcinogenic testing program (approximately 10 years).

Approximately 30 percent of some 3,000 known cosmetic ingredients are approved for use as an ingredient in, or associated with, the manufacture of foods and/or drugs. To minimize duplication of effort, the CIR Procedures permit the Expert Panel to defer from priority consideration those ingredients that are subject to other existing safety reviews. Thus, over-the-counter-drug active ingredients, food flavors, GRAS food ingredients, food additives, fragrance ingredients, ingredients in new drug applications, and all substances that are subject to a final regulation of the FDA are either deferred or excluded from the priority list of ingredients for CIR attention. Similarly, all color additives are excluded because their safety is determined by the FDA under section 706 of the federal Food, Drug, and Cosmetic Act. The color additives will be reviewed separately by the CIR to determine whether the existing safety evaluations and/or government approval for use in drugs and foods are applicable for cosmetic use. Ingredients whose use in cosmetics is higher than that approved by the FDA are not deferred. Instead, they are reviewed with the needed additional data to support a safety recommendation at the higher levels currently being used in cosmetics.

V. SCIENTIFIC REVIEW

The scientific review of each cosmetic ingredient goes through several stages. A flowchart outlining the review procedure as it relates to report development is presented in Fig. 1. The technical staff of the CIR prepares a Scientific Literature Review that summarizes the published literature for an ingredient. All known bibliographical references are included in this report. The document is then publicly an-

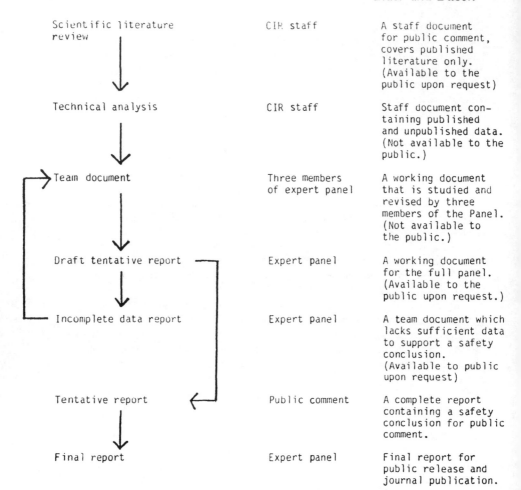

FIGURE 1 CIR report development.

nounced and made available for comment. At this time, a request is
made for any data, published or unpublished, that the literature review does not include. At the end of 90 days, any comments that may
have been received (including all unpublished data submitted by industry) are incorporated into a Technical Analysis for review by a
three-member team of the Expert Panel. The report goes through a
series of revisions called Team Documents before a Draft Tentative Report is presented to the full panel. One or more public sessions are
normally needed before the Expert Panel issues a Tentative Report for
a 90-day public comment period. The Tentative Report provides a
thorough review of the available data and includes the Expert Panel's
tentative conclusions. The panel may make one of three conclusions:

(1) the ingredient is safe as currently used; (2) the ingredient is unsafe; or (3) there is insufficient information for the panel to make a determination. After consideration of the public comments on the Tentative Report, the Expert Panel then issues a Final Report. The Final Reports are eventually grouped and published in a scientific journal. Should the Expert Panel decide at any stage of its review that additional data are required before they can issue a final advisory opinion, an Insufficient Data Report is issued. This latter document states the type of information that must be provided before a final decision on the safety of the ingredient can be made.

A key feature of the review process is the fact that a lack of documented information is not permitted to stand as an adequate basis for a determination of safety. Thus, ingredients that have had a long history of use in cosmetics must bear the same documented burden of proof of safety as would new and more recently adopted chemicals.

VI. AVAILABILITY OF CIR DOCUMENTS

All CIR documents that are issued by the full Expert Panel are available to the public. CIR staff documents and drafts of reports being studied by a Team of the Expert Panel are not available. A Public Documents Room has been established at the CIR for interested persons to review and study available reports, including the published and unpublished literature substantiating the reports and their conclusions. The official minutes of all Expert Panel meetings, and the formalized written procedures under which the CIR operates, are also available to the public.

VII. INFORMATION DEVELOPMENT

Many of the chemical ingredients used in cosmetics are frequently used in other consumer products. A program designed to collect this information, in advance of the CIR program review, was established during 1980. A data base for ingredient use in foods and drugs has also been established. The CIR staff maintains an ongoing surveillance program for cosmetic ingredients that are under toxicological testing by other groups.

The CIR also maintains access to a number of government and commerical data bases through an on-line computer system. The online system provides chemical and biological data for thousands of chemicals, as well as information on the published scientific literature for cosmetic materials. Thus, the CIR program maintains a capability to conduct its own literature surveys, as well as an ability to maintain a continual surveillance on new data, which may become available on ingredients for which a safety recommendation has already been made.

VIII. DISCUSSION

Carcinogenesis testing of chemicals to which the consumer is exposed is of concern to the Expert Panel, as well as to industry and the public. This national issue extends beyond cosmetics to foods, drugs, and other industrial chemicals. The Expert Panel has not hesitated to limit its final recommendation where data of this type were not available. Yet the Expert Panel's recommendation is a judgment of safety on the basis of the type of data that should be available from all good industrial product safety evaluations. When the chemical structure of the ingredient is of concern, the mutagenesis data are lacking, further testing would be requested. This role is not different from the judgment of the scientific peer groups who recommend to the National Toxicology Program (National Institute of Environmental Health Service, DHS) what chemicals should or should not be tested for carcinogenesis. Cosmetic ingredients, like all other chemicals in use today, are screened by this government program. Experience with the CIR review effort has already shown that the careful and methodical review given by the Expert Panel, and the documentation of the specific need for more testing, has prompted industry to do the necessary testing to prove safety. It is in the best interest of the consumer and the industry to complete the needed studies to prove an ingredient safe or unsafe. This holds for all safety tests, routine or specialized mutagenesis and carcinogenesis testing. Full public discussion of the scientific rationale for requesting further testing, rather than a "checklist" type of approach, places a heavy burden on the panel members. The panel has not hesitated to discuss and document its opinions when making this scientific as well as societal decision.

The agreement by the cosmetic industry to fund a program that by its very nature may limit industry's future choices of ingredients to be used in their formulations is unique. It was initially recognized that in order for this review to be effective, it must be scientific and must not be prejudiced by any interest groups. The procedures used in preparing an Expert Panel recommendation must be documented and rigidly followed. This was accomplished by the CTFA's acceptance of the formal CIR Procedures. The separation and independence of the Expert Panel as well as the openness of the deliberations were planned by the CTFA to permit a full public debate on the safety of cosmetic ingredients by scientists and other interested persons.

The CIR Program is one of the most extensive and progressive industry endeavors to evaluate the safety of consumer products. The program is designed and structured to bring all data, published and unpublished, on cosmetic ingredients into a open public forum for review and scientific evaluation. The cosmetic industry, in supporting the effort, challenged itself to prove the safety of its products made from ingredients that had a long history of use. In doing so, it also accepted the responsibility to use this same philosophy and criteria for open public safety review on all new ingredients that are or will be developed. In establishing the program, it was recognized that it would be a long and continuing effort.

The results to date are impressive from two points of view. First, the needed unpublished scientific data have been made available both from industry and govermental agencies. This cooperation demonstrates that both groups believe that it is in the public's interest to support the program. Second, the Expert Panel's careful scientific deliberation as documented in the CIR's Final Reports is being accepted by both consumer and scientific peer groups who have reviewed and commented on the panel's decision. The latter is of absolute necessity, and has set a level of performance that the CIR must sustain.

13

THE INDUSTRY RESPONSE PROGRAM TO THE COSMETIC INGREDIENT REVIEW

GERALD N. McEWEN, JR.

Science Department, The Cosmetic, Toiletry and Fragrance Association, Inc., Washington, D.C.

THEODORE WERNICK

Compliance Coordinator, Gillette Medical Evaluation Laboratories, Rockville, Maryland

ROBERT P. GIOVACCHINI

Corporate Product Integrity, The Gillette Company, Boston, Massachusetts

I. INTRODUCTION

The Cosmetic Ingredient Review (CIR) was initiated in September 1976 by the Cosmetic, Toiletry and Fragrance Association (CTFA) to ensure that ingredients, as used in cosmetic products, were safe, and to eliminate the need for wasteful duplicative testing of these ingredients. When CIR began, approximately 2,500 ingredients were listed in the *CTFA Cosmetic Ingredient Dictionary*. The magnitude of the CIR program was apparent to CTFA member companies from the beginning; even so, industry was willing to make the necessary long-term commitment to ensure its success. Six years following the inception of the CIR, this commitment was reaffirmed by the CTFA Board of Directors, which acknowledged the CIR as one of the industry's most important science programs.

In the early planning stages for the CIR, industry scientists recognized that their active participation would be necessary to ensure that the CIR operated efficiently and effectively. Two of the major reasons for needing this whole-hearted participation were the relative unfamiliarity of the CIR Expert Panel regarding cosmetic formulation and ingredient use characteristics and the limited amount of published information on many cosmetic ingredients. The panel's inexperience with cosmetic formulations was a result of the stringent nonconflict of interest requirements for panel members mandated by the CIR procedures.

The limited amount of published data on cosmetic ingredients was, in part, a result of the relative innocuousness of most of these chemicals, which resulted in little attention being directed toward their testing by the academic scientific community. Additionally, widespread safe use of these ingredients, since before safety testing was routinely conducted, has led to their recognition generally as being safe, and scientific journal space being limited, editors rarely publish results from tests in which no effects are observed. The lack of published safety data requires the submission of unpublished data from individual companies that have tested the ingredients to evaluate their safe use in the companies' products.

In order to maximize CTFA member company efforts in assisting the Expert Panel and to establish a cooperative rather than adversarial association between industry and the CIR, the CTFA organized the Industry Response Program. The Industry Response Program was also developed to make sure that all companies interested in the CIR, whether or not they were members of the CTFA, would be kept abreast of what the CIR was doing and would be given the opportunity of participating in industry activities relating to the CIR. This participation affects the quality and completeness of the review process.

The major goal of the Industry Response Program is to provide a coordinated framework for ensuring that the CIR Expert Panel gets the most relevant information possible to assist in its deliberations. Coordinating industry efforts decreases wasteful duplication and increases the level of expertise that industry can provide to the panel by encouraging a combined approach of industry scientists to issues of concern.

While the procedures of the CIR, as outlined in the preceding chapter, are relatively static in order to provide continuity and to protect the review process from ill-conceived change or outside pressure, the procedures for the Industry Response Program are meant to be dynamic. This allows the program to adjust readily to changes in issues of concern to the Expert Panel, individual company interests, and changes in assignments of participating industry scientists. The dynamic nature of industry's procedures also encourages the development of new methods to increase efficiency and effectiveness.

The remainder of this chapter is devoted to a brief history of the development of the procedures of the Industry Response Program and a more detailed look at the current procedures with emphasis placed on the purposes of the various steps. A copy of the CIR/Industry Response Program Procedures, followed in 1982, is included in the appendix at the end of the chapter. Further information on the Industry Response Program is available from the Industry Liaison Representative to the CIR, CTFA, 1110 Vermont Avenue, N.W., Washington, D.C. 20005.

II. HISTORICAL PERSPECTIVES OF THE INDUSTRY RESPONSE PROGRAM

In 1976, while initial steps were being taken by the CIR Steering Committee to review recommendations for Expert Panel members and develop a priority list of chemicals to review, the CTFA Board of Directors charged its Scientific Advisory Committee (SAC) with developing procedures for industry to supply information to the CIR. The SAC appointed a Task Force Coordinating Committee to develop those procedures. It was understood from the beginning that the Industry Response Program must encompass information flow in two directions to be effective. Information had to flow from the CIR to industry so that companies would know what types of data were necessary and what issues concerned the Expert Panel, and information had to flow from companies to the CIR to support the Expert Panel's review efforts. How best to organize this information exchange was the first consideration of the committee.

Early in 1977, the Task Force Coordinating Committee completed work on a set of procedures that set the framework for the Industry Response Program. The responsibility for information flow from the CIR to industry was that of the Industry Liaison Representative to the CIR (ILR), mandated by the CIR Procedures. Originally the ILR made reports to the SAC. Later, the ILR developed an information bulletin that was mailed to the official representatives of companies to convey CIR information.

To prepare information to be submitted to the CIR from industry, the Task Force Coordinating Committee decided that interested companies should form a working group (Ingredient Task Force) for each ingredient. Representatives from those companies would meet and organize the data each had to supply, and then prepare an outline of that information for submission to the CIR.

Early application of these procedures demonstrated several critical problems. Information from the CIR was not always reaching the appropriate persons in each company. Frequently, because of this, companies that had information to supply did not attend working group meetings so that their data were not available for the Expert Panel's review. Because the assembly of data began with these working group

meetings after the CIR requested data, there was only a short time
available to assemble the data and prepare industry submissions. And
finally, since each industry submission was prepared by a separate
working group, the members of which in some cases had limited toxicologic expertise, there was not a consistent presentation in terms of
format or quality.

Realizing that the resources of industry were not being efficiently
utilized, the SAC, in 1980, recommended two significant changes in
the Industry Response Program to increase the efficiency and effectiveness of the system. One change was to transfer responsibility for
preparing industry submissions to the CTFA's Pharmacology/Toxico-
logy (P/T) Committee, which promptly appointed a CIR Subcommittee
for this purpose. The second was to recommend that the CTFA provide a staff person whose primary responsibility would to be coordinate the Industry Response Program.

III. CURRENT PROCEDURES FOR THE INDUSTRY
RESPONSE PROGRAM

Many changes and refinements in technique have occurred since the
initial development of the Industry Response Program; however, the
basic goals of the program remain the same: that is, to provide a
framework for all interested parties to participate in the CIR review
process if they so desire, and to ensure that the CIR Expert Panel
receives all available relevant data necessary to complete a valid scientific safety assessment of each ingredient as used in cosmetic products.

The organizational framework for the Industry Response Program
includes the P/T Committee, made up of senior company personnel
directly responsible for product safety within their companies, and a
CIR Subcommittee comprised of members of the P/T Committee. The
CIR Subcommittee is charged with the day-to-day oversight and management of the Industry Response Program and with interfacing with
the CIR Expert Panel through the ILR. The CIR Subcommittee also
serves as the core for individual Ingredient Task Forces that include,
as members, other interested industry personnel and invited experts.
Each of these groups is supported by appropriate CTFA staff.

The first step in the Industry Response Program is ensuring that
all segments of industry are advised of the actions and concerns of
the CIR Expert Panel. This is done through the *CIR Developments
Bulletin*, a newsletter published by the CTFA, that not only reports
on the actions of the CIR, but also gives an evaluation of the impact
of these actions, alerts industry to upcoming reviews, points out
problems and concerns, and reports on the activities of various task
forces and the P/T CIR Subcommittee. The *Bulletins* are also used
to relay information on ingredients under review by foreign countries
that might impact on the CIR Review. *CIR Developments Bulletins* are
mailed to both CTFA member and non-member companies. Companies

involved in making finished cosmetic products or producing raw ingredients may submit the names of more than one staff member to receive the *Bulletins* directly, if necessary, to ensure that each individual within the company who has the responsibility of interacting with CIR has his or her own copy. Currently, about 800 individuals receive their own copies of the *Bulletins*.

A starting point in determining what unpublished information should be submitted to the CIR Expert Panel for its review of a particular ingredient is to determine what information is available. It would obviously not be efficient to flood the panel with all available unpublished data; however, it is important to make certain that representative types of tests that ensure all use conditions can be adequately evaluated is available to the panel. Choices of data to be submitted to the panel are made on the basis of test type and product type evaluated, *not* on the basis of results of the test. For instance, if an ingredient is found to have been tested in a variety of product types at different concentrations but with essentially the same test, those tests that are representative of each use condition, rather than all tests available, are submitted. Of course, if the panel requests further data, these are supplied.

To identify the available safety data, the CTFA issues a Notice in the *CIR Developments Bulletin* of ingredients to be reviewed by the Expert Panel. This notice requests that each company fill out a form that lists the type of test data available, the type of product tested, and the concentration of the ingredient in the product. Results are *not* included on the form. At the same time, interest in serving on a task force for preparation of the industry submission and review of CIR documents is requested. When these forms are submitted to the CTFA, CTFA staff blind the company identifiers before copying them for review purposes so that the source of the information will remain confidential. The forms are then reviewed by the ILR and members of the P/T CIR Subcommittee to determine which tests would be valuable to the Expert Panel for its review. The forms, marked with the tests requested by the subcommittee, are then returned to CTFA staff for ordering from the individual companies.

Generally, the types of safety data not published in the scientific literature, and therefore those most requested from industry, include tests most appropriate to the evaluation of safety in cosmetics, that is, tests for phototoxicity and photoallergenicity. Most often, academic research is designed to produce an adverse biologic effect so that the mechanism of toxicity can be studied, and if no effects are observed, the information is rarely published. This distinguishes the types of testing most often conducted by academic scientists and published from that research conducted by companies to evaluate safety. For safety evaluation, the aim is to ensure safety under conditions of normal use and anticipated misuse of a product. As the results of this type of testing rarely demonstrate adverse effects, because the prior experience of product formulators in the industry generally precludes the development of prototypes that would be hazardous, these studies are not usually published.

Before the CIR begins its preparation for the review of an ingredient, the CTFA sends letters to individual companies requesting that they send in specific test data noted from the available safety data forms they had previously submitted. As these reports are received, company identifiers are removed to maintain confidentiality and they are collated and stored for future use.

When the CIR issues its Scientific Literature Review (SLR) on a particular ingredient, this SLR, the individual test reports submitted by industry companies, and a draft industry summary of these tests, termed the Cosmetic Ingredient Safety Analysis (CISA), prepared by CTFA staff, are submitted to an Ingredient Task Force and all members of the P/T Committee for their review and evaluation. The Task Force includes all individual company personnel who experessed an interest in the ingredient plus invited experts and the members of the P/T CIR Subcommittee. By encouraging anyone with an interest to participate, the procedures ensure the most open and representative evaluation. By including all members of the subcommittee, industry ensures a high level of quality and continuity, and a consistency of presentation for the CISA.

The subcommittee, after final preparation of the CISA, makes a report to the P/T Committee that outlines the data available on the ingredient, both published from the CIR's SLR and unpublished summaries in industry's CISA, and highlights issues and makes recommendations for further data development if necessary. The P/T Committee addresses all recommendations and determines the type of response most appropriate to answer any anticipated concern. The P/T Committee, being composed of senior safety research personnel from major CTFA member companies, brings to bear a great deal of expertise in the area of safety substantiation on any problem. If the appropriate response involves preparation of a discussion document on a particular issue, an ad hoc task force is set up to prepare the report and it is reviewed by the P/T Committee. If the appropriate response involves additional testing or hiring a consultant, the P/T Committee reports this to the SAC for their review and approval. The SAC, as the major CTFA scientific committee, is made up of representatives invited from each active member company and therefore is most suited to consider technical questions that might result in expenditure of association funds.

When additional testing is necessary, the P/T Committee identifies or develops a protocol that will answer the question, identifies testing facilities that can perform the research, and secures proposals from these facilities. The SAC determines if the ingredient is of sufficiently widespread interest to warrant the financial support for the testing being taken out of general funds, and with the approval of the CTFA Board of Directors authorizes the research. The P/T Committee, with CTFA staff, then monitors the research and reviews the final report before it is submitted to the CIR. If the SAC finds that the ingredient is not of sufficiently widespread interest to warrant industry-sponsored testing, individual companies are notified and they may organize to conduct the needed research. The P/T Committee and CTFA staff offer assistance to these companies at their request.

During Expert Panel review of an ingredient, questions may arise or additional testing may be requested that was not anticipated by the Task Force and the P/T CIR Subcommittee. These are reported to industry and the subcommittee by the ILR and are considered in the same manner as previously stated for industry-identified concerns.

Finally, the Task Force and P/T CIR Subcommittee review reports on deliberations of the CIR Expert Panel as they assess the safety of the ingredient until a CIR Final Report is issued.

It is important to note that while the Industry Response Program was initiated and is constantly being refined to aid companies to interact with the CIR, both the CIR and CTFA Procedures recognize the right of individuals to present their own views or information before the CIR Expert Panel. While this has not been a common occurence, there have been instances in which individual companies have decided to participate directly in the deliberations of the Expert Panel. Coordination of these activities with CTFA staff and the ILR has ensured that the individual companies and the CTFA Technical Committees have maintained adequate communications and coordination of effort.

IV. THE FUTURE OF THE INDUSTRY RESPONSE PROGRAM

In order to aid companies increase their effectiveness in the Industry Response Program, the P/T Committee has initiated a three-part program to increase awareness of the program and the procedures for individual company participation. The first part of this program is a "how to" booklet for those persons within individual companies who have the responsibility to answer requests for data from the CTFA. An article containing the text of this document was published in the *CTFA Cosmetic Journal* in September 1982. The second part of the program is a sourcebook of questions and answers about the CIR and the Industry Response Program designed to present all of the information on each program necessary for a full understanding. The sourcebook became available from the CTFA in November 1982. The third part of the program consists of workships for persons involved with the Industry Response Program designed to aid them in functioning within their companies.

The Industry Response Program, by aiding the CIR, is one of the most important efforts of the industry to demonstrate to consumers, by assisting a third party of experts, that the industry is committed to ensuring that the ingredients used in cosmetic products are safe for the consumer under the label conditions of use. The CIR review process and the Industry Response Program are unending tasks. It will, of course, take considerable time to complete the initial CIR review of all of the individual ingredients, but, in addition, updating these evaluations with new data, reviewing conclusions in light of new scientific thinking, and so on, ensures that both programs will continue indefinitely. Still, the CIR Reports on individual ingredients, based in large part on unpublished data submitted through the Industry Response Program, will continue to be a source of important toxicological information that can be used not only by the industry, but also by the public and by various health regulatory agencies worldwide.

V. APPENDIX: CIR/INDUSTRY RESPONSE PROGRAM PROCEDURES

Part A. General

Section 1. Definitions

(a) "Industry Response Program" means any actions taken by the Cosmetic, Toiletry and Fragrance Association, Inc. (CTFA) on behalf of industry as a whole, authorized by the CTFA Board of Directors and conducted pursuant to these procedures, in response to the Cosmetic Ingredient Review.

(b) "Cosmetic Ingredient Review" (CIR) means the CIR program is defined by the CIR Procedures, revised February 25, 1979. CIR is considered synonymous with the CIR Expert Panel.

(c) "Ingredient" means any chemical substance used as a component in the manufacture of a cosmetic product, but shall not include a proprietary mixture.

(d) "Cosmetic Product" means a finished cosmetic, the manufacture of which has been completed.

(e) "Cosmetic Ingredient Safety Analysis" (CISA) means a document summarizing unpublished safety data prepared in response to the CIR request issued with notification of availability of a Scientific Literature Review for submission to CIR.

(f) "Report" means a document prepared by the CTFA Pharmacology/Toxicology Committee, concerning a specific technical issue or question from CIR, for use by the Industry Liaison Representative to the CIR.

(g) "CIR Testing" means any testing undertaken by CTFA as a whole or by groups of interested companies through CTFA to answer questions on ingredients under review by CIR.

(h) "CTFA Pharmacology and Toxicology Committee" (P/T) means the committee composed of cosmetic industry scientists responsible for CTFA technical activities in the areas of pharmacology and toxicology.

(i) "Vice Chairman—Ingredient Safety" means the Vice Chairman of the P/T who has the responsibility as delegated by the Chairman of P/T for coordination of all P/TC ingredient safety activities.

(j) "Vice Chairman—Research Programs" means the Vice Chairman of the P/T who has the responsibility as delegated by the Chairman of P/T for coordination of all P/TC testing activities.

(k) "P/T-CIR Subcommittee" means a Subcommittee composed of P/T-CIR members for the purpose of interfacing with the Industry Liaison Representative to the CIR Expert Panel to ensure an efficient flow of data from industry to the CIR Expert Panel.

(l) "P/T-CIR Subcommittee Chairman" means the chairman of the P/T-CIR Subcommittee who has the responsibility as delegated by the Chairman of P/T for coordination of the activities of the P/T-CIR Subcommittee.

(m) "CIR Ad Hoc Ingredient Task Force" (Ingredient Task Force) means a Task Force of the P/T-CIR Subcommittee that participates is particular document development activities of the Industry Response Program.

(n) "CIR Ad Hoc Safety Testing Task Force" (Testing Task Force) means of P/T Task Force that participates in safety testing activities of the Industry Response Program.

(o) "CIR Contact Person" means the individual designated by his or her company to communicate with CTFA Staff for the purpose of maintaining an awareness of CIR developments and supplying safety data and other relevant assistance to the Industry Response Program when required.

(p) "Interested Cosmetic Industry Persons" means any individuals representing companies engaged in commerce to produce cosmetic products or ingredients, not necessarily members of CTFA, who demonstrate an interest in participating in the Industry Response Program.

(q) "Industry Liaison Representative" (ILR) means the individual appointed by the CTFA Board of Directors to represent and serve as industry liaison to CIR.

(r) "CTFA Staff" means any individual CTFA staff person responsible for supporting the Industry Response Program.

Section 2. Purpose of the Industry Response Program

The purpose of the Industry Response Program is to identify, acquire, develop, and submit information on ingredients and technical issues to CIR for its use in assessing the safety of ingredients under their conditions of use.

Section 3.   Interpretation and Amendment of Procedures

(a) If any dispute arises as to the proper interpretation or application of these procedures, a majority vote of the P/T, subject to approval by the CTFA Board of Directors or its assignees, shall be final
and binding with respect to such matter.

(b) These procedures may be amended by a two-thirds vote of
the P/T, subject to approval by the CTFA Board of Directors of its
assignees.

Section 4.   Items Excluded from Coverage by These Procedure

Responses to CIR either directly or indirectly through the ILR from
individual companies, or from groups of individual companies, as distinct from the CTFA as a whole, interested in a specific issue or ingredient, are not subject to these procedures at the option of the
respondents.

Unless specifically requested by the sponsor or sponsors, safety
testing performed by individual companies or by groups of individual
companies that is not funded through CTFA General Funds is not subject to *Part E—Additional Data Developments* of these procedures.

Part B.   Responsibilities of Individuals in the
Industry Response Program

Section 10.   Vice Chairman—Ingredient Safety

The Vice Chairman—Ingredient Safety is responsible for the coordination of all document development aspects of the Industry Response
Program, and for providing written reports to P/T on its status and
on CIR developments.

Section 11.   P/T-CIR Subcommittee Chairman

The P/T-CIR Subcommittee Chairman is responsible for coordinating
the activities of the Subcommittee, for the preparation of all CISA's,
for tracking all ingredients under review by CIR, for recommending
additional data development when appropriate, and for the preparation of written status reports for the Vice Chairman—Ingredient
Safety. The Subcommittee Chairman calls all meetings of the Subcommittee and presides over these meetings.

Section 12.   P/T-CIR Subcommittee Members

P/T-CIR Subcommittee Members are responsible for participating in the
preparation of all CISA's, for reviewing all documents produced by
CIR and following the activities of the CIR Expert Panel, and for participating in the preparation of recommendations for additional data
development when appropriate. Subcommittee Members are responsible for attending all meetings called by the Subcommittee Chairman.

Section 13. Ingredient Task Force Chairman

The Ingredient Task Force Chairman is responsible for coordinating the activities of the Ingredient Task Force, for the preparation of the CISA assigned to that Ingredient Task Force, for reviewing all activities of the CIR regarding the ingredients assigned, and for preparing written reports to the P/T-CIR Subcommittee Chairman on the status of the ingredient group and any issues of concern affecting that ingredient group. The Ingredient Task Force Chairman calls all meetings of the Ingredient Task Force and presides over these meetings.

Section 14. Ingredient Task Force Members

Ingredient Task Force members are responsible for participating in the preparation of the CISA on the ingredients assigned to that Ingredient Task Force, for reviewing documents produced by CIR on those ingredients and following the activities of the CIR Expert Panel regarding those ingredients, and for participating in the preparation of recommendations for additional data development on those ingredients when appropriate. Ingredient Task Force members are responsible for attending all meetings called by the Ingredient Task Force Chairman.

Section 15. Vice Chairman—Research Programs

The Vice Chairman— Research Programs is responsible for the coordination of all testing aspects of the Industry Response Program and for providing written reports on testing status affecting CIR.

Section 16. Testing Task Force Chairman

The Testing Task Force Chairman is responsible for coordinating the activities of the Testing Task Force, for identifying and/or designing protocols for testing, for reviewing proposals from testing facilities, for monitoring work in progress, for reviewing reports received on testing conducted, and for preparing written status reports on testing activities to the Vice Chairman— Research Programs. The Testing Task Force Chairman calls all meetings of the Testing Task Force and presides over these meetings.

Section 17. Testing Task Force Members

Testing Task Force Members are responsible for participating in the identification and/or design of protocols, the review of proposals, the monitoring of work in progress, and the review of reports received on testing conducted. Testing Task Force members are responsible for attending all meetings called by the Testing Task Force Chairman.

Section 18. CIR Contact Persons

CIR Contact Persons are responsible for maintaining a current awareness of the status of CIR ingredient reviews, for submitting notification of available safety data on ingredients under review and on products containing those ingredients on forms designed for that purpose to CTFA staff, for sending that data to CTFA staff when requested, and for supplying other relevant assistance when requested.

Section 19. ILR

The ILR is responsible for transmitting information from industry to CIR and for reporting to CTFA staff on CIR deliberations. The ILR represents the interests of industry in all CIR deliberations. The ILR complies with the CIR Procedures, Section 23, outlining rights and responsibilities.

Section 20. CTFA Staff

CTFA staff is responsible for giving administrative and technical support to the Industry Response Program. CTFA staff is also responsible for informing industry in writing of CIR developments on a regular basis, for identifying sources for unpublished data and acquiring that data for use in preparing documents for submission to CIR, and for coordinating all activities associated with the identification of laboratories, solicitation of bids, and negotiation of contracts associated with CIR testing. In all contacts with individual companies, CTFA staff is responsible for maintaining strict confidentiality as to source of information.

Part C. P/T-CIR Subcommittee and Task Force Membership

Section 30. Membership of P/T-CIR Subcommittee

Subcommittee membership shall include a Subcommittee Chairman and other members of P/TC. The Vice Chairman—Ingredient Safety shall serve *ex officio*. CTFA staff shall provide administrative and technical support.

Section 31. Selection of P/T-CIR Subcommittee Members

Upon recommendation of the Vice Chairman—Ingredient Safety, the P/TC Chairman shall appoint the P/T-CIR Subcommittee Chairman. The term of office shall be two years to coincide with the term of the Vice Chairman—Ingredient Safety. Volunteers from P/T will be solicited for other positions on the Subcommittee and in the absence of sufficient volunteers, will be appointed by the P/T Chairman, upon recommendation of the Vice Chairman—Ingredient Safety.

Section 32. P/T-CIR Subcommittee Meetings

P/T-CIR Subcommittee meetings shall be held at the call of the Subcommittee Chairman subject to the availability of a majority of the members. Meetings will be held as infrequently as is practical without interfering with the purpose and function of the Subcommittee. Subcommittee meetings shall be presided over by the Subcommittee Chair-

man or his assignee, and minutes shall be prepared by CTFA staff
and distributed to the Subcommittee and the Vice Chairman—Ingredient Safety. The substance of all meetings will be reported to the P/T
by the Vice Chairman—Ingredient Safety or his assignee.

Section 33. Membership of Ingredient Task Force

Ingredient Task Force membership shall include all the members of the
PT-CIR Subcommittee plus other industry personnel or consultants
who may serve on the Ingredient Task Force through notification of
intent or invitation, respectively. The Chairman of the Ingredient
Task Force shall be a member of the P/T-CIR Subcommittee appointed
by the Subcommittee chairman.

Section 34. Membership of Testing Task Force

Testing Task Force membership shall include members of P/T designated by the Vice Chairman—Research Programs plus other industry
personnel or consultants who may serve on the Testing Task Force
through application or invitation, respectively. The Chairman of the
Testing Task Force shall be a member of P/T appointed by the Vice
Chairman—Research Programs.

Section 35. Ingredient and Testing Task Force Meetings

Task Force meetings shall be held at the call of the Task Force Chairman subject to the availability of a majority of the members. Meetings
will be held as infrequently as is practical without interfering with
the purpose and function of the Task Force. Task Force meetings
shall be presided over by the Task Force Chairman or his assignee,
and minutes shall be prepared by CTFA staff and distributed to the
Task Force. The substance of all meetings will be reported to the
P/T by the appropriate P/T Vice Chairman or his assignee.

Part D. Preparation of Industry Documents

Section 40. Identification and Acquisition of Available Safety Data

(a) CTFA staff shall issue a notice to all companies on the *CIR
Developments Bulletin* mailing list that CIR intends to review a particular ingredient, at least 6 months prior to the commencement of that
review, and shall request notification from the companies of the availability of data for use in preparing a CISA. Upon receipt at CTFA of
notices of available safety data from individual companies, CTFA staff
will ensure that company identifiers are removed from these notices
and summarize the responses for review by the Ingredient Task Force.
Based on data needed and appropriateness of the testing conducted,
individual requests for specific data will be sent to the responding
companies by CTFA staff.

(b) Following receipt of data, CTFA staff ensures to the extent
possible that company and product identifiers are removed from the

company submitted reports and collates the information for use in preparing the CISA.

Section 41. Preparation of the CISA

(a) The Ingredient Task Force Chairman designates one or more members of the Task Force to prepare a working draft of the CISA. CTFA staff provides the documents necessary for the preparation of the draft to the responsible individual. The Ingredient Task Force member, with the assistance of CTFA staff, then prepares a working draft. CFTA staff transposes the draft to word processing equipment for formatting, editing and review.

(b) The working draft and all reports cited therein are submitted to the Ingredient Task Force and the P/T, along with CIR's Scientific Literature Review, for technical review and comment. The technical review pays particular attention to accuracy, consistency, presentation, and potential safety issues and data gaps in the safety information available.

(c) A written report regarding the ingredient group is prepared by the Ingredient Task Force Chairman and submitted to the P/T-CIR Subcommittee Chairman. Comments received on the draft and the Task Force Chairman Report are reviewed by the Subcommittee and incorporated into the draft as appropriate. The Subcommittee Chairman then prepares a written report to the Vice Chairman—Ingredient Safety, indicating approval of the CISA by the Subcommittee, pointing out any existing salient safety issues and potential data gaps, and recommending further data development where appropriate.

(d) The Vice Chairman—Ingredient Safety reviews the P/T-CIR Subcommittee Chairman's report and the CISA and, with the concurrence of the P/TC Chairman, forwards the approved CISA to CTFA staff for transmittal to the ILR for submission to CIR. The Vice Chairman—Ingredient Safety or his assignee presents a written report on CIR Ingredient Group status at the next regular meeting of P/T, and presents any recommendations from the Subcommittee.

Section 42. Preparation of Reports

(a) Reports are documents on technical issues being addressed by CIR. Upon recommendation by the P/T-CIR Subcommittee that a Report be prepared, and approval by P/T, the P/TC Chairman, upon recommendation of the Vice Chairman—Ingredient Safety, will designate a P/T member to chair an Ad Hoc Task Force to prepare this Report. The Ad Hoc Task Force Chairman will be responsible for soliciting from industry personnel whatever aid is necessary for him to carry out his assignment. CTFA staff will provide administrative and technical support to the Ad Hoc Task Force.

(b) When completed, the Report will be submitted to the Vice Chairman—Ingredient Safety for review and acceptance and then distributed by CTFA staff to the members of P/TC, for comment and approval. Following approval, the Report will be transmitted by CTFA staff to the ILR, for his or her use in the deliberations of CIR.

Part E. Additional Data Development

Section 50. Determination That Additional Data Is Needed

The P/T-CIR Subcommittee Chairman shall include in his written report on each ingredient group, to the Vice Chairman—Ingredient Safety, the P/T-CIR Subcommittee recommendations concerning the need for additional data. In addition, the Subcommittee will review all requests for additional data from CIR and make appropriate recommendations through its Chairman to the Vice Chairman—Ingredient Safety. Upon receipt of such recommendations, the Vice Chairman—Ingredient Safety will present a written report on these recommendations to P/TC for its consideration.

Section 51. Determination of Test Methodology Required

(a) If the P/TC determines that additional data are required it shall prepare a report for the consideration of the CTFA Scientific Advisory Committee (SAC). The Vice Chairman—Research Programs shall have the responsibility for coordination of the preparation of the report to SAC.

(b) The P/TC Chairman, upon the recommendation of the Vice Chairman—Research Programs, shall appoint a Chairman for the Testing Task Force to prepare the report to SAC. The Testing Task Force Chairman will prepare a written report to the Vice Chairman—Research Programs outlining the basis for and nature of the additional data needed, identifying the testing recommended, listing research facilities capable of performing the testing, and presenting approximate costs for the research. CTFA staff will provide administrative and technical support to the Testing Task Force.

(c) The Vice Chairman—Research Programs or his assignee will present the report to the P/TC for its approval.

(d) Upon P/TC approval, the report will be transmitted by CTFA staff to SAC for its consideration.

Section 52. Clearance for Testing

Approval for funding and actual financing of testing will be addressed by SAC, the CTFA Board Subcommittee on Research, and the CTFA Board of Directors.

Section 53. Solicitation of Bids for Testing

Following testing approval, CTFA staff with the guidance of the Vice Chairman—Research Programs and the Chairman of the Testing Task Force shall solicit bids from qualified facilities.

Section 54. *Identification of Facility to Perform Testing*

(a) The Testing Task Force with the assistance of CTFA staff shall review proposals received in response to the request for bids, assign a ranking to each proposal based on an objective evaluation, and prepare a written report for the Vice Chairman—Research Programs. The Vice Chairman or his assignee will present that report to the P/TC for its consideration.

(b) Following approval of a testing facility, CTFA staff will negotiate a contract with the facility chosen.

Section 55. *Monitoring the Study*

The Testing Task Force with the assistance of CTFA staff will monitor the study. The P/TC Chairman, upon recommendation of the Vice Chairman—Research Programs, will appoint additional members to the Testing Task Force for this purpose as needed.

Section 56. *Testing Reports*

(a) The Chairman of the Testing Task Force will prepare written status reports on the testing for submission to the Vice Chairman—Research Programs. All reports produced by the test facility will be reviewed by the Testing Task Force and checked for accuracy, consistency, and compliance with the protocols. Written comments on these reports will be submitted to the Vice Chairman—Research Programs and to CTFA staff.

(b) The Vice Chairman—Research Programs or his assignee shall keep P/TC advised of the status of the testing by submission of regular written reports. The Vice Chairman—Research Programs or his assignee shall submit the final report of the testing facility along with written comments and recommendations from the Testing Task Force to the P/TC for approval. Following approval, the final report will be transmitted by CTFA staff to the ILR for submission to CIR.

14

THE RIFM STORY

DONALD L. OPDYKE*

Research Institute for Fragrance Materials, Inc., Englewood Cliffs, New Jersey

I. HISTORY

On November 25, 1964, A. L. Van Ameringen, Chairman of the Board of Directors of International Flavor and Fragrances, Inc., sent a letter containing the following paragraph to seven industry leaders representing companies supplying fragrance raw materials to the fragrance, toiletry, and cosmetic industries:

> I am inviting you to attend a meeting to be held at our offices on January 7, 1965, at 10:00 A.M., on a matter which I consider of great importance to our entire industry, the safety of fragrance ingredients. This invitation includes one man of your organization, preferably technical, whom you feel is most knowledgeable on this subject. . . .

This letter was sent to Charles C. Bryan, President, Firmenich, Inc.; Ernest Durrer, President of The Givaudan Corporation; Her-

---

*Retired

mann J. Kohl, President, Norda Essential Oil Chemical Co. Inc.;

Kenneth Voorhees, President, Ungerer & Co., Inc.; Frederick
Leonhardt, Jr., President, Fritzsche Brothers, Inc.; Alfred
Moeller, President, Noville Essential Oil Co., Inc.; and Emanuel
Poons, President, Universal Oil Products Co.

This historically important meeting was the beginning of the Research Institute for Fragrance Materials, Inc., also known as RIFM.
The farsightedness of these founding fathers in establishing an organization for the safety of fragrance materials is now apparent to
everyone in the industry.

On March 10, 1966, the following Certificate of Incorporation was
signed by the members of the first Board of Directors of RIFM. It
was filed April 12, 1966, with the New York State Division of Corporations and State Records, and RIFM became a nonprofit corporation
under the laws of the state of New York.

CERTIFICATE OF INCORPORATION OF
RESEARCH INSTITUTE FOR
FRAGRANCE MATERIALS, INC.
Incorporated April 12, 1966
Pursuant to the Membership Corporations Law
of the State of New York

WE, THE UNDERSIGNED, for the purpose of forming a membership corporation pursuant to the Membership Corporations Law
of the State of New York, hereby certify:

1. The name of the proposed corporation shall be

RESEARCH INSTITUTE FOR
FRAGRANCE MATERIALS, INC.

2. The purposes for which it is to be formed are:

(a) To gather and analyze scientific data and other information from members of the Corporation and others relating to the
properties, preparation and use of fragrance ingredients and to
engage in a testing and evaluation program relating to existing
fragrance ingredients and newly discovered and manufactured
fragrance ingredients taking into account historical experience
data with respect to existing fragrance ingredients, and to review
and evaluate, on a continuing basis, the standards and methods
employed by the Corporation and others of the testing and evaluation of fragrance ingredients.

(b) To distribute from time to time, in the manner prescribed
by the Board of Directors of the Corporation, to the members of
the Corporation, and to other individuals, firms and corporations
as determined by the Board of Directors of the Corporation, reports, descriptions and other information relating to the uniform
standards of testing and evaluation employed by the Corporation
of fragrance ingredients and the results thereof and to such other

activities in which the Corporation may, from time to time, be engaged.

(c) To cooperate with the Food and Drug Administration and other governmental bureaus, agencies, departments, committees and other bodies, federal and state, in connection with any testing or evaluation program, regulation or legislation carried on or adopted, or proposed to be carried on or adopted, by such Administration or any such bureau, agency, department, committee or body, which relates to the use of fragrance ingredients or end products containing such ingredients; and, in that connection, to appear for the members of the Corporation before such Administration or any such bureau, agency, department, committee or body.

(d) To encourage the members of the Corporation and other individuals, firms or corporations in the fragrance industry to adopt and adhere to uniform standards in the testing and evaluation of fragrance ingredients and to recognize and take into account the testing and evaluation standards and program adopted and carried on by the Corporation and the results thereof.

(e) To do everything and anything reasonably necessary, suitable, proper, convenient or incidental to the foregoing purposes or which may properly be done by a membership corporation organized for such purposes under the laws of the State of New York and to possess all powers, rights and privileges permitted to such a membership corporation by such laws.

The Corporation shall not be conducted or operated for profit, and no part of the net earnings, if any, of the Corporation nor any of its property or assets shall inure to the benefit of any member or any individual or be used otherwise than for the purposes of the Corporation. Employees of the Corporation shall be entitled to reasonable compensation for services rendered to the Corporation in effecting one or more of its purposes, as shall be determined by the Board of Directors of the Corporation.

Nothing contained in this Certificate shall authorize or empower the Corporation to perform or engage in any acts or practices prohibited by Section 340 of the General Business Law or other anti-monopoly statutes of the State of New York.

The signatories to this Certificate of Incorporation, the first RIFM Board of Directors, were Charles Bryan, Saul Chodroff, Ernest Durrer, Frederick Leonhardt Jr., Emanuel Poons, Kenneth Voorhees, and Henry G. Walter Jr.

By-laws were soon prepared, and by May 23, 1967, a Board of Directors was elected consisting of 10 company presidents, and the newly chosen President of RIFM. The following officers were also elected: a President, two Vice Presidents, a Treasurer, and a Secretary. The meeting of the members was held at the New York Athletic Club.

The officers were Thomas D. Parks, President; Henry Retailliau and Donald Francis, Vice Presidents; Hans Baenninger, Treasurer, and Eugene Grisanti, Secretary.

At this time there were fewer than 30 member companies.

II. PHILOSOPHY

While the purposes of the organization were clearly defined in the Certificate of Incorporation, there were several guiding principles to which the members subscribed. These were as follows:

1. The first principle was protection of the small companies from larger-company domination of the institute. This was accomplished by having by-laws which assured that four seats on the 11-member board could be filled by the presidents of small companies only—that is, companies having fewer than 110 employees. Six seats were established for the larger companies, who naturally paid a larger share of the costs, and the eleventh seat was reserved for the elected President of RIFM.

2. Another guiding philosophy was the separation of the powers. While decisions respecting dues, policy, business, and relations with other organizations fell properly to the administrative structure, the Board of Directors, it was considered prudent that judgments in scientific—that is, safety matters—should be rendered by an independent academic Expert Panel, consisting of dermatologists and toxicologists who had no involvement with the industry in any way. The RIFM President was and is a member both of the Board of Directors and the Expert Panel. Figure 1 shows the actual anatomical structure of RIFM.

3. Another principle of great importance was that the institute should devote its research efforts to establishing the safety of the *raw materials only* used by the industry. No fragrance compounds or mixtures would be examined, nor would captive chemicals.

4. It was decided that once the Expert Panel had made a determination, the membership would be notified by an advisory letter that an item had a potential for producing an undesirable effect, but that *no recommendations* would be made. The decision of how to go about implementing the decisions of the Expert Panel was left to the individual member companies in the light of their own years of experience with a given material.

III. CRITERIA FOR SAFETY

Actual perfumes such as might be purchased in a boutique may consist of as many as 600 individual raw materials derived from many geographical areas from natural sources such as plants and animals, and from chemical products. They are extremely complex mixtures and are

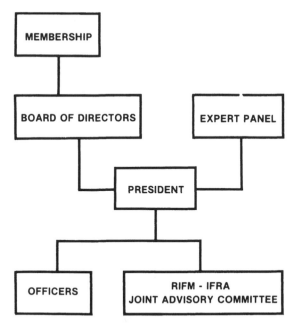

FIGURE 1 Organization of RIFM.

designed to conform to changes in fashion, but always to make the user more attractive both to herself and to companions.

It would be virtually impossible to test each of these combinations thoroughly enough to give an acceptable marketable risk, and it is not at all difficult to see that an ingredient approach to safety in this industry was the only way to go about the task. This is especially true when one considers that up to 4,000 individual materials may be in use.

If one examines several fragrance formulas, it becomes apparent that there are about 800 items in very common occurrence and 200 or 300 more that appear less frequently in formulas but nonetheless may be important to a particular blend. While it is apparent that *volume of usage* per year may tell us which items go into the most fragrances and in the highest percentages and consequently have the greatest exposure level in a certain area, volume of usage is by no means the only criterion of importance from a perfumer's viewpoint. Some items might be very low in volume because of their extreme strength and yet be very important to the perfumer. I refer here to items such as natural musk, civet, tuberose, costus, and ethyl acrylate. All of which make their contributions by imparting special characters to a fragrance in very very low percentages.

Perfumes, colognes, toilet waters, aftershaves, cosmetics, hairsprays, and so on, are applied to the skin, and naturally most of the complaints about fragrances have been cutaneous responses—irritation, photoirritation, allergy, and photoallergy. Consequently, it was decided that the cutaneous area should be investigated first of all with fragrance raw materials, and that two classes of raw materials should have priority, those used to the extent of 500,000 pounds or more per year and those whose reputations had been impugned in the past, regardless of their usage volumes.

An industrywide survey was made of the usage levels of items to be tested.

Consequently, in January 1971 a program was initiated to do the following "package" on as many items as possible in as short a time as possible:

1. A literature search for the 10 preceeding years
2. An acute oral LD_{50} to determine how toxic a material might be
3. An acute dermal LD_{50} to determine if toxic effects could be produced by percolation through the skin
4. A repeated insult patch test or Kligman maximization test on human skin to determine irritation and allergenicity using a 10 X maximum use level
5. A test for phototoxicity, more recently called photoirritation

Additional items were added to the list for specific materials: for example, tests for photoallergenicity, 20-day percutaneous toxicity tests, and 90-day percutaneous toxicity tests. Tests have since been added to include the formation of nitrosamines in complex mixtures by some of the nitrogen-containing aroma chemicals.

Fragrances are ubiquitous; they occur in almost all household products, cosmetics, oven cleaners, garden sprays, and so on. Many of the raw materials date from ancient times. Numerous references occur in the Scriptures, both Eastern and Western. When one considers their ubiquity and their length of usage in human history and when one considers the millions of applications per day of some of these items, one can only conclude that the fragrance industry is indeed one with a good safety record.

Another factor in safety considerations is that many of the items used in compounding perfumes are also flavor ingredients and have been ingested by people for many years, a fact that may have contributed to the low incidence of injury to the skin.

The problems and implications involved in the safety testing program were reviewed in the *Toxicology Annual*, [1].

IV. NOTIFICATION PROCEDURES

One of the major tactical problems in the entire RIFM operation has been the establishment of a uniform method of informing its membership about unfavorable results. Whereas several companies would be involved on committee functions of various sorts, it was of paramount importance that no company received advanced knowledge about an impending action by the Expert Panel since that would automatically give that firm unfair economic advantage. In addition, it became necessary to have a notification system that would also inform the member customers.

There have been three major procedures used to notify both members and customers. These have been the Pre Test Letter, in which information is sent out that a number of items are about to be tested by the RIFM battery of test procedures. This letter also invites the receiver to submit any data in his or her possession about the safety of any of the listed materials.

The second notification procedure, tried experimentally, is the Testing Information Letter, which advises the membership that a given test with a given material has produced some unfavorable results and asks for additional data either via telephone, personal conference, or in writing. It informs the membership that the material will soon be presented to the Expert Panel for its consideration. Members are invited to present their data in person to the Expert Panel, if desired.

The third notification procedure consists of the Advisory Letters, which emanate from the Expert Panel and represent what amounts to a final judgment of the then-available test data plus a literature review. Naturally, nothing is engraved in stone, but there has been an atmosphere of finality surrounding the dissemination of these letters, and the materials that have been the subjects of the letters have usually been dropped.

The fourth notification, of course, was the publication of the finished monographs in *Food and Cosmetics Toxicology*. The first monograph appeared in volume 11 (1973) [2].

Of interest is the observation that among members of the industry there have always been those who feel that all undesirable properties of raw materials were attributable to those materials derived from chemical sources and others who felt that all of the undesirable qualities came from the natural fragrance materials. In actuality, the list of allergens, photoirritants, photoallergens, and other subjects of the Advisory Letters shows a distribution of possible offenders in both the natural and synthetic categories.

V. PROCEDURES IN GENERAL

In May 1978 a Joint Advisory Committee (JAC) was formed consisting of four representatives of RIFM and four from the International Fragrance Association (IFRA), headquartered in Geneva. There were also two alternates, one from each group, in the event of an indisposition of one of the regular members. These were all technical people, industrial people representing the disciplines of perfumery, analytical chemistry, and organic synthetic chemistry—people trained in the processing of natural raw materials. This group meets four times a year and has the following functions:

Selecting materials and concentrations (10 X highest application to the skin) for testing

Reviewing *unfavorable* results and selecting additional grades and/or sources of the material and possible quenching mixtures for testing

Reviewing the results of that testing

Reviewing and editing monographs for publication

The formation and function of this group has provided industry input by providing excellent scientific participation in the program by technical members of the fragrance industry without compromising the deliberative functions of the Expert Panel.

The meetings of the JAC have always been confidential.

Figure 2 provides a handy flowsheet to explain the procedures used by RIFM.

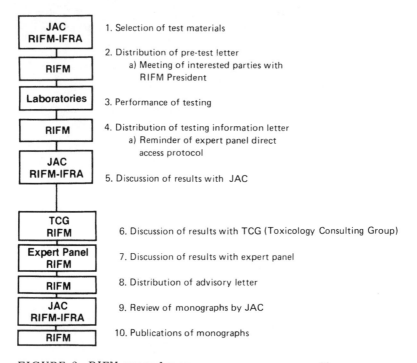

FIGURE 2   RIFM procedures

In 1978 there was felt a need for an additional committee to provide industrial toxicological input for consideration of unfavorable results prior to the deliberations by the Expert Panel. In this committee, for the first time, participation was not limited to the membership but was expanded to include members from finished goods houses. Membership was limited, however, to persons having the best of qualifications as professional toxicologists. This committee was called the Toxicology Consulting Group (TCG). By means of this group, no unfavorable results on a raw material used in fragrances would be presented to the academic Expert Panel without prior deliberation of the significance therof by industrial expert toxicologists who could then interface, if need be, with the Expert Panel prior to the issuance of Advisory Letters.

Meetings of the Expert Panel have usually taken place three or four times per year, with the program divided into two portions. First, an academic portion provided by a lecture by an invited guest on a topic of concern to the experts, for instance, photocarcinogenicity or nitrosamines, or occasionally a presentation by the TCG, or a committee from the Soap and Detergent Association. This academic lecture has been followed by a discussion and deliberation by the panel of monographs on fragrance raw materials together with data developed by RIFM, data contributed by the members of other associations, and an up-to-date literature search. While minutes are kept of these meetings, they have always been confidential since they contain data that are proprietary in nature and cannot be compromised.

On a few occasions, a joint dinner between the Expert Panel and the Board of Directors has been held, and at least once a year, during a meeting of the Expert Panel, the Chairman of the Board of Directors has been invited to lunch with them. Aside from this, contacts between the administrative, commercial divisions of RIFM's hierarchy and the scientific deliberative body, the Expert Panel, have been kept to a minimum.

VI. RELATIONS WITH OTHER ASSOCIATIONS

In a paper "International Coordination in the Flavor-Fragrance Industry" [3], a review of RIFM activities and how they relate to the International Fragrance Association was presented. The paper was divided into portions prepared by the leadership of each organization. In order to describe the activities of IFRA, Dr. F. Grundschober stated in the portion of the paper called "International Fragrance Association (IFRA)":

> The purpose of the International Fragrance Association (IFRA) is to ensure the safety-in-use of fragrance materials and to collaborate in the development of international practices and regulations as to the use of fragrances in consumer products. IFRA was

founded in Brussels, in October 1973, as an international organization with scientific aims. The members of IFRA are the national
associations of fragrance manufacturers from, at present, 11
countries: Belgium, France, Germany, Italy, Japan, Mexico,
the Netherlands, Spain, Switzerland, the United Kingdom and the
USA. Individual companies are members of the national associations of these countries, e.g. British Aroma Chemicals Manufacturers Association for the U.K., Essential Oil Association (EOA)
for the USA [now the Fragrance Materials Association, FMA].

By an unanimous decision of its members, IFRA established a
self-regulatory discipline within the fragrance industry. In October 1975, IFRA issued a Code of Practice for the fragrance industry, to which all the members adhere. This code regulates the
control which the fragrance industry itself exercises over its activities in all domains not dealt with in national or international
regulations. It contains basic standards of good manufacturing
practice dealing with personnel, hygienic requirements in the
manufacturing areas, storage, operations of manufacture, labeling and packaging, as well as quality control. An important part
of the Code concerns the use of fragrance ingredients. They
must only be used in conditions where they present no risk to
health.

Referring to RIFM's published monographs, Dr. Grundschober
continued:

The results of these studies are regularly published. These publications summarize the studies done through RIFM and those published in the scientific literature; however, they do not include
practical recommendations which can be used by perfumers. It is
necessary to have these scientific results in the form of practical
recommendations concerning the use of fragrance materials. The
Technical Advisory Committee of IFRA was established for this
purpose and is continuously updating industry Guidelines for the
use of fragrance materials. The Committee is collecting all data
which are relevant to the safety of fragrance ingredients. Such
data may be drawn from the scientific literature, or may be the
result of testing programs made available by the sponsors of
these tests, as well as from reports of adverse reactions due to
fragrance materials placed at the disposal of IFRA by fragrance
material manufacturers. IFRA cooperates with RIFM; mutual exchange of information enables the RIFM Expert Panel to study data
collected by IFRA, and the results of the RIFM testing program
are made available to IFRA. In the Industry Guidelines for the use
of fragrance materials IFRA advises, in principle, against the use
of fragrance materials under conditions that might provoke irritation and sensitization reactions, or phototoxic effects. Consequently, based on the currently available data, IFRA recommends

that certain fragrance ingredients not be used at all, while certain restrictions of use are recommended for other substances.

From the foregoing it is seen that while RIFM is composed of individual member companies, IFRA is constituted of national associations. RIFM reports data, IFRA interprets data and makes recommendations. A very close association exists between these two organizations.

Another organization with which RIFM had had a close relationship has been the U.S. Soap and Detergent Association, whose scientific committee has gathered retrospective data on certain of the materials that are in use by that industry and have been the subject of RIFM Advisory letters. These data have been presented in person to the RIFM-IFRA Joint Advisory Committee and also to the RIFM Expert Panel and are reported in a series of publications [4–6].

With respect to the CTFA, meetings with the Fragrance Task Force of the Ingredient Safety Subcommittee of the Pharmacology and Toxicology Committee have been held on an irregular basis as occasioned by needs and in order to deal with current problems.

When the CIR Expert Panel first surveyed the scope of its operations, it was decided that it would not include fragrance raw materials since these were being dealt with by RIFM. At its meeting on October 3 and 4, 1977, Dr. Opdyke made a presentation to the CIR Experts describing RIFM activities and accomplishments followed by an intensive question-and-answer period. It was mutually agreed that the door would be left open for additional appearances.

On three occasions when RIFM delegations have made presentations to the Food and Drug Administration, observers from the CTFA have been invited to the meeting and have generously counseled the RIFM presenters in prior briefing sessions with its legal and scientific staff.

In general, then, the IFRA, SDA, and CTFA have received all of the mailings that RIFM distributes to its membership, including Pre-Test letters, Advisory Letters, pre-publication copies of the monographs when they have been accepted for publication, and of course, reprints of all publications.

The relationship with the Flavor and Extract Manufacturers Association (FEMA) has always been a close one, with mutual exchange of data and some jointly supported studies. This was a very natural relationship since, as stated before, both associations deal with many raw materials in common.

The Essential Oil Association (EOA) was one of the earliest organizations to relate to RIFM with its Scientific Committee, under the very capable leadership of Frank Boyd of Fritzsche Dodge & Olcott, supplying RIFM with the materials on which it performed its testing program. The EOA assured quality, identity, and specifications for each item examined.

When the EOA was dissolved, it was replaced by the Fragrance Materials Association of the U.S., which was structured to take a more active role in regulatory affairs while assuring RIFM's position as a strictly scientific body.

VII. RELATIONS WITH REGULATORY AGENCIES

From its inception, RIFM has kept the U.S. Food and Drug Administration informed about its program with regular mailings and frequent informal meetings. On three occasions the RIFM has requested meetings at the FDA to present data on unfavorable test results with fragrance raw materials.

In a similar fashion, RIFM has also kept the Health Protection Branch, Health & Welfare of Canada, informed by mailings and an invited guest lecture at Ottawa by Dr. Opdyke.

The Editorial Board of Food & Cosmetics Toxicology has representatives of several health ministries. The Japanese government is kept informed by the Japanese member of our Expert Panel.

RIFM data are regularly reported to the American Association of Poison Control Centers and to the U.S. Public Health Service for its Registry of Toxic Effects of Chemical Substances.

VIII. GROWTH

RIFM had its beginning with less than 30 American companies and has expanded to represent more than twice that on a world wide basis. It is believed that more than 90 percent of the industry in the United States, Europe, and Japan is represented in the membership.

In April 1978 RIFM expanded to include finished goods houses as active members and now includes some of the largest manufacturers of cosmetics and toiletries in the world.

REFERENCES

1. D. L. J. Opdyke, in *Toxicology Annual* (C. Winek, ed.), vol. 2, Marcel Dekker, New York, 1977, p. 61.
2. D. L. J. Opdyke, *Food Cosmet. Toxicol.* *11*:95 (1973).
3. D. L. J. Opdyke, *Perfumer & Flavorist* 3(April/May):53 (1978).
4. R. J. Steltenkamp, K. A. Booman, J. Dorsky, T. O. King, A. S. Rothenstein, E. A. Schwoeppe, R. I. Sedlak, T. H. F. Smith, and G. R. Thompson, *Food Cosmet. Toxicol.* *18*:407 (1980).
5. R. J. Steltenkamp, K. A. Booman, J. Dorsky, T. O. King, A. S. Rothenstein, E. A. Schwoeppe, R. I. Sedlak, T. H. F. Smith, and G. R. Thompson, *Food Cosmet. Toxicol.* *18*:413 (1980).

6. R. J. Steltenkamp, K. A. Booman, J. Dorsky, T. O. King, A. S. Rothenstein, E. A. Schwoeppe, R. I. Sedlak, T. H. F. Smith, and G. R. Thompson, *Food Cosmet. Toxicol.* *18*:419 (1980).

15

COSMETIC INDUSTRY RESEARCH PROGRAMS

GERALD N. McEWEN, JR., H. JOSEPH SEKERKE and ANITA S. CURRY

Science Department, The Cosmetic, Toiletry and Fragrance Association, Inc., Washington, D.C.

I. INTRODUCTION

A. Background

The cosmetic industry, as represented by the Cosmetic, Toiletry and Fragrance Association (CTFA), includes companies with diverse interests relating to the types of products or services (e.g., hair dyes and hair care products, lipsticks and makeup products, personal cleanliness products, ingredients used in formulating cosmetics, packaging materials, etc.) supplied to the marketplace. Even so, there are substantial areas of interest or concern that are common to all members and are therefore suitable areas for cooperative research efforts. In addition, there are many questions that, while not being of broad interest to the entire industry, are of sufficient interest to a particular industry segment so that these too are suitable for cooperative effort. The simplest definition of an association, in fact, is a group that forms to further common goals and interests. This is especially applicable to research programs.

A common misconception when one speaks of "research" programs is that they must involve some form of testing. This is not the case, however, as much of the research conducted by a trade association like the CTFA involves surveys, problem analyses, systems development, or simply design and evaluation of programs. These research programs may eventually lead to testing, or they may lead instead to the preparation of information documents. The *CTFA Cosmetic Ingredient Dictionary* is a particulary good example of a "research" program that does not involve any actual "testing."

To be considered for sponsorship by the CTFA, research must meet two general criteria. First, the research must be of interest and benefit to the industry as a whole or to all members of the particular industry segment affected by the results; that is, it must not be of a competitive nature designed simply to procure information on a particular company's products or to give advantage only to a particular product type within a class of products. Second, the results of the research must be made publicly available; they cannot be held for the advantage of a particular group of companies or even limited in distribution to members of the CTFA.

Most of the research efforts of the CTFA involve work to confirm the safety of cosmetic ingredients in order to help ensure the safety of cosmetics as a class of consumer products, or work in support of the development of reasonable industry self-regulation or in support of federal, state, and foreign legislation and regulation, where necessary.

B. Benefits of Industry-Sponsored Research

Industry-sponsored research has several important benefits for individual companies. First, of course, is the benefit of cost savings realized by those companies sharing the costs associated with the research. This cost savings could range from as little as 50 percent, if for instance only two companies are interested in participating in the research, to more than 99 percent for those projects of interest to the entire industry and paid for from association general funds.

Another benefit of industry-sponsored research is the broad expert technical and administrative skills available to coordinate and manage such efforts. By utilizing technical staff from various member companies who sit on the CTFA's committees, and CTFA staff who have experience in contract research organizations and in coordinating extramural research projects, expert guidance is available for any given project. This minimizes potential problems, especially common with novel research, by ensuring adequate review and planning prior to project initiation, and by ensuring that, if questions arise during the conduct of the study, they are addressed by professionals with the appropriate technical expertise. In addition, for research involving testing, this broad base of technical expertise ensures adequate scientific monitoring of the study while it is being conducted. It is unlikely that any single company would be able to match the breadth of expertise and experience available to an association through its committee structure.

Individual companies also benefit from industry-sponsored research by having the credibility and prestige of the entire industry supporting the results of the studies. This also allows particularly sensitive issues, those an individual company might not wish to address directly because of publicity or other factors, to be studied and reported on.

C. Classification of Industry-Sponsored Research

There are three basic types of industry-sponsored research as classified by method of funding. Funding type also determines the extent of project supervision exercised by CTFA technical committees.

One type of research is of general industry interest and is funded from CTFA general revenues. The costs for each company amount to the same proportion of the total cost as is its investment in the association. Management of this type of research is under direct control of CTFA technical committees supported by CTFA staff.

Another type of research may be of general industry interest, but it is funded by individual companies on the basis of the relationship of the research to their particular products or services. The costs for this type of research are generally apportioned to each company on the basis of market share for the products or services involved. This type of research is guided in the same manner as that funded from general revenues.

The third type of research is of interest to a limited segment of the industry, and funding apportionment is developed by the concerned companies. Technical review and management is generally in the hands of staff from those companies involved, with assistance from CTFA staff. For this type of research, the companies participating may request technical input from CTFA committees, but this is not required.

Two characteristics that are common to all industry-sponsored research are cost savings through cooperative funding and CTFA staff involvement in an advisory and coordinative capacity. In addition, all industry-sponsored research, regardless of its classification, is subject to legal review regarding antitrust implications and to approval by the CTFA Board of Directors. Contractural authority for industry-sponsored research resides with CTFA management.

There is another type of research, conducted for the benefit of industry, that does not fit into one of these three categories of industry sponsorship. This research is conducted by individual companies, and control remains with the company performing the work. Review and approval of the research being performed, while advantageous if the association will be expected to support the results, is encouraged but not required, and the association does not enter into a contractual relationship with the company performing the work.

II. INITIATION OF RESEARCH PROGRAMS

The first major industry-sponsored research program began in the late 1930s in response to the "coal-tar dye" sanctions of the U.S. Food, Drug and Cosmetic Act of 1938. At this time, the CTFA was known as the Toilet Goods Association (TGA). The Act gave the responsibility of determining which of these dyes should be allowed in cosmetics and what criteria of purity should be required for them to the U.S. Department of Agriculture. In order to assist this endeavor, industry members carried on informal discussions and reached general agreement on a list of colors of prime importance in cosmetics manufacture, and on their desirable properties.

Industry-sponsored research is initiated in response to either internal or external interests or pressures. Programs initiated because of internal interests or concerns (programs suggested by member companies) are usually characterized by general goals and a relatively long planning period that includes reviews and discussions by CTFA committees of not only the central issues but also ancillary concerns.

Committee review and planning for a proposed research program begins by addressing the following questions: Is the issue of sufficiently widespread industry interest to warrant the expenditure of resources? What type of research will best address the issue? How should the research be organized and managed? Are there sufficient industry resources available to make the program successful? Is this research appropriate for sponsorship by the association?

Assuming that all of these questions are answered affirmatively, an organizational structure is developed. If the research has a general goal and will be of a continuing nature, the appropriate industry personnel are formed into a new committee, or the research is assigned to an existing committee. If the research is to have a specific short-term goal, personnel are formed into a task force under an existing committee. Appropriate CTFA staff support the activities of both committees and task forces.

Research in response to external pressures (pressures from groups outside the industry) is generally characterized by a much shorter and more intense planning period. The first question asked is: Should industry respond to the issue, and is it appropriate to address the issue as an association? If this is the case, planning then focuses on finding the appropriate method of responding and on enlisting the appropriate industry personnel, and CTFA staff for support, to address the issue. These programs generally have well-defined short-term goals, and are therefore addressed by task forces.

III. RESEARCH PROGRAM ORGANIZATION

All functions of the CTFA, including research programs, are subject to the review and approval of the CTFA's Board of Directors. The Board sets the goals and limits the resources that can be expended on any particular issue, but is not concerned directly with day-to-day management of program affairs. The Board's Subcommittee on Research has the responsibility of reviewing recommendations for research programs and tracking ongoing research for the board. It reviews overall goals, accomplishments and priorities, and budgets for all research programs.

Reporting to the Board on research programs is the CTFA Scientific Advisory Committee. This committee is made up of the senior research personnel from member companies, commonly the vice presidents of research and development or technical services, and is responsible for reviewing and making recommendations to the board on all research projects.

Reporting to the Scientific Advisory Committee are a number of technical committees and task forces developed to address specific areas, such as pharmacology and toxicology, microbiology, chemical nomenclature, hair coloring, antiperspirants, and so on. Each committee or task force may have subcommittees or task forces reporting to it, and each individual group is supported by a CTFA staff person chosen on the basis of expertise in the committee's or task force's primary area of responsibility.

The personnel making up the technical committees that report to the Scientific Advisory Committee are generally the key individuals— senior scientists—within each company who have the direct responsibility for the particular technical area concerned. These committees and task forces, with CTFA staff assistance, are responsible for the day-to-day management of the research programs.

A few industry-sponsored research programs are in the form of grants to other independent organizations. Examples of these are the Cosmetic Ingredient Review and the Johns Hopkins University Center for Alternatives to Animal Testing. While these constitute industry-sponsored research programs, they are not managed by the CTFA, but by their own separate Steering Committees or Boards of Directors. In these cases, CTFA maintains only a liaison relationship with the program.

IV. MAJOR INDUSTRY RESEARCH PROGRAMS

As was previously noted, the first major coordinated industry research concerned specifications for the coal-tar dyes. From this beginning, the association went on, in the early 1940s, to institute a Board of Standards with the purpose of developing specifications for cosmetic ingredients other than certified color additives. Early work of this

Board included testing samples of raw materials and developing test
methods that would adequately evaluate each material. The TGA
Standards, which have now evolved into the *CTFA Compendium of
Cosmetic Ingredient Composition*, are an example of research conducted by individual companies but reviewed and coordinated by
CTFA technical committees with the assistance of staff. The *Compendium* is valuable in helping to identify the raw material items of commerce used in cosmetics. This is especially useful to the formulator
dealing with new ingredients.

In 1944, the association began a series of annual scientific meetings designed to present a forum for the presentation and discussion
of scientific issues of interest particularly to members of the cosmetic
industry. Reports of research from these sessions were originally
published as *Proceedings of the Toilet Goods Association*. Publishing
these articles allowed those unable to attend the meetings to benefit
from the research conducted. With the growth of specialized journals
for the presentation of cosmetic-related research in the late 1960s, including the *TGA Cosmetic Journal*, the importance of reporting on this
type of research at scientific conferences and publishing the *Proceedings* diminished. In the early 1970s therefore, the thrust of the scientific conferences changed to keep pace with the changing information needs of the industry. The CTFA Scientific Conferences held
now emphasize more the broad issues facing the industry, and present
capsule overviews of the research conducted under the direction of
individual CTFA technical committees.

Many other programs that were essentially informational also began in the late 1930s with the *TGA Bulletin* and continue today. Some
examples of these are the *Executive Newsletter, Legislative Bulletins,
Trademark Bulletins*, and so on. These programs are supported from
general funds and coordinated by CTFA staff.

The CTFA Technical Guidelines, also begun in the early 1970s are
an example of research that included test development, testing, and
information exchange. The Guidelines supply information on various
aspects of product development and production that is especially valuable to some of industry's smaller cosmetic companies. They give
manufacturers information designed to provide a framework that can
be easily modified to fit various in-house programs.

The first Guidelines were published in the *TGA Cosmetic Journal*
in 1968 and concerned microbiological control. Subsequent guidelines
have addressed other aspects of quality assurance as well as safety
substantiation. Support for development of these Guidelines comes
from general funds. Subjects for guidelines are proposed by CTFA
technical committees. Proposals are reviewed annually by the Scientific Advisory Committee, which also must approve the draft Guideline. Guidelines also must receive legal and Board of Directors' approval before being approved for publication. Public comment is invited on draft Guidelines.

Specific test method development studies have been conducted, and are ongoing, that are designed to increase the efficiency and effectiveness of individual company testing. Two recent efforts have involved development of a protocol for testing the effects of inhalation exposure to fine particulate matter, and work to improve the alternative methods for conducting studies to ensure the occular safety of substances. The goal for each of these programs is to develop new methods that will supply the needed amount of valid scientific data while decreasing the costs of testing, the number of animals that must be used, and time for conducting the testing.

Many other programs addressing specific concerns of the industry such as microbiological preservation, nitrosamines in cosmetics, aerosol propellants, talc, cosmetic drugs, hair coloring, and so on, began in the 1970s. These programs were sponsored primarily by individual groups of members, with direction given by committees or task forces composed of personnel from interested companies. As with the development of the Guidelines, these programs generally involved test method development, testing, and information exchange. Some of these programs primarily benefited a particular industry segment, such as studies done on hair coloring ingredients and products; however, it would be naive to assume that any factors affecting one segment of the cosmetic industry do not also have some impact on other segments in the eyes of consumers. In addition, as with other industry research programs, the results of these research programs are available to all, thereby increasing the information base of each individual company.

Two of the major industry research efforts over the years can be classified broadly as dealing with color additives and with voluntary industry self-regulation.

Color additive programs can be said to date from the early work on ingredient specifications in the late 1930s; however, a structured program first began in earnest in the early 1960s. In response to the 1960 Color Additive Amendments to the U.S. Food, Drug, and Cosmetics Act, the industry embarked on a safety testing program. This program continues today, due in large part to changing scientific standards for the evaluation of safety over the years and increasing information demands from the Food and Drug Administration (FDA). Although this program has been extremely expensive, more than $8 million in direct testing costs, the alternatives, either individual companies conducting the necessary testing or not being allowed to use color additives in cosmetics because of a lack of required safety substantiation data, would not be acceptable. The advantage to industry, therefore, is that color additives can be used to formulate cosmetic products. The importance of this to industry is self-evident.

Within the color additive program are three general areas of activity. The major area, in terms of resources expended, is the area of research to support the "permanent listing" (regulatory allowability by

FDA of color additive use) of cosmetic color additives in use at the time the Color Additive Ammendments were issued. A second area of activity involves the development, testing, and petitioning of the FDA for listing of new color additives so that the range of colors available to industry to use in preparing cosmetics is increased. The third area of activity involves testing color additives for new uses, that is, testing to support the proposed use of D&C color additives in eye-area products. This activity is necessary because there are differing requirements for listing color additives dependent on the type of use anticipated: whether the product may be subject to incidental inges-tion, as with lipstick color additives, exposed only to the skin, or might accidentally enter the eye, as with eye-area products.

Programs for self-regulation are the other most important re-search effort of the cosmetic industry. These programs, too, can be said to date from the initial efforts to prepare specifications for raw ingredients; however, a more logical beginning point would be the development and publication of the first *CTFA Cosmetic Ingredient Dictionary* in 1973. The *Dictionary* for the first time presented a source for standard nomenclature for cosmetic ingredients that could be used to standardize ingredient labeling of finished products. Fol-lowing publication of the *Dictionary* and acceptance of its terminology by the FDA, all companies had ready access to uniform raw ingredient terminology. This was especially important for those ingredients not already listed in some other chemical compendium.

Other programs instituted in the early 1970s included voluntary registration of product formulations and voluntary reporting of con-sumer complaints. The major advantages of these programs to indus-try are that they demonstrate to the consumer and to the legislative branch of government that industry is committed to self-regulation involving reasonable public oversight of its activities, and that gov-ernment resources and tax dollars need not be expended to ensure an effective program.

The most recent program in the area of voluntary self-regulation is the industry-sponsored Cosmetic Ingredient Review (CIR), begun in 1976. This program was instituted with an independent Expert Panel whose members are publicly nominated and who must meet the same nonconflict-of-interest requirements stipulated by the FDA for their scientific advisory panels. The purpose of the CIR Expert Panel is to evaluate critically all available information on each cosmetic in-gredient, and to conclude whether it is safe to use that ingredient in cosmetic products. The data reviewed and the Panel's conclusions are then published and made publicly available at a nominal cost. It has been estimated that it could take up to 20 years just to review the ingredients currently being used, and will cost over $10 million.

Supporting the CIR Expert Panel is an Industry Response Pro-gram that searches for unpublished information to present for the Panel's consideration, and reviews and comments on CIR documents

in response to public requests for comments. In addition, there is a
CIR testing program conducted by industry to perform specific tests
on ingredients in response to requests for additional data from the
panel. Examples of types of testing that have been conducted, or are
currently under consideration, are tests for allergenic and photosensitization potential, subchronic toxicity tests, skin depigmentation
tests, and lipstick ingestion studies.

These programs supply a valuable service to industry in addition
to demonstrating industry's commitment to self-regulation. That is,
they bring together information on cosmetic use, typical chemical
specifications, and safety data on ingredients. This information can
then be used by companies when assessing ingredients for use. This
saves valuable industry resources, since each company need not develop the information for itself.

Finally, another independent program supported by industry general funds is the Johns Hopkins University Center for Alternatives to
Animal Testing, begun in 1981. This program is designed to investigate alternatives to some of the common animal tests required to substantiate the safety of consumer products.

Of course, there are many other research programs sponsored by
industry that are of benefit, such as the CTFA workshop program
that has seen workshops since 1970 conducted on microbiology and
quality assurance, occular safety testing, risk assessment, and CIR
and the Industry Response Program, various epidemiology studies on
cosmetologists, programs dealing with foreign cosmetic regulations,
and so on. Further information on the programs discussed in this
chapter and others may be obtained on request from the CTFA Science
Department.

V. FUTURE OF INDUSTRY RESEARCH PROGRAMS

As was previously noted, the initiation of industry-sponsored research
is dependent on many factors, but primarily on individual company
interests and concerns and from pressures from external sources. Although any prediction as to what these concerns or pressures will be
in the future is highly speculative, a few general observations can be
made by reviewing recent history.

First, because of the growing specialization in testing and increasing costs of conducting research, it is safe to predict that more
and more companies will be turning to the association to organize for
cooperative research.

Testing to support the use of new ingredients of wide general interest, such as new cosmetic color additives, where that testing would
be prohibitively expensive for a single company, is a prime example
of cooperative effort that will undoubtedly increase.

Other major areas will probably involve safety test method development and information exchange. Safety is the area that has re-

ceived the most attention in the past, and will become of increasing importance in the years ahead as testing costs increase and as more information is gathered allowing a greater certainty of extrapolation from test results to use exposure conditions.

Test method development is another area of research that will gain importance in the years ahead. Again the impetus will undoubtedly be given by increasing development costs and the realization that the cost savings from better test methods will benefit all members of the association.

Finally, information exchange, whether it involves readily available test results, specifications on cosmetic ingredients, ingredient use information, regulatory and legislative information, or any of a number of other areas, will be of increasing importance because of the difficulty of an individual company tracking all the possible pertinent information on a given subject of general interest. As more and more information on cosmetics becomes available, the pressure will build to develop an industry-sponsored program to assess and consolidate that information and to make it readily available.

16

INDUSTRY PROGRAMS FOR DEALING WITH TRACE CONTAMINANTS

GEORGE POLLACK

Cosmair, Inc., Clark, New Jersey

IRA ROSENBERG

Research and Development, Clairol, Inc., Stamford, Connecticut

I. INTRODUCTION

Trace materials have always existed as the result of side reactions in the preparation of chemicals, but never have they come under such close scrutiny as in today's products. The principal reason for this new-found fame is the greater precision provided by the builders of today's analytical instruments and the improved skills of the analytical chemists using these instruments. The field of analytical chemistry is changing rapidly, and through the use of computers chemists are now capable of doing repeated analyses in a few hours compared to the weeks of analysis of a few years ago. The instruments now being used enable an analytical chemist to achieve levels of precision that were previously impossible. It is now common to report analytical findings at the parts-per-billion (ppb) level. Many of the findings reported by the Food and Drug Administration (FDA) and other researchers are in the parts-per-billion to parts-per-million (ppm) range. These numbers raise many questions regarding the consumer's risk when infinitesimal quantities are present in finished products.

Presently there are three areas of concern to the cosmetic industry. They are the class of compounds known as nitrosamines, the chemical substance 1,4-dioxane, and trace impurities that might be found in color additives used by the industry.

Nitrosamines as a class have been shown to be carcinogenic to animals when administered orally, and 1,4-dioxane has been reported to be an animal carcinogen on the basis of several ingestion studies. Likewise, the impurity p-toluidine, which is found in D&C Green No. 6, has been shown to be a carcinogen in mice. However, many questions still remain concerning the carcinogenicity of these materials when administered topically. Research recently has given some insight into this area[1].

II. NITROSAMINES

The cosmetic industry first became aware of the presence of nitrosamines in some cosmetics on March 22, 1977, with the presentation of a paper at an American Chemical Society meeting[2]. Immediately following the disclosure that traces of N-Nitrosodiethanolamine (NDElA) had been found in cosmetic products, the cosmetic industry, through its association the Cosmetic, Toiletry and Fragrance Association (CTFA), established a Nitrosamine Task Force. It has been the responsibility of this technical group to define the problem(s) and make information and methodology available to be utilized in efforts to measure, control, and/or eliminate this potentially hazardous substance in cosmetic products.

One of the initial programs of the Task Force was to identify, review, and collate pertinent information on the chemistry of nitrosamines. This work culminated in the publication of a major review paper in the fall of 1978[3]. Subsequently, several papers from industry research laboratories have appeared, including HPLC-UV[4—7] and HPLC-photoconductivity[8] methods for NDElA determination, methods for determining nitrite in cosmetic ingredients[9], a method for total nitrosamine analysis[10], and exploration of the roles of 2-bromo-2-nitropropane-1,3-diol[11] and peroxides[12] play in nitrosamine formation.

Concurrent with the preparation of the review paper, the Nitrosamine Task Force launched a research program to (1) use the thermal energy analyzer (TEA) to analyze the more widely used cosmetic ingredients that have potential for containing or forming NDElA; (2) investigate the development of simpler analytical techniques approaching the TEA in sensitivity; and (3) analyze frequently used ingredients for presence of nitrosating agents (e.g., "nitrites"). A number of independent contract laboratories were evaluated, and the Midwest

Research Institute (MRI) in Kansas City, Missouri, was selected. A
TEA was purchased by the CTFA and placed in MRI's laboratory.

The initial goal of the research was to define the scope of NDElA
and nitrite contamination in cosmetic raw materials. Ingredients were
selected and ranked using the following criteria:

Chemical structures capable of forming NDElA or of being nitro-
    sating agents
Frequency and/or volume of use in cosmetic products
Use in products reported to contain NDElA
Number of suppliers offering each ingredient
Synthetic route used for manufacture
Potential for contamination in shipment or handling

The collection of the selected ingredients was programmed to provide
MRI with typical production-quality raw materials being used by cosmetic manufacturers. Anonymity was accomplished by providing the
data to a certified public accountant (CPA), who was the sole individual with access to all information. It was the responsibility of the
CPA to notify any manufacturer whose product contained "high" levels
of contamination. Seven analytical methods using high-pressure liquid
chromatography (HPLC) coupled to a TEA were developed and validated
for the determination of NDElA in the following classes of cosmetic
ingredients:

1.   Ethanolamines
2.   Ethanolamides
3.   Monoethanolamine salts
4.   Triethanolamine
5.   Amphoteric compounds
6.   Quaternary ammonium compounds
7.   Morpholine

The required sample preparation, together with basic methodology
and validation data, have been reported[13]. Detection limits of the
method are about 0.5 nanograms (ng) of NDElA injected on the HPLC
column, allowing the detection of levels as low as 25 ppb and quantification of NDElA levels greater than 50 ppb in cosmetic ingredients.

Analysis for NDElA in 99 samples of 17 different ingredients found
that 70 percent contained no detectable level of this nitrosamine. An
additional 9 percent showed only trace levels (<50 ppb), and six samples were found to contain in excess of 1,000 ppb. These six samples
were all condensation products of ethanolamines with fatty acids and
can be considered "nitrosamine-susceptable" cosmetic raw materials.

Alternative methods were explored, in order to provide less expensive tools for quality control screening of certain cosmetic ingredients. HPLC coupled with ultraviolet (UV) detection (254 nm) has

proved to be a useful procedure[14]. Two methods utilizing a reverse-phase octadecylsilane (ODS) column and methanol/water eluent were developed at MRI for NDElA in the ethanolamines and diethanolamine amide matrices. The detection limits in the HPLC-UV determinations using direct sampling were typically in the range of 0.1 to 1 ppm NDElA, depending on the nature of the sample matrix.

It can be concluded from this program that the NDElA levels reported in the literature for finished cosmetic products cannot be attributed solely to contamination of ingredients used in the cosmetics. Not only are the absolute levels of NDElA in cosmetic ingredients below the levels reported in some products, but their potential for contribution to total product contamination is further diminished by the low formula amounts used in cosmetic products.

From its inception, the Nitrosamine Task Force was concerned with the potential problem that nitrosating agents could be present in cosmetic ingredients and be the precursor for the formation of NDElA in an amine-containing cosmetic product. Therefore, a program was initiated at MRI to analyze 14 frequently used organic cosmetic ingredients. This was followed by a corresponding program at Raltech Scientific Services, Inc., in Madison, Wisconsin, for the analysis of nitrite in seven inorganic chemical ingredients.

The chemistry for the determination of "nitrite" in the organic cosmetic ingredients was a modification of the Greiss reaction wherein a sulfanilic acid is diazotized with "nitrite" and then coupled with N-1-naphthylethylenediamine dihydrochloride to form a colored compound[15]. The intensity of the absorption at 500 nm is directly proportional to the "nitrite" concentration in the sample. Methods of analysis sensitive to 25 ppb "nitrite" were developed for four cosmetic ingredients classified as "neutral or acidic water-soluble" (citric acid, glycerine, propylene glycol, and sodium lauryl sulfate), and for 10 ingredients classified as "water-extractable" (castor oil, cetyl alcohol, dimethicone, isopropyl myristate, lanolin, methylparaben, mineral oil, musk, stearic acid, and stearyl alcohol).

Of the 114 organic cosmetic ingredients analyzed at MRI, 78 contained no "nitrite" (<25 ppb), 20 contained trace levels (25 to 50 ppb), and only three samples of sodium lauryl sulfate contained more than 1,000 ppb (1 ppm). Based on these findings, "nitrite" precursors for the formation of NDElA are not introduced by these ingredients. Sodium lauryl sulfate is the exception in this group of samples.

Similar chemistry was selected for the determination of nitrite in the inorganic cosmetic ingredients. Based on industry-developed methodology with detection limits down to 1 ppm, seven cosmetic ingredients—bentonite, iron oxide, kaolin (hydrated aluminum silicate), magnesium aluminum silicate, mica, silica, talc (hydrous magnesium silicate), titanium dioxide, and ultramarine blue—were analyzed.

Of the 63 samples analyzed at Raltech, 20 contained less than 1 ppm, 39 contained 1 to 5 ppm, and 4 contained greater than 5 ppm. The findings of low parts-per-million concentrations of nitrite in the inorganic cosmetic ingredients and sodium lauryl sulfate could classify these materials as nitrite-susceptible and contributing precursors for nitrosamine formation in finished cosmetic products.

The "nitrite" results indicate that the organic raw materials, with the possible exception of sodium lauryl sulfate, are not significant sources for the precursor "nitrite" necessary for nitrosation to occur with ethanolamines and their condensation products which are present in cosmetic formulations. However, 60 percent of the inorganic raw materials analyzed (including samples of bentonite, iron oxide, magnesium and aluminum silicates, mica, talc, and titanium dioxide) contained levels of "nitrite" greater than 1 ppm. If these materials were used in a finished product that also contained ethanolamines, measurable quantities of NDElA could be formed.

To date the Food and Drug Administration (FDA) has analyzed 317 cosmetic finished products for NDElA. A majority (60 percent) of the products analyzed contained no detectable levels of NDElA, while 7 percent contained trace levels (10 to 30 ppb). Almost all of the samples containing greater than 2 ppm NDElA (7 percent) were formulated with 2-bromo-2-nitropropane-1,3-diol, a known nitrosating agent. The remaining 26 percent of the samples analyzed by the FDA contained low levels of NDElA contamination (30 to 2,000 ppb). Such levels of NDElA could be formed from cosmetic raw materials containing nitrite levels corresponding to those found in the inorganic cosmetic ingredients analyzed.

The FDA published a notice in the *Federal Register*[16], on April 10, 1979, inviting interested parties to submit comments, scientific data, or other information regarding nitrosamine contamination in cosmetics. The FDA notice also strongly suggested that the cosmetic industry take action to solve any nitrosamine contamination that may exist. The cosmetic industry has responded with research programs intended to reduce the level of nitrosamine contamination to the lowest possible levels.

NDElA was the nitrosamine of concern. Little information was available in the literature on its rate of formation as a function of pH, with the exception of studies that showed formaldehyde acting as a catalyst in this reaction. Therefore, a study of the rate of NDElA formation as a function of pH (4 to 9) with and without formaldehyde as a catalyst was undertaken. HPLC with UV detection was used to monitor the reaction.

As expected, the rate of NDElA formation in aqueous solution over a typical product pH range (4 to 9) qualitatively decreased with increasing pH. These rate studies served as the initial model for the inhibition studies conducted at Hazleton Raltech, Inc., (formerly Raltech Scientific Services).

The studies of rate of formation in the presence of formaldehyde showed catalysis to be constant over the pH range 4 to 9, and the rate for formation appears to be four to six times greater in the presence of this material.

The industry now had a clearer understanding of the scope of the problem. Methodology had been developed for the analysis of NDElA in several classes of cosmetic raw materials, and possible sources of NDElA and nitrite had been identified. During this time, industry research had revealed that other nitrosamines (N-nitrosodimethylamine and N-nitrosomethyldodecylamine) were present in lauramide oxide. These findings were confirmed and expanded by the FDA with the identification of several other nitrosamines in cosmetic finished products[17]. With the facts indicating that cosmetics had a broader-based nitrosamine problem, the focus of the Task Force turned toward investigating ways of inhibiting the nitrosation of oil-/and water-soluble amines in the presence of nitrite in model cosmetic systems representing various categories of cosmetic finished products. A program was designed to achieve these goals was conducted at Hazleton Raltech.

Most of the goals of CTFA's inhibition study are completed, or should be shortly. Model cosmetic emulsion systems have been defined and checked for stability. Rates of nitrosation for diethanol-amines, dicyclohexylamine, and methyldodecylamine have been measured in both nonionic and anionic oil-in-water emulsions, and the effectiveness of inhibitors toward blocking nitrosation have been studied in these systems. The inhibitors studied were three water-soluble compounds (ascorbic acid, sodium bisulfite, and potassium sorbate) and three oil-soluble compounds (butylated hydroxyanisole, ascorbyl palmitate, and α-tocopherol). The research also focused on model systems designed around iron oxide as the pigmented material. The pigment has been shown to contain low parts-per-million levels of NO_x (reported as NO_2^-) and therefore could contribute low levels of NO_2^- to formulas in which they are used [18]. The ability of this pigment to nitrosate amines in both aqueous and nonaqueous systems is being investigated.

Since the report which showed that cosmetic products could be contaminated with trace levels of NDElA, both the industry and the FDA have reported finding other nitrosamines, at trace levels, in cosmetic products. Both the FDA and the Nitrosamine Task Force feel that to continue to develop methodology for each newly found nitrosamine would be an arduous, if not lifelong, task. The present thinking of the FDA is toward the development of a total nitrosamine assay for various classes of cosmetic products. However, the Task Force felt that methods of inhibiting nitrosamine formation in cosmetic systems were needed before the levels of total nitrosamines in cosmetic products were generated.

With viable methods of inhibiting nitrosamine formation identified, attention has been turned to developing a total nitrosamine method. The final goal of the cosmetic industry should be to reduce nitro-samine formation to the lowest feasible levels. With both methods of inhibition and a total nitrosamine assay available, inhibition studies on products of interest could be conducted by the member compan-ies. This methodology could be used by companies developing in-hibitors for their nitrosamine-susceptible products, regardless of the nitrosamine involved.

In conclusion, the goals of the CTFA Nitrosamine Task Force—to develop methodology by which member companies can determine the nitrosamine susceptibility of their products, and inhibit nitrosamine formation—are being realized.

III. 1,4-DIOXANE

1,4-Dioxane can be present, at trace levels, in some types of ethylene oxide condensates, and this broad class of compounds is widely used in both the food and cosmetic industries. The FDA became concerned with this issue after a study showing that dioxane is capable of pene-trating skin. In response to this concern the CTFA moved quickly and formed the Dioxane Task Force in July 1979.

At that time, the Birkel procedure, which consists of vacuum distillation followed by gas chromatographic analysis, was the only ac-cepted validated procedure [19]. The average analysis time per sample is 2 to 3 hours, with a 0.5-ppm limit of detection. The Birkel method is too time-consuming for routine quality control analysis, and for this reason the CTFA Dioxane Task Force conducted a round-robin method of analysis to identify alternate methodology[19]. Samples of widely used cosmetic ingredients (i.e., sodium laureth sulfate, Polysorbate 60, and PEG-8) were chosen for study. Twenty-two laboratories, in-cluding the FDA, agreed to participate, and a total of seven generally different analytical techniques were employed, including the Birkel and modified Birkel procedures. Other methods generated used gas-chromatography-mass spectrometry (GC/MS) with perdeuterotoluene as an internal standard [21], purge-and-trap procedures followed by GC, direct GC injection [22,23], headspace GC, and atmospheric azeotropic distillation followed by GC. The CTFA study showed that these alternate procedures yielded results comparable to the Birkel method, except for the purge-and-trap technique. It was felt that with some additional development work, the purge-and-trap technique could be satisfactory. As a follow-up, the producers of these materials were encouraged to take as much precaution as possible to effectively eliminate this contaminant.

IV. TRACE CONTAMINANTS IN COLOR ADDITIVES

The last couple of years have seen an increased concern, within the industry and government, in establishing the safety of the dyes and color additives used in cosmetic products. This concern has been highlighted with respect to D&C Green No. 6 (1,4-bis[(4-methylphenyl)-amino]9,10-anthracenedione), which is listed in the Code of Federal Regulations (CFR) for use in externally applied drugs and cosmetics. Because D&C Green No. 6 is a certified dye, each commercially-prepared batch of this color is subject to FDA certification. This material is formed by reacting 1 mole of quinizarin with 2 moles of p-toluidine. D&C Green No. 6 has been shown to be safe for external use; however, literature reports have indicated that p-toluidine is a carcinogen in mice [24]. Residual amounts of reactants such as p-toluidine are commonly found in this color additive, and trace levels are unavoidably present even in highly purified reagent-grade material. Until recently, there were no validated methods available for detecting trace levels of p-toluidine in D&C Green No. 6, which explains the FDA reluctance at the time to "permanently" list this color additive for external use in consumer products. In response to this need, the CTFA asked member companies to develop the needed methodology. Several p-toluidine detection methods have been reported [25,26] at 500 ppb and probably at lower levels. The methods involve separation by HPLC followed by UV or fluorescene detection. Submission of this methodology and supporting data resulted in permanent listing of D&C Green No. 6 for external uses.
    The regulatory agencies' approach of requiring analytical methodology for trace components is becoming common. The cosmetic analytical chemist will become more involved in developing trace analytical techniques for the quality control of color additives in the future.

V. CONCLUSION

With improved analytical instruments becoming commonly available, challenges were thrust upon the cosmetic industry to identify and quantify diverse trace constituents and assess their significance in cosmetic ingredients and products. The cosmetic industry, through its trade association, the CTFA, proved its ability to respond in a timely and authoritative way to these challenges. The CTFA's scientific committees and task forces appear to be an ideal mechanism for developing appropriate programs in response to future challenges resulting from advances in analytical technology.

REFERENCES

1. E. G. Letheo, W. C. Wallace, and E. Brouwer, The Fate of N-Nitrosodiethanolamine After Oral and Topical Administration to Rats, *Food Chem. Toxicol. 20*:401–406 (1982).

2. T. Y. Fan, U. Goff, L. Song, D. H. Fine, G. P. Arsenault, and K. Biemann, N-Nitrosodiethanolamine in Cosmetics, Lotions and Shampoos, *Food Cosmet. Toxicol.* 15:423—430 (1977).
3. M. L. Douglass, B. L. Kabakoff, G. A. Anderson, and H. C. Cheng, Inhibition and Destruction, *J. Soc. Cosmet. Chem.* 29: 581—606 (1978).
4. P. Rahn and W. Mitchell, High Performance Liquid Chromatography in Cosmetic Analysis, *Drug Cosmet. Indus.* 123:55—66, 126, 130 (1978).
5. I. E. Rosenberg, J. Gross, T. Spears, and U. Caterbone, Methodology Development for the Determination of Nitrite and Nitrosamines in Cosmetic Raw Materials and Finished Products, *J. Soc. Cosmet. Chem.* 30, 127—135 (1979).
6. I. E. Rosenberg, J. Gross, T. Spears, and P. Rahn, Analysis of Nitrosamines in Cosmetic Raw Materials and Finished Product by High Pressure Liquid Chromatography, *J. Soc. Cosmet. Chem.* 31:237—252 (1980).
7. I.E. Rosenberg, J. Gross, and T. Spears, Analysis of N-Nitrosodiethanolamine in Linoleamide DEA by High Pressure Liquid Chromatography and U.V. Detection, *J. Soc. Cosmet. Chem.* 31:323—327 (1980).
8. R. H. Bennett and E. S. Peterson, Determination of Nitrosamines by Liquid Chromatography Using a Photoconductivity Detector, Paper presented at the 21st Rocky Mountain Conference on Analytical Chemistry, Denver, Colo., August 2, 1979.
9. R. R. Gadde and B. Patel, A Simple Nitrite Assay Method for the Screening of Raw Materials Commonly Used in Creams and Lotions, *J. Soc. Cosmet. Chem.* 30:385—391 (1979).
10. V. H. Baptist and R. Brown, Nitrosamine Determination by the Use of Conventional Equipment, *J. Soc. Cosmet. Chem.* 31:219 (1980).
11. J. T. H. Ong and B. S. Rutherford, Some Factors Affecting the Rate of N-Nitrosodiethanolamine Formation from 2-Bromo-2-nitropropane-1,3-diol and Ethanolamines, *J. Soc. Cosmet. Chem.* 31:153 (1980).
12. J. T. H. Ong, B. S. Rutherford, and A. G. Wich, Formation of N-Nitrosodiethanolamine from Peroxidation of Diethanolamine, Paper presented at the Annual Scientific Meeting of Society of Cosmetic Chemists, New York City, December 11—12, 1980.
13. MRI Reports NFT-1 through NFT-7, available through the Cosmetic, Toiletry and Fragrance Association, 1110 Vermont Ave., N.W., Washington, D.C. 20005.
14. MRI Reports NF-8 and NF-9, available through the Cosmetic, Toiletry and Fragrance Association, 1110 Vermont Ave., N.W., Washington, D.C. 20005.
15. MRI Reports NFT-10 and NFT-11, available through the Cosmetic,

Toiletry and Fragrance Association, 1110 Vermont Ave., N.W., Washington, D.C. 20005.

16. *Fed. Reg.* *44*(70):21365—21367 (April 10, 1979).
17. S. S. Hecht and J. B. Morrison, N-Nitroso-N-methyldodecylamine and N-Nitroso-N-methyltetradecylamine in Hair Care Products, *Food Chem. Toxicol.* *20*:165—169 (1982).
18. CTFA Report on Mechanisms for Nitrosamine Formation and Inhibition, Available through the Cosmetic, Toiletry and Fragrance Association, 1110 Vermont Ave., N.W., Washington, D.C. 20005.
19. T. Birkel, C. Warner, and T. Fazio, Gas Chromatographic Determination of 1,4-Dioxane in Polysorbate 60 and Polysorbate 80, *J. Assoc. Offic. Anal. Chem.* *62*:931—936 (1979).
20. CTFA Final Report on the Dioxane Round Robin Program, Availthrough the Cosmetic, Toiletry and Fragrance Association, 1110 Vermont Ave., N.W., Washington, D.C. 20005.
21. B. A. Waldman, Analysis of 1,4-Dioxane in Ethoxylated Compounds by Gas Chromatography/Mass Spectrometry Using Selected Ion Monitoring, *J. Soc. Cosmet. Chem.* *33*:19—25 (1982).
22. J. J. Robinson and E. W. Ciurczak, Direct Gas Chromatographic Determination of 1,4-Dioxane in Ethoxylated Surfactants, *J. Soc. Cosmet. Chem.* *31*:329—337 (1980).
23. M. L. Stafford, K. F. Guin, G. A. Johnson, L. A. Sanders, and S. L. Rockey, Analysis of 1,4-Dioxane in Ethoxylated Surfactants, *J. Soc. Cosmet. Chem.* *31*:281—287 (1980).
24. E. K. Weisburger, et al., Testing of Twenty-one Environmental Aromatic Amines or Derivatives for Long-Term Toxicity or Carcinogenicity, *J. Environ. Pathol. Toxicol.* *2*:325—356 (1978).
25. E. Cox, High Performance Liquid Chromatographic Determination of Quinizarin, p-Toluidine and D&C Violet No. 2 in D&C Green No. 6, *J. Assoc. Offic. Anal. Chem.* *62*:1338—1341 (1979).
26. *Fed. Reg.* *47*(64):14140 (1982).

17

THE JOHNS HOPKINS CENTER FOR ALTERNATIVES TO ANIMAL TESTING

ALAN M. GOLDBERG

Department of Environmental Health Sciences, The Johns Hopkins University School of Hygiene and Public Health, Baltimore, Maryland

I. WHY A CENTER

All commercial products have the potential for producing toxicological insults. The scientific and industrial communities have a special interest and a responsibility to assess the nature of toxic effects and to evaluate the safety of products.

Although many approaches to safety evaluation have been developed, most methods use animal models and extrapolate from animals to humans. Such methods are often cumbersome and expensive. Approaches using *in vitro* methods would be preferred, and with the many recent advances in biomedical research there are opportunities, as never before, to explore such possibilities. Moreover, during the past several years there has been increasing pressure to decrease or to prevent the use of animals in toxicity testing and safety evaluation. Thus, it is in everyone's best interest to develop appropriate alternative methods to diminish, as rapidly as is feasible, the need for animals in testing.

The Cosmetic, Toiletry and Fragrance Assoication, Inc. (CTFA), recognizing its responsibilities in this area, established in 1981 the Johns Hopkins Center for Alternatives to Animal Testing. The primary mission of the center is to develop the fundamental knowledge base that will provide alternative methods to whole-animal testing.

II. GOAL

The goal of the center is to develop appropriate basic scientific knowledge for innovative methods to evaluate fully the safety of commercial and/or therapeutic products.

III. PURPOSE

The specific purposes of the center include:

1. To encourage fundamental research needed for the development of *in vitro* test procedures or other non-whole-animal test procedures to examine the toxicity of chemicals and chemical compositions
2. To develop and validate methodology that will provide alternative approaches to whole-animal studies for the evaluation of safety
3. To encourage and promote the acceptance of applicable methods of non-whole-animal safety testing
4. To provide the best available practical methodological approaches for safety evaluation.

IV. RESEARCH APPROACH

The research program focuses on developing fundamental knowledge to provide alternative methods to animal tests. The approach examines the cellular and biochemical mechanisms responsible for tissue damage and then will develop appropriate *in vitro* correlative tests as substitutes for whole-animal studies.

The research programs are both intramural and extramural programs of grants and projects that are reviewed scientifically and awarded on the basis of quality and appropriateness to the mission of the center.

V. SYMPOSIA

In order to shape the research approach and identify the research program, symposia examine the currently available alternatives to whole-animal testing. Each symposium has several objectives:

1. Understanding the state of the art of appropriate research areas
2. Targeting specific research areas for development
3. Stimulating research in alternative approaches to whole-animal testing
4. Identifying ways to reduce or replace animals in testing.

The proceedings of the symposia are published as monographs which will serve to disseminate knowledge on mechanisms of toxicity and methods of safety evaluation and to stimulate further research.

The symposium mechanism was chosen as an initial and an on-going activity because it will identify the important areas of research needed and thereby establish the focus of the research program of the center. Symposia will in addition provide a mechanism to evaluate the progress of the center's programs.

VI. THE ADVISORY BOARD

A distinguished Advisory Board of voting and nonvoting members has been established. The voting members of the Advisory Board are charged with setting and approving the policies of the program, including the selection of research projects to be funded. Nonvoting representatives from the animal rights movement and the U.S. government have agreed to serve. In addition, representatives from the CTFA and Bristol-Myers serve on the Advisory Board as liaison representatives. Current members include:

Voting members:

Leon Golberg, M.B., D. Sc., D. Phil
Professor, Division of Community & Occupational Medicine
Duke University Medical Center
Durham, N.C.

Alan M. Goldberg, Ph.D
Department of Environmental Health Sciences
Director, Center for Alternatives to Animal Testing
The Johns Hopkins University School of Hygiene & Public Health
Baltimore, Md.

Gareth M. Green, M.D.
Chairman, Department of Environmental Health Sciences
The Johns Hopkins University School of Hygiene & Public Health
Baltimore, Md.

Donald A. Henderson, M.D. M.P.H.
Dean, The Johns Hopkins University School of Hygiene &
    Public Health
Baltimore, Md.

Paul Kotin, M.D.
Adjunct Professor of Pathology
University of Colorado
Denver, Colo.

Franklin M. Loew, D.V.M., Ph.D.
Dean, School of Veterinary Medicine
Tufts University
Boston, Mass.

James P. McCulley, M.D.
Professor and Chairman, Department of Ophthalmology
University of Texas
Dallas, Tex.

Henry N. Wagner, Jr., M.D.
Professor, Departments of Medicine Radiology & Environmental
    Health Sciences
Director, Divisions of Nuclear Medicine & Radiation Health Sciences
The Johns Hopkins University School of Hygiene & Public Health
Baltimore, Md.

Peter A. Ward, M.D.
Professor and Chairman, Department of Pathology
The University of Michigan Medical School
Ann Arbor, Mich.

Nonvoting members:

James M. McNerney, M.P.H.
Vice President—Toxicology
The Cosmetic, Toiletry and Fragrance Association, Inc.
Washington, D.C.

W. Gary Flamm, Ph.D.
Associate Director for Regulatory Evaluation Division of
    Toxicology, Bureau of Foods
Food and Drug Administration
Washington, D.C.

Thomas E. Hickey, D.V.M.
Bristol-Myers Pharmaceutical Research & Development Division
Evansville, Ind.

VII. THE INTRAMURAL PROGRAM

The Center Director coordinates and manages a program of focused
research areas in which expertise exists at The Johns Hopkins Univer-

sity. The center integrates ongoing activities and provides additional funds to target, expand, and create new programs. Applications are reviewed by the solicited grant review process. It is anticipated that through the identification of existing expertise and programs with The Johns Hopkins University, effective utilization of the funds for new research will be achieved.

VIII. THE EXTRAMURAL PROGRAM

Once the intramural program was established, the areas not covered therein were identified and targeted for funds in the extramural program. Specific scientists with proven accomplishments and expertise were then requested to submit proposals for review and at the same time applications were solicited by advertising in professional journals. The research programs are chosen on the basis of quality and appropriateness to the mission of the center.

IX. THE DIRECTOR'S SMALL GRANT PROGRAM

An additional mechanism to stimulate research is the development of the director's research initiative small grant program. This program will fund very specific and highly original ideas so that preliminary data can be obtained. It is anticipated that these programs will then seek additional funds either from the center or other sources to develop fully their research programs. The flexibility offered by such an approach will generate a maximum of research with a minimal budget. Having these funds available at the discretion of the Center Director will allow a more rapid development of these programs.

X. REVIEW OF GRANT PROPOSALS

Written reviews of grant proposals* are solicited by the Center Director from qualified scientists without conflict of interest. Outside reviewers are requested to provide detailed evaluations, critiques, and suggestions for improvement of each grant. The Advisory Board evaluates the proposals and the solicited reviews and makes recommendations for funding.

XI. GRANTS MANAGEMENT

The monitoring and quality control of the funded applications is through the grants management and evaluation process. Renewal and review

---

*Requests for proposals, guidelines for proposals, and instructions for reviewers are available on request from the center

mechanisms evaluate quality and stimulate conceptual and productive development. As part of this process, the regularly scheduled symposia allow grantees to present progress reports to members of the Advisory Board and the sponsors of the center. Further, written reports of program progress are required of all grantees at six-month intervals and progress reports are submitted by the Center Director to its sponsors.

XII. FACILITIES AND RESOURCES

The field of public health is evolving from a primary focus on infectious diseases to an emerging focus on the environment and its effects on the health of populations. The Department of Environmental Health Sciences, the academic home of the center, addresses this focus in its Divisions of Occupational Medicine, Environmental Physiology, Toxicology, Chemistry, Radiation Health Sciences, and Health Engineering and Biology. The department is larger in faculty and student size than most schools of public health and is involved with a truly interdisciplinary faculty throughout the School of Hygiene and Public Health and the university.

The department is organized to offer the science base, technical and professional skills, and the educational and research resources for knowledge development, training, and consultant services to graduate students, industrial corporations, and governmental agencies for the solution of health problems stemming from the environment. An outstanding faculty of scientists, physicians, and engineers has been created to study risk factors of disease and mechanisms of injury; to educate environmental health scientists and practitioners; and to provide risk assessments, preventive services, and environmental control technology for the promotion of public health.

XIII. SUMMARY

The guiding principle of the center and its programs will be scientific excellence and credibility.

The approach presented seeks to develop a highly visible and creative International Center to examine alternatives to whole-animal testing. This research program is developed through a process of familiarization with the needs of its members; identification of areas of research need; development of intra- / and extramural research programs; and dissemination of its research findings.

The Johns Hopkins Center for Alternatives to Animal Testing was established by a grant from the Cosmetic, Toiletry and Fragrance Association, Inc. Additional funding for organ-specific toxicity was provided by the Bristol-Myers Corporation.

18

COMMUNICATION AND COOPERATION—
THE DERMATOLOGIST AND THE COSMETIC
INDUSTRY

WALTER LARSEN

Portland Dermatology Clinic, Portland, Oregon

I. INTRODUCTION

Until recently there has not been much communication and cooperation between the dermatologist and people in the cosmetic industry. Dermatologists generally perceived the industry as one that markets mysterious products with overblown promises. In addition the dermatologist had a skewed viewpoint in that he or she saw only the adverse reactions that occured as a result of using cosmetics.

Recently, however, a dialogue has been established between dermatology and the cosmetic industry on a local and international level. A number of courses have been conducted around America on cosmetics for the dermatologist. Witness the popularity of the cosmetic symposium at the annual American Academy of Dermatology (AAD), which generally attracts hundreds of dermatologists. Patients are seeking advice regarding cosmetics from dermatologists, so we must know which cosmetics work and how they work.

II. COSMETIC LABELING

I believe that a significant breakthrough in understanding cosmetics was made when cosmetic labeling was instituted several years ago. This put the understanding of cosmetics on a more scientific level and broke through the so-called mystery of cosmetic ingredients. With the publication of the *CTFA Cosmetic Ingredient Dictionary*, dermatologists can find in one place all relevant information on a cosmetic ingredient. (A comprehensive third edition was published in 1982.) Now when a patient has sustained an allergic reaction to a specific ingredient, he or she can avoid future reactions by carefully reading labels.

III. DIRECTORY OF COSMETIC INDUSTRY RESOURCE PERSONNEL

Another helpful aid was the CTFA publication, *Directory of Cosmetic Industry Resource Personnel for Physicians* (3rd ed., 1982). This handy booklet lists the chemist or research person in each company so that a dermatologist can call and receive information or specific ingredients of a cosmetic.

IV. STANDARD PATCH TESTS

Additionally, dermatologists patch-test with a standard patch test series, put out by the American Academy of Dermatology and monitored by the North American Contact Dermatitis group. This group of chemicals (Table 1) screens for the most common allergens found in cosmetics, such as fragrances, preservatives, and lanolin derivatives. Some chemicals used in cosmetics are difficult to obtain, so it is most helpful for the dermatologist if a company is able to supply such chemicals. Moreover, some cosmetic companies supply a complete patch-testing kit for each of their cosmetics as another helpful tool.

Fragrances are a unique problem in that they cause most cases of allergic contact dermatitis from cosmetics and are listed on the label only as fragrance. In case of fragrance allergy it is most helpful when the fragrance manufacturer can help by giving partitions of the fragrance for patch testing. When a partition is positive it can be further broken down ot its complete components until the offending specific allergen is determined. The fragrance mixture contained in the AAD screening series consists of eight common fragrance materials that cause allergic reactions (Table 2). By eliminating these chemicals or reducing their concentration in perfumes, future reactions to cosmetics can be reduced. I believe that when known fragrance sensitizers are used in perfume they should be identified parenthetically following the term "fragrance" on the label. This would be helpful in assisting patients to avoid those fragrance materials to which they are allergic.

TABLE 1   AAD Standard Test Series

Medicaments, preservatives, and fragrances

    Neomycin sulfate, 20%
    Ammoniated mercury, 1%
    Ethylenediamine dihydrochloride, 1%
    Parabens (methyl, ethyl, propyl, and butyl, 3% each), 12%
    Wool alcohols, 30%
    Caine mixture, 8%
    Lanolin, 100%
    Benzyl alcohol, 5%
    Quaternium 15, 2%
    Imidazolidinyl urea, 2%
    Fragrance mix, 16%

Metals

    Potassium dichromate, 0.5%
    Nickel sulfate, 2.5% (optional)

Rubber Chemicals

    Mercapto mix, 1%
    MBT, 1%
    Thiuram mix, 1%
    Black rubber *p*-phenylenediamine mix, 0.6% (PPD mix, 0.6%)
    Carba mix, 3%

Others

    Formaldehyde (in water), 2%
    p-phenylenediamine, 1%
    Epoxy resin, 1%
    p-tert-Butylphenol, 2%
    PCMX, 1%

V. DETECTING ADVERSE REACTIONS

The incidence of adverse reactions to cosmetics is unknown. In the typical dermatologic practice I do not believe that such reactions are common; but when a reaction does occur, it is important for the dermatologist to determine the cause. The dermatologist has a responsibility to determine if the reaction is due to misuse of the cosmetic or an irritant reaction, allergic reaction, or phototoxic reaction to the cosmetic. If it is an allergic or photoallergic reaction, the specific allergen must be determined so that it can be avoided in the future.

TABLE 2   The AAD Fragrance Mixture

This mixture consists of 2% each of the following in petrolatum:

Alpha amyl cinnamic alcohol
Cinnamic aldehyde
Cinnamic alcohol
Oakmoss absolute
Hydroxycitronellal
Eugenol
Isoeugenol
Geraniol

The difficult problem faced by dermatologists is that approximately one-half of cosmetic reactions are unsuspected by the patient and the physician. The incidence of irritant contact dermatitis versus allergic contact dermatitis is unknown. However, most workers believe that irritant dermatitis due to cosmetics is more common than allergic contact dermatitis. It is relatively easy to test for allergic contact dermatitis or allergic photodermatitis, whereas testing for irritant dermatitis is difficult.

For instance, many reactions around the eye are due to irritant dermatitis and not allergic contact dermatitis. Most studies of allergic contact dermatitis due to cosmetics have demonstrated that the fragrance component is the most common cause of allergy. The fragrances most commonly implicated include cinnamic alcohol, hydroxycitronellal, isoeugenol, eugenol, oakmoss, and geraniol.

The second most common cause of allergic dermatitis is due to preservatives such as Quaternium-15, imidazolidinyl urea, etc. Lower down the list are the lanolin derivatives. The paraben derivatives, which have been loudly touted as sensitizers, are rarely implicated in cosmetic dermatitis. In fact, most patients who are allergic to parabens can use cosmetics containing parabens and not have any adverse reactions. The North American Contact Dermatitis Group has generated epidemiologic data showing the relative occurrence of reactions to various components of various cosmetics. This type of data can be valuable to cosmetic chemists when formulating new products.

VI. CONCLUSION

A new era of cooperation has developed between dermatology and the cosmetic industry, and I believe this will continue in the future. Planning is now underway to conduct a regular program of information exchange between physicians and the cosmetic industry, through publications and through representation at major industry and AAD meetings. Such efforts will help dermatologists better serve our patients.

19

INDUSTRY PERFORMANCE AND PUBLIC OPINION: SOME REFLECTION ON PUBLIC RELATIONS

VICTOR R. HIRSH

Consultant, Chevy Chase, Maryland

Some 1,700 trade and professional associations are headquartered in the metropolitan Washington area, according to the American Society of Association Executives. That makes the trade association business the second-ranking private industry in the nation's capital, exceeded only by tourism. Naturally, industry trade groups did not settle in such profusion along the banks of the Potomac because of the grandeur of the capital city alone. There are some practical reasons, for example, the growth of federal government.

U.S. Department of Labor figures show us that between 1900 and 1980 the nonmilitary federal workforce grew from 94,893 to 2,866,000. Government, at both the federal and state levels, became the leading growth industry in the United States. While there has been a cutback in federal personnel and certain agency budgets since President Reagan came to office, Washington remains a factory town, and government runs the factory.

As the center of government, Washington is also the news media capital of the country. The 1983 edition of *Hudson's News Media Contacts Directory* lists 3,014 Washington news-gathering organizations and free-lance writers. This practical guide to the news business bears a dedication to the Washington press corps that reads ". . .who ever knew the Truth put to the worse in free and open encounter?" In this question lies the *raison d'être* for the livelihoods of many who work for trade associations. Some might say all because, one way or the other, trade association staff people are the voices of the industries that hire them.

The lines of communication run principally to Congress, federal agencies, and the news media, including both general and trade press.

There is also a steady flow of news to association members from their trade groups in Washington. Ideally, all trade associations operate by consensus of their memberships. Thus, one major association responsibility is to inform its membership faithfully of Washington developments.

Federal legislation and regulation profoundly influence the general economy, the way businesses operate, and the public welfare. Much new legislation and regulation follow public opinion, reflecting public demand on Congress for action in a given field of interest. The balance of consumer and industry interests as determined by law and regulation is a subject of constant public debate. Product safety particularly, taken up as a topic in other chapters of this volume, is a primary concern both to government and to the cosmetic industry. An objective of this chapter is to consider the relationship between public information and education and public policy, whether expressed through legislation or regulation.

One mission of public information activities by the Cosmetic, Toiletry and Fragrance Association (CTFA) and by CTFA member companies is to place accurate information about products, their ingredients, and their proper uses into the hands of the public and government. Another aim is to support industry positions on regulatory and legislative affairs.

To examine how these goals are pursued, it will help to outline briefly some key elements of the CTFA's operation.

CTFA membership has remained steady for some years at just under 500 companies, about half of them manufacturers of consumer products and about half suppliers of raw materials and packaging to product manufacturers. A well-established structure of CTFA committees, involving member representatives who are expert in their special fields, develops CTFA positions on the entire range of industry issues.

The committee system provides the foundation of industry science, especially with respect to ingredient safety, and policy development on regulatory affairs. Under direction of the CTFA Board of Directors, CTFA members and staff do not merely react to pressures created by the public or government. On the contrary, one hallmark of CTFA's durability through the years has been its capacity to recognize industry's responsibilities and to anticipate consumer concerns and take responsive initiatives.

While the consumer movement is popularly thought to be a development of the 1960s, consumerism and federal law to protect the public date back to the turn of the century. Congress passed the Interstate Commerce Act in 1887 to protect consumers against big business and the railroads. The Sherman Antitrust Act of 1887 dealt with large corporations. The National Consumer League was founded in 1899. Upton Sinclair's muckracking novel, *The Jungle*, attacking abuses in the meat packing industry, appeared in 1906. That was the same year Congress enacted the Food and Drugs Act to prohibit poisonous and other harmful additives to food and drugs, later repealed with enactment of the Food, Drug, and Cosmetic Act of 1938. In 1960, the Food, Drug, and Cosmetic Act was amended to require the safety substantiation of all color additives used in cosmetics.

Since that time the CTFA has undertaken on behalf of the cosmetic industry a continuing scientific research program to demonstrate the safety of cosmetic color additives that industry seeks to use.

The CTFA's $8 million color research program is not the result of voluntary action: It is required under the color additive safety regulations. But it serves to illustrate that the industry had the resources and expertise to establish a sophisticated, long-range scientific enterprise.

More to the point, the industry had the foresight to establish several other respected programs voluntarily. These demonstrate the industry's willingness and capacity to mount a program of responsible self-regulation.

In establishing these programs the cosmetic industry sent several important messages to the public, government, and the scientific community, all targets of public information activities describing these programs. One fundamental aim of the industry's public information efforts is to create a climate of acceptance for the industry and the products it markets. Doing so requires the industry to take action to earn credibility and to publicize the actions it takes. Effective public relations are based on solid accomplishments, not on exaggerated claims for lightweight enterprises.

This simple reality is amply recognized by members of the CTFA Public Information Committee (PIC), comprised of a core of corporate communications executives, as well as a few scientists and an attorney who are liaisons to association scientific and legal committees. The PIC plans the CTFA public information programs. It also steers staff activities and places emphasis on hitting the industry theme that cosmetics, toiletries, and fragrances are essential to contemporary lifestyle, not merely luxuries or afterthoughts to the process of good grooming.

Industry has gone to great cost and effort to document the role that good grooming and the use of grooming products play in our society and throughout human history.

In 1979, the CTFA published *The Cosmetic Benefit Study*. The study found that while "cosmetic products might touch only the skin, . . .their beneficial effects on the whole are profound."

For more than 26,000 years men and women have used cosmetics. The Cro Magnon people painted their faces red. The Egyptians colored their hair, painted their eyelids and brows, and rouged their lips and cheeks. They moisturized, washed, and deodorized.

The Romans used hair dyes, night creams, dentifrices, and mascaras.

In India, cosmetic use was largely a male domain. Men rubbed their bodies with sandlewood powder to cool them in the tropical heat and painted their lips, their nails, and the soles of their feet.

Throughout history men and women have believed that outward beauty is a reflection of inward beauty. That belief, CTFA's study found, has not changed much since ancient times. At least 87 percent of those studied placed personal grooming and attractiveness in the context of total well-being and the care and growth of the whole person — physically, psychologically, socially, and emotionally. Use of personal care products begins in childhood with lotions and toothpaste and continues through adulthood to old age with products to help the elderly feel more productive and to look and feel more alive, well, and happy about life.

Anthropoligist Harry Shapiro, former chairman of the anthropology department of the American Museum of Natural History in New York City, stated: "[S]o universal is this urge to improve on nature that one is almost tempted to regard it as an instinct. Aside from such fundamental drives as those for food, love, security, and the expression of maternal solicitude, I can think of few forms of human behavior that are more common to mankind as a whole."

These projects to strengthen the case for cosmetics as essential tools in the quest for personal self-esteem support other association initiatives to maintain and build industry credibility in regulatory, scientific, and media circles.

Perhaps the most prominent effort by industry to underscore its commitment to self-regulation is its establishment of the Cosmetic Ingredient Review (CIR), a wholly industry financed, independent organization to evaluate the safety of ingredients used in the entire range of cosmetic products.

Founded by industry through the CTFA in 1976, the CIR is a nonprofit organization that evaluates data concerning the safety of cosmetic ingredients. It operates under procedures that closely parallel those used by advisory panels of the U.S. Food and Drug Administration (FDA) for the Over-the-Counter Drug Review Program. These procedures include a public evaluation of available data on ingredients by a panel of expert scientists and clinicians.

The CIR's seven-member Expert Panel must meet rigid conflict-of-interest requirements, and three persons participate in a nonvoting capacity in panel discussions, a consumer organization liaison, an industry liaison, and an FDA contact person.

The CIR is truly a "sunshine" operation, sponsored by industry and predating federal sunshine laws.

To illustrate further that industry can act effectively both in its self-interest and in the public interest, one need look no further than the Johns Hopkins Center for Alternatives to Animal Testing. It was established in 1981 under a $1 million grant by the CTFA from a fund created by voluntary contributions from CTFA member companies.

As its name implies, the Center has launched a search for test-tube alternatives to live-animal safety testing of cosmetic ingredients and products. In its two short years of operation at the time of this writing, the Center—under administration of the Johns Hopkins University School of Hygiene and Public Health—has become the leading institution seeking ways to reduce, perhaps eliminate, all ingredient and product safety testing using animals.

The Center is the focus of worldwide attention by animal rights advocates and scientists concerned with the humane and scientific issues involved in animal testing. Its establishment by the CTFA is a study in how an industry elects to meet a public opinion and scientific challenge in one stroke.

The cosmetic industry had been concerned with the problems of animal testing long before advocacy of animal rights became a strong international cause. Other chapters of this book discuss the regulatory and scientific requirements for animal testing as a means of assuring human safety. Only the most extreme animal rights activists would deny the necessity for at least some animal testing, and the most prominent animal rights leaders praise the industry and the Center for their work to date. In fact, most animal rights leaders are partners rather than adversaries of the cosmetic industry in supporting CTFA efforts to meet both industry and public interest goals.

It is true that industry took action in part because animal rights leaders singled out the cosmetic industry in their first organized national campaign to address their policy goals. It is equally true that the industry listened and responded with sensitivity and strength.

So there are many winners and no losers in the continuing outcome of industry's response to the animal rights issue. In only two years the Center has forged a leading role in the scientific community as an indispensable resource in the search for alternatives. It publishes a regular newsletter to scientists and the science press, including science writers in the general press. The CTFA augments these communications with an information flow that reaches opinion makers inside and outside the industry, both directly and through press coverage of Center affairs. Additionally, the CTFA and the Center are in regular touch with animal rights leaders reporting on the Center's progress.

Good news, as the Center demonstrates, can travel as fast as bad news, despite the more conventional view that bad news grabs all the headlines.

Though not as dramatic a venture as the Center, another public-spirited project of the industry illustrates industry's commitment to voluntary self-regulation and helps substantiate the accurate picture of an industry concerned both with its reputation and its customers.

In 1971, the CTFA petitioned the FDA to establish at the FDA a program that would gather information about the location of cosmetic manufacturing plants, information about ingredients used in cosmetic products, and data concerning adverse consumer reactions to products. Implemented in 1973, it is called the Voluntary Reporting Program. As the name implies, cosmetic manufacturers participate on a voluntary basis, in keeping with industry's view that self-regulation is more effective than mandatory regulation.

The value of the program from industry's viewpoint is that it provides the FDA with information that helps the FDA monitor product performance among consumers. Data obtained under the program may also serve to alert both the FDA and manufacturers to problems associated with ingredients or class of formulations. Further, a successful voluntary program argues strongly against legislation that would require mandatory reports and generate costs to government and industry typical of mandatory programs.

As effective as the program has been, it is not well known to the general public, though it is respected by government officials and legislators interested in the cosmetic product field.

All three of the many CTFA self-regulatory programs discussed thus far have established that an industry can earn the public trust by taking innovative steps to meet emerging challenges. Doing so successfully and using effective means to inform opinion makers and the public about these developments builds the climate of industry credibility for which there is no substitute.

This chapter has reviewed only the most prominent programs to make the point that acceptance by government and the public is related to institutional performance, not to wall-papering editorial offices with self-serving news of little consequence.

The means by which the CTFA disseminates news about its achievements to the public and specific target groups are, for the most part, conventional, though effective.

At the basic level the CTFA issues press releases to the trade press and supplies these periodicals with copies of legal and scientific submissions to government agencies, principally the FDA. Doing so is one more way to inform CTFA members of association actions of which they should be aware. Further, and sometimes overlooked, is the practice of general press reporters to pick up information and story leads for popular consumption from the trade press. Most media relations professionals working for trade associations have more than a few tales to tell about how a trade press story touched off general press inquiries on sensitive issues that otherwise might never have surfaced.

Another communications tool is publication of a series of brochures that explains in lay language the goals and operations of the CTFA's self-regulatory programs, such as those described here. These are distributed in quantity to member companies, related trade groups, and special interest groups such as the American Academy of Dermatology, whose members have an obvious concern with cosmetic product safety and efficacy. For the same reason, issue brochures—in effect are "white papers"—are distributed to contacts at key federal agencies.

A continuing public information project is the development of a briefing book on major industry issues for use by members and staff for press and public inquiries on industrywide concerns. The briefing book—constantly being updated—is particularly helpful to small businesses that do not have large technical and legal staffs to field the range of inquiries cosmetic companies receive.

Obviously, press and public contact work focuses on two major day-to-day CTFA enterprises, safety science, and regulatory affairs. To these efforts the CTFA, on behalf of industry, has dedicated its greatest resources. The CTFA's dealings with regulatory agencies, especially its submissions to the FDA, meet the highest legal standards. Such strong routine performance does as much to influence the professionals in the world of regulation as do the star programs described earlier.

It is worth noting that this competence is reflective of industry on the whole. While CTFA members sell about 90 percent of all cosmetics, toiletries, and fragrances sold in the United States, 60 percent of the association's members are small businesses. The organizational strength behind the CTFA derives from diversity, the strong leadership of the Board of Directors, and the thoroughly professional operation of CTFA's 50 committees and task forces. Without this cohesion and integrity, the CTFA would not stand where it does today as an organization vital to the future of the industry.

Section B: As Individual Companies

20

PREPARING FOR COSMETIC GMPs

ALLAN T. REEVES

The Procter & Gamble Company, Cincinnati, Ohio

I. INTRODUCTION

The U.S. Food and Drug Administration (FDA) may be issuing Current Cosmetic Good Manufacturing Practices (CCGMP) regulations or guidelines in the future to convey their manufacturing expectations to

the cosmetic industry. The FDA is responsible for seeing that cosmetic products are not adulterated or misbranded. As pointed out in
Chap. 1, "Regulation of Cosmetics in the United States-An Overview," the FDA is authorized to check the sanitation of a cosmetic
Packaging and Labeling Act. The FDA is also interested in the competence of the personnel, the material control systems used, and record systems maintained in the production of cosmetics. It should be
noted that the FDA currently does not have authority to examine records for cosmetics and cosmetic drugs which are also "old drugs" (anticaries toothpastes, antidandruff shampoos, etc.).[1] The FDA does
have authority to examine records for cosmetics which are also "new
drugs,"[2] to check compliance with the drugs' New Drug Application
(NDA). Even though the FDA does not have authority to examine
records, control systems, personnel qualifications, etc., it will ask
questions during an establishment inspection relating to these subjects. In fact, the FDA has recently published a revision of its Inspectors' Operations Manual, which contains "GMP Guidelines for Cosmetics" (see the Appendix to this chapter).

The FDA compliance program for the 1980 cosmetic manufacturers'
also had 32 questions which indicated the areas of interest to the FDA.
This chapter is written with the FDA's questions concerning each area
listed at the beginning of the area. While each firm should obtain its
own legal advice, this chapter provides the FDA's legal requirements
and areas of interest. It must be kept in mind that this guidance is
intended to illustrate principles and not necessarily to specify the only
acceptable ways to meet these principles.

II. ORGANIZATION AND PERSONNEL

One of the keys to CCGMP is personnel. The organizational structure,
personnel qualifications, personal sanitation, and health are discussed
in the following sections.

A. Organization Structure

A cosmetic manufacturer is expected to have an organization where the
quality assurance function is separated from the production function.
It is desirable to have a quality unit which sees that appropriate steps
are taken to approve or reject raw materials, packaging materials,
labels, finished product, and investigate production errors. It is obvious that at some level of every company, the responsibility for the
quality function and production function resides in one individual.
The FDA's position seems to be that the closer that person is to the
chief operating officer of the company, the better. In a small company, people have responsibilities for several areas; if possible, the
quality and production responsibilities should reside in different people.

In larger companies, it is desirable to have a corporate manager of quality assurance as well as a corporate manager of manufacturing.

B. Personnel Qualifications

FDA Compliance Program Questions

>   Is there a standardized training program for personnel directly involved in the manufacture or control of cosmetic products?

A cosmetic manufacturer is expected to have an adequate number of qualified people to produce quality cosmetics. The qualifications can come from any combination of education, training, or experience. A suggested way to meet this expectation is to have an initial training program for new employees and periodic ongoing training. Suggested areas to cover for an initial training program are listed in Table 1. Possible areas for ongoing training are listed in Table 2.

TABLE 1 Suggested Initial Training Program

Individual training on "how to do the job"

Food and Drug Administration (FDA) authority and responsibilities for cosmetics

Company policies concerning compliance with FDA requirements

Overview of the FDA's Good Manufacturing Practices (GMPS)

Individual responsibilities for compliance with GMPs

TABLE 2 Suggested Ongoing Training Programs

Company quality assurance program

Personal sanitation

Building design and maintenance

Equipment design and maintenance

Material specifications and standards

Material control systems

Material and product sampling and testing

Production and process controls

Warehousing and distribution

C. Personal Sanitation

FDA Compliance Program Questions

> Are personnel properly attired when directly involved with the
> manufacturing process?

A cosmetic manufacturing company is expected to have people with
good sanitation habits so as to avoid adversely affecting product quality. Clean clothing is expected, as well as adequate hair and skin
coverings (hair, beard, arms, and hand if appropriate). All personnel are expected to wash their hands after toilet use. A policy stating
this should be established and signs should be posted at appropriate
locations.

D. Health

FDA Compliance Program Questions

> Does the firm have a policy for dealing with illness or lesions
>   among employees who might come in contact with raw materials
>   or cosmetic products?

Employees who produce cosmetics are expected to practice good
health habits. Although this is somewhat nebulous, it is obvious that
unhealthy people should be kept from contact with cosmetic products
and packages. Employees should be instructed to report illnesses to
their supervisors; if they are ill, the supervisor should decide if they
should be removed from contact with cosmetic products and packages.

III. BUILDINGS AND FACILITIES

FDA Compliance Program Questions

> Does the building (or buildings) provide adequate space for or-
>   derly placement and storage of equipment, materials and fini-
>   shed product?
> Are there adequate toilet, locker and workroom facilities?
> Are eating and smoking areas separated from the manufacturing
>   facility?
> Is the manufacturing and processing equipment properly stored
>   when not in use?
> Is there adequate lighting and ventilation in the manufacturing
>   area?
> Are floors, walls and ceilings free from loose dirt, chipped, and
>   peeling paint?
> Are hand-washing facilities available near the manufacturing area?
> Are waste receptacles adequate in number and utilized properly?

Buildings and facilities should be of suitable size and construction
to produce cosmetic products. If possible, they should be laid out for

logical flow of materials (receiving area, quarantine area, raw material and packaging material warehouse, preweigh area, process area, and finished product warehouse). All buildings and facilities should be designed to facilitate cleaning, maintenance, and proper operation. Floors, walls, and ceilings should be smooth, hard surfaces to facilitate cleaning. Ventilation should be adequate in all areas of the building and, if appropriate, temperature and humidity controls should be used. The building can be under positive pressure to help minimize intake of foreign material. An environmental monitoring system can be used to measure foreign materials. The building should have adequate toilet, locker, and washroom facilities near the manufacturing area and should have designated eating and smoking areas. The building should be designed to handle waste disposal efficiently; waste can be stored in closed containers to help control pests.

IV. EQUIPMENT

FDA Compliance Program Questions

> Are scales, balances and other measuring devices in the manufacturing establishment checked for accuracy according to an established schedule?

Equipment used to manufacture cosmetics should be of suitable size and construction to produce the products. The equipment should be designed for easy cleaning, sanitation where necessary, and proper maintenance. The equipment should be constructed of materials which will not affect the cosmetic products or packages. The equipment should be cleaned at appropriate intervals and should be sanitized periodically; records of the cleaning and sanitization are suggested. Equipment such as scales and meters should be calibrated periodically; records of the calibrations are also suggested.

V. CONTROL OF RAW MATERIALS AND PACKAGING MATERIALS

FDA Compliance Program Questions

> Are raw materials identified (either by chemical or proprietary names)?
> Are all raw materials sampled and tested to meet established standards of identity, quality and purity?
> Are tests performed for microbial contaminations of all susceptible raw materials?
> Are raw materials rotated to follow a "first-in, first-out" inventory control procedure?
> Do raw materials carry lot identification?
> Are storage conditions adequate to protect raw materials from adverse environmental conditions, filth, vermin, or other sources of contamination?

Are raw material shipments, not yet approved by the quality control department, stored under quarantine in a location separate
from approved raw materials?

Is the process water treated prior to use in cosmetic products?

A. Control Systems

Raw materials and packaging materials should be controlled from the
time they are received until they are used. The materials should be
received, quarantined, sampled, tested, released or rejected, and
dispensed for processing when needed. This control can be accomplished several ways; the following are three examples.

Physical Quarantine System

A physical quarantine system depends on physical separation of quarantined and released materials. As materials are received, they are
placed in a separate quarantine area and are sampled. If the sample
test results meet the acceptance criteria, the materials are physically
moved to a "released" area. If the sample tests do not meet the acceptance criteria, the materials are physically moved to the "reject"
area. Materials from the "released" area are dispensed for processing
when needed. The main advantage of the quarantine system is its
reliability. The main disadvantage of the system is the cost of multiple material handling.

Tag Quarantine System

A tag quarantine system depends on each container of materials having
a tag which describes the status of the material. When the materials
are received, they are tagged appropriately (Quarantined, Held, etc.)
and stored. The materials are sampled and if the sampling test results
meet the acceptance criteria, the materials are tagged with an appropriate tag (Released, Acceptable, etc.). If the materials do not meet
acceptance criteria, they are tagged with appropriate tags (Reject,
Unacceptable, etc.). The released materials are dispensed to processing as needed and are used after the operator checks the "released" tag. The main advantages of the tag system are the low materials handling cost and ease of determining the status of each container of material. The main disadvantage is the cost to tag each container several times.

Computer (or Paper) Quarantine System

Some warehouses are controlled by computer (or paper) systems. In
such a system, the status of each unit of materials can be stored in
the computer system. When the materials are received, they are
stored in specific locations in the warehouse; the computer system
should indicate that the materials are "quarantined." The materials
are sampled, which can also be indicated in the computer, and tested.

If the sampling test results meet the acceptance criteria, the status of the materials in the computer system is changed to "released." If the sampling test results do not meet the acceptance criteria, the status of of the materials in the computer system is changed to "reject." The "released" materials are dispensed to processing on an "as needed" basis. The main advantages of the computer quarantine system are the low materials handling cost and ease of changing the materials status. The main disadvantage of the system is that the status of the material is not easily determined without a computer terminal or print-out.

B. Sampling and Testing

When the materials are received, they should be stored in an area and controlled to prevent their use as described. The materials should be sampled and tested by the supplier's lot number for appropriate chemical, physical, and microbiological tests; specifications should be established for all tests. The Cosmetic, Toiletry and Frangrance Association (CTFA) provides guidance for microbiological limits and process water, and also provides asssitance in establishing analytical methods to analyze the materials.[3] The results of the tests should be recorded and retained until the product has been distributed and used by consumers. If the materials meet all the specifications, they should be released for use in cosmetics. If the materials do not meet all specifications, they should be rejected. Rejection will require proper disposition, which may mean returning to the supplier, scrapping, reworking, or accepting as is with concurrence of appropriate management. The rejected materials should be controlled (stored separately, tagged, etc.) to prevent their use. It should be noted that analyses from suppliers can be used if the reliability of the supplier is established and results are monitored periodically; identity and purity of each lot of materials should be established upon receipt.

C. Storage

Raw materials should be stored properly to prevent contamination. Usually this means appropriate temperature control, humidity control, and pest control. Professional pest control firms usually do a good job of controlling pests if the building is "tight" and surrounding areas are kept clean and well lighted. The materials should be used on a first-in, first-out basis and should be retested at appropriate intervals; six-month intervals are suggested.

VI. PRODUCTION AND PROCESS CONTROLS

FDA Compliance Program Questions

Does the batch record accurately represent the master formula by percent and batch quantities?

Does the batch record accompany the product through the entire
   manufacturing process?
Is the batch record signed or initialed at each of the following
   steps:
   1. Ingredient weighing or measure?
   2. Each process step?
   3. Filling?
   4. Quality control sampling?
   5. Product release to storage?
Does the batch record provide adequate processing, sampling, and
   filling instructions?
Does the batch record make provision for in-process controls
   where needed?
Is the manufacturing and filling equipment adequately cleaned
   (and sanitized, where appropriate) to preclude product contamination?
Are the lot numbers of raw materials recorded on the manufacturing batch record?

There should be instructions on how to formulate and process the
product. The formulation instructions should specify the raw materials to be used, the quantity of materials to be used, and the order of
addition of the materials. Equipment, temperatures, mixing times,
etc., should also be specified for the batch.

When the batch is made, the first thing to do is to check the formulation for accuracy. As the batch is built, the operator should record the quantity of material added, the identification of the lot of material being used, and initial the batch record at the completion of
each ingredient addition. When the batch is built and is being processed, the operator should record key processing variables (times,
temperatures, pressures, etc.) and initial the batch record at the
completion of each step. When the batch is finished, it should be
sampled and tested for appropriate chemical, physical and microbiological properties; specifications should be established for each test.
Batches that meet all specifications should be released for packaging.

Batches which do not meet specifications should be rejected (re-
worked, scrapped, or used as is with appropriate management approval). It is a good idea to check the quantity produced versus the
quantity intended to catch weighing errors, leaks, etc. The batches
should be stored properly to prevent deterioration or contamination,
and should be tested at appropriate intervals against the specifications.

The above is in the context of a batch process. With a continuous process, the principles must be applied with specifics relevant to
that situation.

VII. PACKAGING AND LABELING CONTROLS

FDA Compliance Program Questions

Are storage and disbursement of labels adequately controlled to prevent mixups?

Are labels checked at this establishment for compliance with regulatory requirements?

Do labels bear warning or caution statements where required?

Do finished cosmetic products bear a manufacturing code number?

Are samples or finished product batches retained?

Are all finished cosmetic products sampled and tested to meet established standards of identity and quality?

Are susceptible finished cosmetic products tested for microbial contamination?

As stated previously, packages, including labels, should be inspected and released before they are put on a packing line. Labels should be controlled particularly well to prevent labeling mixups. One of the key items to check is the compliance of the labels with their legal requirements (see Chap. 23). Instructions should be available detailing the packaging and labeling operation. A coding system should be available on the packing line to put a code on each unit which will allow tracing the history of the product batch and each packaging material.

Before the finished product units are put in a shipping container, samples should be inspected for labeling. Representative samples of the finished cosmetic should be taken and tested against preestablished packaging and product specifications; the tests should include microbial content if the product is susceptible to microbes. Products should not be shipped until the results of these tests are known and the products are released. Samples of each finished cosmetic batch should be retained until the product has been distributed and used by consumers.

VIII. WAREHOUSE AND DISTRIBUTION

FDA Compliance Program Questions

Does the firm have an effective system for recall at the wholesale and/or retail level?

Instructions should be available detailing how the product should be warehoused (temperature limits, humidity limits, maximum length of storage time, stacking heights, etc.). A system should be established to control distribution of the product until it is released. When possible, a first-in, first-out system should exist which will show where the product is shipped to facilitate its recall, if necessary. The following systems can be used for recalls:

Direct Recording Recall System

The direct recording recall system consists of recording the number of cases shipped from each batch code to each customer. If a recall is necessary, the records can be retrieved and the customers contacted to return the product. The system does take time but is very fast and accurate if administered properly.

Time Block System

The time block system consists of setting aside one particular batch and recording its starting and ending shipping dates. If a recall becomes necessary, the customer shipments for the shipping period can be traced to return the product.

IX. SUMMARY

To summarize, the FDA may issue CCGMP regulations or guidelines for the manufacture of cosmetic products. Whether the FDA's actions take the form of regulations, guidelines, or inspections, the FDA expects cosmetic manufacturers to have adequately trained personnel, well-designed buildings and equipment, good control of the materials used in cosmetics, adequate instructions for the manufacture of cosmetics, records of the production of cosmetics, and an adequate warehouse and distribution system.

X. APPENDIX: U.S. FOOD AND DRUG ADMINISTRATION TRANSMITTAL NOTICE OF 8-2-82

Cosmetic Establishment Inspection Instructions

---

653   COSMETICS

653.1   Cosmetics

653.11 Establishment Inspection

A. General: The purpose of cosmetic establishment inspections is to determine whether the inspected firms manufacture, hold or deliver for introduction into interstate commerce cosmetics that are adulterated or misbranded and to prevent these and other practices violating Sec. 301 of the FD&C Act. Sec. 704(a) of the Act provides the authority to enter establishments in which cosmetics are manufactured or held and to inspect such establishments as well as all pertinent equipment, finished and unfinished materials, containers and labeling therein.

A cosmetic may be deemed adulterated for essentially four reasons, namely:

1. It may be injurious to users under conditions of customary use because it contains, or its container is composed of, a potentially harmful substance.
2. It contains filth.
3. It contains a non-permitted, or in some instance non-certified, color additive.
4. It is manufactured or held under insanitary conditions whereby it may have become injurious to users or contaminated with filth (Sec. 601).

A cosmetic may be deemed misbranded for reasons of:

1. False or misleading labeling.
2. Failure to state prominently and conspicuously any information required by or under authority of this act.
3. Misleading container presentation or fill (Sec. 602).

Follow the instructions in this section unless directed otherwise by your supervisor or your assignment.

IOM Exhibit 653.1-A provides background information, regulatory requirements and specific health hazard concerns associated with certain types of cosmetic products. Review this information prior to, and refer to it during, EI's.

Prior to any inspection, review:

1. Current Cosmetic Compliance Program.
2. 21 CFR 700 to 740.
3. District files of the firm to be inspected, including EIR's of previous inspections, consumer complaint files, and files on regulatory actions.

B. *Abbreviated Inspections*: see IOM 502.2, Abbreviated Inspections. Use the critical factors as follows:

1. Check processing and filling areas for smooth, easily cleanable floor, wall and ceiling surfaces, adequate facilities for washing equipment, utensil and hands, and adequate floor drainage.
2. Check working and utensil storage areas for adequacy, cleanliness, and sanitary condition.
3. Check processing, filling, finished bulk holding and transfer equipment, and utensils for absence of dust, dirt, incrustation or other contamination or insanitary condition.
4. Check for personal cleanliness (garments, protective devices) and conformance to hygienic practices (washing hand, no eating, drinking or smoking in processing, filling or storage areas) of persons handling raw materials, equipment or finished products in bulk.
5. Check for maintenance of proper batch records (formula, manufacturing instructions, quality control testing, etc.).

6. Determine whether raw materials are tested for identity and contamination with filth, harmful microorganisms and extraneous chemicals.
7. Determine whether finished bulk and/or products are tested to meet established acceptance criteria, including microbial limits and adequacy of preservation under use conditions.
8. Check for use of non-permitted color additives.
9. Review a number of product labels for compliance with the requirements of the Act & regulations.

C. *Comprehensive Inspections*

1. *Building and Facilities*. Check whether:
   a. Buildings used in the manufacture or storage of cosmetics are of suitable size, design and construction to permit unobstructed placement of equipment, orderly storage of materials, sanitary operation, and proper cleaning and maintenance.
   b. Floors, walls and ceilings are constructed of smooth, easily cleanable surfaces and are kept clean and in good repair.
   c. Fixtures, ducts and pipes are installed in such a manner that drip or condensate does not contaminate cosmetic materials, utensils, cosmetic contact surfaces of equipment, or finished products in bulk.
   d. Lighting and ventilation are sufficient for the intended operation and comfort of personnel.
   e. Water supply, washing and toilet facilities, floor drainage and sewage system are adequate for sanitary operation and cleaning of facilities, equipment and utensils, as well as to satisfy employee needs and facilitate personal cleanliness.

2. *Equipment*. Check whether:
   a. Equipment and utensils used in processing, holding, transferring and filling are of appropriate design, material and workmanship to prevent corrosion, build-up of material, or adulteration with lubricants, dirt or sanitizing agent.
   b. Utensils, transfer piping and cosmetic contact surfaces of equipment are well-maintained and clean and are sanitized at appropriate intervals.
   c. Cleaned and sanitized portable equipment and utensils are stored and located, and cosmetic contact surfaces of equipment are covered, in a manner that protects them from splash, dust or other contamination.

3. *Personnel*.
   a. Obtain the names, titles and other information on key personnel as indicated in IOM 525.

b. Check whether:
   (i) The personnel supervising or performing the manufacture or control of cosmetics has the education, training and/or experience to perform the assigned functions.
   (ii) Persons coming into direct contact with cosmetic materials, finished products in bulk or cosmetic contact surfaces, to the extent necessary to prevent adulteration of cosmetic products, wear appropriate outer garments, gloves, hair restraints etc, and maintain adequate personal cleanliness.
   (iii) The eating of food, drinking of beverages and using of tobacco is restricted to appropriately designated areas.

4. *Raw Materials*. Check whether:
   a. Raw materials and primary packaging materials are stored and handled in a manner which prevents their mix-up, contamination with microorganisms or other chemicals, or decomposition from exposure to excessive heat, cold, sunlight or moisture.
   b. Containers of materials are closed, and bagged or boxed. Materials are stored off the floor.
   c. Containers of materials are labeled with respect to identity, lot identification and control status.
   d. Materials are sampled and tested or examined in conformance with procedures assuring the absence of contamination with filth, microorganisms or other extraneous substances to the extent necessary to prevent adulteration of finished products. Pay particular attention to materials of animal or vegetable origin and those used in the manufacture of cosmetics by cold processing methods with respect to contamination with filth or microorganisms.
   e. Materials not meeting acceptance specifications are properly identified and controlled to prevent their use in cosmetics.

5. *Production*. Check whether manufacturing and control have been established and written instructions, i.e., formulations, processing, transfer and filling instructions, in-process control methods etc., are being maintained. Determine whether such procedures require that:
   a. The equipment for processing, transfer and filling, the utensils, and the containers for holding raw and bulk materials are clean, in good repair and in sanitary condition.
   b. Only approved materials are used.
   c. Samples are taken, as appropriate, during and/or after processing, transfer or filling for testing for adequacy of mixing or other forms of processing, absence of hazardous microorganisms or chemical contaminants, and compliance with any other acceptance specification.

d. Weighing and measuring of raw materials is checked by a second person, and containers holding the materials are properly identified.

e. Major equipment, transfer lines, containers and tanks used for processing, filling or holding cosmetics are identified to indicate contents, batch designation, control status and other pertinent information.

f. Labels are examined for identity before labeling operations to avoid mix-up.

g. The equipment for processing, holding, transferring and filling of batch is labeled regarding identity, batch identification and control status.

h. Packages of finished products bear permanent code marks. Obtain and report key to the code.

i. Returned cosmetics are examined for deterioration or contamination. Review the acceptance specifications.

6. *Laboratory Controls*. Check whether:

a. Raw Materials, in-process samples and finished products are tested or examined to verify their identity and determine their compliance with specifications for physical and chemical properties, microbial contamination, and hazardous or other unwanted chemical contaminants.

b. Reserve samples of approved lots or batches of raw materials and finished products are retained for the specified time period, or stored under conditions that protect them against contamination or deterioration, and are retested for continued compliance with established acceptance specifications.

c. The water supply, particularly the water used as a cosmetic ingredient, is tested regularly for conformance with chemical-analytical and microbiological specifications.

d. Fresh as well as retained samples of finished products are tested for adequacy of preservation against microbial contamination which may occur under reasonably forseeable conditions of storage and consumer use.

7. *Records*. Check whether control records are maintained of:

a. Raw materials and primary packaging materials, documenting their handling, storage, laboratory control, and usages as well as disposition of rejected material.

b. Manufacturing of batches, documenting the:
(i) Kinds, lots and quantities of materials used.
(ii) Processing, handling, transferring, holding and filling.
(iii) Code marks of batches and finished products.

c. Finished products, documenting sampling, individual laboratory controls, test results and control status.

d. Distribution, documenting initial interstate shipment, code marks and consignees.

8. *Labeling.* Check whether the labels of the immediate and outer container bears:
   a. On the principal display panel:
      (i) In addition to the name of the product, the statements of identity and net contents,
      (ii) The statement "Warning - The safety of this product has not been determined" if the safety of the respective product has not adequately been substantiated. Determine whether and what toxicological and/or other testing the firm has conducted to substantiate the safety of its products. See 21 CFR 740.10.
   b. On an information panel:
      (i) The name and address of the firm introducing the product into interstate commerce.
      (ii) The list of ingredients (only an outer container) if intended for sale or customarily sold to consumers for consumption at home.
      (iii) The warning statement(s) required at 21 CFR 740.11 and 740.12.
      (iv) Any other warning statement necessary or appropriate to prevent a health hazard. Determine the health hazard or other basis for a warning statement.
      (v) Any direction for safe use of a product.
      (vi) In the case of a hair dye product, the caution statement of sec. 601(a) of the Act and appropriate directions for preliminary patch testing. This warning only applies to coal-tar hair dyes which, if so labeled, are then exempted from the adulteration provision of the Act.
9. *Complaints.* Check whether the firm maintains a consumer complaint file and determine:
   a. The kind and severity of each reported injury and the body part involved.
   b. The product associated with each injury, including the manufacturer and code number.
   c. The medical treatment involved, if any, including the name of the attending physician.
   d. The name(s) and location(s) of any poison control center, government agency, physician's group etc., to whom formula information and/or toxicity data are provided.
   e. Whether the firm voluntarily files Cosmetic Product Experience Reports (21 CFR 730).
10. *Other.* Check whether the firm is:
    a. Participating in the program of voluntary registration of:
       (i) Cosmetic manufacturing establishments (21 CFR 710).
       (ii) Cosmetic product ingredient and cosmetic raw material composition statements (21 CFR 720).

(iii) Cosmetic product experiences (21 CFR 730).
  b. Using a color additive which is not listed for use in cos-
     metics (21 CFR 73, 74, and 82) or which is not certified
     (21 CFR 80).
  c. Using a prohibited cosmetic ingredient (21 CFR 700).

653.12 Sample Collection

A. *Compliance Samples*

  1. *General*. Collect compliance samples if, during an inspection,
      adulteration or misbranding is noted or suspected, or on spe-
      cial assignment, including a follow-up to an investigation of a
      cosmetic-related adverse reaction.
      Obtain the key to code at the time of sampling (or from file
      at district office) and explain in C/R under "Remarks."
      Collect the following quantities (these include quantities for
      702(b) samples) and ship for testing as indicated below.
  2. *Chemical Analysis*. Collect at least 3 units and not less than
      12 oz. total quantity in duplicate and ship to the Division of
      Cosmetics Technology (HFF-440), except for:

Aerosol Products	Not less than 24 oz. total quantity duplicate		
Bath Salts	24	"	"
Bubble Baths	24	"	"
Eye Make-ups	2	"	"
Facial Make-ups	8	"	"
Mouthwashes	24	"	"
Nail Preparations	6	"	"
Perfumes	6	"	"
Pressed Powders	6	"	"

  3. *Microbiological Analysis and Filth Examination*. Collect at
      least 10 individual retail units (20 units if less than 1/2 oz.
      each) or 1-4 oz. subsamples from each of 10 containers of bulk
      material and ship to the microbiological laboratory which nor-
      mally services your district, unless otherwise directed. Col-
      lect subsamples in duplicate.
  4. *Color Analysis*. Collect at least 3 units and not less than 8 oz.
      total quantity in duplicate and ship to the regional color ana-
      lyst.
  5. *Label Examination*. Collect at least 3 intact units with all
      available labeling for examination at district office.

6. *Toxicity Testing*. On assignment or clearance by the Division of Cosmetics Technology (HFF-440), collect at least 3 units and not less than 6 oz. total quantity and ship to the Division of Cosmetics Technology for testing by the Division of Toxicology.

B. *Complaint Samples*

1. *General*. Refer to the current Compliance Program and IOM 926 for guidance in the collection of samples in connection with investigations of cosmetic related adverse reactions. Collect samples only if directed by the Division of Cosmetics Technology (HFF-440). Attempt to collect a sample of the remainder of the product associated with the injury.
2. *Sample Quantity*. When directed to collect compliance samples, collect quantity indicated in Section A.2 above.
3. *Sample Submission*. Ship all samples to the Division of Cosmetics Technology (HFF-440).

C. *Surveillance Samples*

1. Refer to current Compliance Program for guidance in the collection of surveillance samples. Collect surveillance samples only if so directed.
2. *Sample Quantity*. Collect the total quantity indicated in section A.2 above.
3. *Sample Submission*. Ship samples for examination to the offices or laboratories indicated in section A.

NOTES

1. An "old drug" is a drug product which is generally recognized as safe and effective by a panel of experts.
2. A "new drug" is a product which is not recognized as safe and effective. The company marketing a "new drug" must file a New Drug Amendment (NDA) with the FDA containing safety data, efficacy data, and instructions on how the new drug will be manufactured. The NDA must be approved by the FDA before marketing the "new drug."
3. Charles R. Haynes and Norman F. Estrin, eds., *CTFA Technical Guidelines*, 2nd ed., The Cosmetic, Toiletry and Fragrance Association, Inc., Washington, D.C., 1982.

21

EVALUATING YOUR MICROBIOLOGY PROGRAM

JOHN L. SMITH

Microbiological Research, Chesebrough-Pond's Inc. Research Laboratories, Trumbull, Connecticut

I. INTRODUCTION

A. Regulatory Responsibility

Many ingredients used in the manufacture of cosmetics can be metabolized by microorganisms. This is true particularly in the case of water-containing cosmetics. Microorganisms may render the cosmetics unacceptable by causing visible degradation in the cosmetic or may pose

a potential health hazard to the user if certain microorganisms which
are pathogenic proliferate in these products. If not handled properly,
certain dry ingredients such as talcs or color additives may also contain objectionable microorganisms. These microorganisms are not
usually able to grow in dry cosmetics or dry ingredients, but their
viable presence may be considered a potential health problem given
the proper circumstances.

Cosmetics and toiletries are not, nor do they need to be, produced
from raw materials which are free of microorganisms. It can be expected that they may contain numbers of inocuous microorganisms.
A microbiological survey of Swedish nonsterile drugs showed that
203 out of 460 products tested were contaminated with microorganisms
in 1969. The microorganisms were identified as *Bacillus subtilis*,
Staphylococcus epidermidis, yeasts, and molds. Many products also
contained Gram-negative bacteria such as *Klebsiella, Pseudomonas*,
and some coliforms. A survey conducted in Canada in 1967—1971
found that approximately 5 percent of nonsterile drugs and cosmetics
were contaiminated to an extent great enough to warrant recall. A
1971 survey in the United Kingdom showed that 50 percent of the
items were essentially sterile and that approximately 90 percent contained fewer than 1,000 microorganisms per gram of product. Actually,
more than 80 percent of the 172 preparations contained fewer than
300 microorganisms per gram, and only 5 percent contained more than
10,000 per gram [1].

An extensive study conducted by the Cosmetic, Toiletry and Fragrance Association (CTFA), testing products that account for more
than 90 percent of the U.S. national sales volume, showed that 99.5
percent of the 3,967 samples tested were within the CTFA microbial
limit guidelines [1].

The U.S. Food and Drug Administration (FDA) considers its role
a legal responsibility to ensure that cosmetics entering interstate
commerce will not present health hazards to the consumer. The potential hazard criteria developed by the FDA include the pathogenic and
invasive properties of particular microorganisms, the intended use of
the product, and the actual numbers of microorganisms present. The
FDA also considers preservation of cosmetics and the potential health
hazards posed by a contaminated product to debilitated individuals.

The FDA has been charged by Congress with the responsibility of
enforcing the Food, Drug, and Cosmetic Act, as amended in October
1976. This act is intended to assure the consumer that cosmetic products shipped in interstate commerce are safe and made from appropriate ingredients. Safety includes the absence of injurious chemical
ingredients and microorganisms and their metabolities. The applicable
sections of the Food, Drug and Cosmetic Act are listed in Table 1 [2].

This Act, in effect, gives the FDA wide-ranging responsibility
which may be construed to encompass the manufacturing process as

TABLE 1   Food, Drug, and Cosmetic Act, As Amended October 1976
(Parts That Pertain to Microbial Contamination)

Section 601.   A Cosmetic shall be deemed to be adulterated

(a) If it bears or contains any poisonous or deleterious substance
which may render it injurious to users on the conditions of use
as are customary or usual . . . .

(b) It contains in whole or in part of any filthy, putrid, or decomposed substance.

(c) If it has been prepared, packed, or held under insanitary conditions where it may have become contaminated with filth, or
whereby it may have been rendered injurious to health.

---

well as the microbiological quality of the cosmetic on the retail
shelf.

The awareness of potential problems posed by the presence of
certain microorganisms in cosmetics and the necessity for their control during the manufacturing process attained major importance,
in the late 1960s and early 1970s. Even though some microbial problems had been reported in cosmetics in the past, the large variety of
cosmetics, ever-expanding sales, and the increased use of new cosmetic ingredients such as proteins, gums, and resins resulted in increased opportunities for spoilage and other associated microbiological problems [3].

The experience of the FDA and the cosmetic industry in the United
States during this time suggested that production standards in some
sections of the cosmetic industry needed improvement. There were
recalls of some obviously contaminated cosmetic preparations [4]. In
addition, instances of nonapparent contamination were discovered in
surveys of the microbial content of some cosmetics.

As a result of information developed by individual companies, the
CTFA and the FDA became acutely aware of potential problems in this
area and the industry began preparation of technical guidelines and
methods that addressed production, microbial content, and preservation
of cosmetics in a more definite fashion [5].

In this chapter, an attempt has been made to cover all major microbiological checkpoints of cosmetic manufacture. Many of these concepts have been obtained from the author's contacts and participation
in the CTFA as well as from his own in-house experiences. It is
hoped that cosmetic industrial microbiologists, particularly those who
have just entered the field or who work in areas where resources are
limited, will benefit from this compilation of information and experience.
It should be used as an aid in developing and evaluating microbiological control programs suited to particular needs.

B. Staffing

If a real commitment is to be made to maintain adequate microbiological control, it is important to have persons on the staff, or available as consultants, who are trained microbiologists and able to view operations from a microbiologist's aspect. Although persons trained in other professions may have an understanding of sanitation processes, etc., it is usually the microbiologist who must provide the proper interpretation, particularly in trouble-shooting operations. It is also usually the microbiologist who has an overall picture of the entire production process, because it is that person's responsibility to help in the development of all cleaning and sanitizing procedures; testing of raw and in-process materials, finished products, components, and process water; review of microbiological sampling plans for all these items, and setting up equipment monitors to verify that proper sanitization has been achieved.

II. THE PRODUCTION PLANT SCENE

A. General Housekeeping and Plant Design

Although only processing equipment usually comes in direct contact with cosmetic products, the importance of general housekeeping cannot be overemphasized. See the CTFA Guidelines on plant housekeeping and cleanliness [6]. Proper ventilation is important in controlling mold spores in the plant environment. Accumulation of dirt, debris, and spilled product may serve as a reservoir for molds and bacteria. It is a short trip for these agents from such an area to a compounding ramp or other sanitary area when carried by workers' shoes, mops, carts, or other portable equipment. Essentially, housekeeping should be considered part of an overall cleaning and sanitization program with a critical focus on the actual production equipment. A good sanitization program is one where housekeeping duties are clearly delineated and understood by those who are to carry out the program. It is important to have an active training program for housekeeping staff which includes monitoring the effectiveness of the entire plant housekeeping program. The housekeeping staff should be trained in sanitary approaches which include regular cleaning of the plant production areas as well as external equipment surfaces and overhead pipes and fixtures which may accumulate dust and dirt. Unused raw material and equipment or supplies should be removed from production areas. An adequate ventilation system should be included to prevent condensation on ceilings and overhead fixtures. In those areas where dust control has been determined to be a problem, an adequate program should be instituted. An effective waste disposal program should also be included. All staff toilets and washing facilities should be maintained in a sanitary manner, cleaned and checked regularly. Talc production areas should be segregated from

other production to prevent excessive airborne contamination being
carried in on the talc particles. Often this will be included in the
overall dust management program.

Since bacteria and molds can be carried through the air, particularly when facilitated by the presence of dust particles, all raw
materials which may be subjected to contamination by this source,
particularly during storage, should be protected by proper covering.
Components should preferably be stored in plastic bags rather than
in the cardboard containers in which they may be shipped. If filters
are used in air systems, they should be changed periodically. Any
sweeping or cleaning operations which occur during processing or
packaging operations hould be performed so as to minimize chances
that airborne contamination will affect the product in question. Vehicles which are used to transport material from one area, or to and
from compounding and filling areas, should be inspected and cleaned.
All raw material containers should be cleaned before they are brought
into compounding or filling areas. During any construction, care
should be taken to protect compounding and filling areas from contamination.

Part of the overall program of good housekeeping includes allowing
adequate space for various factory functions, for washing, changing,
and toilet facilities which may be located close to but separate from
the working areas, and for placement of equipment to facilitate ease
of carrying out good housekeeping functions.

Proper housekeeping is also facilitated by choosing the proper
finish for walls and floors. Preferably these surfaces should be smooth.
Floors should be nonslip and provided with adequate drainage. They
should be sloped to provide for easy washing and thorough run-off
to drains. In particular, floor areas under and around production
equipment ramps should be constructed so that they can be regularly
maintained and sanitized with suitable chemical disinfectants. A complete housekeeping program includes inspection and disinfection of
all floor and sink drains on a routine basis. These areas can serve
as loci of contamination which may be spread throughout the plant.

It can be easily comprehended that to implement an adequate housekeeping program, the talents of a trained microbiologist are essential,
whether the person is a permanent member of the plant staffing or
is obtained through a consulting agency. This policy, coupled with
an adequate training program and monitoring system, should result
in a comprehensive and effective houskeeping program which will
serve as the first line of defense against microbiological problems.

B. Cleaning, Sanitization, Validation, and Microbiological Monitoring of Equipment

All equipment used in the production of cosmetics should be subjected
to appropriate cleaning and sanitization processes in order to assure

that the manufactured product is not injured by heavy challenges of microorganisms. During the development program for cosmetics, it is often necessary to add preservatives in order to prevent microbial proliferation once the cosmetics are in the hands of the user. This is the real purpose of adding preservatives to a cosmetic. Preservatives are not intended to compensate for an inadequate cleaning and sanitization program. In fact, continued use of improperly cleaned and sanitized equipment can cause problems with the preservation of a cosmetic either during production or once it is in the hands of the consumer. Every preservative system has what might be called a finite capacity to destroy microorganisms. If this capacity is used up in correcting insufficient manufacturing procedures, the shelf or in-use life of the cosmetic may be very short indeed. In addition, improper cleaning and sanitization may lead to the development of strains of microorganisms which are very resistant to the preservative system and which will proliferate in these products during production and cause unsalable products to be produced. This is particularly true in instances where product is not entirely removed from equipment and where it may reside for even a short time in equipment where it has been partially diluted with water or cleaning solutions.

For theoretical purposes, cleaning and sanitization should be described as two separate operations; that is, it is normal practice first to clean equipment and then to subject it to a sanitizing procedure. In certain instances these may be synonomous. It is generally recognized, however, that most sanitizing procedures are effective if there is no residual product left inside the equipment and that all surfaces are free from product scale or build-up. This is particularly true where chemical sanitizers are employed.

Cleaning equipment prior to sanitization usually involves rinsing equipment to remove residual product with a surfactant. If necessary, equipment surfaces should be physically scrubbed to remove product. In all areas where it is not possible to scrub equipment surfaces, such as line interiors, these should be inspected regularly to determine if product buildup has occurred and at some point a determinaation must be made as to whether the buildup will interfere with the product integrity or cause problems with sanitization.

After cleaning procedures are performed, the equipment should be subjected to a suitable sanitizer. There are many sanitizers from which to choose. They generally fall into the following categories: (1) steam or hot (180°F) water; (2) alcohol; and (3) chemicals, such as quaternaries, iodophors, or hypochlorite compounds. There are other sanitizers available, but they are probably not in such wide use. They tend to be employed only in specific situations where the more popular sanitizers are not available. These include formaldehyde-type sanitizers and ozone. These are more difficult to deal with because of equipment incompatability or safety considerations in the production plant.

All equipment which comes in direct contact with products must be cleaned and sanitized. Pumps, filters, valves, tubing, lines, fillers, compounding and holding tanks, storage vessels, spatulas, and brushes used in cleaning must be processed and held under sanitary conditions which will not permit microbial contamination to occur. Often it is important to designate a specific period of time after cleaning and sanitization after which equipment must be resanitized even though it has not been exposed to product. Equipment used for producing various products must be cleaned and sanitized between product changes. Insofar as it is important to maintain batch integrity, the same may also apply to different batches of the same product. Equipment should preferably be constructed of suitable material, usually stainless steel with smooth and nonpitted surfaces. A grade of steel should be used which is able to withstand repeated cleaning and sanitization of the chosen agents without significant deterioration. The surfaces of equipment should be maintained in good condition so that product residue cannot become encrusted in splits and cracks. All valves and fittings should be sanitary in nature. Covered tanks and vessels are recommended.

All integrated equipment designs should be reviewed by the microbiologist when they are in the blueprint stage to avoid any serious shortcomings in design. The microbiologist should work carefully with the design engineers to chose equipment that is sanitizable and compatible with easy monitoring. All equipment and lines should be designed so that they will drain dry after cleaning and sanitization procedures are carried out. Extensively long lines through which product is pumped should be inclined so that they will gravity-drain. Any dead ends should be eliminated, and any connections which will act as reservoirs for accumulating stagnant product should be removed from the system or redesigned. Sanitary fittings should be used wherever possible, and any areas which are not designated as sanitary should be dismantled, inspected, and cleaned and sanitized thoroughly at every cleaning and sanitization sequence.

Validation of all cleaning and sanitization procedures is probably the most important facet of a successful program. It should be carefully determined by inspection that the procedures developed for a certain product in certain equipment do, in fact, accomplish the intended objective. That is, the cleaning procedures must effectively remove residual product from equipment, and if it is not possible to accomplish this *in situ*, such equipment must be broken down and processed manually to accomplish proper cleaning. Since different products may be produced using the same items of equipment, it is advisable to revalidate this process with each product and with any new products which may be made in the future on the same equipment. It may be practical to set down written procedures for cleaning and sanitization of all plant equipment. This document should detail the exact procedures as well as describe the various agents which are used to accomplish the process. A system should be set up for approving alterations in equipment. Any changes should be reviewed with the microbiologist as well as any other persons concerned with such changes. Although routine equipment disassembly may be performed during cleaning and sanitization operations, it is wise to institute a program which requires periodic special disassembly and inspection of equipment for wear and for product buildup over the long term. Most procedures are predicated upon observations made during start-up with a new product, and it is not possible to judge the long-term effect of that product on equipment without actual inspection after a large amount of product has passed through the equipment and numerous cleaning and sanitizations have been carried out.

Equipment monitoring is an ongoing part of the entire program. It is usually part of the function of the microbiology laboratory to perform routine equipment monitoring after cleaning and sanitization procedures have been carried out. Since actual microbiological readings are usually obtained after the equipment has been used in production, they must not be relied on totally to spot problems. Observations and visual inspection by operators must not be overlooked as a means of verification that the equipment has been cleaned and sanitized. Equipment cleaning and sanitization logs should be kept, in which it is recorded that the operators in charge have themselves inspected the equipment and found it to be at least visually suitable for use and that the prescribed procedures have been carried out.

Microbiological monitoring should be centered around actual sampling of all critical points on the equipment. This may entail breaking down equipment if necessary, as long as it does not compromise the sanitary condition of the equipment. It must be realized that most microbiological monitoring checks only the condition of the several designated points on the equipment at a particular time and cannot easily monitor the internal condition of a long pipeline or other complex systems such as a filler unless special methods are employed such as flow-through monitoring with a liquid carrier.

All accessories used in cleaning and sanitization, such as pots, brushes, spatulas, etc., should be also cleaned and sanitized in an area suited for that purpose. This area should be set aside specifically for that purpose, and the maintenance of the area should be included in the entire program. A proper place should be provided for storage of all utensils and brushes in these areas and on the ramps where they are used. Sink areas are particularly notorious for harboring problem microorganisms, and all sinks and drains should be routinely sanitized. Taps can also act as microbial contamination loci. Even city water remaining overnight in a spigot may become contaminated with microorganisms. All taps should be flushed with hot water before commencing cleaning and sanitization in those particular areas.

C. Raw and In-process Materials and Components and Finished Products

In addition to establishing an adequate microbiological program for sampling incoming raw materials, bulk in-process components, and finished goods for microbiological quality, it is necessary to determine the conditions of storage and susceptibility to contamination. All susceptible materials should be identified. An adequate monitoring program for these materials should be developed. It is suggested that the CTFA *Microbiological Sampling Guidelines for the Cosmetic Industry* [7] be consulted when developing a program. These Guidelines describe microbiological techniques for sampling raw materials, bulk in-process materials, packaging components, and finished goods based on years of industry experience. All materials, whether ingredients, packaging components, bulk in-process materials, or finished goods, differ in their susceptibility to microbial growth. For practical purposes the Guidelines categorize these materials into different susceptibility categories.

Each shipment of raw materials should be inspected for damage to the shipping containers as indicated by actual injury to the container or leakage of the stored materials. Containers which do not maintain their integrity can expose the contents to outside contamination. Tank car shipments should be inspected through the entry ports for contamination or damage. The presence of water condensate in tank cars containing waxy or oily materials may render the shipment unacceptable, since many microorganisms found to be a problem in cosmetics may proliferate under these conditons.

Sampling techniques must be performed with microbiologically susceptable materials. Specific sampling procedures are described in the CTFA Guidelines. The nature and properties of individual products will help to determine the necessary sampling. Since homogeneous samples require less sampling than heterogeneous materials, the extent to which the material can be considered homogeneous should be determined. A decision on the number of samples to be taken should be determined by either the supplier lot number or the quantities received for raw materials. The best sampling plan is one which can be related to batches of materials as separate units. Typical sampling plans can be found in the CTFA *Quality Assurance Sampling Guidelines* [8,9]. Usually 30 to 100 grams of sample are aseptically transferred from each container chosen by the selected sampling plan. It is also practical and feasible to composite certain samples of raw materials from the same batch or lot which have been determined by previous testing to be essentially homogeneous. If composite samples are found unacceptable, then individual containers of the materials should be retested. It may be necessary to perform more extensive testing in these instances. Compositing should be done only when absolutely necessary.

In addition to initial testing, stored materials should be retested to determine if they continue to adhere to microbiological specifications. This is particularly true with susceptible materials. Materials which have been opened or partially used should be retested if they have been stored between uses.

The following CTFA categories are based on the necessity and frequency determined by historical profiles of numerous batches of materials. Also, susceptibility may be determined for water-containing materials by performing a microbial challenge test.

Category 1: High Susceptibility. Aqueous or semiaqueous materials, emulsions, geriatric and pediatric preparations, cream lip preparations (water-based), all other water-based products, and raw materials of natural origin.

Category 2: Medium Susceptibility. Pressed powders, compact powders, blushing powders, etc., makeup sticks, loose powders, batch powders, dusting talcs, aerosols, eye powders (pressed and loose), and eye stick preparations.

Category 3: Low Susceptibility. Alcoholic preparations (more than 25 percent alcohol), deodorants and antiperspirants, bath salts, certain aerosols, and raw materials with antimicrobial activity.

Category 4: Nonsusceptible. Materials which by the nature of the components, not including preservatives, will not support the survival of vegetative microorganisms.

Properly taken samples are necessary for accurate microbiological test results. Devices such as sampling thieves can be used for liquid or powder preparations; sterile scoops are used for powders where it is not necessary to obtain a deep sample; sterile cups can be used for both liquids and powders. All sampling personnel should be thoroughly trained in aseptic sampling techniques by a qualified microbiologist familiar with these techniques.

Packaging components should be inspected upon receipt for dirt, insects, molds, etc. If any of these items are found unacceptable, a microbiological sampling should also be done. Any components which have been previously found to be a source of microbiological problems should be placed on a routine monitoring program. Suppliers should be notified if problems exist, and an attempt should be made to correct the situation.

All sampling should be done in a prescribed manner and accurate records dept. All materials should be held in quarantine until they have been released by the microbiology laboratory.

Bulk and in-process product should be sampled to determine the suitability for filling into the final container. This is particularly true if they are stored for long periods of time. As usual, a historical performance record will help to determine the need and frequency for sampling bulk and in-process materials. It is necessary to test preserved in-process materials such as adjusting color solutions which, incidentally, should be subjected to evaluation of preservation if they are to be stored and used at a future date.

Finished goods should be tested via a well-developed microbiological sampling program. Samples should be taken at the beginning, middle, and end of each shift, filling run, etc., whichever is appropriate. To determine the number of samples, the number of filling lines, container size, extended down time, and product susceptibility should be determined. Unopened samples should also be retained for future reference. Samples of finished goods should be taken as soon as possible after manufacture. With few exceptions, all susceptible products should be tested, and it is prudent to use the same sampling frequency for all these susceptible products. This procedure will help in analysis of microbiological problems arising from formula changes or problems in production. A classic example which emphasizes the importance of taking beginning, middle, and end samples from production runs is that if a filling device is implicated in sanitization failure, the initial samples will usually be highly contaminated while the samples at the end of the run will show much lower counts. If contamination has occurred in the compounding process, usually all samples will show the same levels of contamination.

Although a vast majority of hospital infections are from environmental and resident microflora, contaminated topical and oral preparations have occasionally been implicated in the medical literature as responsible for infection in the hospital environment. These types of products have been topical antiseptics, eye medicaments, saline solutions, and various creams or jellies which contain medicaments [1].

Specifications for filled cosmetic products should be determined by the character and intended use of the preparation. As far as the number of innocuous microorganisms which might be tolerated in a preparation, any set number is somewhat arbitrary. The numbers in any ethical drug product should be as low as the content of raw materials permits, using sanitary manufacturing practices [1].

Microbial limits for raw materials have not been set by the CTFA. Except for the USP test for microbial content of gelatin and certain other substances, there are no clearly defined limits. The USP limits are actually based on consumption of the gelatin for food purposes and not for its use in cosmetics. The CTFA has set microbial limit guidelines for finished cosmetics and toiletries [10]. The Guidelines set forth general content guidelines which finished cosmetics should not exceed. These are separated into four categories as follows:

1. Baby products: not more than 500 microorganisms per gram or milliliter
2. Products used about the eye: not more than 500 microorganisms per gram or milliliter
3. Oral products: not more than 1,000 microorganisms per gram or milliliter

4. All other products: not more than 1,000 microorganisms per gram or milliliter

The Guidelines also state that: "In addition to the microbial limits specified above, no product shall have a microbial content recognized as harmful to the user as determined by standard plate count procedures."

It is generally understood that the limits apply primarily to Gram-positive spore-forming microorganisms of a ubiquitous nature. The experience has been that the most commonly occurring microbial contaminants in cosmetics, particularly the aqueous and semiaqueous types, are Gram-negative bacteria, primarily of the genus *Pseudomonas*. A list of potentially pathogenic bacteria isolated from cosmetic preparations by the FDA laboratories was presented by Madden [2]. It includes a wide variety of Gram-negatives and is reproduced in Table 2 [2].

It may be seen from Table 2 that given the proper conditions, almost any microorganism can be a cosmetic contaminant. It is much simpler to decide on a responsible and effective preventive program than to repeatedly attempt to determine if the microorganisms one isolates from one's products are potential pathogens. The situation in a susceptible aqueous product is so dynamic from a microbiological viewpoint that little reliability can be given that contamination levels determined at one point in time, if the product is contaminated, will remain at the same level for any extended time. Bacterial numbers may change drastically, and certain microorganisms originally below detection limits may blossom forth to change the initial character of the contamination. It is generally accepted that vegatative bacteria can be controlled in most creams and lotions to insignificant plate count levels. This is not always true for dry products, and for this reason finished-product specifications should reflect this fact.

There are no standard methods prescribed for checking microbial content. Suggested sources are the CTFA *Microbiological Limit Guidelines for Cosmetics and Toiletries* [10], the FDA bacteriological analytical manual, *Microbiological Methods for Cosmetics* [11], and the USP *Microbial Limit Tests* [12], for microbial content of certain materials.

To recap, it should be mentioned that microbiological sampling and specifications must be based on susceptibility of the materials and their historical performance with no tolerance allowed for potential pathogens within the detection limits of the cited guidelines. Generally speaking, dry materials will not support microbial growth, but they may contain microorganisms in a static state. Materials from natural sources may contain high levels of various microorganisms, some of which may be objectionable. In these instances it may be necessary to have the material treated before it can be used in production. Aqueous raw materials, particularly certain surfactants,

TABLE 2 Potentially Pathogenic Bacteria Isolated from Cosmetic Preperations

Acinetobacter calcoaceticus	*Escherichia coli*	*Providencia rettgeri*
Citrobacter diversus	*Hafnia alvei*	*Providencia stuartii*
Citrobacter freundii	*Klebsiella oxytoca*	*Pseudomonas cepacia*
Clostridium sp.	*Klebsiella ozaenae*	*Pseudomonas fluorescens*
Enterobacter aerogenes	*Klebsiella pneumoniae*	*Pseudomonas putida*
Enterobacter agglomerans	*Morganella morganii*	*Serratia liqueufaciens*
Enterobacter clocae	*Proteus mirabilis*	*Staphylococcus aureus*
Enterobacter gergovae	*Proteus vulgaris*	*Staphylococcus epidermidis*
Enterobacter sakazakii		

must be preserved. This is usually done by the supplier and should be confirmed by the cosmetic manufacturer. In-process solutions which are stored and used for batch adjustment purposes also should be preserved or prepared in a solution of a hostile agent such as alcohol or propylene glycol. Components such as glass or plastic bottles and tubes usually have very low microbial counts. It is most important to keep these from becoming contaminated during storage and handling in order to maintain this quality. Cardboard and paper items often contain mold spores, which may cause them to mildew if not held under sufficiently dry conditions.

All raw materials should be evaluated during the course of product development to determine the acceptability for use in cosmetics and what problems might be encountered if they are used. Preferably several batches of the same raw material will be available for evaluation in order to determine if the microbiological profile is consistent from batch to batch. Microbiological specifications should be negotiated with the supplier to ensure a consistent quality of raw material. Raw materials should be used on a first-in, first-out basis; this applies particularly to susceptible materials. Susceptible materials in particular should be quarantined in order to prevent cross-contamination in the plant. Attention should be given to pest and dust control in warehouse storage areas.

D. Process Water

Many factors influence the quality of water used to manufacture cosmetics. Potable water may be generally suitable for food production purposes, but the levels and type of microorganisms which may be found in potable water are not necessarily suitable for cosmetic manufacture. They may contaminate the equipment and the entire manufacturing system. In particular, *Pseudomonas* and other Gram-negative bacteria which are commonly found in potable water, taps and spigots, and sink drains have served as the bane of the cosmetic industry. These microorganisms are extremely adaptable to changing environmental conditions and are able to adapt to grow in many types of water-containing cosmetics. Many of these microorganisms may be of public health significance. The microorganisms found in cold process water present a severe problem in cosmetic manufacture, particularly in cold-mix products, if steps are not taken to remove them prior to using the water. The other primary source for these microorganisms is filtered, deionized water. All water which contains organic and/or inorganic material may be undesirable for cosmetic manufacture, because these materials can supply nutrients to sustain the growth of microorganisms and may have an adverse effect on the physical properties of the product itself. These materials can cause other problems, such as cloud formation, or impart an objectionable odor to the water. Water with a high mineral content may cause scale to form on equipment and can also inactivate certain disinfectants used to sanitize equipment. For these reasons process water should be purified and tested microbiologically to verify its quality.

The microbiological methods for testing water are numerous. The best method is one which gives an accurate answer in a short time. In most instances, standard plate counts are not available until after the water has been used. It is therefore necessary to develop a good critical process control program for water sanitization of resin beds on a routine basis.

Any process water system should be designed to minimize the occurrence of elbows and traps or dead ends in the lines. These stagnant areas allow for microbial proliferation and are major sources of microorganisms in any water system. The system should be examined regularly; this should be the responsibility of a competent individual who has familiarity with the equipment and the system. Any new systems or changes in process water lines should be reviewed with a microbiologist to determine if the changes are within sanitary guidelines.

Water purification may be achieved by placing filters and/or ultraviolet lamps in lines. There may still be a problem unless they are installed at point-of-use locations throughout the plant. Chemicals such as chlorine or ozone have been suggested, but their use is occasionally offset by considerations of equipment deterioration which may develop. Hot recirculating water systems are very effective. Hot water may be used as an agent of sanitization for many types of equipment. It may be used to flush out water meters and heat exchangers which produce cold process water. This type of system is, of course, very expensive to operate, but its simplicity is an important factor to consider.

Although the CTFA has no microbial limit guidelines for process water, it is necessary for each manufacturer to determine the limits suited to his or her products and conditions. The level should be the lowest practicable for a given situation.

Since process water may pose so many problems to the cosmetic manufacturing plant, the areas listed in Table 3 are suggested for monitoring a demineralized water system. They are taken from the CTFA *Microbiological Guidelines for Process Water* [13].

If a charcoal bed is included in the system, samples must be taken there also. Samples should also be taken after the water has passed through any ultraviolet irradiation device. If a filtration unit is used, samples should be taken after water has passed through the unit, if practical. It is also possible to perform a filter integrity test *in situ*. These procedures are usually available from the manufacturers of filter units.

Besides routine monitoring of the water systems, trouble-shooting necessitates extraordinary inspection procedures. If heat exchangers are used, they should be put through appropriate checks to determine if any leakage from the cooling water system has occurred. If a

TABLE 3   Suggested Sampling Points for Demineralized Water Systems

---

1. The incoming water feeding the unit

2. The water emanating from the ion exchange resin beds

3. Each station downstream from the demineralizing unit (point of use)

4. Any other point in the system where information on the microbio-
    logical status is deemed necessary

---

buildup of Gram-positive microorganisms is noted, the system should
be examined thoroughly for dead ends and the incoming water monitored to ascertain if it is source of entry of these bacteria.

All modern ultraviolet systems are equipped with devices which
measure output of these lamps at the germicidal wavelengths; they
should be equipped with alarm devices should the output of the lamps
fall below the required levels.

III. MICROBIOLOGICAL SUPPORT IN THE PRODUCT DEVELOPMENT STAGE

The focus of microbiological support to production centers around the
testing of new raw materials and cosmetic formulas for microbial content and preservation. Preservation of new developmental formulas
is the primary aspect of any support program. The main objectives
of preservation are (1) to control microbial content of cosmetic products in order to prevent proliferation of microorganisms that could
adversely affect the health of cosmetic users, and (2) to prevent
deterioration of the product on the shelf or in the hands of the consumer [14].

The support given in the form of microbiological approval and
qualification of raw materials has already been covered. The cosmetic
industry uses many raw materials that contain indigenous microorganisms at significant levels. Raw materials such as certain color additives, gums, proteins, talcs, and other ingredients from natural
sources may contain microorganisms indigenous to these sources or
introduced during their manufacturing or mining processes. In addition, many cosmetic products can support growth of microorganisms
because they are essentially biodegradable. As a result, preservatives must often be added to prevent spoilage. Such materials as
proteins, resins, and ethoxylated and propoxylated emulsifiers may
render certain preservatives ineffective. Preservation of developmental formulas is another matter and will be reviewed further.

Although there are many publications dealing with the preservation of cosmetics, there are as yet no standard methods or set groups
of microorganisms which are required to evaluate any given formula

TABLE 4 CTFA Guidelines for Preservation of Aqueous Liquid and Semi-Liquid Eye Cosmetics

Challenge levels:	1×10^6 bacteria per challenge
	1×10^4 yeast and fungi per challenge
Bacteria:	99.5% reduction for remainder of the test
Yeast and fungi:	90% reduction in 7 days—continued reduction for the remainder of the test
Bacterial spores:	Static activity

Source: Ref. 12.

or set of formulas. Most preservation tests involve several challenges to the test formulas at a level of $\log_{10}$ 6.0 for bacteria and $\log_{10}$ 4.0 for molds and yeasts. The microorganisms are tested for at various time intervals following challenge.

Although suggested reductions are described in the CTFA *Guidelines for Preservation of Aqueous Liquid and Semi-Liquid Eye Cosmetics* [15], these criteria are considered the minimal qualifying activity for any cosmetic falling into this water-containing category (see Table 4).

Other suggested criteria may be found in the American Society for Testing and Materials (ASTM) ANSI/ASTM E-640-78 [16].

These Guidelines reflect activity which is minimal and therefore may not be entirely suitable for certain cosmetics which are packaged in containers to which the consumer has direct and frequent access. Many schools of thought exist on this subject, and it is possible to encounter almost any opinion. It is for this reason that any microbiological program should take several things into account:

1. ATCC microorganisms may be used if they are shown to be resistant to preservative systems. Nonresistant cultures are of very limited value in predicting actual performance of any particular preservative system in usual production/field situations.

2. The microorganisms used in these tests should be those found to be resistant to other in-house cosmetics. These cultures should be held and maintained in a condition which perpetuates their resistence. Accurate records should be kept of field experience for all susceptible cosmetics. This includes an accurate assessment of all microbiological problems which arise in the field, including those verified consumer complaints of a microbiological nature.

3. Combinations of preservatives are often better than single preservatives because they can be tailored to make up for deficiencies that single preservatives may possess.
4. Any stability program should include evaluation of the preservative system after the product has been in the package for a period of time. This is particularly important in the case of certain plastics or other components to which the product will be intimately exposed. If long-term stability decisions are based on elevated-temperature storage, it may be advisable to test actual retained production samples after the product is on the market to verify earlier predictions.
5. In-house panels should be performed on cosmetics which have passed initial stages of preservation or which are already on the market to verify that the laboratory tests are predictive of field experience. Preservation challenge testing should include some sort of stability evaluation of the activity of all preservative systems with aging and at elevated temperatures.
6. These systems should be followed during scaleup and verified after the product is on the market with either pickups from the retail shelf or tests of line retains which are several years old.

IV. SUBCONTRACTORS

Any cosmetic made or filled by a subcontractor should meet the same requirements as an in-house cosmetic. If the cosmetic is purchased from the supplier and formulated by the supplier, assurances must be had that the cosmetic meets reasonable microbiological criteria. The manufacturing practices of the subcontractor should be reviewed to determine if the subcontractor is meeting industry guidelines regarding raw material testing and whether or not the subcontractor has developed adequate cleaning and sanitization procedures for the product.

If a product is manufactured in bulk and then shipped to the supplier for filling, it is recommended that the microbiological quality of the bulk be determined before shipment in order to avert any unnecessary cost and inconvenience which would be incurred by shipping unfillable product.

The microbiological quality of the filled products should be verified by mutually agreeable methods and sampling plans, and materials should not be shipped until completion of microbiological testing.

V. CONCLUSION

In summary, a comprehensive microbiological program for cosmetic production requires careful integration of microbiological product development and quality assurance. Emphasis must be placed on

keeping accurate records in order to track down and rectify microbiological problems. Because the entire subject of cosmetic microbiology involves control of living organisms which are usually not visible
to the naked eye, ubiquitous in the environment and thrive in raw
materials and moist or damp places such as equipment and storage
areas, a good microbiological program must involve all aspects described in this chapter. Failure to track and control any one or a
combination of these will likely result in loss of time and money. Review and approval of all procedures should be performed by persons
who are thoroughly familiar and with formal training in microbiological concepts and techniques. The contents of this chapter should
serve as a useful guide to achieving these goals.

VI. ACKNOWLEDGMENTS

The author would like to express his thanks to the Research and Development Division of Chesebrough-Pond's, Inc., for the many years
of continued support of activity in the CTFA, ASTM, and other groups
which permitted development of a comprehensive understanding of
cosmetic microbiology.

In particular, the author also wishes to recognize the invaluable
professional guidance derived through the CTFA Microbiology Committee and its members over many years in addressing the microbiological
aspects of cosmetic microbiology.

The opinions expressed in this chapter are those of the author,
not necessarily those of Chesebrough Pond's, Inc.

REFERENCES

1. Tennenbaum, S., Considerations Leading to the Development of
   the Microbial Limit Guidelines of the CTFA, *Cosmet. Toilet. 92*
   (March):79—83 (1977).
2. Madden, J., Cosmetic Microbiology: Viewpoint of the FDA, in
   Developments in Industrial Microbiology, vol. 21, Society of Industrial Microbiologists, Pittsburg, Pa., 1980, pp. 149—156.
3. Boehm, E. E., and D. W. Maddox, Problems of Cosmetic Preservation, *Manufacturing Chem. Aerosol News 42* (April):41—43
   (1971).
4. Parker, M. S., Some Factors in the Hygienic Manufacture of Cosmetics and Their Preservation, *Soap, Perfume, Cosmetics 43*:483—
   485 (1970).
5. Smith, J. L., Preserving Cosmetics, *Cosmetics and Toiletries 96*
   (March).39—41 (1981).
6. C. Haynes and N. Estrin, ed., *Plant Housekeeping and Cleanliness*, 2nd ed., CTFA Technical Guideline, The Cosmetic, Toiletry
   and Fragrance Association, Washington, D.C., 1982.

7. C. Haynes and N. Estrin, eds., *Microbiological Sampling Guide-
    lines for the Cosmetic Industry*, 2nd ed., CTFA Technical Guide-
    line, The Cosmetic, Toiletry and Fragrance Association, Washington,
    D.C., 1982.

8. C. Haynes and N. Estrin, eds., *Microbiological Quality Assurance
    Guidelines for the Cosmetic Industry*, 2nd ed., CTFA Technical
    Guidelines, The Cosmetic, Toiletry and Fragrance Association,
    Washington, D.C., 1982.

9. C. Haynes and N. Estrin, eds., *Quality Assurance Sampling Guide-
    line Part I and II*, 2nd ed., CTFA Technical Guideline, The Cos-
    metic, Toiletry and Fragrance Association, Washington, D.C.,
    1982.

10. C. Haynes and N. Estrin, eds., *Microbiological Limit Guidelines
    for Cosmetics and Toiletries*, 2nd ed., CTFA Technical Guide-
    line, The Cosmetic, Toiletry and Fragrance Association, Washington,
    D.C., 1982.

11. Food and Drug Administration, Bureau of Foods, Division of Micro-
    biology, *Microbiological Methods for Cosmetics*, FDA, Washington,
    D.C., July 1979, chap. XXIII.

12. USP XX NF XV, *Microbial Limit Tests*, 874-879 (1980).

13. C. Haynes and N. Estrin, eds., *Microbiological Guidelines for
    Process Water*, 2nd ed., CTFA Technical Guideline, The Cosmetic,
    Toiletry and Fragrance Association, Washington, D.C., 1982

14. Smith, J. L, Product Testing for Preservative Efficacy, *Cosmet.
    Toilet. 92* (August):30—34 (1977).

15. C. Haynes and N. Estrin, eds., *A Guideline for Preservation
    Testing of Aqueous Liquid and Semi-Liquid Eye Cosmetics*, 2nd
    ed., CTFA Technical Guideline, The Cosmetic, Toiletry and Fra-
    grance Association, Washington, D.C., 1982.

16. American Society for Testing and Materials, *Standard Test Method
    for Preservatives in Water-Containing Cosmetics*, ANSI/ASTM E-
    640-78, May 1978.

22

ESTABLISHING A TRAINING PROGRAM FOR PLANT PERSONNEL

JOHN L. BECK*

Technical Services, The J. B. Williams Company, Inc., Cranford, New Jersey

JOSEPH E. KILSHEIMER

Max Factor and Company, Hollywood, California

I. INTRODUCTION

Industry of any type relies heavily on the skills and creativity of human beings for its success. This is equally true for individual com-

---

Present affiliation: Quality Affairs Department, Berlex Laboratories, Inc., Wayne, New Jersey

panies within any given industry, and it is particularly true for highly
technical, highly regulated companies. This is, of course, a description which applies to the cosmetic industry and to the companies within it.

The development of human skills and the enhancement and nur-
turing of the creativity available in the workforce are among the primary objectives of training programs. Training programs modify
the behavior of people so as to accomplish better the objectives of
the company. This, the enhancement of a company's abilities to meet
its objectives, is the overall goal of any good training program. Regardless of the type of industry, size of company, or specific phase
of training, the *prime* consideration must be the company's overall
objectives or the training program is poorly conceived. Training,
therefore, can be defined as a *tool* which can help an organization to
meet its objectives. Differences in training programs arise from the
following factors:

1. The type of industry
2. The size and character of the company
3. The particular sets of company objectives chosen as the subject matter
4. The format chosen to present the subject matter

For example, training programs conducted by the automobile industry would undoubtedly have different requirements, and very likely
different formats, than those conducted by the cosmetic industry.
On the one hand, the automobile industry is a heavy industry, highly
concerned with critical machining and tooling tolerances, and although
heavily semiautomated, also very labor-intensive, with relatively un-
skilled workers. On the other hand, the cosmetic industry deals
with many raw materials in their natural or synthetic bulk chemical
state, is concerned with compounding procedures, cleanliness and
microbial contamination, and packaging. Except for packaging operations, the cosmetic industry utilizes relatively skilled workers and is
not very labor-intensive. One would not expect the training programs of these two industries to be very similar. Second, training
programs conducted by a large company with a dedicated training
staff would certainly differ from those conducted by a smaller company without such a staff, even though both companies might be in
the same industry. Also, the training program of a company with a
broad and diverse product line might differ from the training program
of a company with a very narrow product line because their requirements differ. Third, the training program and its format are dependent on the specific objectives. For example, one set of objectives
might be concerned with improving productivity in the manufacturing
and/or packaging areas. Another might be to improve operating
techniques and safety in the warehouse and shipping areas. Yet another might involve reenforcing good manufacturing practices (GMPs)

in specific areas of the company. The cosmetic industry has proposed, through the Cosmetic, Toiletry and Fragrance Association (CTFA), a set of voluntary guidelines for GMPs in the industry [1]. We are also governed, insofar as we manufacture so-called cosmetic drugs, by the drug GMPs [2], which are not voluntary but have the force of law. These require specific training, particularly with regard to procedures and documentation. In fact, one of the requirements of the drug GMPs is that each company maintain an ongoing training program for its particular operations and that this program include training in the GMPs. Finally, training programs which are designed for new employees may differ significantly from those designed for older employees or from those designed to reenforce earlier training.

II. TRAINING CATEGORIES

Training can generally be broken down into three broad categories:

1. Orientation training
2. Ongoing training
3. Qualification training

A. Orientation

Orientation training is that training which is given to new employees. Such training may cover any introduction to the industry, the company's product lines, its quality policies, the regulations imposed on it by federal and state agencies, and similar topics to which all employees should be exposed. This phase may also teach individual job skills required in various areas of the company, such as manufacturing, packaging, shipping, and quality control. Normally, this training is best accomplished in a classroom atmosphere. Audiovisuals can be very effective for this training phase, especially if they are tailored to reflect the needs of the particular company.

All companies should have some type of orientation training. Larger companies may have formal orientation training, whereas smaller companies may have very informal training, perhaps simply having supervisors indoctrinate new employees. However, if new employees are simply shown where to work and what to do and are left without an explanation of the company's policies, its goals and expectations, and the operating procedures that apply to the particular work area of the employee, the best interests of the company and of the employee have not been served.

B. Ongoing

Ongoing training is that which is given to reenforce skills acquired, refresh training previously given, and update training to reflect current regulatory requirements, technological advance, product line

expansion, etc. Normally this category of training is fairly job-specific.
As opposed to orientation training, ongoing training is frequently,
but not always, conducted on the job (OTJ training).

Normally ongoing training does not require large blocks of time,
except in those cases where product line extensions may carry a company into an entirely new area, for example, when a contract packaging company decides to venture into contract manufacturing as well.
In these situations extensive training may be necessary, but this is,
in actuality, a combination or orientation, ongoing, and qualification
training.

C. Qualification

Qualification training, sometimes termed validation training, is that
which is given to ensure that employees have obtained the knowledge
and skills to perform specific jobs in accordance with the goals set
for those jobs. These, of course, should comply with departmental,
company, and regulatory agency requirements. This type of training
is often given to older employees who may not have received earlier
formal training of any kind. It can also be given when a company is
expanding into new areas which may have requirements differing from
those for which employees have previously been trained. Such a
need might be created when, for example, a company specializing in
in perfumes and colognes decides to add skin-care emulsion products
to its line.

III. ESTABLISHING A TRAINING PROGRAM

Having covered some of the more general and philosophical aspects of
training, we will now go on to discuss the steps necessary for establishing an effective training program. This section will stress those
criteria which are important and must be considered whether we are
dealing with an entire program or simply a phase of any of the categories, that is, orientation, ongoing, or qualification training. These
are as follows:

1. Analyze the needs.
   A. Determine goals.
   B. Determine performance necessary to attain goals.
   C. Evaluate current performance.
2. Plan the program.
   A. Evaluate training resources.
   B. Design a training program which will change behavior so
      as to result in the level of performance desired.
3. Execute the program.
   A. Schedule training.
   B. Implement training program.

4. Assess the results.
   A. Test to evaluate effectiveness of the training.
   B. Audit to evaluate behavioral changes desired and results.
   C. Determine if the primary goals desired have been achieved.
5. Follow up.
   A. Exit if goals have been achieved.
   B. Reevaluate the training program and retrain if goals have not been achieved.

A. Analysis

Analysis of the situation requires a knowledge of the company's objectives, its quality policies, the regulatory requirements, and any job-related problems before one can determine the goals of the training program. Only when there is a thorough understanding of these factors can we proceed to evaluate the changes in behavior which the training should produce to accomplish these goals. In addition to the general parameters mentioned above, the analysis should include an evaluation of the current level of knowledge of the employees and any particular obstacles to reaching the desired level of behavioral change. In this regard, one must consider such factors as management support, educational level of employees, language barriers, etc.

B. Planning

Planning the program involves an evaluation of your training resources. Will you establish a professional training staff, or will the training be done by supervisors? If the latter, who will train the supervisors? Will you do it yourself, or will you employ an outside agency? Do you have an appropriate "classroom" situation—quiet, large enough for the groups you would like to train, etc.? Are there audiovisual materials available to you? If so, these can help to alleviate much of the training burden from your staff or the supervisory personnel charged with this responsibility. It helps if they are customized to your company, and this can be done by modifying existing programs available from the regulatory agencies, the CTFA, or private companies specializing in this field. Once prepared, they need only be updated as technology, regulatory requirements, or the company's changing product lines require. However, do not fall into the trap of presenting audiovisuals in a "sterile" manner but, rather, augment them with interaction between the trainer and the personnel being trained. Many different kinds of audiovisuals can be employed, such as videotape, motion pictures, and slide-tape presentations. Although these can be excellent training aids, do not despair if they are not available to you. Good programs can be conducted around "classroom" training, in an OJT situation, or a combination using visual aids such as overheads, blackboards, flip charts, etc. OJT has certain advantages in that actual job conditions and equipment are

used. If the company is not overburdened and has good supervisory
personnel, this type of training can be very effective. However, if
the company is very fast-paced, one may discover that this training
tends to be superficial. It also places a heavy burden on first-line
supervisors, since a professional training staff is not often used for
this type of training. When these options have been evaluated, one
can plan the program, whether it be the company's overall training
program or a program designed for one specific area.

In order to be effective the program must have management approval and should, if possible, involve management and the operational
supervisory staff. This will ensure that the program will not fail for
lack of monetary support, time spent training, or goals which are not
in line with management's objectives.

C. Execution

Now it is time to execute your training program. Prepare the training
materials or arrange for an outside professional training organization
to do the job, schedule time off the job for training, and implement
the program. Time off work must be made available for the program
to succeed. Normally, initial training, orientation, or qualification
will require several blocks of time, whereas ongoing training may require only one block on a relatively infrequent schedule.

D. Assessment

If the program is concise, you may want to wait until it is finished before beginning an assessment of its effectiveness. If it is long and/or
complicated, you may want to evaluate progress at various discreet
stages of the program. This can be done by testing, either singly or
in groups. Group participation in testing has the advantage of re-
moving stress from the "exam" and can be a useful tool to enlist the
cooperation of a union in the acceptance of tests. It may also help to
reenforce the training program, "on line" so to speak, since the discussions of the group usually tend to have a positive effect on the
weaker members. In the final analysis, ongoing audits will tell you
if the program has been successful. Remember, success is not measured by whether the participants enjoyed the course or not, although
the learning process is greatly facilitated if the experience is enjoy-
able. It is not measured by the amount of money or energy spent,
nor is it measured by an increase in the level of activity. The measures of success are the results that have been achieved on meeting
the goals and on how long these results last.

E. Follow-up

Based on the assessments a decision should be made to exit the program and schedule long-range refresher training, to retrain, or to

reevaluate the training program. In any case, the ongoing audit plays a part here, in the follow-up stage, as well as in the assessment stage. Normally, follow-up or refresher training need not be conducted more frequently than once a year. Exceptions to this occur when there is a major change, such as a new product, a new plant, new equipment, or new regulations.

In summary, then , the five critical steps to establishing a training program are: analyze, plan, execute, assess and follow-up.

The next section of this chapter will deal with more specific examples of training in cosmetic industry. Before we proceed, let me remind you that training is extremely important and can be an effective tool. It can also be an ineffective tool if not applied properly and, in the worst case, may have a negative effect. You may believe that any kind of training, no matter how poorly thought out, is better than none. Such is not the case. *Training can hurt* if is is not in accord with the objectives of the company.

IV. TRAINING PROGRAM GUIDELINES

A. General

Up to this point we have reviewed the philosophical aspect of the different types of training. As one gets into the more specific aspects, one should keep in mind that we conduct training at either a conscious or subconscious level. We are familiar with some of the activities that we use to indoctrinate new employees, such as where to punch their timecards, where the restroom facilities are located, and the hours of work, at the very minimum. Most successful companies carry this further to include a variety of skills to help the new employee and the company even if the new employee has been previously employed in a similar position, or in the general cosmetic industry in another type of position.

The CTFA has prepared a number of guidelines that cover the area of general practices that are often useful in preparation in a training situation for newly hired or transferred employees. The following discussion centers about several more specific types of training course outlines that could serve as models for use by the reader.

B. Regulatory Training Guidelines

Many cosmetic companies now find themselves regulated by the drug GMPs as well as the cosmetic requirements of the federal Food, Drug, and Cosmetic act (FDC Act) because they are producing one or more of the so-called cosmetic drugs. Two major classes of products falling in this category are the antiperspirants and sunscreens. For regulatory purposes the definitions of cosmetics and drugs are found in the FDC Act [3], and the distinction between these, including the posi-

tioning of the "cosmetic drugs", is discussed elsewhere in this volume.
Suffice it to say that any training which covers regulatory areas
should consider which of these categories apply to the operation in
question and should use the drug GMPs and/or the CTFA *Technical
Guidelines* [4] or Citizen Petition [1] as primary source materials,
along with other technical matter to be presented.

Because the format, and indeed the basic content of the Citizen
Petition for cosmetic GMPs and the drug GMPs is similar, the following
example of regulatory material to be included in a training program
is modeled after the CTFA's Citizen Petition.

This training must cover the following areas in such a manner as
to ensure that the cosmetics have not been prepared, packed, or
held under unsanitary conditions whereby they may have become contaminated with filth, or whereby they may have been rendered injurious to health [5].

1. Plants and grounds
   A. Must be orderly and clean.
   B. Should have adequate lighting, ventilation, and screening
      to prevent macro and micro (biological) contamination or
      cross-contamination.
   C. Should prevent contamination by separating the various
      functional areas either physically or via some other effec-
      tive method.
2. Equipment and utensils
   Shall be appropriate, clean, orderly, adjustable.
3. Personnel sanitation facilities
   A. Shall be adequate, available, and furnished with appropriate
      facilities.
   B. Signs should be posted for hand washing.
4. Equipment and utensil cleaning and sanitization
   A. Should be cleaned appropriately, at the proper intervals,
      using materials which are stored so as to prevent product
      contamination.
5. Personnel
   A. Supervisory personnel should be of sufficient number and
      have adequate education, *training*, or experience for their
      assigned duties.
      (1) Operating personnel should be trained in good hygiene
          and use of the protective garments, where applicable.
      (2) Personnel must be aware of health problems and ex-
          cluded from product contact, if necessary.
      (3) Personal belongings, smoking, and eating are not to
          be permitted except in prescribed areas.
6. Raw materials
   A. Raw materials should be stored in a clean and orderly man-
      ner.
   B. Raw material identification

(1) Cosmetic materials shall be tested for conformance with specifications before use or the products shall not be marketed until they are tested.

(2) Raw materials may be released without testing if the finished products are tested, a certificate of analysis is supplied, or the supplier's history is satisfactory.

(3) Control of raw materials should include examination for labeling, broken containers or contamination, appropriate sampling, identifiable coding, proper storage and rotation, and reject marking, where necessary.

    (a) Testing specifications, including a description of the methods and proper authorization, should be available.

    (b) Appropriate records, including the name of the supplier, lot numbers, date and amount received, tests performed, and authorized records of disposition, should be maintained.

7. Packaging materials

Packaging material identification.

(1) Cosmetic packaging materials shall be tested for conformance with specifications before use or the products shall not be marketed until they are tested.

(2) Packaging materials may be released without testing if the finished products are tested, a certificate of analysis is supplied, or the supplier's history is satisfactory.

(3) Control of packaging materials should include appropriate handling and storage to prevent mixups and contamination, appropriate marking, and appropriate rotation. Records covering the name of the supplier, date and amount received, tests performed, and authorized records of disposition should be used.

8. Manufacturing record systems

Manufacturing record systems shall include the product formula, manufacturing directions, product specifications and test procedures, and appropriate batch production records, including raw material and packaging records.

(1) The product formula should include the name, code identification, authorizing signature and date, a list including the weights or percents of raw materials, a statement of total weight, measure or percent, and provisions for adjustments, where appropriate.

(2) Manufacturing procedures shall include an authorizing signature and date, product identification, and all significant aspects of the process.

(3) Product specifications and test procedures shall include an authorizing signature and date, product

              identification, and appropriate physical, chemical,
and microbiological test procedures and limits.

(4)   Production records

     (a)   Batch records shall include the formula, product
identification, coding, date, a record of the raw
materials used and their weights, and in-process
checks and endorsements.

     (b)   Processing control records shall include product
identification, coding, results of testing and
endorsements, and disposition of the lot.

     (c)   Packaging records shall include product identi-
fication, dates, coding. and quantity produced.

     (d)   Raw material and packaging material records
shall include records outlined in 5.C and 5.D,
respectively.

9.   Production and control procedures

Production and control procedures shall be sufficient to
ensure that the products meet their specifications. They
should include requirements for qualified personnel, auditing of equipment, appropriate marking and handling
to prevent mixups, withholding from market until released,
appropriate handling of rejected and returned goods, and
collection and retention of samples.

10.   Laboratory controls

Laboratory controls shall include the scientific establishment of raw material and finished product specifications
and testing procedures, as well as provisions for establishing and retaining appropriate records for each lot of
finished product.

11.   Records

A.   Records of examinations of raw materials, packaging materials, and finished products shall be maintained.

B.   Processing and production records shall be maintained.

C.   Distribution records shall be maintained.

D.   All records in A, B, and C shall be maintained for at least
two years.

Examples of other training guidelines for general orientation, specific skills, and higher levels where multidiscipline skills are required
are shown in the following sections. These are presented as only one
of several possible choices. Depending on the size and complexity
of your company, you may choose to employ different formats and
trainers.

C. General Orientation Guidelines

1.   New employee—any position

A.   Company's goals, expectations, policies

B. Regulatory aspects (see regulatory training guideline for specific area of work)
C. Phase of work—specific location
D. Hours of Work
E. Supervision
F. Specific task or tasks expected
G. Rate of pay
H. Performance review period
I. Opportunities for advancement
J. Location of employee services—restrooms, dining facilities, etc.

D. Specific Skills

Example: Lipstick Packager Training Program

Phases I and II—Lipstick Compounding and Molding: Estimated time for this phase is 20 to 30 minutes. The size of the training group should range from four to eight people, and the trainer may be a quality control or production manager/supervisor or designate.

1. Phase I—compounding
   A. Introduction
      (1) Brief description of incoming raw material control
      (2) Brief description of lipstick formulation (general)
   B. Central weighing
      (1) How formulas are received in production area
      (2) How batch numbers are assigned
      (3) How raw materials are weighed and checked per formula (example)
      (4) How weighed raw materials are palletized by batch and forwarded to lipstick compounding room
   C. Compounding
      (1) Methods used in bulk product compounding per formula and manufacturing procedures (use example) (note problems in procedure deviation)
      (2) Methods used in shade matching (including brief description of color blends)
      (3) What is involved for a batch to receive quality control approval, and expiration dating of bulk approvals
      (4) Straining and storage of mass between manufacturing and molding
      (5) Pour pot charging procedures
2. Phase II—molding
   A. How bulk product is maintained in pour pot (note temperature importance)
   B. Pouring techniques required for good sticks and factors affecting yield (quality and economics)

C. Problems arising from variations (and combinations thereof
   in molding techniques)

Phase III—lipstick inspection standards: Estimated time for this
phase is 30 to 45 minutes. The Phase III session should be done off-
line in a meeting room or laboratory. A table, chairs, and a display
board will be required.

1. Explanation of lipstick configurations (shapes) for brands of
   lipstick manufactured. Show drawings or photos of each,
   pointing out differences.
2. Show diagram or photo of a lipstick and point out the three
   areas of inspection and the relative concern in each area.
3. Samples of each of the following defects, graded and marked
   with numbers 1 through 4, are passed around. Those marked
   1 and 2 are good, 3 and 4 are rejectable ("NGs").
   A. Pinholes
   B. Bad face
   C. Mold marks
   D. Debris
   E. Overflame
   F. Miscellaneous (cuts, scrapes, etc.)
4. After each trainee has examined each of the numbered and
   graded samples, a group of 20 test samples (numbered 1 to
   20) are passed around for each to see and record as good or
   bad. The trainees are then informed of the correct results
   and the operation is repeated for one more cycle.
5. There will then be a question and answer period, at which
   time the trainees may seek further information.
6. Each member in attendance will then sign a dated attendance
   sheet indicating that he or she has received this training.

Note: These sessions should be repeated quarterly, or more often
if specifications change.

Any number of specific skill training sessions can be assembled
with this type of outline.

When one first considers the need to formulate specific skill training,
one should create a flowchart and include all areas of knowledge that
a person needs to perform the job. Obviously, for a lipstick packager
you do not have to delve deeply into the various waxes, oils, and
pigments, other than to describe generally the information already
available on an ingredient label. When training a lipstick compounder,
one need not spend a great deal of time on pouring techniques that
affect stick strength and/or appearance. However, both groups will
work more efficiently with a nodding acquaintance of the overall
process. One firm studied demonstrated up to a 20 percent increase in
output by training crews with similar types of training. Two or
three hours per year invested over 2,000 hours worked seems a cheap
investment for that kind of payback.

E. Multidiscipline Skills

At a higher level, that is, professional or supervisory level, several orders of knowledge are required. For example, specific skills which pertain to that particular organization, general skills, people skills (organizational skills, training skills), and communication skills (oral and written) are but a brief list. Supervisory training can be, and may generally be, conducted by supervisors within a particular organization. Production trains production supervisors, quality control trains it own laboratory or packaging area supervisors, etc. However, an area often overlooked is cross-fertilization. Departments which interface closely may never see what other departments are attempting to accomplish for the overall good of the company. Further, they may never understand why certain operations must be conducted in a particular way. This lack of understanding can often lead to friction and mutual disrespect, which can be negative influences on the company's overall goals. How much more valuable for the planning department to understand the problems and solutions needed to solve compounding or packaging supervisors' challenges. This very desirable training is often achieved by utilizing a professional training staff and a comprehensive training program, at least at the higher levels.

Let us examine an outline of a prospective quality assurance supervisor training program. In-depth topics specific to quality assurance will not be reviewed, but we will suggest coordinate types of training by other departments and outside agencies. The end result desired is to help quality assurance supervisors at any level interface effectively to provide assistance to the operations function so as to make better-quality, lower-cost product.

Section 1—Organization and functional responsibilities
Section 2—Quality assurance analytical laboratories
Section 3—Quality engineering
Section 4—Distribution quality assurance
Section 5—Receiving inspection
Section 6—Finished product inspection
Section 7—Field/subcontractor inspection
Section 8—Research and development
    A. Marketplace direction
    B. Company product strengths and product positioning
    C. Functional development laboratories, i.e., hair care, emulsions, makeup
    D. Product evaluation techniques
Section 9—New Product introduction
    A. Company's new product introduction system
    B. Quality assurance interface in product planning
    C. Marketing/quality assurance interface in performance specifications
    D. Communications and feedback

Section 10—Packaging development
    A. Development orientation—brand design managers
        (1) Brand X
        (2) Brand Y
        (3) Brand Z
    B. Art department—production art
    C. Packaging specification systems
    D. Packaging engineering
    E. Ingredient listing integration
Section 11—Purchasing Orientation
    A. Purchasing agents/managers
    B. Reject handling
    C. Vendor certification system
Section 12—GMPs/regulatory affairs
    A. GMPs and CTFA guidelines
    B. GMP inspection of facilities
    C. "The inspection" (FDA)
    D. Safety
Section 13—Manufacturing control system
    A. Paperwork flow
    B. Sample accounting (somebody pays for quality assurance
        samples)
    C. Product release systems
Section 14—Summary and reviews
Section 15—Interim takeover

This may seem "heavy," but most companies really perform these
functions, consciously or subconsciously. Whereas we have presented
a prospective quality assurance supervisor's training outline, this
type of outline is useful for the cross-fertilization training of any
position which should interface with other departments. Daily interface with the production and warehouse functions will provide additional training opportunities to help integrate the quality assurance
supervisor with his or her peers. We would suggest that a production
supervisor spend similar time in the nonproduction areas to understand the quality assurance interrelationship to the compounding or
packaging departments.

V. CONCLUSION

Most company goals include consumer satisfaction, increased productivity, and lower costs. Teaching employees, new or old, exempt or
nonexempt, those skills needed to make them successful in producing
product economically and in compliance with regulatory agency dic-
tates would seem to be a self-evident necessity. However, human
nature is not easily convinced. Edward Deming convinced the Japanese,
and their successes in the 1965—1981 era are well known. There are

many case histories in the literature documenting increased labor productivity as defect levels decrease. Training can be a key to increased profits while helping to increase consumer-perceived quality. Better quality also equals lower exposure to increased regulatory agency activity.

References

1. CTFA Citizen Petition for Proposed Regulation, Part 750 Current Good Manufacturing Procedures in Manufacture, Processing, Packing, or Holding of Cosmetic Products, July 28, 1977.
2. 21 Code of Federal Regulations, Parts 210 and 211.
3. Section 201 of the FDC Act, 21 U.S. Code (321) (G) (1) and (321) (I).
4. C. Haynes and N. Estrin, ed., CTFA Technical Guidelines, 2nd ed., The Cosmetic, Toiletry and Fragrance Association, Washington, D.C., 1982.
5. Section 60L of the FDC Act, 21 U.S. Code (361) (C).

23

COPING WITH COSMETIC LABELING

JAMES M. AKERSON

Regulatory Affairs, Clairol Research Laboratories, Stamford, Connecticut

JAMES M. LONG

Ohio Ethics Commission, Columbus, Ohio

I. INTRODUCTION

Every cosmetic product sold in the marketplace must be packaged in a manner pleasing and acceptable to the potential purchaser or consumer. The information on this package must also be informative and eye-catching in order to help convince the potential buyer that this is precisely the produce that he or she needs. Since some of the information on a cosmetic label relates directly to the safety or economic value of the product, Congress has enacted several pieces of legislation that specifically mandate certain items that must appear on cosmetic products. These acts, the Food, Drug, and Cosmetic Act (FDC Act) and the Fair Packaging and Labeling Act (FPLA), require that cosmetic labels contain the following items:

1. A statement identifying the use or function of the cosmetic
2. The name and address of the manufacturer or distributor
3. The net contents of the product
4. A statement of the ingredients contained in the product
5. Certain relevant warning statements indicating potential hazards related to the use of the product

Before we discuss the specific labeling issues, however, we should first review the general labeling provisions and restrictions of these two laws. Each of these acts has peculiar characteristics which must be clearly delineated in order to follow the twists and turns of cosmetic labeling. Although these acts have been discussed in previous chapters, a few moments should be taken to review just what constitutes a cosmetic product.

A. The Food, Drug, and Cosmetic Act

The FDC Act defines a cosmetic as:

(1) articles intended to be rubbed, poured, sprinkled, or sprayed on, introduced into, or otherwise applied to the human body or any part thereof for cleansing, beautifying, promoting attractiveness, or altering the appearance, and (2) articles intended for use as a component of any such articles; except that such term shall not include soap.

From this definition it can be seen that the status of a product as a cosmetic is determined by the intended use of the product as stated in the labeling and advertising representations. In addition, it is also apparent that several products typically considered to be cosmetics are indeed subject to regulations other than the cosmetic regulations. These products include the true soaps, which are not regulated by the FDA at all, and a variety of cosmetic drugs which, because of their claims, must be labeled in compliance with both the cosmetic and drug requirements of the FDA. Typical examples of these types of products include antiperspirants, sunscreens, antidandruff shampoos, and anticaries toothpastes.

The FDC Act is the basic act empowering the FDA to regulate cosmetics as well as foods, drugs, medical devices, and other related items. Although this act is primarily a safety act, it also directs the FDA to issue regulations governing false or misleading labeling. The act requires that all cosmetic packages contain the net contents, the name and address of the manufacturer or distributor, and certain applicable warning statements. The label information required by the FDC Act must be located on the immediate container of all cosmetic products. If there is both an inner and an outer container, these mandatory statements must appear on both containers. This requirement is found in the definition of "label" in the FDC Act which states:

The term "label" means a display of written, printed, or graphic matter upon the immediate container of any article; and a requirement made by or under authority of this Act that any word, statement, or other information appear on the label shall not be considered to be complied with unless such word, statement or other information also appears on the outside container or wrapper, if any there be, of the retail package of such article, or is easily legible through the outside container or wrapper.

Note that this definition, by referring to the retail package, does not include shipping cartons used to transport the product to the retail outlet. However, the Department of Transportation has also issued regulations that may affect the labeling of these shipping containers. Although these regulations are outside the scope of the present discussion, they should be consulted for relevance to specific shipping questions.

In addition to the definition of "label" quoted above, the FDC Act also defines the term "labeling" as follows:

The term "labeling" means all labels and other written, printed or graphic matter (1) upon any article or any of its containers or wrappers, or (2) accompanying such article.

As can be seen, this is a broader term than "label" and can have significance for those regulations applicable to "labeling" rather than just to the "label."

B. The Fair Packaging and Labeling Act

The second law that affects cosmetic labeling is the Fair Packaging and Labeling Act (FPLA). This act is basically an economic law and was enacted to prevent consumer deception and to provide consumers with information enabling them to make value comparisons in the marketplace. Under this act the cosmetic package must bear a net contents, an identity statement, the name of the manufacturer or distributor, and a list of the ingredients. Since this act is designed for presenting information to the consumer at the retail level, these mandatory items are required only on the outer retail packaging.

In addition, the FPLA, in its definition of "consumer commodity," states that the product must be sold to the consumer for personal use or for services ordinarily rendered in the household. This definition has the effect of exempting free samples and professional products used solely by beauticians or similar commercial cosmetic establishments from the FPLA labeling requirements. This does not, however, exempt these products from FDC Act requirements. In utilizing these exemptions several considerations must be kept in mind. First, free samples must be truly free of cost to the consumer. If there is a tie-in requirement or deal in which the consumer must buy

product A in order to obtain B, product B may not be fully compliant
with the "free" requirement. "Free" products should be carefully
reviewed for hidden consumer costs.

Second, professional products must be used solely in commercial
establishments and cannot be resold through these outlets for use by
consumers at home. The labeling of the container "For Professional
Use Only" may not be sufficient for this exemption if the product is
also promoted for consumer sale in the salon or beauty shop.

One last general feature of the FPLA that should be considered
before the specifics of cosmetic labeling are discussed is the principal
display panel (PDP). Under the FPLA the principal display panel is
defined as follows:

> The term "principal display panel" means that part of a label that
> is most likely to be displayed, presented, shown or examined under
> normal and customary conditions of display for retail sale.

It should be emphasized that the PDP is an FPLA concept and is
not a mandatory feature of the FDC Act. Therefore, the limitations
of the FPLA to the outer package of retail consumer products also
holds for the PDP. In other words, only the outer package has a
PDP, and a regulation requiring placement of information on the PDP
relates only to that outer package.

The PDP must be large enough to accommodate all the mandatory
information in a clear and conspicuous manner. As with all mandatory
copy, the information should not be obscured or crowded by other
printing or by designs. If a package is so designed as to have alternate principal display panels, each PDP must bear all the mandatory
label information. This ensures that no matter which PDP the consumer examines on the retail shelf, the required label information
will be appropriately displayed.

The location and size of the PDP varies with the individual package.
If the package is rectangular, the PDP is considered to be one entire
side. The area of this PDP would then be the product of the height
times the width. If the package is cylinderical, the PDP will constitute an area on the cylinder equal to 40 percent of the product of
the height times the circumference of the container. This area must
be the 40 percent of the vertical surface which is designed to face
the consumer when displayed for retail sale. For all other shapes of
packages the PDP is equal to 40 percent of the total surface area
except for those containers that have an obvious principal display
panel. In determining the area of the PDP, the tops, bottoms, flanges,
necks, and shoulders of the packages can be excluded.

II. BASIC LABELING INFORMATION

As we have seen, there are five basic types of information required
by these federal laws to appear on cosmetic labels. These items are

1. An identity statement
2. The manufacturer's name and address
3. The net contents
4. The ingredient labeling
5. Appropriate warning statements

We will now discuss each of these items individually.

A. Identity Statement

The Fair Packaging and Labeling Act specifies that each cosmetic product subject to the FPLA "shall bear a label specifying the identity" of the cosmetic. The FDA has issued regulations under the Code of Federal Regulations (C.F.R.), 21 C.F.R. 701.11, detailing the steps required to meet this requirement. First of all, the identity statement must appear on the principal display panel as one of its principal features. However, since the identity statement is required only by the FPLA and not by the FDC Act, it need appear only on the outer container.

Second, the regulation provides several alternatives for the identity statement. These alternatives are as follows:

1. The common or usual name of the cosmetic, such as "lipstick," "shampoo," or "hairspray"
2. A descriptive name which explains the product's function, such as a foaming body bath, hair conditioning gel, or skin smoother
3. A fanciful name when the use of the product is obvious
4. An illustration or vignette that clearly identifies the intended use of the cosmetic

As noted earlier, the identity statement must be one of the principal features on the principal display panel, but it does not have to be the principal feature. The regulation does not specify a type size requirement for this identity statement, but it does require that it be in a "size reasonably related to the most prominant printed matter" on the panel. The identity statement must also be in a bold type clearly conspicuous to the consumer. It should not be obscured by other phrasing or design material and it should be printed in lines "generally parallel" to the base of the package.

As can be seen from this requirement, there is much room for variation and interpretation of the identity statement. However, as long as the labeler uses some discretion and assures that the product is not confusing or misleading to the consumer and that the identity statement is clearly conspicuous, the regulatory risk should be minimal.

B. Manufacturer's Name and Address

The name and address of the manufacturer, packer, or distributor is required by both the FPLA and the FDC Act. In addition, it must

also appear on free samples and professional products that are regulated by the FDC Act. As with all mandatory label copy, the name
and address must be conspicuous. Although no type size or location
is specified in the regulations, the statement must be of sufficient
size and legibility to be easily read and should not be obscured by
other printed matter or colored backgrounds. The regulation requires
the labeling of the name and address of the manufacturer, the distributor, or the packer, whichever is relevant. It does not require
the identification of all three. To satisfy this requirement the product must show the corporate name and address. It may also indicate
the relevant division of the corporation at the discretion of the labeler.
If the labeler is an individual, partnership, or association, the name
under which the business is conducted is the prescribed name to be
used. If the label is listed in a current telephone or city directory,
the address may consist only of an identification of the city, state,
and ZIP code. However, the street address must also appear if the
directory requirement is not met.

If the name on the label is not the manufacturer of the cosmetic,
then the name and address must be preceded by a qualifying phrase
that accurately describes the relationship of the labeler to the product. This is commonly done with the phrase "Distributed by," "Manufactured for," or "Packed by." If the address identified on the
cosmetic package is outside the United States, the labeler should also
consult the Department of Commerce regulations concerning the labeling of the country of origin on imported products.

The cosmetic regulations do not specify the amount of manufacturing
control a company must exert in order to represent itself on the label
as the "manufacturer." However, the drug regulations do define the
term "manufacturer" and establish requirements for proper descriptions of the term "manufacturer" on the label of a drug product.
Therefore, labelers of cosmetic products that are also drugs should
consult those drug regulations for more precise information.

C. Net Contents

With the possible exception of the ingredient labeling requirements,
those for the net contents are the most intricate of the mandatory
labeling issues. The net contents is required by both the FDC Act
and the FPLA and, therefore, must appear on all packages. For
products regulated by the FPLA the net contents must appear on the
PDP of the outer package in the lower 30 percent of the panel in lines
generally parallel to the base of the product as it appears on the retail shelf. If the PDP is 5 square inches or less, however, the placement in the lower 30 percent is waived.

The net contents must appear in terms of weight, measure, or
numerical count. However, a product in kit form, containing more
than one component, can also be identified by the number of applica-

tions provided when the consumer follows the label directions. Cosmetics that are solids, semisolids, or viscous materials must express their contents in units of ounces or pounds and must use the phrase "net weight." Products that are liquids may optionally use the terms "net" or "net contents," but must express the contents in terms of fluid ounces, pints, or quarts. Products that contain at least 1 pound or 1 pint must identify their contents in dual terms, such as "16 fluid ounces (1 pint)." In addition, these units cannot be qualified or described by terms which would tend to exaggerate the contents, such as "full" or "giant." Furthermore, the use of drams as a unit is expressly forbidden.

The type size of the net contents on the outer package is determined by the size of the principal display panel or PDP. Although the measurements of the PDP were previously touched upon briefly, the labeler should refer to the labeling regulations themselves, for specific problems. Table 1 summarizes the minimum type size requirements for various-sized cosmetic packages.

If the net contents is embossed on a bottle, the minimum type size must be increased by an additional 1/16 inch. However, embossing of the net content can be done only if all the mandatory label information is embossed.

The ratio of height to width of the letters of the net contents cannot exceed 3:1, but it can be expressed in mixed upper- and lowercase characters. If lowercase letters are used, however, the type size requirement must be met by the size of the lower case "o." When using fractions, each number must be at least one-half the height determined by the size of the PDP.

In addition to all of these requirements, the net contents must be separated from other printed matter above and below by a clear space equal in height to the size of the type used for the net contents. It must also be separated from printed matter to the left and right by a space equal to twice the width of the letter "N" in the type style used for the net contents.

All these restrictions have been enacted under the FPLA. As mentioned earlier, however, the FDC Act also requires the net contents to be prominent and conspicious on any inner container as well, but

TABLE 1   Minimum Type Size for Net Contents

Size of PDP	Minimum type size
5 square inches or loss	1/16 inch
Over 5 and up to 25 square inches	1/18 inch
Over 25 and up to 100 square inches	3/16 inch
Over 100 and up to 400 square inches	1/4 inch
Over 400 square inches	1/2 inch

that act does not specify the type size or the location of the net contents.

The net contents must be expressed in English units and cannot be listed only in metric units. However, metric units may be used as a secondary or supplementary net contents in conjunction with the English units. When both units appear on the label, the English units must be in full compliance with the labeling regulations, while the metric units may appear anywhere on the package with no restriction on type size. If the metric units are used adjacent to the English units, the clear space requirement under the FPLA is waived for the space between the two expressions of the net contents.

D. Ingredient Labeling

Ingredient labeling of cosmetic products has become mandatory only within the last few years and is probably still the least understood of the various components of cosmetic labeling. Enacted under the statutory authority of section 5(c) (3) of the Fair Packaging and Labeling Act, the ingredient statement must appear only on the outer package of all cosmetics sold to the consumer. Since this is strictly an FPLA requirement, free samples and purely professional products are exempted from ingredient labeling. Due to the fact that these regulations are extremely complicated and interrelated, this discussion has been subdivided into four areas: general requirements, nomenclature and order of ingredients, labeling of kits and assortments, and off-package labeling. The subdivisions are purely arbitrary and have been established only for convenience. Individual labeling problems usually will involve more than one of these subdivisions.

General Requirements

The declaration of ingredients must appear in descending order on an "appropriate" panel of the outer package. This declaration must be prominant and conspicuous so that ordinary individuals can read and understand it under normal conditions at point of purchase. Although "appropriate" is not defined in the regulation, it clearly does not restrict the ingredient declaration to the principal display panel and allows the labeler to use whatever part of the package he feels meets the general requirements of clarity and legibility.

The declaration must appear in a type size of at least 1/16 inch except in those cases where the total surface area of the package suitable for labeling is less than 12 square inches. In these cases the type size can be reduced to 1/32 inch. In calculating this area the regulation states that "surface area is not available for labeling if physical characteristics of the package surface, e.g., decorative relief, make application of a label impractical." Thus, a package may have more than 12 square inches of surface area yet still meet the requirements for the reduced type size.

The regulations also provide an optional method for labeling decorative packages or packages that have insufficient label space. Under these conditions the ingredients may be placed on a "firmly affixed tag, tape or card." Although this alternate labeling may be designed to be discarded by the consumer during use, it must be attached to the cosmetic package sufficiently well to withstand shipping and still be accessible to the consumer at point of purchase.

Cosmetic products that are also drugs must be labeled in compliance with both the drug and cosmetic regulations. For these products the active drug components must be identified first in the declaration of ingredients, followed by the remaining cosmetic ingredients. It should also be noted that the drug ingredients must be named using the "established name" procedures in the drug regulations. Although in most cases the cosmetic and drug names for these ingredients will be identical, this is not always true. Since the ramifications of the drug labeling regulations are beyond the cosmetic scope of this article, the drug labeler should carefully review those regulations to identify properly the drug ingredients and to satisfy himself that his cosmetic/drug product is properly drug labeled.

The cosmetic labeling regulations also provide relief for a current or anticipated shortage of a raw material. In these cases the labeler has the option of either inserting the alternative ingredient, preceded by the word "or," immediately after the ingredient it replaces, or listing the alternative ingredients in descending order at the end of the ingredient declaration following the phrase "may also contain." It should be noted that this option cannot be used to replace an ingredient that is highlighted in advertising or identified elsewhere on the label. For example, this exemption would not apply to a substitution for menthol in a mentholated shaving cream.

Nomenclature and Order of Ingredients

The procedures for naming the cosmetic materials in the declaration of ingredients are spelled out in detail in the regulations. Primarily, the FDA has reserved the right to provide specific nomenclature for certain ingredients in 21 C.F.R. 701.30. At present this list consists only of the names for the halogenated aerosol propellants and Ethyl Ester of Hydrolyzed Animal Protein. It should be noted, however, that most of these propellants have subsequently been banned for use in cosmetic products. Aside from this list, which can be amended at the FDA's discretion, the priority of proper nomenclature is as follows: (1) CTFA *Cosmetic Ingredient Dictionary,* 2nd edition, (2) *U.S. Pharmacopeia,* (3) *National Formulary,* (4) *Food Chemicals Codex,* (5) USAN and the USP *Dictionary of Drug Names,* (6) a name generally recognized by consumers, and (7) a chemical or technical name or description.

In utilizing these sources several points should be considered. First, the order of these publications in the list is important. If an

ingredient is identified in more than one source with different nomenclature, the highest source in the listing takes precedence. For example, if the same ingredient is listed in the CTFA *Dictionary* and in
the *Food Chemical Codex* but with differing names, the CTFA *Dictionary* nomenclature must be used for ingredient labeling.

Second, the regulation specifies the 2nd edition of the CTFA *Dictionary*. Recently the CTFA published a 3rd edition to the *Dictionary*
and has petitioned the FDA for recognition of this new edition. However, until such time as the FDA formally recognizes the 3rd edition
in the *Federal Register*, the 2nd edition remains the "official" version.

Third, in the regulation recognizing the 2nd edition of the CTFA
Dictionary, several names were specifically not adopted by the FDA
for purposes of ingredient labeling. These issues were all addressed
during the preparation of the 3rd edition, and it is anticipated that
these restrictions will not be retained in the eventual regulation recognizing the 3rd edition.

Fourth, the only exception to the use of this listing is the identification of drug active ingredients in cosmetic products that are also
drugs. The procedures for these products were discussed previously
under "General Requirements."

Fifth, the term "chemical description" has been defined in the regulations at 21 C.F.R. 700.3 (a) as follows: "The term 'chemical description' means a concise definition of the chemical composition using
standard chemical nomenclature so that the chemical structure or
structures of the components of the ingredient would be clear to a
practicing chemist. When the composition cannot be described chemically, the substance shall be described in terms of its source and
processing." This definition should be carefully considered before
assigning arbitrary names to ingredients not otherwise identified in
the listed compendia.

There are several important exceptions to the requirement for specific identification of all cosmetic raw materials. Those materials that
are used solely to impart an odor or taste to a cosmetic product can
be identified simply by the terms "fragrance" or "flavor." These
terms should be used, however, only when these ingredients have no
other function in the cosmetic product and the materials so designated
are clearly within the meaning of such terms as commonly understood
to consumers.

In addition, the FDA has authorized the use of the phrase "and
other ingredients" to be used at the end of the declaration of ingredients to represent those ingredients that have been accepted by the
FDA as representing trade secrets. This phrase, however, can only
be used when the FDA has actually granted an exemption from public
disclosure and cannot be used simply because a labeler does not wish
to disclose the composition of an ingredient.

In general, the ingredient statement must appear in descending
order of concentration of ingredients. In calculating these usage

levels the exact concentration of each component must be determined.
If the raw material is obtained as a solution or as an intentional mixture with other ingredients, each component must be identified individually and listed in the descending order of its "active" percentages.

There are three variations that can be used to express the descending order of ingredients. The first of these methods involves
the strict listing of all ingredients in descending order. Under this
procedure the percentage of each component is calculated and the ingredients are named in order. The FDA, recognizing the practical
difficulties inherent in determining the relative amounts of lesser
ingredients, has issued two alternative methods. These are (1) the
listing of all noncolor additives in descending order of predominance
followed by all the color additives in random order and (2) the listing
of all noncolor additives at concentrations greater than 1 percent in
descending order followed by the random listing of all noncolor additives at concentrations of 1 percent or less followed by the random
listing of all color additives.

In using these methods it should be noted that there is no regulatory level or concentration below which an ingredient need not be
identified for labeling. However, the FDA has issued regulations
which exempt certain ingredients if they are "incidental" ingredients.
These ingredients are defined as materials present at insignificant
concentrations and which have no technical or functional effect in
the cosmetic product. These materials are further identified as substances that are added to the cosmetic product as either components
of another ingredient or as processing aids in the manufacture of the
final product. Any ingredient present in the finished cosmetic product that does not satisfy either of these requirements must be ingredient labeled regardless of how low its concentration is in the cosmetic.

The FDA has also made provision for the listing of a color additive
that is used occasionally for color-matching certain batches of cosmetics and that is not always present in a given product. These colors
should appear at the end of the ingredient declaration after the phrase
"may contain." Note should be made of the similarity between the
phrase "may contain" as used for color additives and the phrase "may
also contain" as used for raw materials in short supply. The term
"may contain" can be used only with color additives.

Kits and Assortments

The labeling regulations offer some help in developing common labels
for certain lines of products. Although these regulations are complex, the following brief summary should help identify the types of
relief available. In the final analysis, however, the labeler is advised
to consult the *CTFA Labeling Manual* [1] for more detailed information.

First of all, an assortment of different types of cosmetic products
sold in the same kit may have a composite listing of colors. Although

the noncolor additives of each component must be specifically identified for every particular cosmetic product in the kit, the color additives may be declared in a single composite list. This list, however, must be identified as pertaining to all the products in the assortment.

Several regulations relate to the declaration of ingredients used for a line of similar products. If the cosmetic products are "similar in composition and intended for the same use" and are sold as an assortment in the same package, the ingredient declaration can be partly combined. For these products, the noncolor ingredients that are common to all the products may be declared in their cumulative order of predominance. These ingredients are then followed by a listing of those noncolor ingredients that are not common to all products, clearly identifying those components in which they are present. This listing is, in turn, followed by a composite listing of all the color additives present in the assortment, regardless of whether or not they are present in all of the component products. There are several points of value to this provision. First, it allows the use of the cumulative order of the common ingredients in an assortment and does not require that each component have the same descending order of concentration. Second, this provision applies to all ingredients and not just to color additives. Third, it allows the use of a composite listing of colors; and finally, it is not restricted to any particular type of cosmetic product.

In addition, if the assortment package has less than 12 square inches of total surface area available for labeling, the cumulative list of ingredients can be expanded to include those materials not common to all products. For these small assortment packages all the ingredients in the assortment would then appear in a single list in cumulative descending order of predominance. The color additives may either be included in the cumulative listing or appear at the end in a composite list.

Special composite labeling rules are also available for "shaded" products that are sold independently, use the same declaration of ingredients, and are all sold under the same trade name. Shaded products are defined as and limited to eye makeup, other facial or body makeup, lipstick, and nail polish. For these products a color additive not present in all shades in the line can be identified following the phrase "may contain." In addition, the ingredient listing for a line of individually packaged shaded products or a line of assortments of shaded products that are similar in composition and intended for the same use may also be combined. For these products the ingredients common to all shades appear in a single list in cumulative order. This list is followed by those ingredients not common to all shades with an identification of those shades of which they are components. The color additives common to all shades may then be listed with the noncolor ingredients or can be combined in a composite list after the noncolors. Finally, any color additive not present in all shades should appear after the phrase "may contain."

As noted earlier, ingredient labeling need appear only on the outer container of cosmetic packages. However, if an inner container is sold separately, it must also be properly ingredient labeled at that time, since it has now become an "outer" package.

Off-package Labeling

The ingredient labeling regulations make special reference to the use of off-package labeling for certain types of products. The labeling may be on leaflets or padded sheets in a type size of at least 1/16 inch. This labeling must accompany the cosmetic products but need not be attached to the actual package.

In order to qualify for off-package ingredient labeling, several criteria must be met. These are as follows: (1) The total surface area of the package must be less than 12 square inches (note that this is not the same as the 12 square inches available for labeling to qualify for smaller type size); (2) the product cannot be packaged in an outer container; and (3) the product must be in "tightly compartmented trays or racks."

In addition, a difference is made between "display units" and "display charts." A display unit is a retail device containing actual products for sale to consumers. The compartmented racks or trays are part of this display and hold the products "displayed for sale." A display chart, on the other hand, is a visual device used to inform the consumer of the various shades of a cosmetic product that are available. The actual products in their racks and trays may be located remotely and need not be part of the display. The racks or trays of products need only be "held for sale." Although a display unit can be used for any type of cosmetic, the display chart can be used only for "shaded" cosmetics as were discussed earlier (eye makeup, other facial and body makeup, lipstick, and nail polish).

Copies of the ingredient listing for products using display units or charts must be placed in a holder attached to the units or charts. The labeling in this holder must meet one of several criteria. The labeling sheets may be placed on the front of the display unit or chart fully visible to the customer. Variations of this procedure allow the labeling sheets to be placed in the front of the display unit or chart but only partially visible to the consumer or on a side of the unit or chart. However, a clearly visible notice in type at least 3/16 inch must be present directing the customer to the location of the ingredient labeling. The labeling sheets cannot be located on the top, bottom, or back of the display.

There are several further restrictions for off-package labeling which can be summarized as follows. The sheets for each display must all be identical and must provide the ingredients for all products in the display. The ingredient sheets and the display must be shipped to the retailer as part of the same shipment. When the last ingredient sheet is removed, the statement "Federal law requires ingredient lists to be displayed here" must be clearly visible to the consumer. Whenever the formulation of any of the products changes, new labeling must be supplied to the retailer that indicates the changes and clearly identifies the package applicable to the ingredient list. Sufficient copies of the labeling sheets must be supplied to the retailer so that a consumer may obtain a copy with each purchase. Further, the company identified on the products must also promptly send a copy of the ingredient declaration to any person requesting it.

Ingredient declarations for products sold by direct mail may also be located off-package. The ingredient list can either be identified in a catalog retained by the purchaser or may be included in the shipment with the product. This labeling must be clearly conspicuous and at least 1/16 inch in height. The package must also bear a statement indicating the location of the labeling.

E. Warning Statements

Several warning statements have been issued by the FDA for cosmetic products in order to "prevent a health hazard that may be associated with the product." These warning statements must appear on the label prominently and conspicuously in a type size of at least 1/16 inch. At present there are only a few warnings relevant to cosmetic products. These warnings are as follows.

Hair Dye Exemption Warning

The hair dye exemption warning is found in the FDC Act itself and, when placed on the label, exempts coal-tar hair dyes from the adulteration provisions in the FDC Act. It can be used only with respect to coal-tar dyes and does not apply to any non-coal tar colorant that a manufacturer chooses to use on hair. This warning states: "Caution—This product contains ingredients which may cause skin irritation on certain individuals and a preliminary test according to accompanying directions should first be made. This product must not be used for dyeing the eyelashes or eyebrows; to do so may cause blindness."

In addition, this provision also requires the inclusion of directions for pretesting or patch-testing the product for sensitization.

Aerosol Warnings

There are several different warnings for cosmetic aerosol products. First, all aerosol products must state: "Warning—Avoid spraying in eyes. Contents under pressure. Do not puncture or incinerate. Do not store at temperatures above 120°F. Keep out of reach of children." If the product is not a spray, the eye warning can be deleted. If the package is made of glass, the phrase "Do not puncture or incinerate" should read as follows: "Do not break or incinerate." Finally, if the product is for use by children, the last line should be amended to read: "Keep out of reach of children except under adult supervision."

Certain cosmetic aerosols containing halocarbon or hydrocarbon propellants must also have the warning: "Warning—Use only as directed. Intentional misuse by deliberately concentrating and inhaling the contents can be harmful or fatal." This warning is not required for:

1. Foams with less than 10 percent propellant
2. Products of less than 2 ounces that have metered dosage
3. Products packaged with a physical barrier to prevent the escape of the propellant
4. Products of less than 1/2 ounce contents.

Feminine Deodorant Sprays

A cautionary statement is required for all "feminine deodorant sprays," which are defined as "spray products represented for use in the area of the female genitals or for use all over the body." The cautionary statement reads as follows: "Caution—For external use only. Spray at least 8 inches from skin. Do not apply to broken, irritated or itching skin. Persistent, unusual odor or discharge may indicate conditions for which a physician should be consulted. Discontinue use immediately if rash, irritation or discomfort develops."

This caution statement must appear on both inner and outer packaging but need not be on the principal display panel.

Safety Substantiation Warning

The FDA has enacted a general warning statement applicable to any cosmetic that has not been adequately tested for safety prior to marketing. This regulation requires safety substantiation for both the final product and each of its ingredients. If a manufacturer has not substantiated the safety of the product, the following warning must appear: "Warning—The safety of this product has not been determined."

This warning must appear on the principal display panel of the outer package as well as on the label of any inner container.

III. CONCLUSION

This has been a brief highlighting of the various aspects of mandatory labeling required by the Food and Drug Act and the Fair Packaging and Labeling Act. Due to space limitations this presentation has not presented all the variations or inuendos of these regulations. The labeler of cosmetic products is referred to the *CTFA Labeling Manual* [1] in order to obtain fuller details and explanations of these various labeling issues.

In addition to the labeling requirements regulated by the FDA, several other federal laws and agencies also have some impact on cosmetic labeling. These areas include the Federal Trade Commission requirements for the labeling of soap, the labeling of the country of origin under the jurisdiction of the Department of Commerce, the labeling

of products containing alcohol as regulated by the Treasury Department, and the various Department of Transportation requirements
for shipping containers. These facets of cosmetic package labeling
are generally outside the scope of the chapter, and the reader is again
referred to the *CTFA Labeling Manual* for further guidance.

A final comment: Most states have laws and regulations enforced
locally. Although these state labeling requirements are usually identical or at least compatible with the federal regulations, the reader is
cautioned to verify the applicability of any local regulation for the
particular states in which he or she has an interest.

REFERENCES

1. James M. Long, ed., *CTFA Labeling Manual,* 4th ed. The Cosmetic,
   Toiletry and Fragrance Association, Washington, D.C. 1982.

24

THE CTFA *COSMETIC INGREDIENT DICTIONARY*

PATRICIA A. CROSLEY

Information Resources, The Cosmetic, Toiletry and Fragrance Association, Inc., Washington, D.C.

I. INTRODUCTION

In the 10 years since its creation, the *CTFA Cosmetic Ingredient Dictionary* (the *Dictionary*) has grown dramatically in both volume and sophistication, evolving into an essential reference work with worldwide recognition as a source of ingredient labeling nomenclature. Developed originally from a survey which compiled chemical and trade vernacular for materials used in the manufacture of cosmetic products, each edition of the *Dictionary* has demonstrated the cosmetic industry's initiative in providing both a source of uniform nomenclature for use in compliance with ingredient labeling regulations and a comprehensive reference for chemical information on cosmetic ingredients. The success of this scientific effort, as well as its regulatory significance, has benefited the cosmetic industry, the government, the medical profession, and the consumer. In an era of heightened consumer and regulatory scrutiny, the *Dictionary* stands out as an example of the cosmetic industry's continuing commitment to self-regulation and the perpetuation of the highest standards of quality and safety.

In order to comprehend fully the versatility and impact of the *Dictionary* in the scientific and regulatory communities of the cosmetic industry, its development and use must be discussed in terms of its applications as a chemical reference and an aid in regulatory compliance with ingredient labeling requirements. That the *Dictionary* is the most comprehensive collection of data of cosmetic ingredients ever compiled is evidenced by a review of its contents. What gives the *Dictionary* its significance, however, is its emergence at a critical period in the cosmetic industry's regulatory history, and its consolidating effect on the industry's efforts to regulate voluntarily the safety and integrity of its products. This effort cannot be demonstrated simply by discussing the *Dictionary's* contents, but may be inferred by a description of its development, its applications, and those scientific, regulatory, and consumer forces that collectively shape its value and scope.

The following discussion will focus on the *Dictionary* as a *dynamic* compendium, reflecting changes in both its contents and the context within which its information is generated and used. Although the *Dictionary*, by design, is self-consistent in its presentation of ingredient and supplier information, a comparison of its three editions (1973, 1977, and 1982) reveals enormous changes in the *quality* of information provided and demanded by its users. What is less apparent from such a review, yet essential to its significance, is the design, and the commitment to its execution, that has allowed the *Dictionary* to perpetually accommodate change and growth in the cosmetic industry.

II. THE *DICTIONARY* IN HISTORICAL PERSPECTIVE

The transition from the late 1960s to the early 1970s was particularly significant for the cosmetic industry, and nowhere was that change reflected more clearly than in its leading trade association. Growing concern for the industry's commitment to safety precipitated increasing pressure and action within the scientific and regulatory communities that had previously shared a relatively passive relationship with the Toilet Goods Association (TGA). The changing of the TGA's name to the Cosmetic, Toiletry and Fragrance Association (CTFA) in 1971 symbolized an organizational "metamorphosis" that prepared the association, and the industry, to face a new decade of challenge. By initiating new scientific and research programs throughout the early 1970s, the CTFA successfully anticipated and met such challenges as asbestos contamination of talc and regulatory action on mercurial preservatives and hexachlorophene [1]. It was during this period of sudden, heightened concern for the safety of chemicals in general, and cosmetics in particular, that the concept of a dictionary of cosmetic ingredients was conceived.

Events in the early 1970s which affected the development of the *Dictionary* included consumer, regulatory, legislative, and industry initiatives addressing the fundamental issue of a lack of information on the cosmetic industry and its products. The CTFA petitioned the Food and Drug Administration (FDA) in 1971, requesting regulations implementing a voluntary program for registration of cosmetic product establishments, formula disclosures, and product experience reports [2]. During this period the issue of ingredient labeling was emerging as a widespread concern. A consumer petition was submitted to the FDA in May 1972 which requested regulations for ingredient labeling to be adopted by the FDA under the Fair Packaging and Labeling Act [3]. At the petitioner's request, action on that petition was suspended pending the disposition of ingredient labeling legislation previously introduced in Congress. To encourage the implementation of voluntary ingredient labeling practices while legislative consideration continued, the FDA published a general format for voluntary ingredient label declarations [4]. Prompted by the failure of Congress to enact cosmetic labeling legislation, the consumer petition was resubmitted in October 1972 [5]. The FDA expanded the voluntary format, publishing proposed regulations for ingredient labeling in February 1973.

During this period of intense industry and regulatory pressure, the CTFA began a voluntary effort to assess the safety of cosmetic ingredients. A preliminary step in that effort involved the distribution of an industry survey which requested information on cosmetic raw materials. The Battelle Institute was commissioned to compile a master list of cosmetic ingredients, in order to preserve the confidentiality of industry response to the CTFA survey. The resulting list contained not only a wide, yet redundant, array of chemical and trade names for cosmetic ingredients, but also an assortment of noncosmetic materials. It was the CTFA's attempt to organize this information, coupled with the realization that in order to facilitate compliance with proposed ingredient labeling regulations the industry would require a simpler, more uniform method for the meaningful identification of its raw materials, that fostered the idea for a dictionary of cosmetic ingredients.

Work began toward the goal of providing label information on cosmetic products that was meaningful to the consumer and the FDA, while minimizing the burden of industry compliance. The CTFA took an immediate and active role in an effort—beginning with the extensive research conducted by James Akerson (then with the Gillette Medical Evaluation Laboratories) into existing nomenclature systems—to establish distinctive conventions for use in cosmetic ingredient labeling. The results of this initial effort were presented at a meeting with the FDA in July 1972, in the form of a prototype document derived from the initial industry survey [6]. This proposal consisted of two alternative nomenclature systems, which applied both standard and generic conventions to the diverse list of cosmetic raw materials compiled at the Battelle Institute. During this meeting, ground rules were established for the creation of a nomenclature system acceptable to the FDA for its intended use in cosmetic ingredient labeling.

Armed with guidelines for the creation of a comprehensive, unique system for identifying cosmetic ingredients, the CTFA established the Cosmetic Ingredient Nomenclature Committee (the Nomenclature Committee) in August 1972, assigning it the responsibility of converting the results of the industry survey into the foundation for the first edition of the *Dictionary*. A draft of the first edition was published in October 1972, and consisted of an alphabetical listing of chemical and trade names for cosmetic ingredients. This list was expanded from the initial CTFA survey and further identified by corresponding labeling nomenclature. Called the "CTFA Adopted Name," this terminology was intended for eventual use by the cosmetic manufacturers in compliance with ingredient labeling regulations.

In preparation for publication of the first edition of the *Dictionary*, the Nomenclature Committee began the task of supplementing these Adopted Name assignments with definitions, chemical structures, and related data. It was agreed at the onset to cross-reference CTFA nomenclature to the widely used *Chemical Abstracts* number system (CAS numbers), designed to assign numeric identifiers to unique chemical structures. In addition, references were compiled for international compendia and U.S. regulations, and included along with chemical definitions, structures and synonyms for each CTFA Adopted Name. This effort successfully broadened the scope of the first edition beyond that of simply a directory of labeling terminology. The first edition of the *Dictionary* was recognized by the FDA as the preferred source of nomenclature for use in cosmetic ingredient labeling, and by the cosmetic industry as the most comprehensive source of information on cosmetic raw materials ever compiled. Ten supplements to the first edition were issued in response to the growing body of ingredient data precipitated by the release of this first volume. These supplements were subsequently compiled, together with the first edition data, to form the basis for the second edition of the *Dictionary*.

The publication of the second edition in 1977 coincided with the effective date of regulations promulgated by the FDA making ingredient labeling mandatory for cosmetic products [7]. The volume of material compiled by the Nomenclature Committee had nearly doubled by that time—increasing from 1,500 monographs in the first edition to 2,500 in the second. The collection of supplementary information was expanded as well, to include more than 10,000 analogous terms or "Other Names" for CTFA Adopted Names, and references to more than 35 international compendia. The FDA published regulations in January 1980 recognizing, with a few exceptions, the nomenclature contained in the second edition. By then, work on the third edition was well underway.

The third edition departed from the short but successful tradition of the *Dictionary* in several ways. The number of monographs for CTFA Adopted Names grew to 3,400, and a more sophisticated method of cross-referencing chemical and trade names allowed for the inclusion of more than 20,000 Other Names in this edition. Nomenclature used by *Chemical Abstracts* in conjunction with their CAS number system was included for the first time, as were empirical formulas and references to the Toxic Substances Control Act Inventory (TSCA). Less apparent, yet critical to the achievement of this level of sophistication, was the first-time use of a computer to compile and produce the third edition. The use of this technology was intended to expand the CTFA's ability to handle this growing collection of data and increase the potential for augmenting the existing format of the *Dictionary* with information from new sources. The use of a computer will also expedite the accumulation and publication of data for supplements and future editions. In the meantime, the CTFA has again petitioned the FDA to recognize the third edition of the *Dictionary* as the preferred source of nomenclature for the labeling of cosmetic products [8].

III. ESTABLISHING COSMETIC INGREDIENT NOMENCLATURE

The task of compiling nomenclature suitable, from consumer, FDA, and industry perspectives, for the identification of cosmetic ingredients has resulted in 40 rules which have been developed and used by the Nomenclature Committee over the last 10 years in their efforts to establish terminology consistent with the principles outlined during the earliest discussions with the FDA. These rules are highlighted in Sec. VI of this chapter, and preceded here by a discussion of the considerations and activities underlying their development [9].

The Nomenclature Committee's initial efforts involved the evaluation of existing nomenclature systems and their potential for adaptation to cosmetic ingredients. Focus quickly narrowed to food and drug nomenclature systems—neither of which lent itself fully to the labeling of cosmetic products. The nomenclature for food additives proved, in many cases, to be unrelated to cosmetic ingredients, and in other cases too generic in its representation to provide a sufficiently distinctive basis for cosmetic labeling requirements. Nomenclature used by the United States Adopted Name Council (USAN) in the identification of complex drug ingredients with simple, generic names was considered too narrow in its construction for broader application to cosmetic products, in which these ingredients performed a different function. Conversely, nomenclature used by *Chemical Abstracts* to provide unique identifications on the basis of specific chemical structures was determined to be excessively detailed and cumbersome for labeling purposes. Although it appeared initially that the unique requirements for labeling cosmetic products would necessitate the development of an entirely new system of nomenclature, terminology from these and other sources was subsequently adopted for use in the *Dictionary* as the complexity and variety of those requirements became more apparent.

Among the ground rules established by the FDA for the creation of cosmetic labeling nomenclature was the requirement that names for cosmetic ingredients be representative of their chemical composition. Also, where appropriate, names consistently used by cosmetic scientists were to be retained. The FDA acknowledged the distinction of cosmetics from foods and drugs by precluding strict adherance to existing nomenclature systems for either group. In particular, the use of generic identifiers—such as the term "fatty acids" used in food additive labeling—was considered unacceptable. For purposes of ingredient labeling, terminology was to be kept as short as possible, with minimum requirements for punctuation and capitalization.

In transforming these guidelines into the mechanics of creating labeling nomenclature, the Nomenclature Committee soon found it necessary to expand the rules required to maintain the consistency of ingredient names. Between the publication of the first draft of the *Dictionary* in October 1972 and the first edition in May 1973, the list of rules expanded from 16 to 32. Drug nomenclature, such as that used by the USAN, the *U.S. Pharmacopeia* (USP), and the *National Formulary* (NF), became a valuable reference as well as a direct source of nomenclature for the Dictionary. For example, the terms "laureth" and "nonoxynol" were adopted directly from *USAN and the USP Dictionary of Drug Names*.

Inherent in the use of drug nomenclature was the problem of distinguishing an ingredient's presence in a cosmetic from its association with pharmaceutical applications. The Nomenclature Committee, therefore, adopted drug names only where their use in ingredient labeling would not result in the implication of therapeutic value or standards of purity inappropriate to the finished cosmetic product. At the same time, the Nomenclature Committee made every effort to incorporate acceptable nomenclature from related industries, in order to limit the confusion, on the part of consumers and scientists, that multiple terms for identical ingredients implied. Adding to the Nomenclature Committee's concern in this regard was the fact that existing nomenclature was, in certain cases, already inconsistent between the food and drug industries. In those cases, established guidelines for clarity and brevity were relied upon for the adaptation of terminology. For instance, use of the term "EDTA" in the *Dictionary* parallels its use in food labeling to abbreviate the chemical term "ethylenediamine tetraacetic acid." The nomenclature for this material when used in pharmaceutical applications is "edetic acid." Also, the term "PEG," indicative of polyethylene gylcol derivatives, is finding extensive use as an alternative to the USAN's "polyoxyl" nomenclature.

A second area of concern between existing drug and proposed cosmetic nomenclature was the conflicting principles of concise, specific names for complex drug ingredients and the requirement for names broad enough to encompass cosmetic ingredients of similar composition that may have been derived from different materials or synthesis routes. Weighing the problem of exact chemical identity against meaningful terminology, the Nomenclature Committee established several additional guidelines. The first of these stated that nomenclature is intended to reflect only the final composition of an ingredient, and not its derivation. In addition, a distinction was made between natural and mechanical mixtures of raw materials, which resulted in the use of the linking term "(and)" to indicate combinations of materials that did not occur normally. A second differentiation between natural and fortuitous mixtures was created to indicate differences in the distribution of components from natural sources. Retention of the natural stem (for example, "coco-" from coconut, or "tallow") in the nomenclature was intended to reflect the unaltered distribution of component materials in a cosmetic ingredient such as "coconut acid." Deviations from this natural distribution were indicated by using the stem of the component which predominated as a result of significant cutting or enrichment of the original, natural material (for example, "stearic acid").

In general, the Nomenclature Committee's efforts to generate cosmetic ingredient nomenclature attempted to incorporate existing terminology and conventions whenever possible, in addition to complying with the requirements established by the FDA for the purpose of ingredient labeling. The fact that the Nomenclature Committee succeeded in creating specific rules for this nomenclature is significant in that it facilitates, for the user of the *Dictionary*, an understanding of the established terminology and provides a precedent for consistent nomenclature decisions for future editions.

IV. THE CTFA COSMETIC INGREDIENT NOMENCLATURE COMMITTEE

In establishing the Nomenclature Committee in 1972, the CTFA succeeded in assembling a group which satisfied a unique requirement in
its representation. Unlike the majority of the CTFA's committees,
which are comprised exclusively of industry representatives, members
of the Nomenclature Committee were recruited from all the disciplines
whose contributions were felt by the CTFA to be essential to the development of the *Dictionary*. The purpose of this approach was two-
fold. First, it allowed direct input from experts representing the
concerns of regulators, medical professionals, cosmetic producers,
and suppliers. Second, this balancing of the committee's membership
ensured a nonpartisan forum for the evaluation of chemical information
on cosmetic ingredients, and the development of impartial nomenclature.
The CTFA's approach undoubtedly helped to prevent subsequent allegations of misleading nomenclature, as well as many of the difficulties
involved in gaining regulatory acceptance of an industry-sponsored
compendium.

The original membership of the Nomenclature Committee included
five representatives of the cosmetic industry, two members of the
FDA's Division of Cosmetic Technology, and a delegate from the American Medical Association's Committee on Cutaneous Health and Cosmetics. The CTFA also invited a consumer representative to participate
on the committee, but that invitation was declined [10]. Under the
editorial leadership of Dr. Norman Estrin—who conceived the idea
for the *Dictionary*—this group established the assignment and infor-
mation-gathering procedures which have resulted in the publication of
three editions of the *Dictionary*.

The complexion of the Nomenclature Committee has not changed
significantly in its 10 year history. Part of the committee's success
as a group may be attributed to its small size, and the development by
each of its members of key areas of expertise and input. Collectively,
this diversification of contributions is reflected in both the consistency and scope of the *Dictionary's* contents. Despite the replacement
of some individual members, the Nomenclature Committee has retained
its diversity and the balancing of concerns that contribute to the
Dictionary's thoroughness and objectivity.

The work of the Nomenclature Committee is performed on a continuing
basis, interrupted only by the concentration of effort surrounding
the publication of each edition of the *Dictionary*. With the assistance
of CTFA staff and the contributions of industry, the Nomenclature
Committee generates a constant stream of information which, in its
published form, represents state-of-the-art data on cosmetic raw materials. The contributions of the FDA have been of great value in
terms of the *Dictionary's* content and its regulatory significance. The
FDA continues to share the Nomenclature Committee's commitment to
the development of concise, accurate nomenclature for the labeling of

cosmetic products, in the interest of providing complete and valuable information to the consumer.

V. REGULATORY IMPACT ON THE *DICTIONARY*

The incorporation of the *Dictionary* into the regulations for cosmetic ingredient labeling, and the subsequent impact of regulatory scrutiny on its contents, affirmed the cooperative effort on the part of the CTFA and the FDA to increase consumer protection and awareness. In addition, establishing the *Dictionary* as the controlling compendium for labeling nomenclature reduced the burden on the FDA and the cosmetic industry in their respective enforcement of and compliance with labeling regulations, by creating the precedent for use of this consolidated source of nomenclature. The significance of the *Dictionary's* position within this regulatory setting is not that it is a publication generated by industry—this precedent was established by compendia such as the *U.S. Pharmacopeia,* the *Colour Index, USAN and the USP Dictionary of Drug Names,* and the *Food Chemicals Codex.* What is significant is the swiftness of the unqualified recognition by the FDA of the first edition, which had immediate effects on the regulatory environment of the cosmetic industry. Consequently, the impact of the FDA's activity had its first ramifications on the *Dictionary's* status, rather than its contents.

The question of mandatory ingredient labeling for cosmetic products was raised as early as 1971, in response to the FDA's proposed regulations for a program of voluntary registration by industry. Because the labeling issue was determined to fall outside the scope of the ensuing regulations, the Commissioner of Food and Drugs, while concurring with the necessity of ingredient labeling, deferred a proposal for regulatory action until a consumer petition specifying consideration of the labeling issue was filed in May 1972. Prior to that petition the FDA, sensing the inevitability of some form of mandatory labeling requirements, had published a voluntary format for manufacturers wishing to adopt ingredient labeling practices pending the disposition of ingredient labeling legislation then before Congress. That voluntary format became the foundation for the cosmetic ingredient labeling regulations subsequently proposed by the FDA.

Responding to the May 1972 petition, the FDA published proposed rules for ingredient labeling in February 1973, which read in part:

§1.205 Cosmetics: labeling requirements; designation of ingredients. . .

(c) An ingredient shall be identified in the declaration by:

(1) The name specified for that ingredient in any recognized compendium, such as United States Pharmacopeia, National Formulary, Food Chemicals Codex, United States Adopted Names, or any industry compendium; or

(2) In the absence of such a listing, the common or usual name; or

(3) In the absence of such a listing or a common or usual name, the chemical or other technical name or description [11].

It was the phrase "or any industry compendium" that opened the door for the first edition of the *Dictionary* (published three months after the FDA's proposed regulations) to be incorporated into the final regulations for ingredient labeling.

In comments filed with the FDA regarding the proposed regulation, the CTFA presented the first edition of the *Dictionary* as conforming in design to the list of recognized compendia, subject to periodic review and revision, and available as an industry source of labeling nomenclature [12]. Included in the preamble to the final regulation of October 17, 1973, which amended the Fair Packaging and Labeling Act by establishing requirements for cosmetic ingredient labeling, were the following comments by the Commissioner of Food and Drugs:

> The Commissioner has encouraged the establishment of a compendium or dictionary which could serve, inter alia, as a standard reference for determining the names to be used for label declaration of such ingredients, and the Cosmetic, Toiletry and Fragrance Association. . . has developed such a dictionary. . . The dictionary, in large part, retains those names which are the common or usual name and/or those names specified in official or recognized compendia. . . . The dictionary also attempts to translate trade names which would not ordinarily be meaningful to consumers into uniform and more commonly understood names. Accordingly, the final regulation recognizes the CTFA Cosmetic Ingredient Dictionary as the controlling compendium to be used in label declaration of a cosmetic ingredient. In the absence of an applicable entry in the CTFA Dictionary, other recognized compendia, listed in the regulation, will control [13].

The promulgation of this rule documented the industry's ability to provide information voluntarily which, in its published form, warranted preferred regulatory status. The status afforded the *Dictionary* at this significant point in the regulatory history of the cosmetic industry became further evidence of the CTFA's effort to facilitate the industry's and the FDA's *need to know*, while protecting the consumer's *right to know*. Although the *Dictionary* was conceived as a reference tool for the evaluation of cosmetic ingredient safety, as well as a source of uniform terminology, the timing of its introduction by the CTFA into the regulatory environment of ingredient labeling enhanced not only the value of its contents, but also the reputation of the industry for successful self-regulatory efforts.

Having established the *Dictionary's* regulatory significance to cosmetic ingredient labeling, emphasis on the part of both the CTFA and FDA shifted to its contents. The CTFA began to compile the supplements and revisions which would become the second edition of the *Dictionary*, and the FDA begain its evaluation of the effectiveness of

its labeling regulations as nomenclature from the first edition began
to appear on cosmetic products. The FDA also continued to participate in the activities of the Nomenclature Committee until it was officially petitioned by the CTFA in June 1976 to recognize the *Dictionary's*
second edition [14].

Pending official action on the CFTA petition, the FDA responded
to the CTFA's request by reviewing the contents of the *Dictionary*
(which was, by then, in the final page-proof stages of production),
and providing written comments which raised issues critical to the
acceptance of several of the second edition's nomenclature conventions.
In a letter to the CTFA dated September 17, 1976, the FDA questioned
the Nomenclature Committee's approach to establishing, and retaining
from other sources, several types of labeling nomenclature. Although
these issues were raised for the first time in this letter, many of the
conventions questioned by the FDA had also been contained in the
first edition. In addition, the FDA's comments raised fundamental concerns regarding the *Dictionary's* function as the controlling, consolidated source for chemical information on cosmetic ingredients. These
issues, and their resolution, had repercussions on the development
and production processes for both the second and third editions of
the *Dictionary*. The FDA's impact, therefore, on the *contents* of the
Dictionary involved not only the creation and use of acceptable nomenclature, but also its function as a chemical reference.

Nomenclature issues categorized into five areas were reviewed by
the CTFA and the FDA in a series of meetings and correspondence.
The depth of these issues, and the lengthy mechanics of their resolution, precipitated a decision by the CTFA to publish the second edition
prior to the completion of all the changes and clarifications requested
by the FDA. Because the CTFA placed equally high priorities on resolving questionable nomenclature issues and making available the
large volume of new, unquestioned nomenclature which comprised the
majority of the second edition, this edition was published in early 1977
and incorporated as many changes as were practicable without delaying
the *Dictionary's* availability [15]. Those modifications requested by
the FDA that were not reflected in the published second edition were
exempted from the proposed regulation of October 28, 1977 [16], and
the final regulation of January 18, 1980, which otherwise recognized
the 1977 edition of the *Dictionary* as the controlling source of ingredient labeling nomenclature [17].

The CTFA persued the clarification of exempted nomenclature
throughout the development of the third edition of the *Dictionary*.
Through discussions with the FDA, cosmetic producers, suppliers,
and other CTFA committees, the Nomenclature Committee systematically evaluated each remaining issue, implementing changes and modifications, or justifying the retention of certain conventions, where
appropriate. This effort culminated in the publication of the third
edition in March 1982, and the submission by the CTFA of a petition

in May 1982 to gain FDA recognition of this edition [18]. In addition
to the purpose of providing current and uniform labeling nomenclature,
the third edition serves as documentation of the resolution of the
issues raised by the FDA six years before its publication. The FDA's
action on the CTFA's petition to recognize the third edition will determine whether the questions raised concerning the acceptability of
several fundamental nomenclature conventions have been satisfactorily
resolved.

A discussion of the sequence of events and activities involved in
the resolution of specific nomenclature issues, while appropriate to
the illustration of the FDA's impact on the contents of the *Dictionary*,
would detract from the observation that those issues represented, in
fact, a small segment of the second edition's contents. More significant is the extent of the impact that those questions produced—in the
practical application of CTFA nomenclature and the principles underlying its development. With the exception of the deletion of prohibited
ingredients from the second edition, resolution of the issues enumerated by the FDA required deliberations spanning two editions of the
Dictionary. An evaluation of those issues and their impact follows.

A. Ingredients Whose Composition Is Undisclosed

The FDA cited second edition monographs for 34 coal-tar dyes for which
recognition was withheld due to the failure of the *Dictionary* to provide information on their composition in the form of a structural representation or chemical definition. Information for those monographs
was obtained from the *Color Index*, as was the nomenclature retained
for use in the *Dictionary*. In addition to indicating that terminology
such as "acid black 58" was not chemically informative, the FDA stated:

> Reinstatement of a name of a chemical substance listed in another
> source. . . without appropriate description of the substance's
> chemical composition, is contrary to the intent of making the dic-
> tionary a reference to other, more comprehensive sources of infor-
> mation. . . . The adoption of a nondescriptive name of an ingre-
> dient without disclosure of its chemical identity, origin, or compo-
> sition in a monograph would be meaningless to the consumer and
> defeat the purpose of cosmetic ingredient labeling [19].

Additional information on the chemical composition of these ingredients was not available for inclusion in the second edition. However,
the Nomenclature Committee concurred with the FDA's assessment of
the benefits of adding this information. In preparation for the third
edition, the CTFA issued a list of these ingredients and requested
information on their chemical composition or structure. Those ingredients for which no response was received were subsequently omitted
from the third edition.

Although the CTFA complied with the FDA's request to delete un-
defined nomenclature from the *Dictionary*, it also emphasized that the

removal of this terminology contradicted the purpose of the *Dictionary* in providing a consolidated and uniform source of nomenclature for cosmetic labeling. The CTFA pointed out that the FDA's request to remove *Colour Index* nomenclature from the *Dictionary* necessitated the use by manufacturers of other, more diversified sources for the purpose of ingredient labeling. In the absence of appropriate terminology in the other sources included in the labeling regulations (which, in the case of coal-tar dyes, did not exist), the manufacturer would be compelled to use a common chemical or technical name for the ingredient declaration. Because nomenclature from the *Colour Index* was so widely recognized, although an unofficial source of nomenclature, the CTFA suggested that manufacturers would ultimately use the same terminology to label coal-tar colors not covered by regulation, whether that terminology appeared in the *Dictionary* or exclusively in the *Colour Index*. The CTFA felt that to remove this widely accepted nomenclature from the *Dictionary* would detract from its purpose as an information source, as well as a collection of commonly recognized terminology. To create new nomenclature exclusively for the purpose of cosmetic ingredient labeling would also detract from the *Dictionary's* function as a compilation of uniform terminology for chemicals used not only in cosmetics but in related applications as well.

B. Ingredients Whose Composition Is Partially Disclosed

In their September 1976 comments to the CTFA, the FDA cited several ingredients for which they determined that the composition was not completely dislcosed. Unlike the 34 coal-tar dyes, the nomenclature for this second group of materials was not questioned. Instead, the FDA requested that the monographs for these materials be revised to include a more complete description of each ingredient's derivation or composition, and provide additional chemical information on these materials as available from patents and other scientific and regulatory literature. The FDA's emphasis in this case centered on the inadequacy of certain monographs to disclose chemical information, rather than their acceptability in terms of ingredient labeling nomenclature. The carbomer polymers and specially denatured (SD) alcohols exemplify the FDA's objections in this category.

Carbomers

The FDA stated that the monographs for carbomer polymers lacked important chemical information which had been disclosed in patent literature, and for that reason recognized this nomenclature in the second edition with the provision that this additional information be included in a future supplement [20]. Prior to publishing the second edition, the CTFA made an unseccessful attempt to obtain this information from the manufacturer of these materials. In its response to the FDA's comments on the second edition, the CTFA suggested that its inability to obtain the requested information was secondary to the goal of making available the large volume of new information contained in the second edition to the industry and consumers.

When the FDA provisionally recognized this nomenclature, the CTFA made a second unsuccessful attempt to obtain the missing information. By then, however, the carbomer polymers had been monographed, and recognized, in *USAN and the USP Dictionary of Drug Names*. In comments filed with the FDA after publication of the second edition, the CTFA stated that in the absence of acceptable nomenclature for carbomer polymers in the *Dictionary*, manufacturers could, according to the labeling regulations, refer to two alternative sources of nomenclature for these materials—USAN and NF—each of which contained descriptions of their chemical composition similar to the *Dictionary* monographs. In fact, the NF monograph provided a more generic labeling designation than either USAN or the CTFA. The CTFA pointed out that failure to recognize this nomenclature in the *Dictionary* would do little to increase the amount of chemical information available for these materials, and nothing in terms of providing more valuable information to the consumer in the form of a label declaration.

As in the case of *Colour Index* nomenclature for coal-tar dyes, the CTFA emphasized the value of retaining available information in favor of removal of incomplete monographs. It upheld the *Dictionary's* value as a chemical reference by publishing available chemical descriptions for the carbomer polymers in the third edition, as well as referencing the NF and USAN monographs for these materials. The CTFA maintains that the disclosure of this information should be sufficient, and preferable to the exemption of this nomenclature, for ingredient labeling purposes.

SD Alcohols

The FDA also cited 27 monographs for SD alcohols as containing insufficient information in their respective descriptions. Because the preparation of SD alcohols is regulated by the Bureau of Alcohol, Tobacco and Firearms in Title 27 of the *Code of Federal Regulations,* these regulations were included as information sources in the second edition monographs for these materials. In its comments to the CTFA, the FDA suggested:

> In order to properly serve the needs of dictionary users who do not have ready access to 27 CFR 212 it would be appropriate to list, in the definitions in the monographs, the denaturants of the various alcohols [21].

During discussions of these and other issues, the FDA indicated that a delay in the publication of the second edition was not warranted for the purpose of modifying the SD alcohol monographs. However, these monographs were included in the list of ingredients for which recognition of labeling nomenclature was provisionally accepted by the FDA. That provision, which requested that the specific denaturants

used in the preparation of each of the 27 SD alcohols be included in
the chemical definitions of these materials, was satisfied in the third
edition by citing the specific regulation for each SD alcohol monograph,
and including a complete list of its denaturants in the definitions for
each of these ingredients.

C. Ingredients Commonly Identified as Drugs

The FDA's comments in this category encompassed two areas: ingredients which are found exclusively in drugs, and ingredients found
in both drugs and cosmetics. The FDA's suggestion to remove those
ingredients ordinarily found only in drugs was based on a concern
that including such materials in the *Dictionary* would be misleading to
manufacturers by implying that their use was permissable in cosmetic
products. In order to expedite publication of the second edition, the
CTFA complied with the FDA's recommendation in this case, but objected strenuously to the FDA's rationale for this requirement. The
CTFA emphasized the *Dictionary's* function as a chemical reference
containing ingredient information obtained directly from manufacturers.
Although many of these ingredients may have been retained simply
as a result of the initial industry survey (in which entire product lines
were inadvertently submitted and included in the first edition), the
CTFA noted that if these types of materials were subsequently submitted as cosmetic ingredients, they would necessarily be considered for
future supplements and editions of the *Dictionary*. The CTFA further
stated:

> The Nomenclature Committee does not feel it is within its charge to
> search in such areas as the product category in which the ingre-
> dient is used, or the country in which the product is sold. . . .
> The Dictionary is merely a source document and does not determine
> the legal status of any ingredient. . . . It is designed neither as
> an encyclopedia nor as a substitute for the Code of Federal Regu-
> lations [22].

In keeping with the intention of the *Dictionary* to serve as a reference for materials used or promoted for use in the manufacture of cosmetic products, the language of the Regulatory Information section of
the third edition was strengthened to emphasize the obligation of the
manufacturers of cosmetic raw materials and finished products to comply with all federal regulations concerning the use and label declaration of cosmetic ingredients.

The FDA also cited several ingredients used in both drugs and cosmetics, requesting that the drug nomenclature be adopted in the *Dictionary* to avoid the possibility of confusion, and because the CTFA
failed to "provide any reasonable grounds in [its] petition for the new
names proposed" [23]. The CTFA's response stated that the Nomenclature Committee had indeed considered adopting existing drug

terminology for these ingredients, but concluded that the alternative nomenclature proposed in the *Dictionary* was, in fact, more informative to consumers and physicians, and provided a distinctive basis for identifying different functions of such ingredients in drug versus cosmetic products. The CTFA contacted representatives of USAN, requesting consideration of *Dictionary* nomenclature for the following ingredients:

> Aluminum chlorohydrex
> Amyl dimethyl PABA
> Butyl PABA
> Lime sulfur
> Octyl dimethyl PABA
> Sulfur

USAN subsequently published a monograph adopting CTFA nomenclature for aluminum chlorohydrex. The CTFA nomenclature for the ingredients listed above was recognized by the FDA in the second edition of the *Dictionary*, with the provision that name changes may be requested as a result of USAN's consideration and action on the remaining terminology. Comments in the final FDA regulation included the following:

> FDA stated in the preamble to the proposed regulation that it concurs with CTFA that the. . . six names may be more informative to consumers than their respective drug names. . . . FDA agrees that. . .referencing of alternate names may be useful to interested persons. The CTFA dictionary already lists both the drug and cosmetic names for these ingredients, and both sources disclose the chemical compositions of these substances. These descriptions and representations offer adequate information to determine the identities of the six ingredients [24].

Prior to publication of the third edition, the CTFA was notified by USAN of its intention not to adopt the CTFA nomenclature for the remaining ingredients under provisional recognition. In a meeting with the FDA in March 1980 the CTFA reiterated its position that *Dictionary* nomenclature for these materials was more informative, and suggested that allowing alternative nomenclature for these ingredients in drug and cosmetic applications might prove advantageous. The third edition includes monographs for butyl PABA, octyl dimethyl PABA and sulfur. Amyl dimethyl PABA and lime sulfur were omitted, due to their discontinued use in cosmetic products. The FDA's concurrance with the retention of this nomenclature will be determined in its recognition of the third edition.

D. Ingredients Prohibited by Regulation

The FDA cited 23 monographs in this category, including materials such as chloroform, vinyl chloride, and FD&C Red No. 2—ingredients for which use is restricted or not currently permitted—and carrot juice powder, chlorophyllin, and lamp black—ingredients for which use is prohibited by regulation. In its comments, the CTFA stated that the presence of these ingredients in the *Dictionary* was a result of requests or use by foreign countries in which such prohibitions did not exist. Emphasizing again the function of the *Dictionary* as a chemical reference, rather than a substitute for the regulations regarding use of cosmetic ingredients, the CTFA noted that because it was not in a position to deny any request for a CTFA Adopted Name assignment, nor its inclusion in the *Dictionary*, the restricted ingredients would be deleted from the second edition with the provision that they would be reinstated necessarily as a result of subsequent manufacturer submissions. The CTFA also strengthened references to federal regulations regarding prohibition of ingredient usage, and concurred with the FDA that in the future, banned ingredients would be deleted from the *Dictionary* as soon as practicable.

E. Ingredients with Questionable Nomenclature

Within this category, the FDA distinguished between ingredients with nomenclature reflecting usage or function, and ingredients with nomenclature which failed to provide a chemical representation of the material. Resolution of these issues was accomplished by several routes, which are highlighted below by ingredient category.

UV Absorbers

The FDA objected to the designation of ultraviolet (UV) absorbers on the basis of their function in a finished product. The FDA also pointed out that in cases where these materials were used for purposes other than UV absorption, this designation would be misleading [25]. In order to prevent a delay in the publication of the second edition, the CTFA elected to remove these ingredients from the *Dictionary*, pointing out that manufacturers would be required to use alternate designations in the period between publications. USAN terminology for these materials was subsequently adopted for the third edition.

Fluorescent Brighteners

The FDA made similar objections to the use of the terminology "fluorescent brighteners" to describe function, rather than chemical or structural identity. The CTFA removed these monographs from the second edition of the *Dictionary*. These ingredients were also among those cited by the FDA as being inadequately disclosed by chemical or structural respresentation. Because the CTFA was unsuccessful in obtaining sufficient information to provide either a more complete compositional disclosure or alternative nomenclature which satisfied the FDA's requirements for additional information, these ingredients were also omitted from the third edition of the *Dictionary*. As the CTFA stipulated in its comments, this action did not imply lack of use by the cosmetic industry, but instead required manufacturers using these materials to refer to alternate sources of nomenclature or other commonly recognized terminology, as required by labeling regulations.

Food Colors

Four ingredients with "food" designations were cited as misleading by the FDA due to their restricted use in coal-tar hair dyes. The FDA reasoned that use of this terminology would misrepresent the permitted use of these ingredients to consumers. The CTFA indicated that although this nomenclature was adopted directly from the *Colour Index*, where it is included without use distinctions, these terms would be removed from the second edition. The CTFA subsequently developed alternative nomenclature for these materials, and included them, along with cross-references to *Colour Index* nomenclature, in the third edition of the *Dictionary*.

Vitamins

The FDA termed the use of label declarations of "vitamins" misleading to consumers because this nomenclature "conveys the impression that these ingredients offer a nutrient or health benefit when used in a cosmetic" [26]. As the FDA suggested, this nomenclature was changed to the chemical representations for publication in the second edition of the *Dictionary*. (For example, vitamin C became ascorbic acid, vitamin E became tocopherol.) Subsequent name assignments for vitamins reflected this distinction of function in cosmetic products, and chemical terminology for these ingredients was incorporated into the third edition of the *Dictionary*.

Amphoterics

The 20 ingredients identified in the second edition by the term "amphoteric" plus a numeric designation, were cited by the FDA as being inconsistent with other nomenclature because these terms did not reference chemical structures or classifications [27]. The FDA recognized this nomenclature for the purpose of including these materials in the regulations for labeling, but placed a time limit on its use which would coincide with the development of alternative nomenclature by the CTFA within one year of the second edition's recognition. The CTFA concurred with the requested redisignation. However, the lack of existing acceptable nomenclature alternatives necessitated the development of new terminology for these materials. New monographs for these ingredients, containing structural representations, are included in the third edition. At a meeting in March 1980, the FDA agreed conceptually with the alternative designations for amphoterics [28].

Quaterniums

The FDA objected to the 54 ingredients termed "quaternium" plus a numeric designation, because this nomenclature implied similarities

within this group that were not indicated by their structural representations. Although the quaternium terminology was acceptable for
a smaller class of ingredients, the FDA's "recommencation for renaming
[these] ingredients . . . was prompted primarily by the agency's concern that the similarity in nomenclature of ingredients having non-
related properties and usage may confuse or deceive consumers" [29].
The CTFA agreed to develop alternative nomenclature for these materials.

The FDA accepted the quaternium designation for a period of one
year after the publication of final regulations recognizing the second
edition of the *Dictionary*, to allow for the completion of the redesigna-
tion effort by the Nomenclature Committee. It became apparent, however, that the Nomenclature Committee could not meet the one-year
deadline established by the FDA. Therefore, the FDA extended the
recognition period for quaternium nomenclature, pending publication
of the third edition of the *Dictionary* [30]. In the meantime, a lengthy
series of discussions and correspondance were exchanged between the
CTFA and the FDA regarding the difficulty in establishing alternative
nomenclature for quaternium compounds that was not overly cumbersome for labelers or confusing to consumers. The CTFA argued that
that this nomenclature had been "chosen to provide the consumer with
easily recognizable terminology and yet highlight the primary chemical structure of these materials" [31]. The CTFA received repeated
indications from the FDA that this approach would not be acceptable
for ingredient labeling purposes. The Nomenclature Committee, therefore, developed several solutions which provided alternative types
of nomenclature. One subclass of ingredients, called "polyquater-
niums," was created to denote the common polymeric characteristic
of these materials. Chemical terms were also adopted in cases where
this type of designation was not overly cumbersome to labelers,
or too technical for recognition by consumers. In certain cases, the
"quaternium" designation was retained in order to eliminate the confusion that the most burdensome of these chemical terms would cause,
and because their use had become widespread in scientific literature
and data bases.

Upon receiving additional indications from the FDA that these alternative approaches were considered unacceptable, the Nomenclature
Committee developed chemical designations for 64 quaternary ammonium
compounds to be included in the third edition of the *Dictionary*. In
a February 26, 1981, letter to the FDA, the CTFA summarized its
position with regard to the proposed alternative nomenclature in the
following statements:

> We believe that the use of these [quaternium] names is consistent
> with the practice of providing concise, informative names for the
> purpose of ingredient labeling while at the same time, providing
> in the Dictionary definitions, chemical structures and other names

to further describe the ingredients. . . . Although the alternative nomenclature has been developed to address fully the FDA
objections to the Quaternium nomenclature now in use. . . neither
the Nomenclature Committee nor CTFA's Scientific Advisory Committee considers the new approach to be in the best interests of
the consumer, FDA or the cosmetic industry [32].

In its response, the FDA recognized the efforts of the CTFA to
provide meaningful terminology which did not result in additional
confusion or burden on the part of consumers and industry:

> [T]he agency supports your decision to petition for the use of the
> alternative nomenclature when it is clearly in the interest of the
> consumer and to work to otherwise limit the total number of com-
> pounds that may fall within this category [33].

The resolution of this issue, as well as others raised by the FDA
in 1976, represented the CTFA's effort to obtain unconditional recognition of the third edition of the *Dictionary*. That the number of ingredients initially questioned by the FDA in terms of their nomenclature or definition involved only a small percentage of the monographs
contained in the second edition, was a testament to the industry's
sound approach in its development and adherance to the nomenclature
system presented in the *Dictionary*. The satisfaction of these limited
qualifications on the *Dictionary*'s contents in a manner acceptable to
the FDA and the industry, and meaningful to the consumer, will be
much more significant to the *Dictionary*'s credibility and success as
an industry-generated publication worthy of preferred regulatory
status.

VI. THE THIRD EDITION: CONTENTS AND APPLICATIONS

Despite changes in the *Dictionary*'s contents, reflecting the FDA's
impact and the consequences of fluctuations and trends in the cosmetic
ingredient marketplace, the format of the *Dictionary* has changed little
in its 10 year history. The *Dictionary* was designed to provide a concise, convenient body of ingredient information, presented in a format
which maximizes its meaningful use by a wide variety of scientific,
regulatory, and consumer groups. The success of this design, as
evidenced by its consistency through three editions, must be credited
to Dr. Norman Estrin—the *Dictionary*'s founding editor. Dr. Estrin's
assessment of industry, FDA, and consumer requirements for chemical
information on cosmetic ingredients, coupled with his perception of the
most effective organization of this information, resulted in the nine-
section format into which the *Dictionary* is divided. A description of
the contents and functions of each of these sections follows [34].

A. Foreword and Preface

Section I: Foreword

This section provides a brief history of the *Dictionary*'s development and a description of its intent to provide a uniform source of nomenclature that is informative to the industry, the FDA, and consumers.

Section II: Preface

This section highlights the scope of subsequent sections of the *Dictionary* and describes the activities and intentions of the Nomenclature Committee in their systematic evaluation of submissions of cosmetic ingredient information for the purpose of developing concise, uniform ingredient labeling nomenclature.

B. Introduction

The Introduction (section III) provides information essential to the use and evaluation of subsequent sections of the *Dictionary*. Although this section is too often overlooked, it is the key to the interpretation of ingredient information and its use in industry, regulatory, medical, and consumer applications.

Nomenclature Rules

The 40 rules listed in this section document the conventions established by the Nomenclature Committee. This is a cumulative list, which reflects the original guidelines established for development of the first edition as well as subsequent rules established by the Nomenclature Committee as a result of FDA impact and new industry issues.

In addition to describing the mechanics of establishing nomenclature for the *Dictionary*, these rules offer explanations of policy with regard to conventions developed in conjunction with regulatory requirements and the appearance in subsequent sections of terminology for which additional qualification is required. For example, rule 28 summarizes the use of name/number designations in the *Dictionary* and explains the necessity for this type of nomenclature (including the use of quaternium designations in limited cases, as recognized by the FDA). Rule 33 describes the use of the term "ampho" as a combining term in the nomenclature developed as an alternative to the amphoteric designations contained in the second edition. Rule 34 defines the term "And Other Ingredients," appearing in section VI of the *Dictionary*, as indicative of the trade-secret status granted certain materials by the FDA. Finally, rule 39 offers an explanation of the absence, in some monographs, of chemical or trade names for the CTFA Adopted Name. In these cases, the use of an ingredient is usually limited to one product or manufacturer and not commercially distributed. Its appearance in the *Dictionary* is indicative of the assignment of a generic CTFA Adopted Name, for use by such a manufacturer in ingredient labeling compliance.

Abbreviations

This section contains a list of abbreviations used in CTFA Adopted Name
assignments appearing as monographs in section V of the *Dictionary*,
and in the definition portions of those monographs. This list is intended to serve as the key to the appearance of abbreviated nomenclature throughout the text.

Registry Numbers

This section notes the occurance of CAS numbers and recognized
disclosure (RD) numbers in the text, and supplies explanations of
their derivation and use. In particular, the use of these numbers is
described in conjunction with the voluntary disclosure of formulation
data to the FDA. The conditions of this application are provided in
detail, and qualify the use of CTFA nomenclature, accompanied by
appropriate registry numbers, in the context of the Voluntary Disclosure Program.

RD numbers are assigned by the FDA for the purpose of identifying
materials which have not yet been assigned a CAS registry number.
In the Voluntary Disclosure Program, these numbers are used to provide additional identification to the CTFA Adopted Names included in
formulations. The FDA assigns a second type of registry number under this program—the cosmetic raw material composition statement
(CRMCS) number—to trade name ingredients whose composition has
been voluntarily disclosed. These numbers appeared in the second
edition as a separate section, and were not included in the third edition.

Information Sources

Section V of the *Dictionary* contains abbreviated references to additional sources of information for monographed ingredients. Part 4 of
the Introduction contains an explanation of their use, and an alphabetical listing of the complete references for each of the 58 Information Sources cited in the Monograph section.

Changes

This section is an important guide to those who use the *Dictonary* for
ingredient labeling purposes. It contains a two-column list of nomenclature which has been revised from the previous edition, and, in the
case of the third edition, includes 35 quaternium compounds for which
new nomenclature has been created as a result of FDA requirements.
In other cases, nomenclature changes reflect modifications of terminology intended to provide clarification or abbreviation of previously
established nomenclature.

This list serves as a convenient summary of revisions which, with
FDA recognition of the third edition, will become mandatory for ingredient labeling declarations.

Deletions

This section lists CTFA Adopted Names which have been removed from the current Monograph section of the *Dictionary*. In the third edition, this list includes the coal-tar colors exempted from recognition by the FDA for incomplete disclosure of their composition, and the 20 amphoteric materials for which new designations were also requested by the FDA. In the case of the amphoterics, the revisions in nomenclature required a reclassification of these compounds by chemical structure. The result of this revision prevented the renaming of existing monographs, requiring instead the development of a larger number of new monographs. To emphasize the extent of this revision, the amphoterics are listed as deletions from the third edition, and the new, revised monographs are included in section V.

This list is another convenient source for required changes in labeling nomenclature. Manufacturers may refer to this list to determine those ingredients for which alternate sources of nomenclature must be obtained, as stipulated in the labeling regulations.

Conventions

Several features of sections V and VI of the *Dictionary* are described in this section. Included is an explanation of the term "(and)", used as a linking term for materials having multiple components and corresponding CTFA Adopted Names, and a caution against including this term in ingredient labeling statements. A new feature of the third edition, the *Materials Containing* portion of the Monograph section, is also discussed. This section also identifies terms enclosed by parentheses, which are associated with "Other Names" as suppliers of these raw materials, or, in the case of certified colors, the country of usage.

C. Regulatory Information and CTFA Nomenclature Assignment Procedures

Section VI A: Regulatory Information

This section contains important information concerning the responsibility of manufacturers of cosmetic ingredients and products to comply with regulatory requirements for safety substantiation and ingredient disclosure. Emphasis is placed on the responsibility of the manufacturers of raw materials to provide safety substantiation for their ingredients, and the manufacturers of finished products to consult and comply with labeling regulations included in the *Code of Federal Regulations* and the *Federal Register*.

This information also disqualifies the *Dictionary* as a substitute for regulations regarding the use or labeling of cosmetic ingredients, and stipulates that the inclusion of an ingredient in the *Dictionary* does not constitute FDA approval or CTFA endorsement of a cosmetic raw material, or any supplier of that material.

Section IV B: CTFA Nomenclature Assignment Procedures

This section contains a summary of the procedures required to request
a CTFA Adopted Name assignment or an amendment to a previous
assignment. Guidelines pertaining to the types of information required
for consideration by the Nomenclature Committee are included, as well
as a statement of CTFA's intent to periodically submit new and revised nomenclature for formal recognition by the FDA. This section
also states the assumption by the CTFA that all information submitted
by manufacturers for review by the Nomenclature Committee is accurate,
and serves exclusively as the basis for all nomenclature assignments.
Unacceptable for review is information submitted by one manufacturer
which pertains to the product(s) of another manufacturer, or information which is submitted as confidential. Manufacturers of cosmetic
raw materials or finished products wishing to obtain a CTFA Adopted
Name assignment for any ingredient should consult this section prior
to making their submission.

D. Monographs

The Monograph section (section V) is the largest section of the *Dictionary*, and that from which all subsequent sections are derived.
This section consists of an alphabetical listing of monographs for each
CTFA Adopted Name assignment, accompanied by additional information
pertaining to each ingredient. This information is presented in a
standard format, consisting of the categories described below.

CTFA Adopted Name

The name assignment given to a cosmetic raw material by the Nomenclature Committee is the monograph heading, and the terminology intended for use in ingredient labeling, as recognized by the FDA.

Registry Numbers

Ingredients for which *Chemical Abstracts* or the FDA has assigned
numeric identifiers list CAS and/or RD numbers directly beneath
the CTFA Adopted Name. In cases where an ingredient has been
assigned registry numbers by both groups, the CAS number is retained due to is widely recognized usage. Some monographs list
multiple CAS number assignments because components of their structural representation are common to more than one material previously
identified by a CAS number. For groups of ingredients possessing
identical CAS numbers (a second circumstance of common structural
classification), the RD number is also retained to further distinguish
each ingredient within the group. The first and second editions of
the *Dictionary* included the CAS number prefixes MX (mixture) and
PM (polymer). The use of these prefixes has been discontinued by
Chemical Abstracts, and they have been eliminated from the third
edition of the *Dictionary*.

These registry numbers are included in ingredient monographs
for several reasons. First, as mentioned in section III of the *Dictionary,* these numbers are required when using CTFA Adopted Names
in compliance with the Voluntary Disclosure Program for ingredient
and formula declarations. The second function of these registry numbers, in particular the CAS designation, is their value in literature
searching and retrieval of information from scientific data bases. In
addition, CAS numbers provide a common and uniform identification
to chemicals of concern to other federal agencies whose activities may
also be of interest or concern to cosmetic producers or suppliers.
CAS numbers enable manufacturers to cross-reference their materials
to lists such as the TSCA Inventory, compiled by the Environmental
Protection Agency (EPA), or compendia such as the USP, USAN, and
the *Merck Index.* In summary, these registry numbers serve as a
valuable tool in the identification of chemicals for a wide variety of
functions and applications. Their inclusion in the *Dictionary* serves
to perpetuate the recognition of cosmetic ingredients as one component
of the larger chemical sphere.

Empirical Formula

This category is included in the Monograph section for the first time
in the third edition. These formulas traditionally identify chemicals
at the atomic level, and provide a second method for literature searching
and information retrieval for ingredients in the *Dictionary.* Empirical
formulas also provide additional assistance in cross-referencing cosmetic ingredients to pharmaceutical compendia.

Definition

Every ingredient in the *Dictionary* is defined in terms of its chemical
composition, structural representation, or, in the case of natural
materials, its species of origin. Chemical formulas are included whenever possible to provide the most accurate and concise description
of the ingredient. In the third edition of the *Dictionary,* a number
of definitions were expanded to accommodate the FDA's requirements
for more complete chemical disclosures, including the 26 SD alcohol
monographs specified in the FDA's final regulations. The appearance
of the word "generally" in conjunction with the disclosure of a chemical
formula in the definition of an ingredient denotes the lack of a clear-
cut structural representation for that ingredient.

Information Sources

This portion of the monograph contains the abbreviated references
listed in section III of the *Dictionary.* This information is provided
as a tool for obtaining additional data on a specific ingredient. In
some cases, an information source also indicates pertinent regulatory
references or the original source of nomenclature adopted from other

compendia for use in the *Dictionary*. References to the TSCA Inventory are also included in this section for the first time in the third edition. One unique feature of this category is its reference to international sources of information. As the worldwide availability of information on cosmetic ingredients expands, the value of these references will increase—particularly in the area of safety substantiation.

Other Names

This category provides alternative nomenclature for CTFA Adopted Names. The majority of these additional names are trade names submitted to the Nomenclature Committee by manufacturers or users of cosmetic raw materials. Other types of nomenclature include chemical terms which denote the unabbreviated form of a CTFA Adopted Name, more commonly recognized or functional terminology, *Chemical Abstracts* nomenclature which corresponds to the CAS registry number(s) for a given monograph, and nomenclature included in other compendia (for example, USAN and USP). In monographs for color additives, additional terminology includes *Colour Index* designations (CI, followed by a five- or six-digit numeric identifier), and nomenclature used in foreign countries (for example, Germany: C-Blau 17).

"Other Names" are intended to represent, in their entirety, the ingredient defined by a particular monograph, with the exception of extracts in which the solvent is exempted from declaration—as listed in nomenclature rule 30. Certain ingredients may include a natural distribution of materials within which one component predominates. The Nomenclature Committee evaluates each submission for indications of the alteration of this type of natural distribution, and in cases where such distribution is not altered, will include these trade name materials as "other names." When modification of this natural distribution is indicated in a submission, the material cannot be listed as an "other name." This distinction is important to formulators who must incorporate combinations of CTFA nomenclature into ingredient labeling declarations.

Materials Containing

This category was created to make the cross-referencing more comprehensive in the third edition by listing additional trade-name materials in the Monograph section of the *Dictionary*. The Materials Containing category encompasses ingredients submitted as blends or mixtures of two or more monographed ingredients, and materials within which the natural distribution of components has been modified. The second edition of the *Dictionary* included these types of ingredients in its alphabetical listing of chemical and trade names, but omitted them from the Monograph section. Compounds listed in the Materials Containing section of a monograph are included to denote the presence of the monograph ingredient as one component of a blended material. The trade name for this type of material will appear under each mono-

graphed ingredient indicated in the mixture. This category distinguishes materials for which the name assignment provided by the
Nomenclature Committee consists of multiple CTFA Adopted Names.
In section VI of the *Dictionary*, these multiple name assignments appear in their descending order of predominance. The Materials Containing category has greatly expanded the Monograph section of the
Dictionary, and provides a more accurate reflection of ingredient
usage as a result.

E. Chemical/Trade Names

Section VI of the *Dictionary* consists of an alphabetical listing of chemical and trade names for all CTFA Adopted Names monographed in
section V. This section is of key importance to individuals responsible
for preparing ingredient labeling declarations for cosmetic products.
Due to the computerization effort by the CTFA, this section is especially comprehensive in the third edition.

Section VI lists all Other Names and Materials Containing terms
contained in the Monograph section, and also incorporates all CTFA
Adopted Names to facilitate the evaluation of information within this
section. Qualifying terms indicated by parenthesis accompany CTFA
nomenclature, as well as chemical and trade terminology, within this
list. These terms include supplier designations, modifying terms for
chemical nomenclature [for example, polyoxyethylene (4) cetyl ether],
and, in the case of color additives, the country of usage. This qualifying terminology is carried over from the section V monographs.

Two additional distinctions are unique to section VI, and consist
of the "(and)" and "(or)" designations. These terms are found in
the CTFA Adopted Name column and are used to indicate name assignments consisting of multiple CTFA Adopted Names, and common terminology which is applicable to two CTFA Adopted Name monographs.
As mentioned in section III of the *Dictionary*, the (and) designation
is not part of the CTFA Adopted Name assignment for a trade name
material, but is included to separate CTFA nomenclature terms and
to indicate that these terms appear in their descending order of predominance. Manufacturers using an ingredient with such a multiple
name assignment are required, as part of the labeling regulations,
to incorporate individual terms into the complete ingredient statement,
retaining the descending order of component terms as they occur in
the complete formulation.

The (or) designation is provided for information purposes and does
not denote interchangeability of CTFA nomenclature for ingredient
labeling purposes. It is used primarily in conjunction with nomenclature for color additives, but also for commonly recognized terminology
that may be applicable to more than one CTFA monograph. In the
case of color additives, the (or) term reflects the association of *Colour
Index* designations (in the form of a five digit CI number) with the

certified and noncertified nomenclature for a color additive material.
The *Colour Index* designation is included in CTFA monographs for
reference purposes only. Manufacturers using color additives should
consult Title 21 of the *Code of Federal Regulations* for information on
the use and ingredient declarations for these materials.

F. Registry Numbers and Empirical Formulas

Sections VII and VIII of the *Dictionary* are included as convenient
compilations of information included in the Monograph section. Their
intent is to serve as aids in literature searches and information retrieval of cosmetic ingredient data.

Section VII: Registry Numbers

This section consists of a numeric listing of registry numbers found
in the Monograph section of the *Dictionary*, and cross-references
these numbers to their corresponding CTFA Adopted Names. This
list is designed to facilitate the identification of CTFA nomenclature
by its association with CAS and RD number designations. As a consequence of numerical listing, this section also groups ingredients by
their common CAS identifiers, and is indicative of chemical or structural similarities between ingredients.

Section VIII: Empirical Formulas

This section consists of an alphanumeric listing of empirical formulas
for monographed ingredients, cross-referenced to their corresponding
CTFA Adopted Names. This list provides a second method of identifying cosmetic ingredients on the basis of their chemical structure,
and also denotes structural similarities between monographed ingredients. This section may also be used as an aid in qualifying empirical
formula information obtained from scientific data bases or other chemical compendia, such as USAN and USP.

G. Supplier Index

Section IX of the *Dictionary* provides an alphabetical list of raw material manufacturers and suppliers, including the complete company
name, address, and telephone number, plus a comprehensive list for
each company of all trade name materials monographed in the *Dictionary*.
It is important to note that only those trade name materials submitted
to the Nomenclature Committee and assigned a CTFA Adopted Name
are included in this section. Raw material purchasers may use this
section as a convenient reference to sources of cosmetic ingredients,
but should note that this information does not necessarily constitute
a company's complete product line. Section IX also includes listings
for companies serving as distributors of raw materials manufactured
overseas. These listings serve only as cross-references to foreign

manufacturers. The name of a U.S. distributor does not appear in conjunction with trade name materials in sections V or VI of the *Dictionary*.

H. Additional Applications

The contents of the *Dictionary* are constructed to provide information and to serve the needs of cosmetic manufacturers in their preparation of ingredient labeling declarations. The use of the *Dictionary* for this purpose, in turn, provides the consumer with consistent statements of ingredient content on finished products that facilitate value comparisons and the assessment of the suitability of a cosmetic product for individual use. An important function of the *Dictionary* is its use by physicians to evaluate their patients' sensitivity to a particular component of a cosmetic product. With information provided by the *Dictionary*, a consumer may avoid further contact with specific ingredients through the technique of label comparison. This being one of the ultimate goals of cosmetic ingredient labeling requirements, the value of the *Dictionary*'s design and content becomes more apparent. Through the concise, yet comprehensive presentation of chemical information on cosmetic ingredients, applications of the *Dictionary*'s contents span the life of a cosmetic product—from its formulation and production to its use by consumers. That the *Dictionary*'s design serves this wide variety of applications through its integrated and convenient format is one of the greatest measures of its utility and success as an industry compendium.

Farther-reaching applications of the *Dictionary*'s contents include its incorporation into scientific data bases for the purpose of information retrieval. The CTFA is preparing to enter selected *Dictionary* information into the EPA's Chemical Information System (CIS) data base, where it will become part of a vast collection of chemical data and literature. This effort will greatly expand the availability of data on cosmetic ingredients, and the association of these data with chemical information from related industries.

The *Dictionary*'s application to the evaluation of cosmetic ingredient safety may be exemplified by its role in the Cosmetic Ingredient Review (CIR) program. Not only does the *Dictonary*'s format facilitate the accumulation of chemical data and literature on cosmetic ingredients, but the Adopted Names listed in the *Dictionary* define the realm of ingredients to be evaluated by CIR's Expert Panel [35]. The definitive list of cosmetic ingredients that the *Dictionary* provides is weighted according to data on frequency of use to establish CIR's Priority List for safety evaluation. Without the *Dictionary*, and its function as a consolidated source of uniform ingredient terminology, the CIR program might not have been possible [36]. With the *Dictionary*, this industry-sponsored program has become the center of the cosmetic industry's voluntary effort to ensure the safety of its products.

An essential prerequisite to the availability of cosmetic ingredient information for purposes including, and surpassing, the scope of ingredient labeling compliance involves the maintenance of up-to-date data on cosmetic raw material through the continuing evaluation and publication of new information in supplements and revisions to the current edition of the *Dictionary*. This effort remains the foundation of the *Dictionary*'s impact and its variety of applications. Each publication serves as the tangible basis for the utility and credibility of the industry-developed nomenclature system embodied in the *Dictionary*. In recent years the *Dictionary*'s utility has expanded to a worldwide scale, based, in part, on its contents and their international relevance. The *Dictionary* has consistently provided data in a format that is appropriate to international interest and use, through its integration of registry numbers and information sources with chemical and trade terminology. CTFA nomenclature is also finding increasing use and recognition, as exemplified by its appearance in publications such as Italy's *Cosmetic Index* [37]. This expansion of the *Dictionary*'s applications into the international sphere will become further evidence of the versatility of its format and consistency, as the cosmetic industry is made increasingly aware of this collection of data through future publications.

VII. CONCLUSION

From its beginning as a collection of chemical and trade terminology, the *Dictionary* has had a widespread impact on the cosmetic industry. As the vehicle for compliance with ingredient labeling regulations, the *Dictionary* has become one manifestation of the industry's commitment to self-regulatory practices. The *Dictionary*'s service as a comprehensive source of information on cosmetic ingredients satisfies demands for ingredient data by scientists, regulators, physicians, and consumers. The role of the *Dictionary* in the evaluation of ingredient safety is fundamental in its provision of a definitive assembly of cosmetic raw material data and references to worldwide ingredient information.

The dedication of CTFA scientists and staff to the perpetuation of a uniform nomenclature system for cosmetic ingredients cannot be overemphasized. Their efforts have resulted in the increasing recognition of CTFA Adopted Names as "the international language of the cosmetics field" [38]. The FDA's contributions to the status and reputation of the *Dictionary* have also been essential to its impact on the cosmetic industry.

The *Dictionary* is dedicated to ". . .the world's consumers, as a continuing expression of the cosmetic industry's appreciation of consumer needs and desires" [39]. In this sense, it is the central component of a larger effort by industry and the FDA to ensure consumer protection and awareness. The *Dictionary* represents the foundation of the cosmetic industry's participation in this effort, and embodies the dedication of the industry as a whole to its continuing success.

REFERENCES

1. Norman F. Estrin, Using the CTFA Cosmetic Ingredient Dictionary, presented at the January 6, 1982, meeting of the New York Chapter of the Society of Cosmetic Chemists, p. 1.
2. Norman F. Estrin, Self-Regulation in the Cosmetic Industry— A View from a Scientific Vantage Point, M. Balsam and E. Sagarin, eds., in *Cosmetics: Science and Technology*, Vol. 3, 2nd ed. John Wiley & Sons, 1974, p. 473
3. *Fed. Reg. 38:* 3523 (February 7, 1973).
4. *Fed. Reg. 37:* No. 156 (August 11, 1972).
5. *Fed. Reg. 38:* 3524 (February 7, 1973).
6. James, M. Akerson, The CTFA Csometic Ingredient Dictionary and Its Uses, *CTFA Cosmet. J. 8* (3): 21 (July—September, 1979).
7. *Fed. Reg. 42:* 18061 (April 5, 1977).
8. Citizen's Petition: Letter from James H. Merritt, President, CTFA, to Dockets Management Branch, Food and Drug Administration, May 28, 1982, p. 2.
9. James, M. Akerson, Labeling Nomenclature for Cosmetic Ingredients, *Cosmet. Technol.* (1): (October 1979). 24—70
10. Norman F. Estrin, The CTFA Dictionary—How It Can Help You in Your Practice, presented as part of the American Medical Association Council on Continuing Physician Education's Postgraduate Course on "Cosmetics in Dermatology," San Francisco, California, 1977, p. 30.
11. *Fed. Reg. 38:* 3525 (February 7, 1973).
12. Letter from James H. Merritt, President, CTFA, to Miss Beryl McCullars, Hearing Clerk, Department of Health, Education and Welfare, April 9, 1973, pp. 8—9.
13. *Fed. Reg. 38:* 28912 (October 17, 1973).
14. Citizen's Petition: Letter from James H. Merritt, President, CTFA, to Ms. Jennie Peterson, Hearing Clerk, The Food and Drug Administration, June 24, 1976, p. 1.
15. Letter from James H. Merritt, President, CTFA, to Joseph P. Hile, Acting Associate Commissioner for Compliance, U.S. Food and Drug Administration, November 2, 1976, p. 1.
16. *Fed. Reg. 42:* 56757—56760 (October 28, 1977).
17. *Fed. Reg. 45:* 3574—3578 (January 18, 1980).
18. Citizen's Petition, May 28, 1982, p. 1.
19. *Fed. Reg. 45:* 3575 (January 18, 1980).
20. *Fed. Reg. 45:* 3577 (January 18, 1980).
21. Letter from Joseph P. Hile, Acting Associate Commissioner for Compliance, Food and Drug Administration, to James H. Merritt, President, CTFA, September 17, 1976, p. 6.

22. Letter from Merritt, November 2, 1976, p. 4
23. Letter from Hile, September 17, 1976, p. 3.
24. *Fed. Reg. 45:* 3575 (January 18, 1980).
25. Letter from Hile, September 17, 1976, p. 5.
26. Letter from Hile, September 17, 1976, p. 5.
27. Letter from Hile, September 17, 1976, p. 5.
28. Minutes of March 7, 1980, meeting between FDA and CTFA, April 1, 1980, p. 2
29. *Fed. Reg. 45:* 3577 (January 18, 1980).
30. Letter from Joseph P. Hile, Associate Commissioner for Regulatory Affairs, Food and Drug Administration, to Norman F. Estrin, Ph.D., Senior Vice President—Science, CTFA, June 25, 1981.
31. Letter from Norman F. Estrin, Ph.D., Senior Vice President—Science, CTFA, to Taylor Quinn, Food and Drug Administration, July 15, 1980, p. 4.
32. Letter from Norman F. Estrin, Ph.D., Senior Vice President—Science, CTFA, to Taylor M. Quinn, Associate Director for Compliance, Food and Drug Administration, February 26, 1981, p. 2.
33. Letter from Joseph P. Hile, Associate Commissioner for Regulatory Affairs, Food and Drug Administration, to Norman F. Estrin, Ph.D., Senior Vice President—Science, CTFA, July 23, 1981.
34. N. Estrin, P. Crosley, and C. Haynes, eds., The CTFA Cosmetic Ingredient Dictionary, 3rd ed. The Cosmetic, Toiletry and Fragrance Association, Washington, D.C., 1982.
35. Robert L. Elder, ScD., Director, The Cosmetic Ingredient Review, personal communication, July 1, 1982.
36. Estrin, January 6, 1982, p. 5.
37. G. Proserpio, E. Berdi, and A. M. Massera, eds., *Cosmetic Index: Dizionario Della Materie Prime Cosmetiche,* Sinerga, Milan, Italy, 1981.
38. *Cosmetic Index,* p. 1.
39. *Cosmetic Ingredient Dictionary,* p. ii.

25

COPING WITH THE COLOR ADDITIVE REGULATIONS

MURRAY BERDICK

Consultant, Branford, Connecticut

I. INTRODUCTION

Color is so fundamental to cosmetics that it is startling to reflect that the cosmetic industry was almost oblivious to government regulation of colors until 1960. Only a handful of people in the industry were aware of increasing uneasiness in the government about the legalistic interpretation of ". . .harmless and suitable for use. . .", which had guided the regulation of coal-tar colors for foods, drugs, and cosmetics since the major overhaul of the Food, Drug, and Cosmetic Act (FDC Act) in 1938 [1]. The infamous batch of Halloween candy in which FD&C Orange No. 1 was misused (as a pigment instead of a dye) led to the delisting of that color, plus FD&C Orange No. 2, and FD&C Red No. 32 in 1955. This still had no impact on cosmetics, but the citrus industry was in distress over loss of means to color the skins of oranges. In a curious parallel to Congressional action on saccharine more than 20 years later, a special law [2] passed in 1956 extended for almost three years permission to continue this limited use of FD&C Red No. 32.

By 1957 the Food and Drug Administration (FDA) view of "harmless and suitable for use" had hardened into "harmless in any amount," and four more food colors (FD&C Yellows No. 1, 2, 3, and 4) were proposed for delisting. So far, action had affected colors used primarily in food, but the first hints came in 1957 that important cosmetic colors would be involved (on the basis that lipsticks are ingested). During 1958 Congress passed the major amendment to the FDC Act on food additives [3]. An important part of the debate centered on the so-called Delaney anticancer clause. Interestingly, the FDA was opposed to the clause, but the alleged dangers of food additives were fresh on the minds of Congress.

By the end of 1958 the Supreme Court created a crisis by endorsing the FDA position that "harmless and suitable for use," with respect to "coal-tar" colors, meant "harmless *per se*" [4]. The FDA had no choice but to propose to delist 17 D&C colors, including most of the important lipstick colors. An emergency law was passed to permit use of a new dye, Citrus Red No. 2, for orange skins, for 18 months [5]. Final action on lipstick colors was delayed during 1959, while Congressional hearing were held on legislation proposed by the FDA. Although less than enchanted with some of the provisions of the proposed Color Additive Amendments of 1959, representatives of the food, drug, cosmetic, and color industries were forced to support the main thrust of the administration bill, to allow tolerances to be set for safe use of colors. Congress insisted on inserting a version of the Delaney clause, despite the fact that it had not been a part of the FDA bill.

The FDA proposal was conveyed to the Senate by Arthur S. Flemming, Secretary of Health, Education and Welfare (HEW). A quotation from his letter of transmittal of May 29, 1959 [6] provides a striking illustration of the universality of the first propensity of Haga's law ("Anxiety begets organizing") [7], which was not discovered till many years later:

> The bill is designed to meet a pressing need for replacing the inconsistent, and in part outmoded, provisions which now govern the use of different kinds of color. . . with a scientifically sound and uniform system for the listing of color additives of any kind. . . [6].

More than twenty years have elapsed since the passage of the Color Additive Amendments of 1960 [8]. This expansion of the bureaucracy, under threat of an emergency, has not protected the public from harm any more than a simple insertion into the law of authority for the FDA to adopt tolerances for the already-regulated "coal-tar" colors. The massive efforts by both industry and government to meet the burdensome requirements of the 1960 law have never produced any evidence that the use of colors in foods, drugs, and cosmetics would have caused harm under a more realistic system of regulatory control, but they do illustrate the second propensity of Haga's law ("Organizing begets anxiety") [7].

Perhaps this Introduction will illuminate the puzzling question of why government authority over cosmetic colorants is so far out of line with any meaningful evaluation of risk to the consumer.

II. SCOPE

This chapter offers guidance to the user (in the United States) of cosmetic colorants. The user, i.e., the developer or manufacturer of cosmetic/toiletry products, is fortunate to have available the guidance of the color manufacturer, whose knowledge of color additives goes far beyond the scope of this chapter. The manufacturer of the color additive must know all the legal niceties of dealing with the FDA about petitions, certification, hearings, and appeal procedures, as well as the technical factors of chemistry, properties, specifications, manufacturing processes, analytical procedures, pharmacology, and toxicology.

Thus the brunt of regulatory compliance is borne by the color manufacturer or the petitioner, but the maze of the regulatory system still leaves a good deal of coping up to the user. One feature of the law and the regulations that will not be dwelt on here is the distinction between the "listing" of color additives and the "provisional listing." This feature of the law, intended by Congress to provide a two-and-a-half-year transition, has been stretched by a dilatory bureaucracy to more than 20 years. It is a monumental illustration of Parkinson's law of delay ("Delay is the deadliest form of denial") [9].

As this is written, the status of many color additives is in flux because the reports from a third major round of safety studies are under evaluation by the FDA. Any attempt to distinguish between "permanent" and "provisional" listing will be outdated by the time of publication. Because provisionally listed additives are at greater risk of sudden delisting, it may be advantageous for the cosmetic manufacturer to have up-to-date information on this point. The best central source for such information in the United States is the Cosmetic, Toiletry and Fragrance Association, Inc. (CTFA), which has played a leading role for more than 25 years in supporting the continued use of color additives by designing safety studies, clarifying chemical compositions, sponsoring animal tests, organizing scientists and other trade associations, evaluating results, and representing color manufacturers and users in countless scores of conferences with the FDA. The two major collaborators with the CTFA in these activities were the Certified Color Manufacturers Association, Inc. (CCMA), and the Pharmaceutical Manufacturers Association (PMA). For some years, the three trade associations joined together in this endeavor as the Inter-Industry Color Committee.

Because a book can never be up-to-the-minute in a changing world, the guidance of this chapter should be supplemented by ref-

erence to the latest available FDA regulations on color additives [10].
Even more current is information available from the CTFA.

In recent years, many countries outside the United States have
adopted some restrictions on the use of colors. The greatest activity
has been in Canada, the United Kingdom, Scandinavia, West Germany,
and Japan. The European Economic Community (EEC) has been working toward uniform regulation of colors, and is getting closer to implementing such rules. This chapter will deal with these international
matters only briefly (see Sec. VII) because space does not permit detailed guidance.

III. DEFINITIONS

Many terms have precise definitions under the laws and regulations
regarding color additives. Of these, the ones in this section have
been selected to help clarify the rules of the game for the cosmetic
product developer or manufacturer, without dwelling on details that
impact directly only on the color additive petitioner or manufacturer.

For the precise legal language, one must consult the formal regulations [10]. In this section, some of the language is paraphrased to
simplify or clarify it. The rules allow some exemptions, which are
described (to the extent that they relate to cosmetics) in the next
section.

A. Color Additive

A color additive is a substance from any source that can impart color
(including black, white, or gray) to a food, drug, or cosmetic (or
to the human body). Thus, a substance which, when ingested, changes the color of human skin would be a color additive. It would also
be a drug (unless it was a food) as well as a cosmetic. A substance
used to color a container for a cosmetic product is not a color additive
unless, by migration, it contributes sufficient color to the product
to be apparent to the naked eye.

B. Listing

"Listing" means establishment of a regulation specifying how, and under what conditions, a color additive may be used in cosmetics (or
foods, drugs, or certain devices), based on an evaluation of safety
for the intended use. "Provisional" listing is the transitional status
of those color additives that were in use before the new law was
passed in 1960. Provisionally listed colors may continue to be used
as they were before 1960, so long as there is a petition pending with
the FDA for "permanent" listing.

C. Straight Color

A straight color is the color additive, as listed in the appropriate part*
of the latest Code of Federal Regulations (C.F.R.), including such
substances (even impurities) as are permited by the specifications.
A lake (v.i.) can be a straight color but not a pure color (v.i.).

D. Mixture

A color additive made by mixing straight colors, or by mixing straight
color(s) with one or more diluents (v.i.), is a mixture.

E. Diluent

A diluent is a component of a color additive mixture that is not of itself a color additive, but has been deliberately added to facilitate use
of the mixture as a color additive. The diluent may have a noncoloring function in the food, drug, or cosmetic product in which the mixture is to be used. FDA regulations list diluents for mixtures for food
and drug use, but none for general cosmetic use. Early attempts by
the FDA [11] to establish diluent lists were challenged by the cosmetic
industry in 1963 as exceeding the authority granted to the agency
by Congress. The provisions of the regulations with respect to diluents for mixtures for cosmetic use were stayed in 1964 [12]. After
a protracted court battle, the decision confirmed the industry position
that Congress had not granted the FDA the right to regulate cosmetic
products and cosmetic ingredients as color additives [13]. The FDA
had to revise its definition of "color additive" and of "diluent" in 1971
[14], but delayed until 1980 another attempt to establish a list of diluents for mixtures for use in cosmetics [15]. At this writing, the matter is still pending, but three color additives (guanine, guaiazulene,
and silver) have been listed with specific diluents for each.

F. Pure Color

The color *per se* in a color additive, exclusive of all other substances
permitted by the specifications or by regulation, is the pure color for
those color additives which have one principal chemical entity, the
minimum pure color concentration is stated in the specifications, but
for those with more than one organic chromophore, the specification
is given as "total color." Specifications for provisionally listed synthetic organic colors are given in terms of "pure dye," with the method of determination stated for each color additive. This term, equivalent to "total color," is also used when stating the tolerances for
certain color additives in cosmetics and in drugs [16].

---

*The reference in the 1983 edition of the regulations is "Parts 71 and
81," but it probably should read "Parts 73, 74 and 82."

G. Certifiable Color

A listed color additive that has not been exempt from certification (see Sec. IV) is a certifiable color. In practice, although not by law or regulation, the certifiable colors are the listed synthetic organic color additives that have been assigned code names (e.g., D&C Red No. 30). Certifiable colors are often mistakenly called "certified colors." Under the law and the regulations, a "color" is not certified; only a batch of color additive can be submitted for certification.

H. Certification

Certification is the formal approval process in which the FDA issues a certificate assigning a lot number to a given batch of certifiable color additive, if it has met all of the specifications for a listed (or a provisionally listed) color additive. A batch is a homogeneous lot of color additive or mixture held in quarantine by someone who has requested certification. The label on the lot will state the pure color content.

I. Lake

A lake is a straight color extended on a substratum (a substance permitted by regulation for this purpose) by adsorption, coprecipitation, or chemical combination. The definition does not include combinations of ingredients made by mixing. At the time of this writing, the status of lakes is in flux. The FDA proposed a regulation on lakes in 1965, but did not pursue it, and finally scrapped the proposal in 1979. At that time, the FDA published a notice of intent to list lakes [17] and requested comments on definition and nomenclature, safety, and specifications. There was no action during the five years after the end of the comment period. In its comments on definition and nomenclature, the CTFA pointed out the variety of pigments currently called "lakes" by the FDA, and recommended dividing them into "lakes," "toners," "resinated toners," "extended toners," and "extended resinated toners" [18].

In order to describe more accurately the current materials of commerce, the CTFA suggested the following definitions:

Lake: The term "lake" means an organic pigment formed on a substratum when a water-soluble free acid or salt form of a straight color is precipitated by a cation such as sodium, potassium, aluminum, calcium, ammonium, strontium, or zirconium. The term "lake" does not include any combination of ingredients made by a simple mixing process.

Toner: The term "toner" means an organic pigment free of substratum. A toner may be an insoluble straight color, a metal salt of a straight color, or may exist in the free acid form.

Resinated toner: The term "resinated toner" means an organic pigment consisting of a salt of a resin and a toner.

Extended toner: The term "extended toner" means a toner to which a diluent has been added.

Extended resinated toner: The term "extended resinated toner" means a resinated toner to which a diluent has been added.

In making the above proposals, the CTFA offered them as a means of distinguishing (for bulk labeling purposes) the various pigments currently designated by the FDA as "lakes." However, the CTFA did not recommend these names for cosmetic product labeling, because they would not contribute information useful to consumers.

Another confusion in the nomenclature of lakes is that, for cosmetic use, the lake of an FD&C color might be designated an FD&C lake or a D&C lake, depending on the substratum and the process. The CTFA pointed out the more general problem that the designations "FD&C," "D&C," and "Ext. D&C" no longer consistently reflect the nature of the use permitted for a color, and that some of them imply suitability for uses no longer permitted by regulation. The CTFA recommended deletion of the prefixes [18].

J. Tolerance

Tolerance specifies the maximum quantity of a color additive which may be used (or permitted to remain in or on the article in or on which it is used). Such tolerance limitations are established for specified uses. "Temporary" tolerances were established by the FDA in 1960, based on very sketchy toxicological data, and were allowed to stay in place for lipsticks, mouthwashes, and dentifrices (as well as certain drug products) essentially unchanged until 1979 [16], even though the safety studies on which the revised tolerances were based were all in the FDA's hands by 1968 (most of them by 1965).

K. Area of the Eye

The area of the eye includes the region enclosed within the circumference of the supraorbital and infraorbital ridges, including the eyebrow, the skin below the eyebrow, the eyelids and eyelashes, the conjunctival sac of the eye, the eyeball, and the soft areolar tissue that lies within the perimeter of the infraorbital ridge.

L. Externally Applied

"Externally applied" means applied only to external parts of the body and not to the lips or any body surface covered by mocous membrane.

M. Deceptive Use

Use of a color additive in a manner that may promote deception or conceal damage or inferiority constitutes deceptive use [19].

IV. EXEMPTIONS

A. Coal-Tar Hair Dye

In the 1938 law [1] and again in the 1960 Color Additive Amendments [8], Congress has provided exemption for hair dyes from listing and certification. Although a literal reading of section 601(e) of the law [20] suggests that all hair dyes are exempt, the courts have held [13] that this exemption is limited to the same "coal-tar hair dyes" eligible for the exemption in section 601(a). That provision spells out the conditions under which "coal-tar hair dyes" may be marketed, with cautionary labeling, directions for preliminary testing, and restriction against dying eyelashes and eyebrows. Prior to the court decision, henna had been listed for coloring hair [21]. Subsequently, the FDA required submission of petitions for other "vegetable substances" and metallic salts [22]. As a result, bismuth citrate was listed in 1978 [23]. Lead acetate was listed in 1980 [24] for coloring scalp hair only.

B. Exclusion from Definition

The law [20] excludes from the definition of "color additive" any material determined to be used "solely" for purposes other than coloring. The FDA regulation [25] laying down the criteria for the exemption states that the material must be used in a way that any color imparted is clearly unimportant insofar as appearance, value, marketability, or consumer acceptability is concerned. Until 1975, many cosmetic ingredients used for thickening, stabilizing, extending, and other functional purposes were provisionally listed as color additives by the FDA. During 1975 and 1976, the FDA responded to the submission of industry usage survey information provided by the CTFA and terminated the provisional listings of the 22 materials shown in Table 1. The result was that these materials are now regulated as cosmetic ingredients, but not as color additives, so long as they are not being used in cosmetic products for coloring purposes.

C. Exemption from Certification

Section 706(c) of the law [20] exempts from the requirement of certification any color additive, or any listing or use thereof, if the requirement is found not to be necessary in the interest of the protection of the public health. Requests for exemption are normally incorporated in the original color additive petition [26], but they may also be made in citizen petitions by certain interested persons [27,28]. The impact of the exemption provision on color additives for cosmetics has effectively been that all those previously referred to as "coal-tar colors" under the 1938 law [1] are still batch certified, whereas all other dyes and pigments swept in by the 1960 amendments [8] have, to date, been exempt from certification. Thus, as of this writing,

TABLE 1   Cosmetic Ingredients Previously Regulated as Color Additives

Aluminum hydroxide	Magnesium aluminum silicate
Aluminum stearate	Magnesium carbonate
Barium sulfate (blanc fixe)	Magnesium oxide
Bentonite	Magnesium stearate
Calcium carbonate	Magnesium trisilicate
Calcium silicate	Silicic acid
Calcium stearate	Silicon dioxode (silica)
Calcium sulfate	Talc
Cornstarch	Tin oxide
Kaolin	Zinc carbonate
Lithium stearate	Zinc stearate

all color additives with code designations (FD&C, D&C, and Ext. D&C) must be batch-certified, while those color additives for cosmetics that do not have such code names are exempt. This distinction may not necessarily continue in the future. Even though the exempt color additives need not be submitted to the FDA for analysis, they must meet detailed specifications given in the FDA regulations [29].

D. Investigational Use

Section 706(f) of the law [20] makes provision for the exemption from the color additive regulations for articles "intended solely for investigational use by qualified experts when in his [the Secretary's] opinion such exemption is consistent with the public health." In the implementing regulation [30], the FDA has interpreted the Secretary's (of Health and Human Services) opinion to mean that the only cases that will qualify are those where the article is being shipped to experts qualified "to determine safety." The regulation prescribes labeling, procedures, and record keeping.

V. NOMENCLATURE

There are many ways of designating colorants: by chemical name, by common name, by generic name, by name assigned by the FDA or the regulatory body of some other country, by name and number, by code, by letters and numbers, by number, by chemical structure,

TABLE 2 Certifiable Color Additives for Cosmetics

FDA name	Pertinent section of 21 C.F.R.	Other names
FD&C Blue No. 1	74,2101; 82.101	Brilliant blue FCF; food blue 2
D&C Blue No. 4	74.2104; 82.1104	Alphazurine FG; Erioglaucine
D&C Brown No. 1	74.2151	Resorcin brown
FD&C Green No. 3	74.2203; 82.203	Fast green FCF; food green 3
D&C Green No. 5	74.2205; 82.1205	Alizarin cyanine green
D&C Green No. 6	74.2206; 82.1206	Quinizarin green SS
D&C Green No. 8	74.2208	Pyranene concentrated
D&C Orange No. 4	74.2254; 82.1254	Orange II
D&C Orange No. 5	74.2255; 82.1255	Dibromofluorescein
D&C Orange No. 10	74.2260; 82.1260	Diiodofluorescein
D&C Orange No. 11	74.2261; 82.1261	Erythrosine yellowish Na
D&C Orange No. 17	81.1; 82.1267	Permatone orange
FD&C Red No. 3	81.1; 82.303	Erythrosine; food red 14
FD&C Red No. 4	74.2304; 82.304	Ponceau SX
D&C Red No. 6	74.2306; 82.1306	Lithol rubin B
D&C Red No. 7	74.2307; 82.1307	Lithol rubin B Ca
D&C Red No. 8	81.1; 81.25; 82.1308	Lake red C
D&C Red No. 9	81.1; 81.25; 82.1309	Lake red C Ba
D&C Red No. 17	74.2317; 82.1317	Toney red; Sudan III
D&C Red No. 19	81.1; 82.1319	Rhodamine B
D&C Red No. 21	74.2321; 82. 1321	Tetrabromofluorescein
D&C Red No. 22	74.2322; 82.1322	Eosin YS; eosine G
D&C Red No. 27	74.2327; 82.1327	Tetrabromotetrachloro-fluorescein
D&C Red No. 28	74.2328; 82.1328	Phloxine B
D&C Red No. 30	74.2330; 82.1330	Helindone pink CN

TABLE 2 (continued)

FDA name	Pertinent section of 21 C.F.R.	Other names
D&C Red No. 31	74.2331; 82.1331	Brilliant lake red R
D&C Red No. 33	81.1; 81.25; 82.1333	Acid fuchsin D; naphthalene red B
D&C Red No. 34	74.2334; 81.1; 82.1334	Deep maroon; Fanchon maroon; Lake Bordeaux B
D&C Red No. 36	81.1; 81.25; 82.1336	Flaming red
D&C Red No. 37	81.1; 82.1337	Rhodamine B stearate
FD&C Red No. 40	74.2340	Allura[a] red; food red 17
D&C Violet No. 2	74.2602; 82.1602	Alizurol purple SS
Ext. D&C Violet No. 2	74.2602a	Alizarine Violet
FD&C Yellow No. 5	74.2705; 81.1; 82.705	Tartrazine; food yellow 4
FD&C Yellow No. 6	81.1; 82.706	Sunset yellow FCF; food yellow 5
D&C Yellow No. 7	74.2707; 82.1707	Fluorescein
Ext. D&C Yellow No. 7	74.2702a; 82.2702 a	Naphthol yellow S
D&C Yellow No. 8	74.2708; 82.1708	Uranine
D&C Yellow No. 10	74.2710; 82.1710	Quinoline yellow WS; food yellow 13
D&C Yellow No. 11	74.2711	Quinoline yellow SS
FD&C lakes	81.1; 82.51	
D&C lakes	81.1; 82.1051	
Ext. D&C lakes	81.1; 82.2051	

[a]Trademark of Buffalo Color Corp.

etc. Most of these systems are incomplete or unsatisfactory in some respect when dealing with color additives for a regulated use. It is not enough to state the name, or structure, or number, unless the details of purity are spelled out at the same time. Different types of nomenclature are described below. The best central source of nomenclature is the CTFA *Cosmetic Ingredient Dictionary* [42].

A. FDA Names

In the United States, certifiable color additives are designated by the abbreviation FD&C, or D&C, or Ext. D&C, followed by a color, followed by a number. There are three exceptions to this system, but none of these is permitted in cosmetics. Unfortunately, the once rational abbreviations designed to convey the permitted usage as part of the name [31] are now misleading [18]. Because they are official in the United States, they are shown in Table 2. In order to designate precisely what that name means in term of composition, purity, permitted usage, and tolerances, the pertinent sections of the Code of Federal Regulations, Title 21 (21 C.F.R.) are also given in Table 2. Alternative names, including some used in other countries, are also given in the table. A more comprehensive set of alternate names may be found in the CTFA *Cosmetic Ingredient Dictionary* [42].

With respect to color additives exempt from certification in the United States, there has been no attempt to rationalize names. They include chemical names, common names, names indicating mixtures, some that do not indicate a mixture, and some that indicate the physical state of the substance. Unfortunately, although the regulated use requires meeting strict specifications, each name is not unique, as are the certifiable code names. As a consequence, materials of the same name, but not suitable for use as color additives, may be purchased. Table 3 gives the official FDA names together with the pertinent (and critical) references to 21 C.F.R. Some alternative common names are also given in the table. Others may be found in the CTFA *Cosmetic Ingredient Dictionary* [42].

B. Chemical Names

For the synthetic organic dyes and pigments not found in nature, the chemical names are too long for convenience. Furthermore, they tend to be more precise about the color additive than the composition of the commercial material. Reference to the specifications (as in the FDA regulations) is necessary to understand the extent of intermediates, subsidiary dyes, and inorganic salts permitted to be present.

Among the color additives exempt from certification, some are fairly pure chemicals, and are most simply designated by the chemical name (e.g., dihydroxyacetone, bismuth citrate, silver, and lead acetate). On the other hand, some are complex naturally occurring mix-

TABLE 3   Color Additives for Cosmetics, Exempt from Certification

FDA name	Pertinent section of 21 C.F.R.	Other names
Aluminum powder	73.2645	
Annatto	73.2030	Annatto extract; Bixin; *Bixa orellana* L. extract; natural orange 4
Bismuth citrate	73.2110	
Bismuth oxychloride	73.2162	Pearl white
Bronze powder	73.2646	
Caramel	73.2085	Burnt sugar; natural brown 10
Carmine	73.2087	Carminic acid lake
β-Carotene	73.2095	Carotene; natural yellow 26
Chromium hydroxide green	73.2326	Cosmetic green
Chromium oxide greens	73.2327	Cosmetic green
Copper powder	73.2647	Copper, metallic powder
Dihydroxyacetone	73.2150	1,3-Dihydroxy-2-propanone
Disodium EDTA-copper	73.2120	Copper Versenate[a]
Ferric ammonium ferrocyanide	73.2298	Cosmetic iron blue; iron blue
Ferric ferrocyanide	73.2299	
Guaiazulene	73.2180	Azulene
Guanine	73.2329	Pearl essence; 2-aminohypoxanthine
Henna	73.2190	*Lawsonia alba* Lam.
Iron oxides	73.2250	Cosmetic black; cosmetic brown; cosmetic umber; cosmetic ochre
Lead acetate	73.2396	
Manganese violet	73.2775	Manganese ammonium pyrophosphate
Mica	73.2496	Muscovite
Potassium sodium copper chlorophyllin (chlorophyllin-copper complex)	73.2125	

TABLE 3 (continued)

FDA name	Pertinent section of 21 C.F.R.	Other names
Pyrophillite	73.2400	Hydrous aluminum silicate
Silver	73.2500	Crystalline silver metal
Titanium dioxide	73.2575	Titanic earth
Ultramarines	73.2725	Cosmetic blue (or green, or violet); ultramarine blue (or green, or violet, or pink); pigment blue 29; pigment green 24; pigment violet 15
Zinc oxide	73.2991	Zinc white

[a]Trademark of Dow Chemical Co.

tures of only partially elucidated chemical structures (e.g., henna, caramel, and ultramarines).

Because the user of the color is concerned more with specifications than chemical structure, this brief review does not give chemical names, except where they are the official name or the common name. Chemical names and structures (where known) can be found elsewhere [32,33,42], along with practical information on solubilities and stability information.

C. *Colour Index* Numbers and Names

Another unique system for designation of dyes and pigments is the well-known international *Colour Index* (CI) [34]. Commercially available dyes and pigments from all over the world are related to each other by chemical composition, which is designated by a five-digit number, sometimes with a suffix following a colon to show relationships between materials. Although these are widely used, and are useful to identify related materials, they are not precise enough to define color additives for food, drug, or cosmetic use. A useful feature of the system is that many of the numbers are identified with CI names which designate the normal use or some unifying feature about a group of colors. For example, FD&C Yellow No. 5 is CI 19140 and is CI Food Yellow 4; β-carotene is CI 75130, and is CI Natural Yellow 26.

The *Colour Index* numbers and names of food, drug, and cosmetic colors may be found in several reference sources [32,33,42].

D. *Chemical Abstracts* Registry Numbers

Another registry of unique numbers for chemical substances is used by the *Chemical Abstracts* Service (CAS). It is a three-part hyphenated number, with two to five digits in the first part, two digits in the second , and one digit in the last. As examples, annatto is 1393-69-1, and FD&C Red No. 40 is 25956-17-6. CAS numbers may be found for many food, drug, and cosmetic colors in the CTFA *Cosmetic Ingredient Dictionary* [42].

E. EEC Numbers

Within the European Economic Community (EEC), activity has been underway for years toward harmonization of the regulation of colors. The "E" numbers from E-100 to E-199 were originally assigned to permitted food colors in 1962 [35]. The list has been revised several times since. Many of the same colors are used in cosmetics, but the numbers do not appear in the EEC Cosmetic Directive on cosmetics [36], where CI numbers are used. Examples of EEC numbers are E-180 for D&C Red No. 6 and E-171 for titanium dioxide.

F. DFG Code Names

The Dyestuff Commission of the German Research Society (DFG) has established a rational system of designating cosmetic colors with a one- or two-digit number following a color designation, following a code describing permitted usage. As examples, D&C Red No. 36 is C-Rot 1, silver is C-Pigment 2, and D&C Orange No. 4 is C-ext. Orange 8. The "C" designation means that it can be used in all cosmetic products, including contact with mucous membranes (lip, eye, oral cavity). "C-ext." means for externally applied cosmetic products without mucous membrane contact. Another class is "C-WR" for use in products with short contact to human skin or which are rinsed off after use (cleansing or washing preparations). This designation is used for a number of colors permitted in Germany but not in the United States. The numbers can be found in publications of the DFG [37,38] and of CTFA [42].

VI. COLORS PERMITTED IN COSMETICS IN THE UNITED STATES

The FDA had planned several modifications in the permitted lists of colors, to become effective January 31, 1981 [39]. These proposals were delayed by the "regulatory freeze" imposed by the incoming administration of President Ronald Reagan. The tables in this section reflect the status as of June 1984.

TABLE 4 Color Additives Permitted in Externally Applied Cosmetics in the United States

FD&C Blue No. 1	D&C Red No. 30
FD&C Green No. 3	D&C Red No. 31
FD&C Red No. 3	D&C Red No. 33
FD&C Red No. 4	D&C Red No. 34
FD&C Red No. 40[a]	D&C Red No. 36
FD&C Yellow No. 5[b]	D&C Red No. 37
FD&C Yellow No. 6	D&C Violet No. 2
FD&C lakes of FD&C colors	D&C Yellow No. 7
D&C lakes of FD&C colors	D&C Yellow No. 8
D&C Blue No. 4	D&C Yellow No. 10
D&C Brown No. 1	D&C Yellow No. 11
D&C Green No. 5	D&C lakes of D&C colors
D&C Green No. 6	Ext. D&C Violet No. 2
D&C Green No. 8[c]	Ext. D&C Yellow No. 7
D&C Orange No. 4	Ext. D&C lakes of Ext. D&C colors
D&C Orange No. 5	Aluminum powder
D&C Orange No. 10	Annatto
D&C Orange No. 11	Bismuth oxychloride
D&C Orange No. 17	Bronze powder
D&C Red No. 6	Caramel
D&C Red No. 7	Carmine
D&C Red No. 8	β-Carotene
D&C Red No. 9	Chromium hydroxide green
D&C Red No. 17	Chromium oxide greens
D&C Red No. 19	Copper powder
D&C Red No. 21	Ferric ferrocyanide
D&C Red No. 22	Ferric ammonium ferrocyanide
D&C Red No. 27	Guaiazulene
D&C Red No. 28	Guanine

TABLE 4 (continued)

Iron oxides	Titanium dioxide
Manganese violet	Ultramarines
Mica	Zinc oxide
Pyrophyllite	

[a]Not under conditions that affect the integrity of the color.
[b]Not permitted in hair straighteners, permanent wave preparations, and depilatories.
[c]Not exceeding 0.01% by weight.

A. In Externally Applied Cosmetics

All the color additives in Table 4 may be used in externally applied cosmetics. In only three cases (shown in footnotes) are there explicit restrictions with respect to concentration in, or type of the cosmetic, or other limitation. However, for all colors, the use must be consistent with "good manufacturing practice." In other words, the user of the color is responsible for avoiding concentrations in excess of those necessary to achieve the desired objective, and for avoiding use under conditions that would be deleterious to the composition of the color or of the cosmetic product.

TABLE 5   Color Additives Permitted in Cosmetics in the United States for Use in the Area of the Eye

Aluminum powder	Ferric ammonium ferrocyanide
Annatto	Ferric ferrocyanide
Bismuth oxychloride	Guanine
Bronze powder	Iron oxides
Caramel	Manganese violet
Carmine	Mica
β-Carotene	Titanium dioxide
Chromium hydroxide green	Ultramarines
Chromium oxide greens	Zinc oxide
Copper powder	

B. In the Area of the Eye

The color additives now permitted in cosmetics for use in the area of the eye are shown in Table 5. There are no specific restrictions, other than use consistent with good manufacturing practice. At present, none of the certifiable synthetic organic colors is permitted, but petitions requesting such use have been submitted for some. Approval will depend on adequacy of the submitted skin and eye safety studies.

C. In Cosmetics Subject to Ingestion (Without Restriction)

The color additives shown in Table 6 include certifiable colors, lakes, and colors exempt from certification. All may be used generally in cosmetics, including those subject to ingestion, and those in contact with mucous surfaces, without any restriction other than use consis-

TABLE 6 Color Additives Permitted in the United States for Use in Cosmetics Subject to Ingestion Without Restriction[a]

FD&C Blue No. 1	D&C Red No. 30
FD&C Green No. 3	D&C lakes of D&C colors
FD&C Red No. 3	Annatto
FD&C Red No. 40[b]	Bismuth oxychloride
FD&C Yellow No. 5	Bronze powder
FD&C Yellow No. 6	Caramel
FD&C lakes of FD&C colors	Carmine
D&C lakes of FD&C colors	β-Carotene
D&C Green No. 5	Copper powder
D&C Red. No. 6	Guanine
D&C Red No. 7	Iron oxides
D&C Red No. 21	Manganese violet
D&C Red No. 22	Mica
D&C Red No. 27	Titanium dioxide
D&C Red No. 28	Zinc oxide

[a] Other than use consistent with good manufacturing practice.
[b] Shall not be exposed to conditions which may affect the integrity of the color.

tent with good manufacturing practice. However., contact with mucous membranes, in FDA regulations, does not include use in the area of the eye. Only the colors listed in Table 5 may be used in the area of the eye.

D. In Cosmetics Subject to Ingestion (With Tolerances)

The major reason for the adoption of the Color Additive Amendments of 1960 [8] was to confer on the FDA the authority to permit use of certain color additives under specified conditions, or for specified uses, or with specified tolerances. This authority is illustrated in Table 7, where several of the certifiable synthetic organic colors are shown with tolerances based on the application of safety factors to the results of lifetime animal feeding studies, coupled with estimates of maximum likely consumption of products. For this purpose, the FDA has divided cosmetics subject to ingestion into three groups: lip cosmetics (lipsticks and other products intended for application to the lips), mouthwashes (including breath fresheners), and dentifrices. The allocation of these amounts of colors to these classes of cosmetics is also dependent on similar tolerances for use of these colors in drugs that are ingested, as well as the mouthwashes and dentifrices that are drugs by FDA definition. The details regarding the product tolerances may be found in the pertinent FDA regulation [16].

TABLE 7   Color Additives Permitted in the United States for Use in Cosmetics Subject to Ingestion with Tolerances

Certified color or lake	Maximum concentration (% pure dye by weight)		
	Lip cosmetic[a]	Mouthwash or breath freshener[b]	Dentifrice[c]
D&C Orange No. 5	5.0	d	d
D&C Red No. 8	3.0	0.005	0.002
D&C Red No. 9	3.0	0.005	0.002
D&C Red No. 33	3.0	d	d
D&C Red No. 36	3.0	d	d
D&C Yellow No. 10	1.0	d	d

[a]Combined total of D&C Red Nos. 8 and 9 may not exceed 3.0%.
[b]Combined total of D&C Red Nos. 8 and 9 may not exceed 0.005%.
[c]Combined total of D&C Red Nos. 8 and 9 may not exceed 0.002%.
[d]In amounts consistent with good manufacturing practice.

TABLE 8   Color Additives Permitted in Cosmetics in the United States for Specific Uses Only

Color additive	Use restriction	Maximum concentration
Bismuth citrate	For coloring scalp hair only	0.5% (w/v)
Dihydroxyacetone	In externally applied cosmetics to impart a color to the human body	Consistent with GMP
Disodium EDTA-copper	In the coloring of shampoos	Consistent with GMP
Henna	For coloring hair (not in the area of the eye	
Lead acetate	For coloring hair on the scalp only (not elsewhere)	0.6% Pb (w/v)
Potassium sodium copper chlorophyllin (chlorophyllin-copper comples)	For coloring dentifrices (only in combination with substances listed in 21 C.F.R. 73.2125 (b) (2)	0.1%
Silver	For coloring fingernail polish	1%

E. For Specific Uses Only

Additional examples of the exercise of the authority to limit type and amount of usage of color additives are shown in Table 8. Seven materials, of considerable variety, are permitted by the FDA for very specific uses, at designated concentrations. In one case, not only are the use and concentration specified, but the other permitted components of the product are given.

F. Certifiable Lakes

Lakes (discussed in Sec. III) are frequently the preferred form of color additives for cosmetic uses such as lip cosmetics and nail lacquers. Table 9 presents a highly simplified designation of what "lakes" are permitted in cosmetics by present FDA regulations. In practice, only a fraction of the numerous variations that are possible are actually available commercially. Guidance in appropriate selection from available materials can often be obtained from cosmetic color manufacturers and suppliers. The user should be aware that the materials

TABLE 9   Certifiable Lakes Permitted in the United States for Use in
Cosmetics, Subject to the Same Restrictions as the Corresponding
Color

Type of lake	Derived from	Extended on	Permitted salts
FD&C	Previously certified FD&C color	Alumina	Aluminum, calcium
D&C	Any listed FD&C or D&C color (not necessarily certified)	Alumina[a]	Sodium,
		Blanc fixe[a]	Potassium,
		Gloss white[a]	Aluminum,
		Clay[a]	Barium,
		Titanium dioxide[a]	Calcium,
		Zinc oxide[a]	Strontium,
		Talc[a]	Zirconium
		Rosin[a]	
		Aluminum benzoate[a]	
		Calcium carbonate[a]	
Ext. D&C	Ext. D&C Violet No. 2	Same substrata	Same salts listed for D&C
	Ext. D&C Yellow No. 7	listed for D&C	

[a]Any one or any combination of two or more.

are generally insoluble in most cosmetic systems, and therefore act
as pigments. The user should be aware that any regulatory restriction on a color additive also applies to any lake of that color, but that
quantitative tolerances must be recalculated, because they apply to
the pure dye content (see Table 7). Although certified batches of
color additives never contain 100 percent pure dye, the deviation is
much more significant in the case of lakes.

VII. THE REGULATION OF COSMETIC COLORS OUTSIDE THE UNITED STATES

Within the next few years, it is likely that the European Economic Community (EEC) Council of Ministers will adopt a "final" version of the

Cosmetic Directive and Annexes [36]. Within a few years after that, each of the 10 member nations must harmonize its national legislation and regulations with the directive. When that happens, a certain degree of uniformity regarding cosmetic colors will emerge in the Common Market, and will undoubtedly affect practice in other countries. In the interim, the status of individual colors under national regulations is too complex to cover in the limited space of this chapter. Instead, attention will be directed at those organizations that are making or influencing the policies regarding color usage, thus suggesting what sources of information are available in addition to manufacturers and suppliers of cosmetic colors.

A. Supranational Organizations

The principal international organization influencing cosmetic colors at this time is the EEC, whose Scientific Committee on Cosmetology continues work on the EEC Cosmetic Directive and Annexes, as amended [36], which includes positive lists of colors for cosmetics. The colors are divided into those for use in contact with mucous membranes, those not for use on mucous membranes, and those for use where the contact with the skin will be brief (e.g., rinsed off). The adoption of the Directive's third amendment (proposed January 20, 1981) will be a complex process. The Scientific Committee on Cosmetology is part of the Environment and Consumer Protection Service, which serves the European Commission. The commission makes recommendations to the Council of Ministers, which, before making its decision, seeks the opinions of the European Parliament and the Economic and Social Committee. If adopted by the council, a Directive is published in the *Official Journal of the European Communities*, notifying the member states that they must introduce conforming legislation in each country within a specified period, such as two years. A member state may, if it wishes, make the national law more restrictive than the EEC Directive. The members of the EEC are Belgium, France, West Germany, Italy, Luxembourg, the Netherlands, Denmark, Ireland, the United Kingdom, and Greece.

Close liaison with the work of the EEC in the cosmetic area is maintained by an international trade association group located in Brussels, Comité de Liaison des Syndicats Européens de l'Industrie de la Parfumerie et des Cosmétiques (COLIPA). Its members are the cosmetic industry trade associations of the EEC member states, plus Spain and Switzerland. Its most important recent activity in connection with cosmetic colors is to insert the essential strontium, zirconium, and barium lakes of certain synthetic organic colors into the appropriate positive list of the EEC Directive. Part of the problem has been past variation in practice from country to country, and part of the confusion has stemmed from the use of *Colour Index* numbers which specify chemical structures but not the salts or lakes, each of which must be designated individually.

An indirect influential role is played by a group not concerned with cosmetics. The Joint Expert Committee on Food Additives (JECFA) of the United Nations Food and Agriculture Organization and of the World Health Organization (WHO) reviews and evaluates food colorants, among other additives. It adopts comprehensive specifications and toxicological monographs. Because many of the food colorants are also used as cosmetic colors, the findings of JECFA inevitably have an international impact on the regulation of all colors. JECFA has normally met once a year, and its actions are published (in English) at Geneva in the WHO Technical Report Series.

The WHO International Agency for Research on Cancer (IARC) in Lyon evaluates the carcinogenic risk of chemicals to humans. The latest cumulative index of its publications [40] lists about a dozen color additives used in foods, drugs, and cosmetics, as well as about a dozen which were formerly used.

B. National Regulatory Agencies

In Canada, the Drugs Directorate of the Health Protection Branch of the Ministry of Health and Welfare has jurisdiction over the colors used in cosmetics.

The regulation of cosmetic colors in the United Kingdom is administered under the Food and Drugs Act by the Ministry of Agriculture, Fisheries, and Food.

In West Germany, the Dyestuff Commission of the Deutsche Forschungsgemeinschaft (German Research Society) is a quasi-governmental agency that reviews and evaluates colors, acting in an advisory capacity to German parliaments and government agencies. The Commission's recommendations [37,38] form the basis for the German Color Regulations adopted December 19, 1959, and revised several times since. (See also "DFG Code Names" in Sec. V.)

In Japan, the regulation of colors is under the Ministry of Health and Welfare [41]. The use of colors in foods, drugs, and cosmetics is often suspended voluntarily by Japanese manufacturers, without formal government action.

In Australia, the Consumer Product Safety Subcommittee of the National Health and Medical Research Council of the Commonwealth Department of Health was studying the usage of colors in cosmetics and toiletries during 1980 in order to implement regulations.

C. Trade Associations

In many countries, trade associations maintain liaison with regulatory agencies, survey members for information to be supplied to government agencies, disseminate information on new regulatory developments to members, and, in some cases, almost fulfill the role of quasi-governmental functions in the self-regulatory sense. The outstanding example is in the United Kingdom, where the British Industrial Biological Research Association (BIBRA) acts as the central focus for industry members in many areas bearing on public health and safety. A working party on colors has for years cooperated with BIBRA scientists to compile information and design safety programs in connection with the use of color additives. Originally, BIBRA was a quasi-governmental agency, funded by both the British government and industry. In recent years, it has been supported solely by industry funds. It conducts in its own highly regarded laboratories many of the studies which the various working parties select, usually taking into account the requests of governmental committees, or of international groups such as EEC or FAO/WHO.

Membership in trade associations is often one of the easiest ways of staying informed of applicable regulatory requirements. Trade associations specifically concerned with cosmetics and related products in major countries are listed in Table 10.

VIII. HOW TO COPE: CAUTIONS AND PRECAUTIONS

The user of colors in cosmetics can avoid pitfalls in the color regulations by staying alert to the summarizing points in this section. The points are referenced back to appropriate sections, tables, or literature sources, where more detail can be found.

TABLE 10   Cosmetic Industry Trade Associations in Major Countries

Country	Name of Association
United States	The Cosmetic, Toiletry and Fragrance Association, Inc.
Canada	Canadian Cosmetic, Toiletry and Fragrance Association
United Kingdom	Cosmetic, Toiletry and Perfumery Association
West Germany	Industrieverband Körperflege- und Waschmittel
Japan	Japan Cosmetic Industry Association
France	Fédération Française de l'Industrie des Produits de Parfumerie, de Beauté et de Toilette
Australia	Cosmetic and Toiletry Manufacturers Association of Australia
Italy	Unione della Profumeria e della Cosmesi

A. General Guidelines

Make certain that the color is permitted by regulation for the intended use. For guidance, see Sec. VI and Tables 4, 5, 6, 7, 8, and 9. For changes that may occur in the future consult the latest FDA regulations [10] and the CTFA. Stay continuously alert to changes in the status of cosmetic colors, which are reported in the *Federal Register*, in the trade press, and by bulletins and newsletters of trade associations. Note effective dates of new regulations, and take appropriate action at that time.

If the cosmetic product is intended for export, learn in advance whether the colors to be used are permitted in the destination country. See Secs. V.E, V.F, and VII.

B. Formulating Products

Do not depend on the FDA nomenclature (FD&C, D&C, Ext. D&C) for guidance to permitted uses of color additives. See Secs. III.I and V.A.

Carefully review the proposed directions for use of a product, to avoid inadvertent instructions to use a product in a way that goes beyond the limits of permitted use for the color in the product. Note that many colors are now limited to use in externally applied products, and therefore may not be used in contact with mucous membranes. See Sec. III.L. Some products, not normally thought of as eye cosmetics, may be labeled in such a way that they will be used in the area of the eye, as explicitly defined by regulation. See Sec. III.K.

Pay close attention to the tolerances limiting the amounts of certain colors for specified uses. See Secs. III.J and VI.D, Table 7, and footnote c in Table 4. The tolerances must be calculated in terms of pure dye content. See Sec. III.I. This is especially important in the case of lakes. See Sec. VI.F.

To understand the proper function and use in accordance with good manufacturing practice, be aware of the solubilities of the various forms of the colors as dyes, toners, pigments, lakes, etc. Consult the references in Sec. V.B.

If the product will be both cosmetic and drug by definition, be certain that any color is permitted in both. See FDA regulations for colors in drug products [10]. For example, Ext. D&C Violet No. 2 is permitted in externally applied cosmetics, but not in any drugs; canthaxanthin is permitted in ingested drugs, but not in any cosmetics.

If a color in an existing product has been delisted by the FDA, reformulate to delete it from the product at the earliest possible date consistant with adequate assurance of safety and stability, but in any case before the effective date of the regulation.

Do not depend on the exemption from color additive regulation by virtue of exclusion from the definition as a noncolor unless you have a very clear understanding of the specifics and the full significance of the regulation. See Sec. IV.B.

For hair dyes permitted under the color additive regulations, see Table 8. Do not depend on the exemption from color additive regulations for coal-tar hair dyes without a thorough understanding of the specifics and the full significance of the applicable law [20], especially sections 601, 602, and 706. See Sec. IV.A.

C. Ordering Colors

Color additives that are exempt from certification must still meet the detailed specifications in the Code of Federal Regulations [10]. Similar ingredients, often under identical names, may be sold for other purposes where the requirements are not as rigid. See Secs. III.C and V.A. For greatest assurance of compliance, purchase from suppliers who specialize in food, drug, and cosmetic colors.

Do not attempt to order color additives that have been delisted by the FDA. Informed reputable suppliers will not ship them.

D. Storage and Stability

It is best to store certified colors in the original containers so that they will not lose their identity (certified lot number) and will be marked with all the cautions and restrictions required by law. See Sec. III.H. In the absence of specific storage conditions recommended by the supplier, store all colors away from heat, light, and moisture. Maintain close surveillance of inventory of color additives to permit immediate removal if delisted by the FDA.

Study the stability of the color in the environment of the product. For guidance on stability and fastness of color, see the references in Sec. V.B, as well as technical literature provided by the color supplier. Be aware that all color additives are subject to the general requirement of use in accordance with good manufacturing practice. Among other things, this means avoiding conditions that would be deleterious to the composition of either the color or the cosmetic. See, e.g., Sec. V.B. Note that one color additive is subject to specific precautions about oxidizing or reducing agents, and another is not permitted in certain alkaline products. See footnotes to Tables 4 and 6.

Study the stability of the product in the container in which it will be marketed. It is important to avoid interaction between the color and the container. It is also important to observe whether any color used in the container will migrate into the cosmetic under storage conditions. See Sec. III.A. This problem is most likely to occur with colored plastic containers.

Maintain close surveillance of inventory so that cosmetic products containing delisted color additives can be removed from inventory to prevent shipment after a date specified by regulation. See Sec. VIII.A.

E. Manufacturing Cosmetics

Be aware of the pure dye (or pure color, or total color) content of each lot of certified color in inventory, as stated on the original label. See Secs. III.F and III.H. These are allowed by regulation to vary from lot to lot, within specifications. The variations are often greatest from one manufacturer to another, but will occur even from one batch to another made by the same company. Such variations should be taken into account when calculating tolerances (see Sec. III.J) and when matching the color of the cosmetic from batch to batch of finished product. The physical state and particle size of a batch are also significant with respect to perceived color. All these factors may require that the composition of some cosmetic products intended to impart color to the skin be varied from batch to batch in order to match perceived color. Similar problems may also be encountered with some of the color additives exempt from certification, especially the iron oxides.

Some types of cosmetic products are manufactured from intermediates (e.g., color extensions or color solutions) to simplify the process and to aid in uniformity. Maintain surveillance of the inventories of such intermediates, which are often retained for extended times, to facilitate removal at appropriate times, if required by subsequent FDA regulation. See Sec. VIII.A.

REFERENCES

1. U.S. Congress, Public Law No. 75-717, 1938.
2. U.S. Congress, Public Law No. 84-672, 1956.
3. U.S. Congress, Public Law No. 85-929, 1958.
4. *Flemming v. Florida Citrus Exchange*, 358 U.S. 153, cert. denied, 358 U.S. 948.
5. U.S. Congress, Public Law No. 86-2, 1959.
6. U.S. Senate, Committee on Labor and Public Welfare, Report to accompany S.2197, Color Additive Amendments of 1959, 86th Congress, 1st Session, August 21, 1959.
7. W. J. Haga and N. Acocella, *Haga's Law*, Morrow, New York, 1980.
8. U.S. Congress, Public Law No. 86-618, 1960.
9. C. N. Parkinson, *The Law of Delay*, Houghton Mifflin, Boston, 1970.
10. Food and Drug Administration, Code of Federal Regulations (C.F.R.), Title 21, Parts 70, 71, 73, 74, 80, 81, and 82,

U.S. Government Printing Office, Washington, D.C., (revised annually).

11. *Fed. Reg.*, *28*:6922 (July 6, 1963).
12. *Fed. Reg.*, *29*:18495 (December 29, 1964).
13. *Toilet Goods Association v. Finch*, 63 Civ. 3349, Southern District, New York, November 26, 1969.
14. *Fed. Reg.*, *36*:16902 (August 21, 1971).
15. *Fed. Reg.*, *45*:26977 (April 22, 1980).
16 21 C.F.R. 81.25.
17. *Fed. Reg.*, *44*:36411 (June 22, 1979).
18. N. F. Estrin, Letter to FDA Hearing Clerk on "Lakes of Color Additives," Docket No. 79N-0043, November 16, 1979.
19. 21 C.F.R. 71.22.
20. Federal Food, Drug and Cosmetic Act, as Amended, January 1980.
21. *Fed. Reg.*, *30*:7705 (June 15, 1965).
22. *Fed. Reg.*, *38*:2996 (January 31, 1973).
23. *Fed. Reg.*, *43*:44831 (September 29, 1978).
24. *Fed. Reg.*, *45*:72112 (October 31, 1980).
25. 21 C.F.R. 70.3(g).
26. 21 C.F.R. 71.1(c).
27. 21 C.F.R. 71.18.
28. 21 C.F.R. 10.30.
29. 21 C.F.R. 73 Subpart C.
30. 21 C.F.R. 71.37.
31. *Fed. Reg.*, *4*:1923 (May 9, 1939).
32. S. Zuckerman and J. Senackerib, in *Encyclopedia of Chemical Technology*, 3rd ed., Vol. 6, John Wiley & Sons, New York, 1979, pp. 561-596.
33. D. M. Marmion, *Handbook of U.S. Colorants for Foods, Drugs, and Cosmetics*, John Wiley & Sons, New York, 1984.
34. *Colour Index*, 3rd ed., 5 vols., The Society of Dyers and Colourists, Bradford, Yorkshire, England, 1971.
35. *Official Journal of the European Communities*, No. 115, 2645, November 11, 1962.
36. EEC Council, *Directive and Annexes Relating to Cosmetics*, 76/768/EEC, July 27, 1976 (and subsequent amendments).
37. DFG Farbstoff Kommission, *Vorläufige Information Ringbuch Kosmetische Färbemittel*, Mitteilung XIII, Bad Godesberg, West Germany, 1976.
38. DFG Farbstoff Kommission, *Kosmetische Färbemittel*, Harald Boldt Verlag, Boppard, West Germany, 1977.
39. *Fed. Reg.*, *45*:75226 (November 14, 1980).
40. IARC, *IARC Monographs on the Evaluation of the Carcinogenic Risk of Chemicals to Humans*, Vol. 26, International Agency for Research on Cancer, Lyon, France, 1981.

41. Ministry of Health and Welfare, Pharmaceutical Affairs Law
     (Law No. 145, as Amended), Tokyo, Japan. August 10, 1960.
42. N. F. Estrin, P. A. Crosley, and C. R. Haynes, eds., CTFA
     Cosmetic Ingredient Dictionary, 3rd ed., The Cosmetic, Toilet-
     ry and Fragrance Association, Inc., Washington, D.C., 1982.

26

ADEQUATELY SUBSTANTIATING THE SAFETY OF COSMETIC AND TOILETRY PRODUCTS

ROBERT P. GIOVACCHINI

Corporate Product Integrity, The Gillette Company, Boston, Massachusetts

On Monday, March 5, 1975, the Commissioner of the U.S. Food and Drug Administration stated that:

> Each ingredient used in a cosmetic product and each finished cosmetic product shall be adequately substantiated for safety prior to marketing. Any such ingredient or product whose safety is not adequately substantiated prior to marketing is misbranded unless it contains the following conspicuous statement on the principal display panel: *Warning* — the safety of this product has not been determined [1].

In considering the requirements of adequate substantiation of safety, it is important first to define what is meant by the term "safe." As all toxicologists already know, all materials can be toxic or hazardous under appropriate conditions and concentrations. Thus, there is nothing that is absolutely safe. Nor is it possible technically to measure safety.

Many years ago I defined "safe" as "freedom from unreasonable risk or significant injury under reasonable, foreseeable conditions of use" [2]. In this definition an attempt was made to recognize that there is no absolutely safe ingredient or product, and that the lack of any substantial or "unreasonable" risk is the best that can be expected. It reflects the fact that, although there are no absolutely harmless substances, there are ways of using substances or products in relatively harmless ways. Thus, a decision about safety must be a judgment based on an assessment of relative risk.

"Risk" is defined as the potential of getting hurt as the result of using a substance or product. A "hazard" occurs when the poten-

tial for harm or unacceptable risk is present [3]. Thus, the concept
of safety deals with the issue of the relative risk of a substance or
product resulting from the amount, site, and route of application,
under the conditions of use. Safety means the absence of significant
risk under these conditions of use.

With these definitions in mind, it is possible to consider the type
of information that constitutes adequate substantiation of safety. In
its 1975 notice, the Food and Drug Administration (FDA) advised that
the safety of a cosmetic could be adequately substantiated through
both (1) reliance on already available toxicological test data on individual ingredients and on product formulations that are similar in composition to the particular cosmetic, and (2) performance of any additional toxicological and other tests that are appropriate in light of
such existing data and information. The FDA recognized that there
may be disagreement among experts concerning the amount of data
necessary to demonstrate the safety of a cosmetic. The agency pointed out that, although satisfactory toxicological data may exist for
each ingredient of a cosmetic, it may still be necessary to conduct
some additional toxicological testing with the completed formulation
to assure adequate safety of the proposed cosmetic product. The FDA
also pointed out that, while a manufacturer would not be held responsible for the safety of the ingredient or product under every possible
condition of use, that company would be held fully responsible for
safety under both the conditions of use recommended in the labeling
and other reasonably expected uses [1].

In 1969 I published a paper dealing with my company's approach
to substantiating the safety of cosmetic and toiletries products [4].
Since that time, additional literature has been published dealing with
the approach of other companies and other scientists in substantiating
the safety of a variety of chemicals and consumer products. The reports of the FDA Over-the-Counter (OTC) Drug Review panels are
particularly helpful in this area of safety. These reports are based
on the combined expertise of various qualified scientists who served
on a series of panels with responsibility to review safety of all OTC
drug products. These experts have recommended the types of studies
they believe are necessary before one can conclude that an ingredient
or product can be considered safe [5-11]. All of these panels have
used basically the same approach. They first determined the safety
of the individual active ingredient and then considered the safety of
the product containing those ingredients.

At Gillette, a product or its ingredients cannot be manufactured
or marketed until it has been deemed adequately substantiated for
safety by the staff of the Gillette Medical Evaluation Laboratories
(GMEL). The first step is what we call the Formula Ingredient Review.
Each operating unit of the company must forward to the GMEL a formula sheet on a proposed new or changed product.

The formula sheet contains a listing of each ingredient, its concentration, the ingredient's supplier, the method of dispensing the product, the proposed directions for use, and the extent of proposed human exposure. Each ingredient must be identified correctly either through the Cosmetic, Toiletry and Fragrance Association (CTFA) *Cosmetic Ingredient Dictionary*, or through the *Chemical Abstracts* Service (CAS) number system. The medical, chemical, and pharmacologic literature is reviewed for all information on the safety of the individual ingredients. Gillette's approach to searching the scientific literature has been published by the GMEL [12].

Each manufacturer of an ingredient to be used in the proposed product is contacted for a Material Safety Data Sheet (MSDS), which will frequently contain important information not found in the published literature. This information will also be of use later to the Industrial Hygiene Unit of the GMEL in considering employee safety under conditions of manufacture.

The purity of each material must be established through the supplier's specifications or other supplier information sheets. The CTFA's *Compendium of Cosmetic Ingredient Composition* is an important reference source to be consulted. It is important that the identity, purity, stability, and physical-chemical integrity of each ingredient be established as much as possible so that later there will be no question about exactly what is being tested and marketed.

From this literature review we try to establish (1) the chemical composition (ingredients, purity of ingredients, stability, effect of intermixing of ingredients, and physical-chemical characteristics), (2) the possible biological effects under various exposure conditions (dose, volume, concentration, route of exposure, frequency of administration, duration of exposure, and rate of absorption), (3) the known toxicological information on each ingredient, and (4) the potential for chemical interaction. Usually we find that the ingredients have been subject to a variety of reviews and that substantial scientific information is available. For example, some ingredients have been subject to the FDA-GRAS Food Ingredient Review or the FDA-OTC Drug Review. There are often published data concerning toxicity under various testing conditions. The supplier usually furnishes information on the toxicity of an ingredient under a variety of testing protocols. For a new chemical, first marketed after July 1, 1979, the ingredient has already been reviewed, prior to marketing, through the regulatory procedures established in the Toxic Substances Control Act. Many of the ingredients also have been used previously in a variety of different finished cosmetic products on which safety tests have been conducted. In addition, we now may also find that the material has been reviewed by the Cosmetic Ingredient Review (CIR) Expert Panel and that a monograph or proposed monograph exists for that category of materials.

After reviewing the published and unpublished literature on the safety of the ingredients, either individually or in a previously tested or marketed formulation, we consider (1) how much of each ingredient is available for possible dermal or mucous membrane absorption from acute and/or chronic use, (2) what might be reasonable foreseeable misuse, (3) what is the potential for adverse chemical interactions among the specific ingredients to be used in the product, (4) what adverse effects the product might produce, and (5) in light of the foregoing, what safety data are lacking for the product.

While we do not believe that it is possible simply to look at a list of ingredients and conclude that the various ingredients are or are not safe for use in a particular cosmetic product, we do believe that, after a detailed review of the ingredient literature, one is in a sound position to make a preliminary technical judgment on what additional studies, if any, are required to substantiate the safety of the proposed new product. If one of the ingredients in the formula has, as shown in the literature, a history of adverse toxicological effects, the chemist will be asked if the ingredient can be replaced with a different material. Just because an ingredient has an adverse toxicological history, it is not automatically discarded if the adverse effect is found at a concentration and a previous use that is different from the proposed use.

We also review the question of mutagenesis, teratology, and carcinogenesis potential as part of our consideration of the potential for chronic toxicity at this stage. First we examine any information from the National Cancer Institute, the Environmental Protection Agency, and any other health regulatory agencies or committees, to see if there are any concerns about the chronic toxicity of the ingredients. Any ingredient for which a mutagenic, teratogenic, or carcinogenic hazard cannot be ruled out with reasonable confidence on the basis of toxicological judgment should either be tested for that specific potential or not used. For example, ingredients whose biological or structural activity resemble those of known carcinogens or may form metabolites similar to those of known carcinogens should either be tested or not used in the product. Chemicals that can effect rapidly growing tissues or may effect mitosis should be examined closely. Ingredients that chemically, biologically, and pharmacologically relate to known or suspect mutagens or show depression of hematopoiesis, spermatogenesis, and ovogenesis, stimulate or inhibit growth of organs, cells, or viruses, or inhibit the immune system must be carefully examined before use in a product.

If an ingredient has been the subject of an FDA review, lacks adverse reactions from its use in a variety of different categories of cosmetic products for a substantial number of years and on a substantial number of humans, has been reviewed as part of the Cosmetic Ingredient Review and found to be safe, one can conclude that the ingredient should not be detrimental to the consumer using the ingredient as part of a total product. One must, however, take into consideration the conditions under which the ingredient will be used. It is important to know whether it will be ingested, absorbed through a mucous membrane, or will come in contact with the eye, or instead will be applied only to normal epidermis. Potential absorption must be considered. The ingredient, as part of a product, may be used daily, more often, or on only an occasional basis. The ingredient review will be valid only to the extent that these actual or reasonably expected conditions of use are in fact correlated with the information obtained from the literature review.

After the ingredient review has been conducted we then consider the need for any confirmatory studies to substantiate our conclusions that the proposed new product is absent of significant risk. We term this second step the Product Review. While a broad armamentarium of toxicological tests are available for complete evaluation of a product, for cosmetic and toiletry products one generally limits consideration of the acute and chronic toxicological potential to (1) oral toxicity, (2) ophthalmic irritancy, (3) dermal irritancy, (4) dermal sensitization, (5) photodermatitic potential, (6) systemic absorption, (7) inhalation toxicity and absorption, (8) mucous membrane irritation, sensitization, and absorption potential, (9) teratology potential, (10) mutagenic potential, and (11) carcinogenic potential. One must consider these areas with respect to the hazard potential to the consumer under conditions of product use and the employee under conditions of manufacture. It is obvious that conducting studies in each of these areas would be neither feasible nor scientifically justified for the normal new product. Thus, one must use sound scientific judgment to discern which further studies, if any, will be required to establish adequate substantiation of safety of the finished product. Sound scientific judgment here refers to the ability to utilize all of the ingredient and similar finished product safety data available and extrapolate those findings to the proposed new product.

In some cases, especially with minor reformulation of existing products or with new products using well-established ingredients, no additional product testing is needed. When additional confirmatory toxicological testing is regarded as appropriate, we usually begin with acute toxicity studies. These include a single-dose oral toxicity test in rats, an ophthalmic irritancy test in rabbits and, if an aerosol, sprayed eye tests in rabbits, a dermal irritancy test in rabbits, a single percutaneous absorption test in rabbits, and, if there is concern over sensitization potential, a guinea pig sensitization test. If there is concern about the results of the ophthalmic study, we have initiated dog, monkey, and in some cases, human ophthalmic studies. All acute study protocols are generally modeled after those first described by Draize et al. [13]. Our approach has been to do in-depth, tightly controlled studies on a few animals as opposed to broad-based studies on a large animal population.

Following any acute animal toxicity studies, the new product, if acceptable, is considered for short-term human use studies. Such additional studies may or may not be needed, depending on the nature of the product involved. When such studies are conducted, direct control of the subjects is important. Thus, we usually conduct these short-term use studies in our own facilities with employees or subjects who have agreed to participate in our medical-use studies. The subjects are examined prior to, during, and at the conclusion of the medical-use study by an experienced medical technologist or nurse. These types of short-term studies give us some medical safety information under conditions of controlled use. The development personnel also obtain cosmetic attribute and efficacy information.

If the proposed new product is found acceptable both from a medical safety and cosmetic attribute standpoint and no changes in formula are necessary, we institute human irritancy, sensitization, and photodermatitic studies. These studies are normally conducted on all new products and reformulations because we usually lack sufficient information on the irritancy, sensitization, and photodermatitic potential of the perfume and its ingredients. Depending on the proposed marketing schedule for the product, the perfume may be evaluated in the product or may be evaluated as a single ingredient at an appropriate dilution in a suitable vehicle. The results of such studies are needed to substantiate the safety of the product from a human irritancy and sensitization standpoint.

If we are satisfied that the proposed new product is safe based on our acute animal and human studies, our literature ingredient review, toxicity information from the supplier, and any additional information that may have been generated from our files from previous tests, and previous use of similar products and/or ingredients, we then evaluate the potential for subchronic toxicity effects. As a general rule, we conduct a subchronic study only in those instances where specific information and use conditions lead us to conclude that it is justified. Subchronic studies are not conducted routinely.

If we conclude from our review that a subchronic study should be conducted, we generally commence a 90-day percutaneous, oral, or inhalation study. The route of administration of the test product in our studies is dependent on whether the proposed product is to be applied topically, taken orally, or is to be applied to the mucous membranes, and also upon its form of dispensing (e.g., aerosol).

It must be kept in mind that our studies are designed to deal with safety evaluation and substantiation. Thus, our interest is to see if a sufficient margin of nonhazard exists for the product. This is based on the relationship of concentration, exposure intervals, and mode of application as they relate to proposed human use. Therefore, our objective is to evaluate the proposed product at dose levels that, while they may be exaggerated, do bear a direct relationship to proposed consumer use. It is extremely important to put into proper perspective the difference between toxic potential of a product and safety evaluation and substantiation of a product. While both concepts use toxicological testing techniques, the former deals with what effects occur in a biological system when the chemical or product, in sufficient concentration, reaches a particular part of the biological system. Safety evaluation, on the other hand, examines the relationship of concentration, exposure intervals, and mode of application to evaluate if a margin of safety is possible with the proposed product and its proposed directions of use. Safety evaluation, while it involves toxicological testing procedures, deals with hazard evaluation of the product at certain doses and concentrations, and under foreseeable conditions of use.

Usually following any subchronic study, additional human evaluations are conducted. These studies may involve several hundreds or thousands of people in consumer use or market research studies. We have devised questionnaires, to be filled out or answered by the product users, which will reflect any adverse effects that the subjects may experience while using the product. Complaints such as stinging and burning are usually found in these types of human use studies but may not be found in human patch testing.

After the first step of reviewing ingredient safety and the second step of reviewing product safety, final approval for marketing must be given unless the data obtained demonstrate otherwise. At present, we do not have a common standard scientific procedure for determining what is an acceptable risk that is agreed to by government, industry, academia, the consumer, and the consumer advocate. Since nothing is absolutely safe or has zero risk, it is difficult to determine what is an acceptable risk. Science can identify and measure risk by assessing the probabilities and consequences of hazard based on toxicity data, conditions of use, amount of exposure, and route of exposure. Risk estimations clearly lend themselves to scientific techniques. Nonetheless, the determination of an acceptable level of risk is not a question that can be answered by scientific technique. Acceptable risk is a philosophical and social question whose answer infringes on the rights of all consumers. Thus, when it comes to deciding what is an acceptable risk, we all speak from our own personal point of view.

As a result, there presently exists no consensus on an acceptable level of risk for consumer products generally or even specific types of products. What is needed is a procedure, accepted by the public, by which unacceptable risk can be distinguished from acceptable risk. For example, if saccharin use is an acceptable risk under certain conditions, then materials that present less risk should also be accepted.

Such a procedure should permit the development of a comparative or relative risk assessment ladder, ranking risks in order of relative seriousness. Under such a system we could ban those products or materials that are unacceptably hazardous. Those materials found to have a significant but acceptable risk could be labeled with adequate directions and warnings. This approach would, of course, require special consideration of the risk to employees under specific manufacturing processes and procedures [14-16]. Industrial hygienists and safety engineers would have to determine a permissible employee exposure to ensure an acceptable employment risk during manufacturing.

In spite of how much data are available on the safety of an ingredient and product, new technical questions may arise from time to time and additional toxicological testing may be required. In the same manner, consumer correspondence may raise safety issues and therefore must be closely monitored. One must be constantly alert to examining the reasonableness of new safety questions when they arise and one must always carefully separate suspicion of toxicity from evidence of toxicity under actual conditions of use. It is, of course, the ultimate responsibility of the toxicologist to ensure that the proposed product is safe under conditions of consumer use and employee exposure.

REFERENCES

1. A. M. Schmidt, Food, Drug and Cosmetic Products, Warning Statements, *Fed. Reg.*, *40(42)*:8912-8929 (1975).
2. R. P. Giovacchini, Old and New Issues in the Safety Evaluation of Cosmetics and Toiletries, *CRC Crit. Rev. Toxicol.*, *1*:361-378 (1972).
3. R. P. Giovacchini, Safe Means the Risk is Acceptable, *Cosmet. Technol.*, *1(1)*:30-33 (1979).
4. R. P. Giovacchini, Premarket Testing Procedures of a Cosmetic Manufacturer, *Toxicol. Appl. Pharmacol.*, *Suppl. 3*:13-18 (1969).
5. A. M. Schmidt, OTC Topical Antimicrobial Products and Drug and Cosmetic Products, *Fed. Reg.*, *39(179)*:33102-33141 (1974).
6. A. M. Schmidt, Aerosol Drug and Cosmetic Products Containing Zirconium, *Fed. Reg.*, *40(109)*:24328-24344 (1975).
7. R. P. Giovacchini, The Significance of the Over-the-Counter Drug Review with Respect to the Safety Considerations of Cosmetic Ingredients, *Food, Drug, Cosmet. Law J.*, *30*:232-227 (1975).
8. R. P. Giovacchini, OTC Drug Safety Evaluation vs. Cosmetic Safety Substantiation, *Drug Cosmet. Ind.*, *117*:34-36 (1975).
9. R. P. Giovacchini, Safety Is the Issue, *CTFA Cosmet. J.*, *5*:6-8 (1973).

10. R. P. Giovacchini, Drug Safety Testing vs. Cosmetic Safety Substantiation, *Cosmet. Toilet.*, *91*:47-49 (1976).

11. H. J. Eierman, Cosmetic Safety Substantiation: Regulatory Considerations, *Drug Cosmet. Ind.*, *118*:32-35 (1976).

12. S. W. Johnson and R. A. Radar, Information Retrieval for the Safety Evaluation of Cosmetic Products, Special Libraries, *69*: 206 (1978).

13. J. H. Draize, G. Woodard, and H. O. Calvery, Methods for the Study of Irritation and Toxicity of Substances Applied Topically to the Skin and Mucous Membranes, *J. Pharmacol. Expt. Therap.*, *82*:377 (1944).

14. R. Scheuplein, Risk Assessment of 2,4-DAA Containing Hair Dyes and Comparative Risk Associated with NDELA-Containing Cosmetics and Saccharin in Foods, Internal memo, Department of Health, Education and Welfare, Public Health Service, Food and Drug Administration, March 27, 1978, pp. 11-14.

15. V. O. Wodicka et al., Proposed System for Food Safety Assessment, Scientific Committee, Food Safety Council, *Food Cosmet. Toxicol.*, *16*, Suppl. 2 (1978).

16. P. Hutt, J. Rodericks, R. Scheuplein, R. Wilson, and R. Giovacchini, The Concept of Risk, *CTFA Cosmet. J.*, *10(4)*: 16-30 (1978).

27

MANAGERIAL AND TECHNICAL RESPONSIBILITIES FOR SELECTING A LABORATORY FOR GLP TOXICOLOGY STUDIES

PHILIP C. MERKER

Pharmacology and Toxicology, Vick Research Center, Shelton, Connecticut

I. INTRODUCTION

The advent of the good laboratory practice (GLP) regulations signaled a new era in the practice of toxicology. For the first time, a governmental agency (the Food and Drug Administration, FDA) formally adopted minimum criteria for doing any type of *in vivo* or *in vitro* toxicology study which would be used to support the safety of a test article in an application for research or marketing permit. In effect, the scientific and technical communities were constrained from doing less than the minimum requirements even though other protocols and procedures may have been "scientifically" appropriate and noteworthy. This GLP constraint was established to assure in an objective manner the quality and integrity of the submitted safety data.

These regulations were of particular concern to those companies (sponsors) that did not have testing facilities of their own or who routinely used outside testing laboratories to accommodate extra heavy workloads. Under these circumstances, sponsors were placed in the position of selecting laboratories which could comply with the new regulations and, once a study had been placed, they still had the obligation to approve the protocol and were responsible for the data submitted to the FDA. Therefore, in addition to concerns about selecting a laboratory, there were responsibilities about the conduct of the study.

Considerable attention has been given to the technical development of study protocols and the procedures to be followed when monitoring ongoing GLP-type studies. However, the mechanisms to be used for the selection of a laboratory have not been thoroughly discussed.

The selection of a nonclinical laboratory to do a good-laboratory (GLP)-type toxicology study depends on a number of factors. Some of these factors are the responsibility of the sponsor; some are those of the laboratory. The importance of interaction between sponsor and laboratory during the time a sponsored study is in progress is obvious. Equally important, but perhaps not so obviously so, is the need for interaction prior to selection of the laboratory.

In a practical sense, each party should be aware of the other's responsibilities so that the sponsor and laboratory alike can readily identify areas of concern and, hopefully, resolve them prior to the hardening of a final decision that would either exclude or include a laboratory in the sponsor's testing plans.

The purpose of this presentation is to discuss from the sponsor's viewpoint those factors that enter into the decision-making process and to emphasize the interlocking role of management and toxicologists, biologists, and other technical personnel within a sponsoring company to accomplish the objective of selecting a laboratory.

II. RESPONSIBILITIES OF THE SPONSOR

Prior to choosing a nonclinical laboratory to do a toxicology study, the sponsor has the responsibility of developing: (1) standard operating procedures for selecting such a laboratory; (2) a protocol for the planned study; (3) technically trained staff or consultants; (4) an estimated budget; and (5) a time frame for starting and ending the study. These five key points are used to build a strategy for selecting a laboratory. Of importance in developing a strategy is the extent of cooperation within a sponsoring company that exists between administrative (directors, managers) and technical staffs (toxicologists, chemists). Proper cooperation implies that at several levels within the company: (1) the full intent of the proposed study is understood; (2) internal checks and balances are operational during the selection process; (3) the technical aspects of the study are known, and (4) cost and timing estimates have been designed to meet company needs.

A. Protocol

The choice of a nonclinical laboratory to perform a GLP-type study is dependent on the nature of the tasks specified in the protocol design. A laboratory must be chosen that can perform the tasks under GLP.

Overall, the protocol will in itself be written under GLP guidelines and therefore, by definition, its specifications will include: the type of study; species of animal; route of administration; animal and maintenance requirements; and observations to be made. Of particular importance, the protocol will contain a broad classification for study duration which will be specified, directly or indirectly, as: acute (1 to 14 days); subchronic (up to 90 days); and chronic (longer than 90 days).

For example, the protocol might specify an oral toxicity study in rats using a single dose, coupled with 14 days of observation for gross pharmacotoxic signs, body weight, and gross examinations of all internal organs at termination of the study.

This brief description indicates that the study will be done in rodents; the route of administration will be oral; observations will be limited to 14 days; animals will have to be observed daily for toxic responses; and the organs will be examined *grossly* at the end of the study. With such a protocol, a nonclinical laboratory obviously need not have long-term holding capabilities for dogs; personnel need not be trained to administer materials intravenously; special electronic instrumentation is not required to monitor daily pharmacotoxic signs; and the laboratory need not have facilities to process organs and tissues for histology.

In short, a protocol automatically defines a study administratively and technically and communicates to the laboratory the sponsor's re-

quirements and needs. The protocol, initially, can thus be considered to be a medium of communication that functions as a "request
for a proposal" (RFP) from the laboratory to do the work specified
in the protocol.

Once selected, the testing facility will in all probability play a
material role in the inevitable subsequent modifications of the protocol.

B. Request for Proposal (RFP)

A RFP is a written request to a nonclinical laboratory to submit a dollar bid to do the work described therein. Therefore, the RFP must
contain sufficient information so that the laboratory can respond in
the most informed manner. On the other hand, caution must be exercised in its preparation so that confidential information is not disclosed.

Depending on circumstances, either a completed protocol or a con-
cisely written paragraph containing all the pertinent information can
serve as the RFP. In either case, whatever its form, the RFP should
provide the following information: (1) title and description of study;
(2) number and description of materials to be tested; (3) animal(s)
to be used; (4) route of administration; (5) dosage regimen; (6) observations required; (7) length of study and deadline for receipt of
final report; and (8) requirements for GLP.

The U.S. government publishes RFP's in the *Commerce Business
Daily*; all contract laboratories thus have an opportunity to respond.
However, for a cosmetic company to announce or forward RFPs to all
contract laboratories with possible interest would cost the company
considerable time and expense; first, to publish the proposal; second,
to screen the responses. Thus, a preliminary laboratory selection
process should be employed in order to develop a more restrictive list.
This preliminary selection process can be based on a number of factors, such as (1) geographic location of the laboratory; (2) defined
education and experience of the study director; and (3) industrywide
experience with the laboratory.

C. Geographic Location

Several "working zones" based on traveling convenience can be drawn
around the company facility and priorities assigned to laboratories
based on closeness. Setting up "work zones" — say at intervals of
10 to 200 miles — can save traveling time and trip expenses.

D. Education and Experience of the Study Director

Education and experience requirements for the Study Director can
be specified on the basis of the proposed needs of the study. In gen-

eral, an advanced degree (M.S./Ph.D) in a vertebrate zoology area should be a minimal educational requirement, coupled with two to three years of experience in toxicological investigations. For such special studies as reproduction/teratology, the director probably should have a Ph.D. in embryology, along with three to five years of experience as an embryologist/teratologist. At the same time, a sponsoring toxicologist with strong credentials in embryology/teratology may feel that the Study Director need not be so highly qualified in this field. However, the best-trained Study Director should always be obtained irrespective of the training and experience of the sponsoring toxicologist.

E. Industry or Personal Recognition

For certain types of studies some laboratories are highly regarded within the industry or by key individuals within the sponsoring company. While such general recognition may be desirable, it should not be a signal to select a given facility automatically. A full inspection and analysis of the laboratory is in order, even if it is highly regarded.

F. Standard Operating Procedures for Selecting a Nonclinical Laboratory

Another important element for a company to have prior to the selection of a contract laboratory is a well-defined standard operating procedure (SOP) describing the selection process to be used. The SOP should detail a company's managerial and technical criteria for the selection process, and it should clearly state the objectives, methods, and procedures to be followed. The SOP will therefore contain those processes to be used by managerial and technical staffs for this purpose — processes which will have been established only after the technical staff and management have agreed to them. It is of paramount importance that these processes be thoroughly understood by management and the technical staffs, who will, in essence, be responsible for assembling the necessary information and recommending the use of a particular nonclinical laboratory by the company. The technical staff must be fully aware that such recommendations are made in the interest of the *company* as a whole and not in the narrow interest of the toxicology department. The development of this type of awareness is the responsibility of management; it must be expressed and articulated within the framework of the SOP for selecting a nonclinical toxicology laboratory.

Under these conditions, therefore, the SOP should not be considered a rigid document, but one subject to change and modification as new ideas are formulated by management.

III. OBJECTIVE VERSUS SUBJECTIVE FACTORS ENTERING
AN EVALUATION

The choice of a contract laboratory is dependent on a mixture of objective and subjective factors. The SOP thus should take into account
criteria that are evaluated subjectively and objectively. The objective
factors are measurable or determinable items that are required by
GLP, such as a physical plant that contains a necropsy area and provides for separation of animal species. The subjective factors are
those that pertain to evaluating the *quality* of the items required by
GLP. Thus, a room which is prominently designated as a necropsy
area would not be acceptable as such if it were not appropriately
lighted and equipped and provided with sufficient space for technicians. Accordingly, the sponsor should use well-trained and experienced technical staff to conduct the inspection and then evaluate information; in the absence of such personnel, the sponsor should be
prepared to engage a consultant to do this phase of the work. The
importance of expertise in performing this function cannot be overstated; simple subjective statements, i.e., "the physical plant is adequate," provide insufficient evidence to justify recommending a laboratory.

A recommendation to use a laboratory should be supported by both
objective and subjective documentation. This type of documentation
will provide information that will help to determine the choice of a nonclinical laboratory; it will also become a permanent record of the quality of the laboratory at the time of inspection. This record can then
be used as a criterion-standard against which the laboratory can be
judged at future times, if it is selected to do work for the sponsor.

A detailed listing of GLP requirements for inspecting a testing
facility is beyond the scope of this presentation. For our purposes,
the following considerations should be given for an initial inspection.

A. Organization and Personnel

Emphasis should be given to the Study Director and technical staff
within the investigative area of sponsor interest based on study type
decided upon earlier. Depending on the study, various levels of
scientific education and experiences are required.

There is no reason for a Study Director not to be both learned
and experienced, primarily because the final report of a study is the
product of the scientific judgment of the Study Director. The director's education and experience obviously bear directly on the conduct
of the study and the validity of its results.

The importance of personnel qualifications has been emphasized
by the FDA, and in the broadest sense laboratory management and
study directors have responsibilities for ensuring that personnel meet
necessary educational, training, and experience requirements. In

this regard, attention should be given not only to the Study Director and senior technicians but also to personnel who may not be directly engaged in supervision and collection of data — animal caretakers and physical science technicians.

B. Plant Facilities

Emphasis should be on that area of the facility which will be used for the intended study. One area of a facility could be of better quality than other areas in terms of physical plant, equipment, and personnel. Attention must therefore be given to the specific area of interest. An acceptable area for one type of study does not automatically qualify the testing facility as a whole to conduct other types of studies. Prior inspection of the particular area within a facility that will be used for a study is mandatory. For this purpose particular attention should be given to the physical areas, equipment, and personnel in the following four areas:

1. Animal facilities: separation (rooms) of species
2. Support facilities: diet preparation and storage; materials receiving and storage (pharmacy); chemistry (unless the sponsoring company agrees to do the chemistry); personnel facilities (rest rooms); cage washing
3. Pathology: necropsy areas; necropsy technicians; hematology and blood chemistry; record keeping
4. Quality assurance unit (QAU): records storage

The above areas should be inspected and evaluated from the following two viewpoints: scientific and managerial.

IV. SCIENTIFIC VIEWPOINT

The facilities, physical plant, and personnel should be viewed from the vantage points of the "scientist", i.e., toxicologist, microbiologist, embryologist, etc. For this the technical aspects of the facilities, the plant, and the personnel can be evaluated in the light of the governmental agencies' requirements to fulfill GLPs. The knowledge and experience of the inspecting scientist is of prime importance for this part of the evaluation. In this regard, it is of particular importance that the scientist understand the nature of the test system that will be used in the planned studies. Only under these circumstances will the test facilities be evaluated adequately.

As an example, a sponsor designs a study to test a material for acute skin irritancy; a laboratory is inspected by the sponsor's trained toxicologist, and the laboratory is found to be satisfactory. One of the elements in this evaluation is the adequacy of the program in the laboratory to train technicians to grade skin irritant responses based on a SOP describing the training procedure. The sponsor's toxicolo-

gist must not only record the fact that a SOP for this activity is available, he or she must also comment on the adequacy of the program.
An improper training program is sufficient grounds to exclude the
laboratory from the study, notwithstanding excellence of the physical
plant and expertise of the Study Director. In this case it is not only
sufficient to document that a standard operating procedure for training technical staff exists within the laboratory; it is also important
to know and understand the *quality* of the training program.

Toxicologists or other scientists who are involved with specific
study projects can be placed in one of three categories: (1) experienced in the specific study area, such as skin and eye; (2) experienced in a related area, but not in the specific study type; (3) not
experienced in the specific or related areas, i.e., a chemist acting
in the role of a biological toxicologist. Depending on the category,
different types of scientific and administrative interactions can be established with Study Directors and administrators of testing facilities.

For the fully experienced sponsoring cosmetic toxicologist, the
interaction can be at the "peer" level, provided that the Study Director is at an equal level of experience and knowledge. Under these
circumstances, the sponsoring toxicologist and the Study Director can
interact to exchange ideas and develop meaningful study protocols
and interpretations of the data. When the sponsoring toxicologist has
more experience and knowledge than the Study Director, then, in a
practical sense, the sponsoring toxicologist assumes a new role: The
sponsor becomes part of the study team.

For the inexperienced sponsoring toxicologist, the Study Director
must be fully experienced, primarily because protocol development
and data interpretation will become in a very practical sense the responsibility of the Study Director. These types of interrelationships
must be fully understood in order for the selection process for a testing facility to be meaningful and for an ongoing study to be conducted
thoughtfully and then finally interpreted. Thus, a fully experienced
cosmetic toxicologist may be confident in the capabilities of a Study
Director in a particular testing facility, but this same Study Director
may be unsuitable for a less experienced cosmetic toxicologist. When
the toxicologist is inexperienced, an outside consulting toxicologist
should be engaged for the study. Management and scientific personnel in cosmetic companies should always appraise their own abilities
to conduct studies realistically and then develop appropriate strategies *prior* to the selection of a testing facility. Thus, outside consultants should be brought in *before*, not *after*, the selection process.

V. MANAGERIAL VIEWPOINT

The evaluation of a testing facility for its managerial competency requires as much expertise as does that for technical evaluation. Therefore, such evaluations should be done by experienced managers. In

certain circumstances, technical personnel also can have managerial expertise, and thus both the scientific and managerial evaluations can be accomplished by a single individual. However, if an otherwise qualified scientist has no managerial expertise, a manager should accompany the scientist to evaluate those aspects of the testing facility which can be clearly placed within a managerial framework. Individuals who can fulfill this function may be business or office managers and regulatory affairs or legal people. These individuals should be thoroughly familiar with GLP requirements prior to the visitation and they, together with the scientist, should develop a plan of action concerning the visitation.

Two basic elements should be considered in the "managerial" evaluation of a testing facility: those relating to actual performance of a study under GLP, and those of the business end, i.e., the financial standing of the laboratory, contracts, costs, and billing. In terms of actual performance of the study under GLP, certain "external" potential or real needs of a cosmetic company should be considered in relation to the "managerial" purposes of a study; for example: (1) FDA submission; (2) claim support; (3) poison control information; (4) screening to select the most appropriate formulation; (5) legal support. Prior understanding of the managerial purpose for doing a study should help to classify decisions concerning the choice of a laboratory.

For example, a Study Director may be wanted, from a managerial standpoint, who can eventually testify in a court of law. Under these circumstances, a given level of expert credibility would be needed; hence the decision to use one or another testing facility would depend on the "legal" credibility of the Study Director. This is a managerial decision, the making of which should be taken into account prior to selection.

Additionally, insight into managerial decisions with respect to the testing facility can affect the choice of a testing laboratory. For example, the FDA will not accept a study done in a nonclinical laboratory that has refused prior FDA requests for an inspection. Therefore, compliance records such as federal and state GLP inspection reports and required licenses should be examined and commented on in any written report concerning the laboratory.

Mention was made above about the business activities of the testing facility. This might appear to be outside the scope of a discussion concerned mainly with GLP. Nevertheless, it is of practical importance in that a testing laboratory should have an acceptable financial base to support building maintenance, attract and retain highly qualified personnel, provide for consultants when needed, maintain and upgrade equipment, etc. These are of special importance for long-term toxicologic studies that extend over two or three years. Therefore, managerial inquisitiveness should extend into business areas to provide evidence of institutional stability during the course of a study.

The managerial inspection should also cover at least four other areas: standard operating procedures, which should describe the managerial organization; technical operation manuals, which should describe the processes carried out by the managerial organization; a data auditing system, which should be examined in terms of management report procedures used for the quality assurance unit; and archives, with attention given to the management decisions that pertain to the operation of the storage and retrieval systems and procedures used in the handling of reports submitted to the FDA.

Managerial decisions regarding the suitability of a nonclinical laboratory should be based on the examinations of these elements in the light of whether or not the study is required to be done under GLP and thus will be done under "regulated" procedures. Those safety studies that are to be submitted to the FDA should be managerially classified as either "pivotal" or "material." This requires both managerial and scientific judgment.

A "pivotal" study is defined as one which is used in applications for research or marketing permits which formed the basis for a decision regarding the safety of a regulated product. This type is in contrast to a "material" study -- one which is of continuing importance in support of safety decisions made earlier by an agency.

In both instances the fact that data will be submitted to a federal agency to support the safety of the product should be taken into account and the laboratory so informed.

VI. LABORATORY EVALUATION

Following an on-site inspection and examination of all relevant documents and information, an evaluation report should be issued in a manner previously covered in the SOP describing the selection process. This report should contain a list of items examined and discussed, with pertinent documentation, if available, to support claims and comments. A final evaluation should contain recommendations for approval, disapproval, or a continuation of the evaluation.

An approval by the investigating team indicates to management that the laboratory is suitable; approval by the next level of management should be required prior to actual use of the laboratory.

A disapproval automatically terminates further consideration; however, the laboratory may be visited at a later date depending on new information. For whatever reason, when a previously disapproved laboratory is revisited, very careful documentation is required to approve such a laboratory.

A recommendation to continue the evaluation indicates a need for further information. This can and should take the form of a planned "simulated" study in which a reference material of known biological activity is tested at the laboratory using full GLP procedures. The final decision on the laboratory can be based on actual performance of this test. A procedure of this type can be used to validate the laboratory.

VII. CONCLUSION

The search for a laboratory to do toxicology studies should begin with clearly defined study objectives and search procedures. The study objectives are contained within a study protocol, and the search procedures should be defined by standard operating procedures. In both instances, management and technical staffs must fully understand their areas of underlying objectives and procedures. The need for an understanding on the part of both management and technical staffs of their respective areas is critical for the successful choice of a laboratory. Management's contribution will consist of determining the competency of the laboratory's management, evaluating the Study Director in terms of the "pivotal" or "material" nature of the study, and weighing the costs and fiscal stability of the laboratory. The technical staff should professionally evaluate those elements that bear directly on the planned study. The combined managerial and technical evaluation should result in the selection of a laboratory that will do the planned work expeditiously, at acceptable cost, and under circumstances that will guarantee good laboratory practices.

BIBLIOGRAPHY

1. A. Thompson, Personnel Elements of Good Laboratory Practice, *Clin. Toxicol.*, 15:527-538 (1979).
2. C. D. Van Honweling, M. A. Norcross, and P. D. Lepore, An Overview of Good Laboratory Practices, *Clin. Toxicol.*, 15:515-526 (1979).
3. K. Burck, The Role of Quality Assurance in Good Laboratory Practices, *Clin. Toxicol.*, 15:627-640 (1979).
4. J. Ward, How the GLP Changed One Drug Company's Research Operations, *Lab Animal*, 9(4):35-43 (1980).

28

OCULAR IRRITANCY: THE SEARCH FOR ACCEPTABLE AND HUMANE TEST METHODS

EDWARD M. JACKSON

Research Services Department, Noxell Corporation, Baltimore, Maryland

I. INTRODUCTION

Although the assessment of ocular irritancy potential culminates with the testing of the finished cosmetic product, it begins with an assessmont of the individual cosmetic ingredients and colors as well as a review of the results from testing prototype formulations. While individual chemicals may themselves be irritating or nonirritating, mixing these chemicals together may lead to the production of a different chemical compound (synthesis), have a cumulative effect (augmentation), negate a deleterious effect (quenching), or permit an effect which an individual chemical alone is incapable of (synergism).

The ocular irritancy potential of a finished cosmetic product is best accomplished in a stepwise fashion, the data from one phase of testing being evaluated before proceeding to the next level of sophistication. Only then can the product be introduced into the marketplace with the reasonable assurance that the vast majority of consumers will be able to use it without undue harm to one of the most precious of human faculties, sight. The important steps on this testing ladder, in increasing order of sophistication, are ocular irritancy testing in animals, human patch testing, and the controlled use test in humans. These are premarket evaluations of the product. Postmarket surveillance by a consumer affairs department closely monitoring complaints and alleged adverse reactions after introduction of the product will either confirm the premarket test results, thus extending the safety data base of the product, or indicate possible areas of reformulation requiring further testing.

This chapter on the assessment of the ocular irritancy potential of cosmetic products is divided into the scientific foundations for such testing, regulatory foundations, humane testing in both animals and humans, and a look at possible alternatives to the use of animal models. The scope of this chapter is limited to the ocular irritancy testing of finished cosmetic products as defined by the Code of Federal Regulations [1], with special emphasis on eye-area cosmetic products (mascaras, eye shadows, eye makeup removers, eyeliners, and eyebrow pencils). The safety of cosmetic colors is regulated by the Food and Drug Administration's Color Additives Program [2], and the safety of cosmetic ingredients is being addressed by the cosmetic industry's Cosmetic Ingredient Review Program [3]. Fragrances are rarely a part of eye-area cosmetic formulations, but in those instances where fragrances must be considered the reader is referred to the safety assessment of fragrance ingredients as conducted by the Research Institute for Fragrance Materials [4].

II. SCIENTIFIC FOUNDATIONS

A. The Draize Eye Irritancy Test

Rather than yield to the temptation of adding yet another description or comprehensive review of the Draize eye irritancy test [5,6], or composing another litany of the ocular effects resulting from exposure to various chemicals [7,8], or even critiquing the Draize eye irritancy test [9,10], we will attempt to unfold the history of the development of the Draize test, define the phenomenon known as inflammation in relation to the eye, and place the Draize test in a safety substantiation program for cosmetic products with special emphasis on the eye-area products.

It is a *scientific event* when a test system appears in the literature and successfully weathers the critical examination of the scientific community through extensive use and modification of the test. It is a *scientific phenomenon* when such a test system endures for more than four decades as the best available means of a particular form of safety assessment. Such is precisely the case, however, with what is known as the Draize eye irritancy test.

This statement does not imply that a better *in vivo* animal model cannot or will not be found, nor does it exclude the possibility that ocular irritancy potential may eventually be initially screened in *in vitro* test systems. It does mean that literally generations of toxicologists have reviewed, used, assessed, and modified this test, even highlighted its weak points, but have never produced a better animal model than the rabbit nor a better test system than the Draize eye irritancy test.

Although the Draize test is currently used to assess the ocular irritancy potential of cosmetic, pharmaceutical, ophthalmic, personal hygiene, detergent, household, and even pesticide products, few people realize that this test originated out of a need to legally indict the ocular hazards attendant upon the use of a hazardous eye-area cosmetic product. The product in question was an eyelash-dyeing product containing *para*-phenylenediamine, which resulted in several serious and even permanent eye injuries to consumers [11]. The marketing of several hazardous cosmetic products like this during the 1930s finally resulted in the extension of the Pure Food and Drug Act of 1906 to include cosmetic products in the Food, Drug, and Cosmetic Act of 1938.

The newly formed cosmetics division of the Food and Drug Administration (FDA) had as one of its first charges the responsibility for generating laboratory data to demonstrate the ocular hazards associated with the use of this eyelash-dyeing product. Dr. John H. Draize and his fellow pharmacologists, Dr. Geoffrey Woodard and Herbert O. Calvery of the FDA's pharmacology division, were able to document in an animal model the damage that resulted from the use of this hazardous eye-area cosmetic product [12].

But, as happens so often in science, the originator of a particular test method is not necessarily the one whose name becomes attached to that test. The rabbit eye assay was actually developed by Dr. Jonas S. Friedenwald, an internationally renowned ophthalmological pathologist on staff at the Wilmer Ophthalmological Institute at the Johns Hopkins University, Baltimore, Maryland [13]. In a kind of ophthalmological Manhattan project, Dr. Friedenwald was charged during World War II with assessing the effects of various chemical warfare agents on the eye and determining the effectiveness of various antidotes to chemical warfare agents as well as providing a means of lot-by-lot certification for the actual manufacture of these antidotes. Dr. Friedenwald published some of the results as determined by the rabbit eye assay in 1944 [14] and again in 1946 [15].

Draize developed Friedenwald's test system beyond the assessment of corneal ulceration to the determination of conjunctival and iridial inflammation. The fact that corneal injury is the most serious type of eye injury accounts for the weighting of the scoring in favor of the corneal effects in the overall Draize scoring system. Many reviews seem to date the Draize test from his own paper, also published in 1944 [16], since it contains the Draize scoring scale. But the 1952 paper by Draize [17] is far more illustrative of the actual description, use, and application of the technique to topical products. These two papers may also reflect the historical sequence of events. Draize initially used Friedenwald's test system to study the effects of neat chemicals on the eye, then modified it and applied it extensively to topical product ingredients and final formulations.

The use of the Draize test for the assessment of the ocular irritancy potential of topical products was initiated by the FDA and soon found its way into academic, industry, and contract laboratories as the test best able to assess the ocular irritancy potential of both ingredients and products. Its current extensive data base makes it the most useful, direct eye contact assessment currently available to the toxicologist. Other advantages are cost, time, ease, availability, whole animal and organ evaluation, versatility (it can be used as a quantitative assessment or a screening test), ability to evaluate neat chemicals or whole products in a concentrated or dilute form, amenability to modifications and information on ocular repair and restoration [18]. Very few toxicological tests combine these numerous advantages either to indicate necessary product reformulation or to clear a product for initial human use. There are, of course, some disadvantages to every test, and the Draize eye irritancy test is no exception. These have been adequately reviewed elsewhere [18,19].

The inflammatory response is a complex response of an organ or tissue to a chemical, biological, or physical stimulus. A chemical in a product, biological toxins, and sunlight are examples of irritant stimuli. The inflammatory response is clinically characterized by the objective parameters of erythema, edema, pain, and heat and the subjective parameters of stinging, burning, and itching. Once the insult has occurred, mast cells release histamine, which produces vasodilation (erythema) and increased vascular permeability (edema). The phenomenon of chemotaxis signals the polymorphonuclear leukocytes (PMNs) to marginate along the walls of the vascular system, diapedese through these walls, and travel to the affected area to aid in the reparative processes [20].

The preceding is merely a broad-brush scenario whose cellular and molecular events we know very little about. The reddening and swelling of the skin in response to an irritant are the most observable and quantifiable events in the complex process known as inflammation. Because these events occur with some constancy, animal and human test systems characterized by rather good predictability and application have been developed. The Draize eye irritancy test is such a test system. It capitalizes on the erythema and edema produced by the palpebral conjunctivum and the iris as well as corneal opacification or edema in response to certain irritants while permitting observation of the restorative and reparative processes from these effects.

The Draize eye irritancy test occupies the place of an initial test in the safety substantiation program for a cosmetic product. The Draize test is particularly suited to this task because it uses an animal model more sensitive than the human eye. A test system must either be equivalent to or more sensitive than the system it is ultimately intended to protect. It is far more acceptable to err on the side of safety in this regard than on the side of potential hazard. The Draize test is more sensitive because the rabbit eye exhibits a cornea thinner than the human cornea and is, therefore, more susceptible to swelling or edema in response to an irritant. The rabbit eye is also more sensitive because it tears less than the human eye, which reduces the natural dilution of the test material. Unfortunately, the fact of less tearing by the rabbit eye has been a subject of much controversy by the uninformed. The rabbit eye does tear, and this is scored as discharge in the Draize test, the third observation scored under the effects exhibited by the conjunctivum.

The rabbit eye is further composed of structures analogous to the human eye itself. Although extremely complex as an organ, the effects of the test material are determined on three key structures of the eye: the conjunctivum, the cornea, and the iris. The lens is not a test structure in the Draize test. These three test structures are precisely those that immediately come into contact with a foreign substance. Qualitative and quantitative observations are made on the effects produced on the palpebral portion of the conjunctival membrane for redness (hyperemia), chemosis (edema), and discharge (tearing). Next, the cornea, the clear circular and anterior portion of the sclera of the eye, is observed for the process of opacification or clouding. The area involved in corneal opacification can further be measured by the use of a fluoroscein dye. Finally, the iris is observed for swelling, vascular injection, and its reaction to light. The weighting of these scores is heavily in favor of the cornea (73 percent of the total Draize score), then the conjunctivum (18 percent), and finally the iris (9 percent), or an 8:2:1 irritation emphasis ratio [6].

The end result is that the Draize test is a quantification of the process of ocular inflammation due to chemical or physical irritation as manifested by conjunctivitis, keratitis, and/or iritis.

In concluding this section on the Draize eye irritancy test, we could do no better than to quote from Grant's classic work, *Toxicology of the Eye:*

> Most of the unpleasant or serious effects of chemicals or drugs, venoms and plants on the human eye or on vision have been learned of through human misfortune. Now the aim is to try and anticipate and avoid undesirable effects on the eye by more comprehensive testing before exposure of human beings. This essentially involves preliminary testing in animals. . . [21].

B. Human Patch Testing

The specific purpose of the Draize eye irritancy test is to determine the ocular irritancy potential of a given test substance in an animal model. The close proximity of cosmetic products to the human eye itself, coupled with the remarkable absorption potential of the eye, requires this type of testing to be performed initially in animals. However, eye-area cosmetics are applied directly to the contiguous eye structures of the eyelids, eyelash, and eyebrow, necessitating the assessment of the dermal irritancy potential of these products. We must note that anatomically, the eyelid or palpebral skin is the thinnest in the entire human body, thus magnifying its irritancy potential.

A complete assessment of the dermal irritancy potential of eye-area cosmetic products again begins with an initial assessment in an animal model, proceeds to a prophetic dermal irritancy test in humans (prophetic patch test), and on to a dermal irritancy assessment through repeated insults (repeat insult patch test) before concluding with a controlled use test in humans.

The test of choice for determining the dermal irritancy potential in animals is the Draize dermal irritancy test [16,17,22]. This test utilizes both intact and abraded skin in assessing the irritancy potential of the test substance. There are, however obvious differences between the skin of the two test species, rabbit and human. As in all initial assessments of irritancy potential in animals, the analogy of a sieving operation or a screening procedure holds true. The test substance must successfully pass through the rather large holes, so to speak, of the ocular and dermal animal test before attempting to pass through the smaller holes of the human prophetic patch test, then on to the even smaller holes in the repeat insult patch test, and finally through even smaller holes, the controlled-use test. At this point, the manufacturer is relatively confident that exposing the product to a limited amount of people in a test market population should not be problematic from a product-safety point of view. If few irritation problems occur in the test market population, then the product is ready for mass distribution or national introduction coupled with close monitoring of alleged adverse reactions to the product through the company's consumer services department.

Having placed patch testing in its proper perspective in a safety substantiation program, it now remains to discuss the purpose and the value of these human tests in detail. Several reviews comparing and contrasting the various types of human dermal testing are available to the reader who requires information on test protocols, results, and intrepetation of results [23 — 26].

The safety substantiation of a new cosmetic product may include human patch testing in the form of the prophetic patch test. There are several methods available: The Schwartz-Peck [27] method is the pioneer of this type of test and is still in use today, although there have been modifications to this procedure, as in the case of the Brunner-Smiljanic [28] and the Traub-Tusing-Spoor [29] modifications. The principal features of the prophetic patch test are single exposures to the test material under closed patch, open patch, and ultraviolet test conditions, a minimum of 100 test subjects, and a single sensitization and/or photosensitation challenge.

Whichever procedure or modification is used, the prophetic patch test can yield information as to the rank irritancy, sensitization, and photosensization potential of a product. This test is particularly helpful with de novo formulations, formulations representing a new chemical combination, or a formulation containing a new ingredient. In these situations, the results from the prophetic patch test should be reviewed prior to conducting the repeat insult patch test.

If the prophetic patch test is the only patch test used to substantiate the safety of a product, its principal disadvantages are a severely restricted induction period for the assessment of either sensitization or photosensitization potential and a lack of discrimination between photoirritation or photoallergenicity in photosensitization. When used prior to a repeat insult patch test, these disadvantages are resolved by the results from the repeat insult patch test. In fact, the prophetic patch test then acts as a clearance procedure, ensuring a degree of safety for the panelists in the repeat insult patch test.

The repeat insult patch test furthers the initial assessment of irritancy, sensitization, and photosenitization potential made in the prophetic patch test by discriminating between photoirritation and photoallergenecity through repeated exposures under closed patch, open patch, and ultraviolet test conditions. These repeated exposures to the test material maximize the possibility of determining the limits of irritation and photoirritation by multiple exposures to the test product mimicking daily or weekly consumer use of a product. Further, a true induction can occur prior to challenge to probe the sensitization and photoallergenic limits of a test product. The literature reflects several schools of thought on the number of test subjects needed to assess the irritation, sensitization, photoirritation, or photoallergenic potential of a test product in the repeat insult patch test. Rather than reflect these divergencies, we can suggest either a single panel of a minimum of 50 to 200 test subjects or multiple panels cumulatively

totaling 200 test subjects. The advantage of the latter approach is
that it allows the safety substantiation program continually to assess
slight formula modifications such as the use of alternative chemicals,
lot-to-lot chemical variations, process changes, and actual manufacturing variations from the initial product formulation.

There are two main approaches to the repeat insult patch test,
that of the Shelanskis [30], who pioneered this application of the
dermatologist's standard patch test, and that of Draize [31]. In fact,
a modification using certain features of these two approaches is now
termed the Draize-Shelanski repeat insult patch test. True modifications of the repeat insult patch test stress irritancy potential alone
and discrimination among minor alterations in a product formula [32]
or a sensitization potential through exaggeration [33] or enhancement
of absorption [34,35]. It is interesting that the human patch test as
a predictive or prognostic tool arose from a dermatological diagnostic
tool. The use of the patch test as a prognostic tool substantiates the
safety of a product, while its use as a diagnostic tool can determine
the liability aspects of that same product.

C.   The Controlled Use Test

Once an eye-area cosmetic product has been successfully tested in
the Draize eye irritancy test and the human patch test phase of a
safety substantiation program, the first human eye-area application
under test conditions is often the controlled use test. This test has
been variously described as an in use test [36], a paired comparison
use test [23], or simply a use test [37]. The controlled use test has
recently been reviewed [38] and is the term adopted by the cosmetics
industry [39]. It will be used in this chapter.

The precise purpose of the controlled use test is the appropriate
assessment of the effects of using an eye-area cosmetic product. This
assessment must necessarily merge toxicology with ophthalmology and
dermatology. The objective of the controlled use test is to define the
limits of the cosmetic eye-area product's irritation potential on the
human eye and its associated structures.

The controlled use test utilizes either 50, 100, or 200 test subjects, depending on the type of product, intended use, quantity and
frequency of product application, and the biostatistical significance
of the sample size. For purposes of clarity, we will describe the
parameters of the controlled use test in terms of 50 test subjects completing the test regimen and assume a test market introduction and
monitoring of alleged adverse reactions prior to an intended national
introduction of the product.

The minimum duration of the controlled use test for an eye-area
cosmetic product is 28 consecutive days of use or four weeks. The
frequency of use of the particular eye-area cosmetic product should
be exaggerated. If the product is a waterproof mascara, for example,

which consumer patterns indicate is used approximately three times per week, then a daily application in a controlled use test would be exaggerated. If the product is an eye shadow, three to four applications per day by the test subjects would be exaggerated. The principle is to apply a large quantity of test product with a high frequency of application. The exaggerated dosing of the test subject is therefore comprised of varying the quantity and frequency of use with the product configuration. Observations should be made periodically by trained professionals using a standard scoring system such as that of the International Contact Dermatitis Research Group or the North American Contact Dermatitis Research Group. As with any test, the study sponsor is charged with the responsibility of monitoring the progress of the study, usually through site visits. These monitoring visits ensure that the protocol is being followed, allows for appropriate and agreed-to adjustments of the protocol, and helps avoid circumstances which could cast a shadow over the results of a particular study, such as the introduction of panelist bias, whether positive or negative.

Variations of the controlled use test can expand the scope of the results obtained. For example, parallel testing with a previously marketed product can serve as a control, as can an untreated eye, if the test unit is an eye-area cosmetic. This may be more desirable in certain instances than testing the product on both eyes and using a standard dermatological scoring scale. Although irritation potential is the primary purpose of this test on an eye-area cosmetic product, sensitization potential can be assessed by having a two-week rest period after the conclusion of the controlled use test, after which time the panelists return for a standard patch test with the product, which serves as a challenge.

The safety substantiation of cosmetic products, including eye-area cosmetic products, should begin with an initial assessment of the ocular irritancy potential of the product in an animal model prior to any human test. Human patch testing is next employed to determine the dermal irritancy of the product. Only when these two types of tests are completed should the product be used as intended on human test subjects in a controlled use study. Acceptable results from these phase of product safety substantiation allow the manufacturer to proceed to a test market introduction of the product while closely monitoring the alleged adverse reaction ratio to number of units sold. When these data have been analyzed and indicate a low incidence of consumer complaint, national introduction of the product, again coupled with monitoring of the alleged adverse reaction ratio, can be attempted with reasonable assurance of marketing a safe product. The final test is that of time — time in the marketplace and successful use of the product by consumers as indicated by repeat consumer purchases. It is really only then that a manufactorer confirms that the product is indeed safe for the majority of the consumer public. Therefore, to ethnically introduce a product into the marketplace, premarket tests in animals and consenting human subjects will allow the necessary prognostic safety assessments.

III. REGULATORY FOUNDATIONS

There are many reasons to substantiate the safety of consumer products, including eye-area cosmetic products. Before discussing these reasons we should state that more people use more cosmetic products, including eye-area cosmetic products, more times every single day than any other type of consumer product. That includes prescription drug products, over-the-counter drug products, household products, personal hygiene, soap and detergent products, just to name a few of the more prominent consumer product categories. More people, more cosmetic products, more times every day! That's really quite an astounding statement. This, coupled with the fact that sight is one of our most precious human senses, would seem sufficient reason alone to substantiate the safety of eye-area cosmetic products.

Another reason for safety substantiating eye-area cosmetic products is to determine the toxicity and irritancy potential of such products for emergencies. This is important to the health professional in an emergency situation to prevent inappropriate treatment and to prescribe appropriate treatment. A cardinal rule of medicine is to treat only what is physically real or what has been competently diagnosed. Cosmetic products as a class are not hazardous products and, therefore, are not packaged with childproof closures as mandated by the Poison Prevention Packaging Act (1970). But this menas that children may well get these products into their eyes through accident and, for this reason many manufacturers voluntarily provide information on their products to national and regional poison control centers on both the ingestion toxicity of their products and the ocular toxicity of these products.

Finally, there is the question of the legal requirements for safety testing. It is the precise purpose of this section to examine in detail these legal requirements. This encompasses the federal law as expressed in the original Pure Food and Drug Act (1906), which was revised as the Food, Drug, and Cosmetic Act (1938) and its subsequent Color Additive Amendments (1960). Product liability is a further legal dimension which extends from the manufacturer of the product to both the person testing that product and the consumer using the product.

We will begin this discussion on the law by asking three pointed questions.

The first question is "Do the federal regulations specifically state that cosmetic products must be substantiated as to their safety?" The answer to this is unequivocally yes. Not only are cosmetic products to be adequately substantiated for safety prior to marketing, but each individual cosmetic ingredient must be substantiated as well. If a company does not perform this safety substantiation, the cosmetic product must prominently display the statement "Warning: The safety of this product has not been determined" [40].

The second important question about the legal requirements relating to testing cosmetic ingredients and products is "Do the federal regulations specifically state that animals must be used to substantiate the safety of cosmetic ingredients and products?" The answer to this question is also yes. The Food and Drug Administration has stated:

> . . .the use of animal tests is generally recognized by regulatory agencies as the principal basis for assessing potential risks from exposures to chemicals. . .this basis has been universally recognized and accepted by the courts [41].

This means simply that the ethical code of the science of toxicology as well as the medical profession prohibits the direct testing of materials on human beings without first testing these products in animals.

The third and final question is "Is the Draize eye irritancy test specifically required to assess the ocular irritancy potential of eye area cosmetic products?" There are many affirmative ways to answer this question. We shall discuss only two: the color additive regulations with their consequent petitions for listing of the colors tested, and the Interagency Regulatory Liaison Group recommendations. First, the color additive regulations.

The passage of the Food, and Drug, and Cosmetic Act in 1938 was brought about because of a need to regulate cosmetics and colors in cosmetics. Batch-by-batch certification for synthetic colors became mandatory in 1940 with the establishment of three lists of certifiable colors: FD&C colors for use in foods, drugs, and cosmetics, D&C colors for use in drugs and cosmetics, and External D&C colors for externally applied drugs and cosmetics.

The Color Additive Amendments of 1960 [42] were comprehensive in their scope and permitted the FDA to establish safety tolerances. In spite of 20 years of extensive testing and monumental industry expenditures of millions of dollars including long-term feeding studies in animals, this process is still not completed.

So-called coal-tar colors, which include *para*-phenylenediamine as well as synthetic organic colors, have not been allowed in eye-area products since 1938 but are now being tested by industry in the Draize test in rabbits and patch testing in humans to support the petitioning of the FDA for their use in eye-area cosmetics.

These regulations clearly demonstrate that animals are required for the safety substantiation of cosmetic colors and that the Draize eye irritancy test is required for the certification of eye-area cosmetic colors.

The Interagency Regulatory Liaison Group (IRLG) was formed September 26, 1977, to standardize as much as possible the testing required by various federal agencies charged with protecting the public and the environment from the adverse effects of certain chemical substances. The IRLG was composed of the Consumer Product Safety Commission (CPSC), the Environmental Protection Agency (EPA), the Food and Drug Administration (FDA) of the Department of Health and Human Services, the Occupational Safety and Health Administration (OSHA) of the Department of Labor, and the Food Quality and Health Service (FQHS) of the Department of Agriculture. Before its dissolution on September 21, 1981, the IRLG issued a set of guidelines [43] standardizing certain acute toxicity and irritancy test protocols, among which is the Draize eye irritancy test. The reasons for the choice of this particular animal test should be quite clear at this point. The Draize test originated in the Food and Drug Administration to assess the ocular irritancy potential of a hazardous eye-area cosmetic product, is now mandated for the ocular irritancy assessment of pesticides under the EPA's Federal Insecticide, Fungicide, and Rodenticide Act (1972), and is further mandated for the ocular safety assessment of household products by the Federal Hazardous Substances Act (1960).

When antagonists outside the scientific and consumer activist groups recently raised objections to the Draize eye irritancy test, the Food and Drug Administration responded that this test:

> . . .is the most reliable method to predict the harmfulness, or safety, of a substance that may enter the human eye. The [Draize] test is needed to assure that ingredients that may come into contact with the human eye will not be harmful. . . [44].

In concluding this section on the legal requirements for assessing the ocular irritancy potential of eye-area cosmetic products in animals through the Draize test, we must also mention the serious legal liabilities incumbent upon manufacturers of all consumer products, including cosmetic products. At the present time it is a fact, documented by innumerable court cases, that a manufacturer of a consumer product is considered legally responsible not only for producing safe products but also for any harm attendant upon their use, blatant misuse, or abuse. A manufacturer has only two defenses at this time against lawsuits involving the use, misuse, or abuse of his product: Warn of the specific harm incurred by the plaintiff and prominently display said warning on the product package or be able to produce scientific evidence as to the inherent safety of the product under normal conditions of use.

The first defense is impossible in most instances, since no manufacturer can specifically caution against every set of use, misuse, or abuse circumstances. Therefore, it should be clear that the data substantiating the safety of consumer products are also the data necessary to demonstrate the safety of these products in liability cases.

Next, we shall discuss the humane treatment of not only the laboratory animals used in the Draize eye irritancy test but the humane treatment of the human subjects who test such products after the irritancy potential of the product has been initially determined in animals.

IV. HUMANE TEST METHODS

The scientific bedrock of humane treatment for both experimental animals and human subjects took nearly 2,500 years to form before it was laid bare in the nineteenth century [45]. The historical event which marked the beginning of modern experimental research was the publication of Claude Bernard's *Introduction a l'Etude de la Medecine Experimental* in 1865. This work not only crystallized the technical sophistication that had been achieved in scientific experimentation by the nineteenth century, but announced the principle of all future experimental research as well: the assessment of a single variable against a background of relative constants.

Like any principle, it is paradoxical but fertile, It is paradoxical because it is simple, yet complex. Choosing the correct variable to yield a result, answer a question, or open the door for further experimentation is sometimes more intuition than a priori reasoning, for example.

The principle is fertile because it can become a beacon guiding us to a desired result which is often just one piece of a very large and complex puzzle. This beacon not only lights the way but may also uncover new roads leading to results which were unanticipated.

And like any true principle, the principle of altering a variable against a background of relative constants has a corollary: serendipity. The discovery of penicillin and numerous other scientific discoveries were all results of serendipitous data obtained from experiments designed to produce other results.

Finally, the principle of scientific experimentation is a touchstone. We can gain knowledge with it or risk living without this knowledge if the experiment is inadequately performed or not performed at all.

The application of this principle to experiments assessing the safety of cosmetic products requires maintaining the test subjects, whether animals or humans, in as natural a state as possible prior to, during, and after experimentation to ensure a background of relative constants against which a single variable can be accurately assessed. It is a scientific necessity, therefore, to maintain laboratory animals in peak physical health. As applied to human subjects in an experimental test with a cosmetic product, the final selection process must exclude those who at least in the area where the product is to be applied, are not whole and healthy.

Before exploring the humane treatment of experimental and human test subjects in detail, it would be well to define precisely what is

meant by these terms. The term "experimental animal" does not include
animals in the wild, livestock, endangered species, naturally diseased
animals, or domestic pets.

Laboratory animals are genetically identified animal species and
strains bred for the sole purpose of experimentation. Laboratory
animals used to test cosmetic products are not exposed to agents used
in biological warfare, riot control devices, disease agents, petrochemi-
cals, physiological stress, suspect food ingredients, or experimental
drugs [46,47]. They are exposed to prototype or final cosmetic formulations.

"Human test subjects" are not prisoners, retarded human beings,
or the sick. Human test subjects are volunteers who are screened
prior to being accepted for testing, have signed an informed consent
and release form, and are remunerated for their time and inconvenience
during the testing. These paid human volunteers assume a legitimate
risk by participating in the testing of these products. It is unethical
to ask these subjects to use a prototype or final cosmetic formulation
which has had no prior safety testing. Once the irritation potential
of a new product has been initially assessed in appropriate animal
models, paying people to test the product further then becomes a
legitimate and reasonable risk to ask them to assume.

The ethical aspects of using animals or people in experimentation
initially surfaced in the nineteenth century in England. These issues
are properly the province of ethicians and have been the subject of
several excellent articles [48—50] as well as a national commission
report [51]. It is paradoxical that the ethics of whether or not animals are appropriate experimental surrogates for people is often discussed, but little is said about practical applications of ethics in this
regard. For example, the humane treatment of laboratory animals as
well as human test subjects is ethics in action. Second, it is ethics
in action to require safety data on a test material in animals before
exposing humans to that same test material. Without such animal
testing, we revert to the unethical practice of direct human experimentation and abandon the ethical codes of both the scientific and
medical communities. Since no alternative test systems to the animal
model currently exist, we must protect human test subjects from undue harm by first screening test materials in animals.

We now turn to the legislative requirements for humane treatment
of laboratory animals. The national Animal Welfare Act became a
part of the U.S. Code in 1966 [52] and has subsequently been amended
in 1970 [53] and again in 1976 [54]. Many critics correctly point out
that rats and mice are exempted from these statutes. However, rabbits
are covered by these federal statutes, and the rabbit is the animal
used in the Draize eye irritancy test.

The federal regulations stipulate the conditions under which the
breeding, housing, buying, selling, and transportation of laboratory
animals used in scientific experimentation are to be carried out.

Licensing, registration, and inspection are the responsibility of the Animal and Plant Health Inspection Services under the division of Veterinary Services of the U.S. Department of Agriculture. Health certificates and records for the identification and disposition of each animal are encompassed by these inspection powers as well as the exact uses of the animals in scientific experimentation.

The standards set forth in these federal statutes encompass the humane handling and treatment of these laboratory animals, housing conditions, the amount of space required for individuals and groups, quality and quantity of feed and water, sanitation, and ventilation. These standards are to be employed not only by breeders, dealers, and transporters of laboratory animals, but also by the government, industry, academic, and contract laboratories where the experimentation is performed. Futher, an annual report is required by law from each of these laboratories, which report must list the number of animals used and a general description of the type of experiments performed as well as whether pain was involved in the experiments and whether or not anesthetics or analgesics were administered.

The humane treatment of laboratory animals guarantees that these animals are not treated like abandoned pets, certain livestock, diseased animals, endangered species, or animals left to their own devices in the wild. Indeed, these laboratory animals are cared for better than many domestic pets. Humane treatment of laboratory animals is, therefore, not only a matter of good business and competent science; humane treatment of these laboratory animals is also a matter of federal law. The National Animal Welfare Act is expressed in further detail in the so-called Good Laboratory Practice (GLP) Regulations [55]. The precise impact of these GLPs is that any animal study submitted to the Food and Drug Adminstration in support of the safety of a food, drug, or cosmetic product must now carry with it the express statement that the study was carried out in compliance with the GLPs.

Specific federal recommendations for the handling and care of laboratory animals is best exemplified by the *Guide for the Care and Use of Laboratory Animals* [56]. This booklet preceded the National Animal Welfare Act, is periodically reviewed and updated, and continues to be the most quoted document on the care and use of laboratory animals.

There are also recommendations for the handling and care of laboratory animals by other interested groups. As the Animal Welfare Institute states in its publication, *Comfortable Quarters for Laboratory Animals*:

> The aim is to provide an environment sufficiently natural that the animals can be maintained as normal individuals, upon whom reliable observations can be made [57].

It is frequently stated, no doubt, in the irrational heat of argument, that the rabbits used initially to assesss the irritancy potential
of cosmetic products suffer either at the instillation of the test product or after its instillation. Those who make such allegations appear
to have been misinformed on the performance of these tests on cosmetic
products. But this clarifies the very purpose of the initial assessment of the ocular irritation of the eye-area cosmetic products in animals. The purpose is not to inflict pain. Indeed, if this occurs the
manufacturer would be ill-advised to continue the development of the
test product without its reformulation. The purpose is to determine
if a product can be used safely and repeatedly by the consumer.

The above paragraph applies to most cosmetic products. However,
the ocular irritancy potential of cosmetic products such as shampoos
must also be assessed either because of their possible accidental introduction into the eye or their possible abuse or misuse by the consumer.
In these instances, dilution of the product before instillation of the
test quantity into the rabbit eye is useful to mimic actual conditions
of use and reduce the frequency of Type I errors, false positive reactions. The use of anesthetics on laboratory animals has not been
proven to yield identical scores as those achieved in experimental animals not given anesthetics. Further, anesthetics decrease or block
the blinking (wink) reflex, which is a sign of irritation. Without this
reaction, the eye could be severely damaged in the test. Validation
and corroboration must take place before the use of anesthetics becomes a feasible approach for the testing of shampoos, other detergent products, or neat chemicals. But the proponents of this approach
must also realize that these corroboration procedures will require
many additional test animals to demonstrate their validity. Even the
National Animal Welfare Act stipulates that the use of anesthetics,
analgesics, or even tranquilizers on laboratory animals in experimentation is not to be employed unless the use of these pain relievers
yields identical test results.

Next we shall explore the possibility of alternative test systems
to animals as the initial phase of the safety substantiation of eye-area
cosmetic products.

V. ACCEPTABLE AND HUMANE ALTERNATIVES

A. Proposed Alternatives

As the title of this section indicates, an alternative to a test system
already in use must, at minimum, be as acceptable and humane as the
test method it is meant to replace. At best, it should offer further
advantages over the current test system. The alternative test must
have the same or a similar purpose or objective, which, in the case
of the Draize eye irritancy test, is the determination of the degree
and type of ocular inflammation resulting from exposure to a given

test material. It should be noted here that animal welfare and animal rights proponents alike accept the same premise as the cosmetics industry, namely, that direct ingredient or product testing on human beings is both unacceptable and inhumane.

The phrase "alternative to animal testing" has been defined as the replacement of animal testing or the reduction in the numbers of animals used in testing [58]. This definition will be used in the ensuing discussion on alternatives. The focus of this discussion will be alternatives to the Draize eye irritancy test or the reduction in the number of Draize tests required to substantiate the safety of cosmetic ingredients and products.

The alternatives proposed to the Draize test have been drawn from other areas of scientific research and medical education. They fall into three main categories: structure-function relationships, models, and *in vitro* techniques.

The proposals centering on structure-function relationships rely heavily on the assumption that molecular relationships based on chemical structure are related to biological activity. The initial information on which such a premise is based is founded on experimental fact or the results of experimentation, however. Given that, all efforts to date to correlate chemical structure with biological effects have failed, whether these biological effects were irritation, allergenicity, or photosensitization. The coumarin class of chemical compounds is a good example of the lack of correlation between chemical structure and biological effect. Coumarin is extracted from the Tonka bean. Some 250,000 pounds of coumarin is manufactured and processed per year for use by the fragrance and pharmaceutical industries [59]. When 6-methylcoumarin, a commonly used fragrance ingredient, was shown to be photoallergenic, many of the more commonly used coumarin derivatives were immediately tested. The results of opening yet another Pandora's box was that coumarin and the internally ingested anticoagulant Dicumarol have not been shown to be irritants, allergens, or photosensitizers: 6-methylcoumarin and 7-methylcoumarin have been demonstrated to be photoallergens; and 7-methoxycoumarin has been shown to be both a photoirritant and a photoallergen. Most 6-methylcoumarin analogs tested to date have been negative for these effects. These inconsistent structure-effect results are further complicated by the fact that practical threshold levels exist for allergens in formulations and quenching of known allergens has been demonstrated for several chemicals [60]. The criticism might be made that the above example wasn't drawn from the group of known chemical irritants. However, acids such as hydrochloric acid, bases such as sodium hydroxide, and alcohols such as ethyl alcohol can all produce dermal and ocular irritation in appropriate concentrations, to mention just a few of the chemical structure-biological effect relationships that do not correlate among the known irritants.

The term "models" can mean mechanical and mathematical models or computer simulations. Once again, these approaches rely exclusively on data from previous experimentation. While computer simulations may be directionally helpful for neat materials such as cosmetic ingredients, they are of little value when the test material is a finished cosmetic product which will be applied to the skin or around the eyes of a living biological test system. They are helpful as instructional aids in certain situations such as the pharmacokinetic or pharmacodynamic effects of giving additional medication to a patient in a defined disease or injury state who is already on a set of prescribed medications. As we said in Sec. I, mixing chemicals together, that is, formulating a cosmetic product, may produce one or a combination of chemical phenomena such as synthesis, augmentation, or quenching.

In vitro alternative test systems are tests performed in test tubes as opposed to *in vivo* or whole-animal test systems. This area of research has produced the most discussion on alternatives to the Draize test. *In vitro* test systems include organ culture, protozoan and bacterial cultures, animal and human cell or tissue cultures. Organ culture employs either whole or partial organs, but at its present stage of development is at best complementary to animal or human tissue culture systems. Outside the diagnostic uses of protozoan and bacterial cultures, the most widely accepted use of these alternatives are the bacterial *in vitro* mutagenicity techniques. But, as their name implies, their end point is an assessment of the mutagenicity of a neat chemical compound, not the assessment of the dermal or ocular irritation potential of a finished cosmetic product. We shall, therefore, concentrate on animal or human tissue culture systems in the following discussion of acceptable and humane alternatives, since these test systems are the most viable in terms of true alternatives to the Draize test.

Before describing tissue culture and its applicability as an alternative to the Draize eye irritancy test, we must first define the criteria for accepting any test as an alternative. Among these criteria are the demonstration of the need or advantages for such an alternative, validation of the test system, and generation of an adequate data base to allow switching over to the new test system [61]. First, the demonstration of the need or advantages for an alternative. The alternative test system must be a significant improvement in the quality and quantity of data generated from the currently used test system or offer a unique advantage in terms of cost, time, or labor savings. Improving the quality or quantity of the data resulting from the alternative test needs no explanation. Savings in cost, time, or labor is exemplified by a reduction in the total cost of using the alternative test, the time spent conducting the test and analyzing the data, or a reduction in the number and type of personnel used to run the test. Validating the test system requires parallel testing of the alternative

and current test systems. A detailed comparison of the results from
both test system reveals the sensitivity of the alternative method and
allows the determination of the limits of false positive or negative results. Generating an adequate data base may be either short-term
or long-term. Short-term permits the use of the alternative in specific
instances, while long-term permits cross-referencing results from
different investigators from different laboratories for a myriad of test
substances.

B. Applicability and Examples of *in vitro* Methodologies — The *Limulus* Amebocyte Lysate (LAL) Assay

Tissue culture is the science, some would say art, of extracting and
maintaining animal or human cells in an artificial environment such as
glass or plastic containers with synthetic growth media. The principal
applications of tissue culture at the present time are the elucidation
of cellular structure and function, vaccine production, mutagenicity
assays, and selected tests in immunology. While the uninitiated may
think this a small list of accomplishments, the initiated are sufficiently
impressed with the accomplishments and applications of this very demanding sciences.

There are certain requirements to be met if a tissue culture technique is to be useful as a test method. Sterile technique is the *sine
qua non* of any reliable tissue culture work. It is not merely enough
to keep bacterial or fungal contaminants out of one's cultures. Nor
is it enough to ward off intracellular viral and mycoplasmal invasion.
One must also assure that other types of cells, such as the prolific
HeLa cells, do not infiltrate and take over the culture. Ideally, cultures must be maintained in defined, serum-free synthetic growth medium, not an insignificant accomplishment in itself. The cells must
further be able to reproduce and be serially passed on to new culture
flasks, while maintaining their inherited range of biochemical responses
and karyotype. The result is that there are really very few primary
cell lines at the present time. To date, there are only fibroblast, kera-
tinocyte, and chondrocyte lines that can be used as true *in vitro* models of their *in vivo* parentage [62].

Tissue culture offers decided advantages. Smaller test quantities
are required, which can be a significant advantage when the test compound is difficult to synthesize or is radiolabeled. The mechanism of
action for certain test substances has been explored and often defined
by the use of tissue culture techniques [61]. Indeed, this may be the
chief use of such techniques at this point in time — the elucidation of
the cellular and molecular events of inflammation.

Like any test system, tissue culture has its disadvantages as well.
Chief among these is the *in vitro–in vivo* gap [61], meaning that *in
vitro* results must be extrapolated beyond the histological and organ

levels of organization to the whole-animal level and then to humans. The target cell may be another disadvantage. What relationship is there between effects of topically applied soaps and detergents on yeast cells, erythrocytes, or buccal mucosa cells [63]? Certain cell lines are resistant to known in vivo effects [64]. Certain other known in vivo effects cannot be reproduced in vitro [65]. Still other cultures reproduce only selected in vivo effects [66]. And the tissue culture system does not exist that can assess responses due to age, sex, or previous exposures to the test material [67]. Tissue culture techniques are generally more sensitive to test substances [68], which could mean a high incidence of false positives. The use of activated and nonactivated test substances has been used to mimic metabolic manipulation of the test material by this organism. Finally, since organ systems such as the eye and skin are complex groups of different tissues and cells, a range or battery of cell cultures may be the appropriate approach [69]. Even if a system or group of tissue culture systems already existed, their most immediate use at the present time would be either as a screening test or battery, or as ancillary to results from whole-animal test systems by providing corroborative and supportive evidence. Indeed, the in vitro mutagenicity test systems currently in vogue are used in these two ways. They are screening techniques for test compounds occuring in series, and when the mutagenicity data is positive it is still only supportive of in vivo carcinogenicity bioassays and epidemiological evidence.

The Draize eye irritancy test is an inexpensive, quick, efficient, in vivo test system capable of determining low-level irritants on up to primary irritants. It is further capable of generating information on the restorative and reparative processes of healing and repair. An alternative in vitro test to replace the Draize test does not exist at the present time. Further, from the foregoing discussion, it should now be clear that developing a single in vitro test to replace the Draize test is unlikely; rather a battery of such tests would be the more realistic target. This battery would probably not be less expensive, quicker, or more efficient than the Draize test. It appears unlikely that the battery would be capable of grading the potency of irritants, and it would not yield information on the restorative processes of repair and healing. Such in vitro tests will not eliminate the need for animals, but they will mean a different way of using animals [68]. They may eventually reduce the numbers of animals used in safety substantiation after an initial rise in the total number of animals used in testing due to validation experiments. But what may come from a commitment to this line of test method development may be a greater understanding of the cellular and molecular events of inflammation which is so desperately needed at this time. With knowledge developed from the true strength of tissue culture science, more precise in vitro test methodologies might then be developed.

Two examples currently exist which may serve to illustrate these points. Various shampoo formulations have been tested using L929 murine fibroblast cells and cultures [70]. Cytolysis was the end point of these experiments, which were able to distinguish a severely irritating shampoo from its less irritating analogs. However, this system was unable to distinguish between the less irritating formulations themselves. Is cytolysis a true molecular event in the process of inflammation? We do not know at the present time. Could this be the first step in developing a screening technique? Yes. And in this respect, it should be explored for cosmetic products such as shampoos. But this type of screening technique is of little value for cosmetic products such as eye shadows and mascaras, which vary from zero to a mild irritancy potential even in their prototype formulation stages.

The second example is an analogy drawn from the field of medical device testing. Over the decade of the 1970s an *in vitro* bacterial endotoxin test was developed by Levin and Bang [71] at the Johns Hopkins University which has now been approved by the Food and Drug Administration as a replacement [72] for the standard U.S. Pharamcopoeia and National Formulary Rabbit Pyrogen Test [73]. The Limulus Amebocyte Lysate (LAL) assay, as it is called, uses blood extracted by syringe from live horseshoe crabs (*Limulus*), which is then processed, freeze-dried, and stored for later use. This processed blood is able to detect bacterial endotoxins from medical device extracts, thus determining the pyrogenicity of the medical devise test lot. Although this test has now been validated, it must still be run in parallel with the rabbit pyrogen assay by the manufacturer, and only if the results from both tests are equivalent can the *in vitro* test then replace the *in vivo* test for the continuing manufacture of that medical device. Such a test is almost ideal because it reduces the number of animals used, the *Limulus* are returned to the sea after puncture and removal of a small quantity of blood, and are often reharvested and used again. The *in vitro* test is inexpensive and quicker than the *in vivo* test but cannot be used until parallel testing shows equivalency of results.

C. New Horizons

The cosmetics industry has been the subject of criticism and increased pressure from animal welfare and animal rights groups during the 1970s for its use of the Draize eye irritancy test in the safety substantiation of its ingredients and products. This industry, however, has been uniquely responsive to these groups and their concerns. In 1976, the Cosmetic Ingredient Review was independently established with cosmetic industry funds to receive and review published and unpublished data on cosmetic ingredients and rule on their safety. This program is already reducing the number of animals used to substantiate the safety of cosmetic ingredients. Individual cosmetic companies are devoting their own time and resources to exploring the modification of the Draize test through the use of anesthetics and smaller test volumes, thus addressing the humane aspects of the current test [74]. An industry-sponsored scientific symposium entitled "Workshop on Ocular Safety Testing: *In vivo* and *in vitro* Approaches" was held in Washington, D.C., October 6—7, 1980, where scientists gathered to share information on current and future test methodologies. The industry position on ocular safety testing was clearly stated at this conference:

> The cosmetic, toiletry and fragrance industry's position on ocular safety testing of products is clearly exemplified by this workshop, i.e., we are ready, willing and able to consider modification of *in vivo* procedures and to address alternative *in vitro* procedures [75].

This commitment to a review of current *in vivo* test systems and a realistic assessment of potential future *in vitro* test systems was further expressed in the News Release issued at the annual meeting of The Cosmetic, Toiletry and Fragrance Association:

> The Board of Directors of The Cosmetic, Toiletry and Fragrance Association (CTFA) announced today it will establish a fund to help support creation of a national center for developing alternatives to animal testing [76].

After a review of the Requests for Proposals (RFPs) to establish such a national center, site visits were conducted before choosing The Johns Hopkins University School of Hygiene and Public Health in Baltimore, Maryland, as the recipient of a $1 million, three-year industry grant to establish a Center for Alternatives to Animal Testing on September 22, 1981 [77]. This foundation grant was quickly supplemented by a $225,000 company grant by Bristol-Myers. The first intramural research grants were let in March 1982, and a call for extramural grant applications was made in June 1982 [78] after the Symposium held its First Annual Symposium on Alternatives in May 1982 [18]. The cosmetics industry supplemented the initial $1 million grant with an additional $700,000 grant in February, 1984.

Several other independent grants to develop animal test alternatives have also occurred. Revlon announced a $750,000 three-year grant to Rockefeller University [79], the New England Antivivisection Society Awarded $100,000 to Tufts University [80], the fund for the Replacement of Animals in Medical Experiments (FRAME) donated $176,000 to the Medical College of Pennsylvania, and The Millennium Guild announced the establishment of $500,000 in awards, $250,000 to be held in escrow for the scientist(s) who first devise an acceptable scientific and regulatory nonanimal alternative to the Draize eye irritancy test or the LD_{50} test [81]. Revlon contributed an additional $250,000 to Rockefeller University in 1984.

VI. CONCLUSION

The effort to establish test method alternatives to the use of animal test methods will take time, perhaps an estimated 10 years. It will further take money, which the cosmetic industry has pledged. It will also take the effort, time, and money of all industries such as the pharmaceutical, detergent, and chemical industries, as well as the animal welfare and animal rights groups. The ideal of an *in vitro* replacement test or battery of tests may not actually be achieved in the end, but if we further reduce the numbers of animals currently used in the safety substantiation of consumer products such as cosmetics and elucidate further the cellular and molecular events of the process of inflammation, the effort will certainly have been worthwhile.

ACKNOWLEDGMENTS

The author is endebted to Dr. Geoffrey Woodard, coauthor of the original Draize paper [16], for his review and technical assistance in preparing this manuscript.

REFERENCES

1. The cosmetic product types used in this chapter will be the cosmetic product categories designated in 21 Code of Federal Regulations (hereafter C.F.R.) 720.4.
2. 21 C.F.R. 70 and 73.
3. Cosmetic Ingredient Review, 1110 Vermont Avenue, N.W., Washington D.C. 20005.
4. Research Institute for Fragrance Materials, 375 Sylvan Avenue, Englewood Cliffs, N.J. 07632.
5. J. H. Kay and J. C. Calandra, Interpretation of Eye Irritancy Tests, *J. Soc. Cosmet. Chem.* 13:281 (1962).
6. T. O. McDonald and J. A. Shadduck, Eye Irritation, in *Dermatotoxicology and Pharmacology* (F. N. Marzulli and H. I. Maibach, eds.), Halstead Press, Washington, D.C., 1977.
7. W. M. Grant, *Toxicology of the Eye*, 2nd ed., Charles C. Thomas Publisher, Springfield, Ill., two vol. 1974.
8. A. M. Potts and L. M. Gonasun, Toxic Responses of the Eye, in *Toxicology: The Basic Science of Poisons* (J. Doull et al, eds.), 1980.
9. A. S. Weltman, S. B. Sparber, and T. Jastshuk, Comparative Evaluation and the Influence of Various Factors in Eye Irritation Scores, *Toxicol. Appl. Pharmacol.* 7:308 (1965).
10. C. S. Weil and R. A. Scala, Study of Intra- and Interlaboratory Variability in the Results of Rabbit Eye and Skin Irritation Tests, *Toxicol. Appl. Pharmacol.* 19:276 (1971).

11. A. M. McCally, A. G. Farmer, and E. C. Loomis, Corneal Ul-
    ceration Following Use of Lash-Lure, *J. Am. Med. Assoc.*
    101:1560 (1933).
12. G. Woodard, personal communication, August 19, 1980.
13. M. E. Rudolph and R. B. Welch, *The Wilmer Ophthalmological
    Institute 1925—1975,* The Williams & Wilkins Company, Baltimore,
    Md., 1976.
14. J. S. Friedenwald, W. F. Hughes, and H. Herrmann, Acid-Base
    Tolerance of the Cornea, *Arch. Ophthalmol.* *31*:279 (1944).
15. J. S. Friedenwald, W. F. Hughes, and H. Herrmann, Acid
    Burns of the Eye, *Arch. Ophthalmol.* *35*:98 (1946).
16. J. H. Draize, G. Woodard, and H. O. Calvery, Methods for the
    Study of Irritation and Toxicity of Substances Applied Topi-
    cally to the Skin and Mucous Membranes, *J. Pharmacol. Exp.
    Therap.* *81*:377 (1944).
17. J. H. Draize and E. A. Kelley, Toxicity to Eye Mucosa of
    Certain Cosmetic Preparations Containing Surface-Active Agents,
    in *Proceedings of the Scientific Session of the Toilet Goods
    Association* (now called the Cosmetic, Toiletry and Fragrance
    Association), Washington, D.C. *17*:1 (1952).
18. E. M. Jackson, Industrial Safety Testing Practices, Chapter 5 in
    Alternative Methods in Toxicology, Vol. 1: Product Safety Evalua-
    tion (A. M. Goldberg, Editor) Mary Ann Liebert Inc., Publishers,
    New York, 1983.
19. K. J. Falahee, C. S. Rose, S. S. Olin, and H. E. Seifreid,
    *Eye Irritation Testing: An Assessment of Methods and Guide-
    lines for Testing Materials for Eye Irritancy,* U.S. Environ-
    mental Protection Agency, Office of Pesticides and Toxic Sub-
    stances, Washington, D.C., EPA.560/11.82.001, October 1981.
20. E. M. Jackson, Cutaneous and Ocular Toxicology, Presented
    at A Review Program for Toxicologists (M. A. Gallo, modera-
    tor), Rutgers University Medical School, Piscataway, N.J.,
    July 8, 1980.
21. W. M. Grant, Testing Methods and Species Specificity, in
    Toxicology of the Eye, 2nd ed., Charles C. Thomas Publisher,
    Springfield, Ill., 1974, Vol. II,. 1120.
22. J. H. Draize, Dermal Toxicity, in *Appraisal of the Safety of
    Foods, Drugs and Cosmetics,* Association of Food and Drug
    Officials of the United States, Austin, Texas, 1959.
23. B. Idson, Topical Toxicity and Testing, *J. Pharmaceut. Sci.*
    57(1):1 (1968).
24. A. A. Fisher, The Role of Patch Testing in Allergic Contact
    Dermatitis, in *Contact Dermatitis,* 2nd ed., Lea & Febiger,
    Philadelphia, 1974.
25. S. Fregert, *Manual of Contact Dermatitis,* Munskgaard Pub-
    lishers, Copenhagen, Denmark, 1974.

26. S. Fregert and H. J. Bandmann, *Patch Testing*, Springer-Verlag, New York, 1975.

27. L. Schwartz and S. M. Peck, The Patch Test in Contact Dermatitis, *Public Health Report* 59:2 (1944).

28. M. J. Brunner and A. Smiljanic, Procedure for the Evaluation of the Skin Sensitizing Power of New Materials, *Arch. Dermatol.* 66:703 (1952).

29. E. F. Traub, T. W. Tusing, and H. J. Spoor, Evaluating Dermal Sensitivity, *Arch. Dermatol.* 69:399 (1954).

30. H. A. Shelanski and M. V. Shelanski, A New Technique of Human Patch Tests, in *Proceedings of the Science Section of the Toilet Goods Association* (now called The Cosmetic, Toiletry and Fragrance Association), Washington, D.C., 19:46 (1953).

31. J. H. Draize, Dermal Toxicity, *Food, Drug and Cosmetics Law J.* 10:722 (1955).

32. B. M. Lanman, W. B. Elvers, and C. S. Howard, The Role of Human Patch Testing in the Product Development Program, in *Proceedings of the Joint Conference on Cosmetic Sciences*, The Toilet Goods Association (now called The Cosmetic, Toiletry and Fragrance Association), Washington, D.C., 1968.

33. A. M. Kligman, The Identification of Contact Allergens by Human Assay III. The Maximization Test: A Procedure for Screening and Rating Contact Sensitizers, *J. Invest. Dermatol.* 47:303 (1966).

34. C. G. Roeleveld and W. G. van Ketel, Allergic Reaction to Hair Dye Elicited by an Ointment Containing DMSO, *Contact Dermatitis* 1:332 (1975).

35. W. G. van Ketel, Patch Testing with Eye Cosmetics, *Contact Dermatitis* 5:402 (1979).

36. M. M. Reiger and G. W. Batista, Some Experiences in the Safety Testing of Cosmetics, *J. Soc. Cosmet. Chem.* 15:161 (1964).

37. A. M. Schmidt, Hypoallergenic Cosmetics, *Fed. Reg.* 39(38): 7288 (1974).

38. E. M. Jackson and N. F. Robillard, The Controlled Use Test in a Cosmetic Product Safety Substantiation Program, *J. Toxicol.—Cutaneous and Ocular Toxicology*, 1(2):109 (1982).

39. Guidelines for Controlled Use Studies, in *Safety Testing Guidelines*, The Cosmetic, Toiletry and Fragrance Association, Washington, D.C., 1981.

40. 21 C.F.R. 740.10.

41. *Fed. Reg.* 44:17085 (March 20, 1979).

42. Color Additives Amendments, Public Law 86-618, 1960.

43. Draft IRLG Guidelines for Selected Acute Toxicity Tests, Interagency Regulatory Liaison Group, Washington, D.C., August 1979.

44. The Draize Test, FDA Talk Paper T 80-30, May 30, 1980.
45. R. D. French, Animal Experimentation I. Historical Aspects,
     in *Encylopedia of Bioethics*, Vol. I, Free Press-Macmillan Pub-
     lishing Company, New York, New York, 1978.
46. D. Nevin, Scientist Helps Stir New Movement for "Animal
     Rights," *Smithsonian* *11*(1):49 (1980).
47. C. Stevens, Humane Considerations for Animal Models, in *Ani-
     mal Models of Thrombosis and Hemorrhagic Diseases*, The Na-
     tional Institutes of Health, Bethesda, Md., 1976.
48. W. Lane-Petter, The Ethics of Animal Experimentation, *J.
     Med. Ethics* *2*:118 (1976).
49. F. R. Heeger, Norms and Good Reasons, Presented to the
     Faculty for Veterinary Medicine in Utrecht, 1979, under the
     title Veterinary Medicine and Society, *Tijidschar Diergeneeskd.*
     105:4 (1980).
50. R. Branson, The Ethics of Dental Research: An Overview of
     Basic Principles, *J. Dental Res.* *59*(C):1214 (1980).
51. *The Beaumont Report: Ethical Principles and Guidelines for
     the Protection of Human Subjects in Research*, The National
     Commission for the Protection of Human Subjects of Biomedical
     and Behavioral Research, DHHS Publication No. (OS) 78—0012,
     1978.
52. Laboratory Animal Welfare Act, Public Law 89-544, August 24,
     1966.
53. Animal Welfare Act, Public Law 91-579, December 24, 1970.
54. Animal Welfare Act Amendments, Public Law 94-279, April 22,
     1976.
55. Nonclinical Laboratory Studies: Good Laboratory Practice Regu-
     lations, *Fed. Reg.* *43*:59986 (December 22, 1978), effective
     June 20, 1979.
56. *Guide for the Care and Use of Laboratory Animals*, Committee
     of the Institute of Laboratory Animal Resources, U.S. Depart-
     ment of Health, Education and Welfare, Public Health Service,
     National Institute of Health, DHEW Publication No. (NIH) 74-
     23, 1972.
57. *Comfortable Quarters for Laboratory Animals*, rev. ed., Animal
     Welfare Institute, Washington, D.C. 1979.
58. A. N. Rowan, Introduction, in *The Use of Alternatives in Drug
     Research* (A. N. Rowan and C. J. Stratmann, eds.), Macmillan,
     London, 1980.
59. D. L. J. Opdyke, Coumarin, *Food Cosmet. Toxicol.* *12*:385 (1974)
60. M. A. Cooke, Sensitizers—Problems in Detection, Usage and
     Labeling, *Cosmet. Toiletr.* *92*:37 (1977).
61. J. A. Bradlaw, Interface Between in Vivo and in Vitro Studies,
     presented at the Workshop on Ocular Safety Testing: *In Vivo
     and in Vitro* Approaches, sponsored by The Cosmetic, Toiletry

and Fragrance Association, Washington, D.C., October 6—7, 1980.

62. R. Ham, Recent Advances in the Growth of Normal Cells for Culture, presented at the Workshop on Ocular Safety Testing: *In Vivo* and *in Vitro* Approaches, sponsored by The Cosmetic, Toiletry and Fragrance Association, Washington, D.C., October 6—7, 1980.

63. F. R. Bettley, The Toxicity of Soaps and Detergents, *Br. J. Dermatol.* *80*:635 (1968).

64. A. J. Paine, M. J. Ord, G. E. Neal, and D. N. Skilleter, Cell Models for the Study of Toxic Mechanisms, in *The Use of Alternatives in Drug Research* (A. N. Rowan and C. J. Stratmann, eds.), Macmillan, London, 1980.

65. D. Gospodavowicz, A. L. Mescher, K. D. Brown, and C. R. Birdwell, The Role of the Fibroblast Growth Factor and Epidermal Growth Factor in the Proliferative Corneal and Lens Epithelium, *Exp. Eye Res.* *25*:631 (1977).

66. H. Wilmsman, The Inhibition of Saccharase Activity by Anionics—A Screening Test for Physiological Compatibility, *Am. Perfumer Cosmet.* *78*:21 (1963).

67. R. M. Nardone, Tissue Culture Systems in Toxicity Testing, in *The Use of Alternatives in Drug Research* (A. N. Rowan and C. J. Stratman, eds.), Macmillan, London, 1980.

68. D. H. Smyth, Tissue Culture as an Alternative, *J. Roy. Soc. Med.* *73*:299 (1980).

69. K. R. Rees, Cells in Culture and Toxicity Testing, *J. Roy. Soc. Med.* *73*:261 (1980).

70. P. J. Simons, An Alternative to the Draize Test, in *The Use of Alternatives in Drug Research* (A. N. Rowan and C. J. Stratman, eds.), Macmillan, London, 1980.

71. J. Levin and F. B. Bang, Clottable Protein in *Limulus:* Its Localization and Kenetics of Its Coagulation by Endotoxin, *Thromb. Diath. Haemorrh.* *19*:186 (1968).

72. Bacterial Endotoxin Test, (85) in *The United States Pharmacopoeia, The National Formulary*, 20th rev. United States Pharmacopoeia Convention, Rockville, Md., 1980.

73. Pyrogen Test, (151) in *The United States Pharmacopoeia, The National Formulary*, 20th rev. United States Pharmacopoeia Convention, Rockville, Md., 1980.

74. The Cosmetic, Toiletry and Fragrance Association Pharmacology/Toxicology Committee's Modified Ocular Safety Testing Task Force.

75. J. M. McNerney, Industry Position Concerning Ocular Safety Testing, presented at the Workshop on Ocular Safety Testing: *In Vivo* and *in Vitro* Approaches, sponsored by The Cosmetic, Toiletry and Fragrance Assoication, Washington, D.C., October 6—7, 1980.

76. News Release, The Cosmetic, Toiletry and Fragrance Association Annual Meeting, Boca Raton, Fla., March 2, 1981.

77. M. Knudson, Cosmetic Firms Hope to End Animal Testing, *The Baltimore Sun*, September 22, 1982.

78. Extramural Funding, *Science* *216*:1350 (1982).

79. Revlon Funds Animal Test Research, *Science* *211*:260 (1981).

80. P. Gunby, Animal Rights Group Awards Research Grant, *J. Am. Med. Assoc.* *18*(245):1803 (1981).

81. Scientific News: $500,000 Award Offered for Efforts to Reduce and Eliminate Animal Testing, The Cosmetic, Toiletry and Fragrance Association *Executive Newsletter* 82-8, April 23,1982.

29

WHEN AN FDA INSPECTOR CALLS

STEPHEN H. McNAMARA

Hyman, Phelps & McNamara, P.C., Washington, D.C.

I. INTRODUCTION

This chapter discusses the Food and Drug Administration (FDA) inspection. The chapter explains the extent of the FDA's authority to
conduct inspections of cosmetic and cosmetic drug manufacturers, and
it also provides suggestions for manufacturers concerning how to cope
most effectively with the FDA inspection.

FDA inspections have serious regulatory purpose. The inspector
comes, usually, to determine if the inspected company is complying
with the requirements of the Federal Food, Drug, and Cosmetic Act
(FDC Act) and FDA regulations. The inspector is *not* a "friend" who
comes to "help." A company must regard the inspector as a police
officer gathering evidence that might be used against the company.
Almost every FDA-initiated recall, civil seizure action, injunction action, and criminal prosecution has as its basis data acquired by an
FDA inspector during an inspection. Accordingly, it is important that
company personnel have a good understanding of the FDA's rights
and the company's rights during an inspection, and that company personnel act to protect their company.

Every FDA-regulated company should have a standard operating
procedure, a written plan, for coping with an FDA inspection. The
plan should explain for affected personnel (1) the FDA's rights, (2)
the company's rights, and (3) company policies and practices to be
followed during the inspection. This chapter is intended to help companies develop such a plan, or to review and refine existing inspection procedures.

A useful way to approach this subject is to discuss a hypothetical
FDA inspection, from start to finish.

II. RECEIVING THE INSPECTOR

Before beginning an inspection, the FDA inspector is required by the
FDC Act to present credentials identifying himself or herself, and a
written notice of inspection (Form FD-482) to the owner, operator,
or agent in charge of the establishment to be inspected[1].

A company's inspection plan should designate the person to receive and accompany the inspector. "Backup" personnel should also
be identified. These persons should be trained so that they understand thoroughly the extent of the FDA's rights, the company's rights,
and the company's policies with respect to the various matters likely
to arise during an inspection.

Upon receiving the inspector, the company representative ("you"; informal address is used hereafter for convenience) should begin immediately to compile a comprehensive record of the inspection. This record should open with the notice of inspection provided by the inspector. Examine the inspector's credentials, to be certain they conform to the signature on the notice of inspection. Record the full name of each inspector. If the FDA should later institute an enforcement action based on the inspection, you would want to know the identity of each FDA inspector, for depositions or other pretrial discovery.

III. WHAT ABOUT INSISTING ON A WARRANT?

The FDC Act provides that FDA inspectors are authorized ". . .to enter, at reasonable times, any factory, warehouse, or establishment in which food, drugs, devices, or cosmetics are manufactured, processed, packed, or held, for introduction into interstate commerce or after such introduction, or to enter any vehicle being used to transport or hold such food, drugs, devices, or cosmetics in interstate commerce; and . . . to inspect. . ."[2]

The FDC Act makes no mention of requiring a warrant from a U.S. district judge or magistrate to authorize the inspection. Furthermore, the FDC Act provides that "refusal to permit entry or inspection" is a criminal offense.[3] Accordingly, most companies permit an FDA inspection without attempting to insist on the presentation of a warrant.

However, in 1978 the U.S. Supreme Court, in *Marshall v. Barlow's, Inc.*, ruled that it is unconstitutional for inspectors of the Occupational Safety and Health Administration (OSHA) to conduct an inspection without a warrant unless the inspected company consents to the inspection.[4] In this decision, the Supreme Court stated that "warrantless searches are generally unreasonable" and that "this rule applies to commercial premises as well as homes."

Nevertheless, the Supreme Court stated that warrantless inspections *are* permitted in the exceptional circumstance of certain "pervasively" or "closely" regulated industries "long subject to close supervision and inspection." The Court identified the "liquor" and "firearms" industries as examples of the exceptional industries in which warrantless inspections are permitted without the consent of an inspected firm.

A decision by a U.S. court of appeals interpreting the FDS's inspection authority in light of the Supreme Court decision in *Marshall v. Barlow's, Inc.*, has concluded that the FDC Act does not authorize the FDA to conduct warrantless inspections without the consent of an inspected firm *but* that a company that refuses to allow the FDA to conduct a warrantless inspection is properly subject to criminal prosecution. (That is, if you refuse to permit an inspection when

the FDA inspector appears without a warrant, the inspector would
be required to obtain a warrant before he or she could proceed to enter and inspect your establishment without your consent, but the FDA
could also recommend that you be prosecuted for failure to permit the
warrantless inspection.)[5]

The FDA does *not* routinely obtain a warrant before attempting
to conduct an inspection. If an FDA inspector should arrive at your
company armed with a warrant, this would be a most unusual and suspicious circumstance, requiring prompt and careful attention. If the
warrant should provide for photographs, for access to manufacturing
records, or for other FDA activity you otherwise would refuse to permit, it is especially important to react immediately; you may find it
necessary to comply with the warrant until you can reach the official
who issued the document.

IV. BEFORE THE INSPECTION BEGINS

Before the inspector begins the inspection, ask "him"/"her" (hereafter
for convenience) why he is there and attempt to determine what he intends to review. It sometimes happens, for example, that the inspector is interested only in a particular subject, and that you can provide
the desired information without "opening the door" for him to wander
generally through your establishment. In such a case, you may want
to provide the information and let the inspector depart as quickly as
possible.

Also, before allowing an inspection to commence, you should tell
the inspector of your company's policies that will control the inspection. For example, you may want to tell the inspector that company
policy prohibits taking cameras into the plant and that he must leave
his camera in his car or in your office, that any questions or requests
for information are to be directed only to you and not to other company employees, etc, (Secs. VIII-XI review several such policies that
you may want to consider for your company.)

V. CONDUCT OF THE INSPECTION—THE FDA'S LIMITED RIGHTS

Suppose the inspector states that his purpose is to conduct a routine
surveillance inspection of your establishment: What is the extent of
his inspection authority?

The FDC Act provides the FDA authority "to inspect, at *reasonable times* and within *reasonable limits* and in a *reasonable manner*,
such factory, warehouse, establishment, or vehicle and all pertinent
equipment, finished and unfinished materials, containers, and *labeling*
therein" [emphasis added].[6]

Note particularly what the Act does *not* state. It does *not* mention,
for example, FDA access to master formula records, batch production

records, results of analyses, or complaint files. You generally are
not required to show such records to the FDA inspector. (Caveat:
The Act *does* authorize inspection of such records in the case of prescription drugs or restricted devices, but *not* in the case of cosmetics
or nonprescription drugs.)[7]

Technically, the law does not require even that you talk to the
inspector. However, when an inspector asks reasonable questions
about the type of products you manufacture, your manufacturing procedures, etc., you probably will want to respond. After all, you may
save yourself a lot of time. It is time out of your productive day during which you accompany an inspector. If he has to stand in your
plant for two weeks to determine the kinds of products you manufacture, he may decide to do just that, when you could avoid such an
extended FDA presence simply by answering reasonable questions.

VI. TAKING OF SAMPLES

The FDC Act provides that the inspector is authorized to collect samples.[8] During an inspection FDA inspectors routinely take samples
of finished and unfinished materials and of labeling, and companies
generally permit the taking of *reasonable* samples of this type. The
courts have recognized that this is an appropriate inspection function.[9] You may insist that the inspector pay for the fair value of samples taken, but many companies do not bother to do so.

VII. "HOLDING" A SUSPECT PRODUCT

While he may take samples of materials in your establishment, the FDA
inspector does *not* have the authority to detain or embargo materials
he believes to be in violation of the FDC Act. The inspector may request that you voluntarily hold a cosmetic or cosmetic drug that he
believes to be adulterated or misbranded, but he cannot require that
you do so.

"Seizure" of an article in your establishment pursuant to the FDC
Act requires the institution of a civil proceeding in a U.S. district
court. In general, before an article can be seized under the FDC
Act, the following chain of events must occur: The FDA district office
recommends to FDA headquarters that a civil seizure be instituted,
and if FDA headquarters agrees, the FDA chief counsel writes to the
local U.S. attorney, *requesting* the initiation of a civil seizure action.
Assuming that the U.S. attorney agrees (which is usual), he or she
files a complaint for forfeiture in the local U.S. district court, and
then a U.S. marshall serves upon the article a "warrant for arrest."
Service of this warrant upon the article accomplishes seizure. There-
after, there is a hearing before the court to determine whether the article is adulterated or misbranded and should be condemned as alleged
by the FDA.

However, the FDA may ask *state* health officials to detain goods until a federal civil seizure action is accomplished. State officials may exercise authority under state law to embargo goods pending FDA action.[10]

VIII. CONDUCT OF THE INSPECTION—PROTECTING YOUR COMPANY'S RIGHTS

Let's now consider several policies or procedures you might consider adopting to control the conduct of the FDA inspection, in order to protect your company's rights and interests.

> Accompany the FDS inspector *at all times.* Do *not* allow him to proceed unattended by the company representative.
> Advise the inspector that any questions or requests for data are to be directed *only* to the company's designated representative.

The FDC Act authorizes only "reasonable" inspections, and, surely, it is not reasonable to permit someone who is not an employee to roam unattended through your establishment asking questions of whomever he pleases. Such activity could be disruptive of production and perhaps even dangerous to someone unfamiliar with your plant.

> Employees other than the company representative should be instructed not to converse with the inspector. They should not volunteer conversation, and if asked a question by the inspector, they should respond that it is company policy not to discuss their work with visitors and that any questions should be directed to the company representative designated to accompany the inspector.
> Keep a detailed record of *everything* the inspector says or does. This information may become important in the future, especially if the FDA should undertake regulatory action based on the inspection.
> Whenever the FDA inspector takes a sample of anything, you also should take a sample of the same article, to be maintained as a part of your company's record of the inspection. For example, if the FDA samples a particular lot of finished product, or a particular label, you want to be certain that you have an identical companion sample in your records, readily available for reference if the FDA subsequently asks questions or undertakes regulatory action.
> Do not sign or initial "affidavits" or other documents. FDA inspectors frequently enter information that they believe to be important on a form entitled "Affidavit" (Form FD-463a) and then ask a company representative to sign or initial the form, thereby acknowledging the accuracy of the statement. There

thereby acknowledging the accuracy of the statement. There
is no obligation for you to sign or initial any such affidavit,
and ther is no good reason to do so. Any admissions in the
statement could be used against you or your company in courts.

Many companies have a standard policy that their manufacturing
establishment employees are not authorized to sign or initial documents
for the FDA inspector. If the inspector asks for written acknowledg-
ment with respect to a particular matter, ask the inspector to submit
a *written* request for the information to your company, for review and
consideration by management and company counsel. In practice, this
usually will be the end of the matter, because FDA inspectors appear
to be loath to request anything in writing.

- If the FDA inspector calls your attention to a violation of law that
  is easily correctable, try to correct the situation during the
  course of the inspection.
- Do not volunteer information. It may be reasonable to provide
  certain information in response to questions from the inspector, but there is no reason to suggest new avenues of interest
  that otherwise might not be investigated.
- Always be honest in everything you say to the inspector. For
  example, it is one thing to tell an inspector that he has no
  statutory right to require production of certain information,
  and to decline to provide it. It is a quite different matter
  to give the inspector a potentially devious, or dishonest, response. The former should be understood and respected.
  The latter just invites trouble.
- Finally, be polite. You may need to be firm in asserting your
  company policies or in protecting your rights in some other
  respect, but you should always remain courteous. Personal
  animosity cannot help you.

IX. PHOTOGRAPHS

Over the years, the FDA has often used inspection photographs as
effective evidence in its enforcement actions (civil seizure actions,
injunctions, and criminal prosecutions). The FDA asserts that it has
the right to take photographs during an inspection, and the inspector
probably will argue with you if you tell him not to bring his camera
into the plant.

The FDA Inspection Operations Manual (IOM) includes a section
instructing the inspector to insist that he has a right to take photographs, and to cite a particular judicial decision if a company refuses
to permit photography.[11] However, a statement appearing earlier in
the Manual, which the inspector is unlikely to mention, explains why
the FDA really wants those photographs: The IOM tells the inspector

that "Good photographs are one of the most effective and useful forms of evidence of violations."[12]

The judicial decision that the inspector will cite to you if you refuse to permit photographs, *United States v. Acri Wholesale Grocery Co.*,[13] involved a criminal prosecution for violations of the FDC Act. The defendants, a wholesale grocery corporation and two senior corporate officers, were convicted of causing the adulteration of food after shipment in interstate commerce. The court concluded that the government could properly introduce as evidence at trial photographs taken during FDA inspections of the defendants' warehouse. However, the defendants had *not* objected during the inspections to the taking of the photographs, and the *Acri* case can be interpreted to stand for the proposition that *if* a company permits an FDA inspector to take photographs without objection during an inspection, the photographs may be used as evidence against the company in a judicial enforcement action. Nevertheless, the *Acri* decision also expresses the court's belief that FDA inspectors would be entitled to take photographs even without consent.

In any event, the *Acri* decision does not finally settle the question of the FDA's authority to take photographs without consent during an inspection. The decision is subject to varied interpretations, and it is that of only one U.S. district court. Furthermore, as of the present time, neither that case nor any other reported FDA case has penalized a company for refusing to permit photographs during a routine FDA inspection.

It appears that many companies do *not* permit photographs during an inspection. Photographs may overemphasize a particular detail in a misleading way or may reveal confidential manufacturing procedures that you do not want to release beyond your control. If you are firm about it, the FDA inspector will put away his camera and proceed with the inspection. (The IOM instructs the FDA inspector to report a refusal to permit photographs as a "refusal to permit part of the inspection," but you may conclude that such a statement in an inspector's report -- the Establishment Inspection Report (EIR), discussed in Sec. XV -- is preferable to having him report with detailed color photographs of your manufacturing establishment.)

X. ACCESS TO COMPANY RECORDS—GENERALLY

You are *not* required to provide access to manufacturing records (master formula records, batch production records, analytical data, complaint files, etc.). The FDA inspector may make repeated efforts to examine and copy such records, but (unless, as discussed in Sec. V, the records concern prescription drugs or restricted devices) he is *not* entitled to require you to let him see or copy these documents.

There is, however, a "middle ground" approach you may wish to consider for responding to requests for such records. If, for example, the FDA inspector states that he wants to see your master formula records and batch production records to verify that you include in your products the ingredients listed on the labels, you may elect to follow a reasonable "look but don't copy" policy. That is, in order that he may verify that your company uses the ingredients in its products that it lists on the labels, you may want to let the inspector briefly see the pertinent records from your files, but *not* permit him to copy the information. This approach has the advantage of allowing the inspector to confirm that you follow appropriate procedures in a responsible manner, without releasing from your control documents that may include trade-secret information.

XI. SHIPPING RECORDS

There is a limited exception to the general rule that the FDA is not entitled to require production of manufacturing records for cosmetics and nonprescription drug products: The FDC Act provides that persons receiving FDA-regulated articles in interstate commerce or holding such articles so received, shall, upon *written* request, permit the FDA inspector "at reasonable times, to have access to and to copy all records *showing the movement in interstate commerce* of any food, drug, device, or cosmetic, *or the holding thereof during or after such movement, and the quantity, shipper, and consignee thereof*" [emphasis added].[14] Accordingly, if the FDA so requests *in writing*, you must provide access to records concerning interstate shipment.

However, the Act provides that evidence obtained in this manner may not be used against you in a criminal prosecution (although it may be used in a civil seizure action or injunction). If the inspector does request such information, it is important that you insist that the request be made in writing before providing the documents, so that you assure yourself of the protection afforded by the statute with respect to any resulting criminal prosecution.

XII. FDA INSPECTION TACTICS RE COMPANY RECORDS

Be especially alert with respect to FDA inspection tactics concerning company records. When faced with a refusal by a cosmetic manufacturer to provide manufacturing records, on the grounds that the FDA is not entitled to demand production of such records for cosmetic products, FDA inspectors have been reported to shift tactics and to assert that a particular product is a drug and that they want to see the manufacturing records for the designated product.

It is important to remember that, insofar as the FDA's rights to require involuntary production of your manufacturing records are concerned, the distinction between "cosmetic" and nonprescription "drug" is a distinction without a difference.[15] Generally, the FDA does *not* have the right to compel production of manufacturing records *either* for a cosmetic *or* for a nonprescription drug.

You may need to be especially careful to protect yourself here. The FDA's drug current good manufacturing practice (CGMP) regulations have been written in a manner to suggest to the unwary reader that he has an obligation to permit inspection and copying of manufacturing records for all drug products. Indeed, the drug CGMP regulations state as follows:

> All records required under this part, or copies of such records, shall be readily available for authorized inspection during the retention period at the establishment where the activities described in such records occurred.[16]

Be careful! The "key" word in this regulation is "authorized." The FDA is *not* "authorized" to require production of manufacturing records for nonprescription drugs. Indeed, the FDA's preamble to the drug CGMP regulations explicitly concedes as much. In the preamble, the FDA Commissioner states that "Congress did not include in the scope of the inspection authority. . .authority to inspect records regarding the manufacture of OTC [i.e., "over-the-counter" or nonprescription] drug products. . ."[17] Furthermore, the FDA's Inspection Operations Manual tells the same story: "In general Section 704 of the Act [i.e., the section of the FDC Act authorizing FDA inspections] does not provide mandatory access to: Formula files. . . Complaint files."[18] Don't be fooled by an inspector's assertion of "drug" status!

Remember, you must know your rights to be protected. When the FDA inspector asks for such records, he does not advise you that you are not required to provide them.

XIII. THE EXIT INTERVIEW

At the completion of the inspection, the FDA inspector will meet with the "owner, operator, or agent in charge." At this time, the FDA inspector provides an FDA form entitled "Inspectional Observations" (Form FD-483), listing observations the inspector believes are violations.[19]

It is prudent to discuss with the inspector this list of observations. If you do not understand an item, ask about it. If you do not agree with a particular observation, explain your position. If you have corrected an observation during the course of the inspection, tell the inspector. Ask the inspector to make any appropriate changes in the list of observations at this time. Also, if you intend to correct certain observations, explain this. Even if the inspector does not amend his list of observations, he should include your comments in his report of the inspection [the Establishment Inspection Report (EIR), discussed in Sec. XV]. Such comments may affect the way he and his superiors at the FDA evaluate the inspection. You want to satisfy the FDA that you are taking all reasonable steps to manufacture safe and properly labeled products.

Also during the exit interview, the FDA inspector will provide a "Receipt for Samples" (Form FD-484) for all samples taken during the course of the inspection (unless he has already provided such documentation when the samples were taken). At this time you should confirm that you have taken companion samples of all articles sampled by the inspector.

Caveat: If during an exit interview you promise the inspector to make certain corrections, be certain to do as you have promised. The next time an FDA inspector visits your plant, he will determine and report whether promised corrections have been made.

XIV. AFTER THE INSPECTION

Promptly after the inspection, appropriate company personnel should meet to discuss the inspection. Was your company in compliance with the requirements of law? If not, what corrective steps should be taken? Were the inspector's "Inspectional Observations" accurate? If you disagreed with the inspector's observations during the exit interview, did he make appropriate changes in his observations? If the inspector noted violations, were they of such significance that some type of follow-up regulatory action might be expected from the agency? Who in corporate management should be advised of the inspection and its outcome? Depending on the nature of the "Inspectional Observations" and the exit interview, after the inspection you may want to send the FDA a written response to the observations, thereby making certain that the FDA record includes a considered statement of your views.

XV. THE ESTABLISHMENT INSPECTION REPORT

After departing, the FDA inspector returns to his resident post or to the FDA district office and prepares a detailed Establishment Inspection Report (EIR). After the EIR is completed, the inspector usually will destroy any notes he may have made during the inspection. The EIR thus becomes the FDA's primary record of the inspector's visit to your firm, and it will be reviewed by FDA compliance officers looking for violations of law.

You will, of course, be interested to know what the inspector has said about your establishment in his EIR, and you may obtain a copy of the EIR when the FDA has closed its file on your inspection.[20] If the FDA refuses to release a copy of the EIR concerning your inspection, the agency still has an "open" file on the matter; that is, the agency is still considering whether to institute some form of regulatory action.

One reason to request a copy of the EIR is that EIRs are subject to release under the Freedom of Information Act to any member of the public, including your competitors. Accordingly, you may want to review the EIR to determine whether the FDA has inadvertently failed to purge the document of trade secret information before release. If you find the FDA releasing an EIR that reveals trade secret information concerning your establishment, you should object to the agency immediately.

XVI. FDA ANALYSES

If the FDA performs analytical work on a sample of an ingredient or finished product taken during the inspection, you are entitled to a copy of the results of analysis upon request.[21] Note that you should be able to obtain such reports of analyses without waiting until the FDA "closes the file" concerning the inspection.[22] Thus you may be able to obtain analytical results before you can obtain the EIR. (If the FDA performs analytical work, you generally can also obtain from the FDA a portion of the sample subjected to analysis, so that you can perform analytical work on the same sample tested by the FDA.)[23]

XVII. CONCLUSION

If the FDA should conclude that an inspection has revealed significant violations of the FDC Act or of FDA regulations, the agency may initiate regulatory action (e.g., request a recall, recommend a civil seizure action, etc.). It is precisely because of the serious enforcement actions that can arise out of an inspection that it is so important for you to understand and to exercise your rights during an inspection, and to keep detailed records of each inspection.

Never forget the potentially serious nature of any FDA inspection. In order to protect your company's rights and interests, you should establish standard operating procedures for your company for the conduct of FDA inspections, and affected company personnel should be thoroughly trained to follow these procedures. This chapter should help you to provide an effective inspection plan for your company.

NOTES

1. §704(a) of the FDC Act, 21 U.S. Code (U.S.C.) 374(a).
2. §704(a) of the FDC Act, 21 U.S.C. 374(a).
3. §301(f) of the FDC Act, 21 U.S.C. 331(f).
4. *Marshall v. Barlow's, Inc.*, 436 U.S. 307 (1978).

5. *United States v. Jamieson-McKames Pharmaceuticals, Inc.* See also *United States v. Roux Laboratories, Inc.*, 456 F. Supp. 973 (M.D. Fla. 1978) (ruling, in a case involving a cosmetic manufacturing establishment, that the FDA may not conduct an inspection without a warrant if the company does not consent, but also citing the FDA's authority to recommend a criminal prosecution pursuant to 21 U.S.C. 331(f) for refusal to permit an inspection) and *United States v. New England Grocers Supply Co.*, 488 F. Supp. 230 (D. Mass. 1980) (ruling that neither a warrant nor consent was required for an FDA inspection of a grocery warehouse).

6. §704(a) of the FDC Act, 21 U.S.C. 374(a).

7. §704(a) of the FDC Act, 21 U.S.C. 374(a).

8. §§702(b), 704(c) and (d) of the FDC Act, 21 U.S.C. 372(b), 374(c) and (d).

9. *United States v. 75 Cases. . .Peanut Butter*, 146 F.2d 124 (4th Cir. 1944), cert. den. 325 U.S. 856 (1945); *United States v. El Rancho Adolphus Products*, 140 F. Supp. 645 (M.D. Pa. 1956), affd. 243 F.2d 367 (3d Cir. 1957), cert. den. 353 U.S. 976 (1957); *United States v. Roux Laboratories, Inc.*, 456 F. Supp. 973 (M.D. Fla. 1978).

10. See, e.g., *United States v. An Article of Food. . .345/50 Pound Bags*, 622 F.2d 768, 769 nt. 1 (5th Cir. 1980).

11. FDA Inspection Operations Manual, subchapter 520, section 523.1, "In-Plant Photographs" (TN 79-22; 10-19-79).

12. FDA Inspection Operations Manual, subchapter 520, section 523, "Photographs—Photocopies" (TN 79-22; 10-19-79).

13. *United States v. Acri Wholesale Grocery Co.*, 409 F. Supp. 529 (S.D. Iowa 1976).

14. §703 of the FDC Act, 21 U.S.C. 373.

15. In many *other* respects the distinction *is* significant. See, e.g., McNamara, "When Is a Cosmetic Also a Drug—What You Need to Know, and Why," *Food Drug Cosmet. Law J.*, 35:467 (Aug. 1980).

16. 21 Code of Federal Regulations (C.F.R.) 211.180(c).

17. *Fed. Reg.*, 43:45066 (Sept. 29, 1978).

18. FDA Inspection Operations Manual, subchapter 500, section 501.1, "Authority to Enter and Inspect" (TN 79-22; 10-19-79).

19. §704(b) of the FDC Act, 21 U.S.C. 374(b), provides that "Upon completion of any such inspection. . .and prior to leaving the premises, the officer or employee making the inspection shall give to the owner, operator, or agent in charge a report in writing setting forth any conditions or practices observed by him which, in his judgment, indicate that any food, drug, device, or cosmetic in such establishment (1) consists in whole

or in part of any filthy, putrid, or decomposed substance, or
(2) has been prepared, packed, or held under insanitary conditions whereby it may have become contaminated with filth,
or whereby it may have been rendered injurious to health."
20.   21 C.F.R. 20.64, 20.101.
21.   21 C.F.R. 20.105(c).
22.   21 C.F.R. 20.105(c).
23.   §702(b) of the FDC Act, 21 U.S.C. 372(b); 21 C.F.R. 2.10(c).

30

HOW AND WHEN TO RECALL

WILLIAM C. WAGGONER*

Johnson & Johnson, Inc., New Brunswick, New Jersey

---

*Present affiliation: Thompson Medical Company, New York, N.Y.

I. INTRODUCTION

Successful handling of potential recall situations requires basic pre-
paredness. Chances are that an individual firm may never experience
the need to review a product problem leading to recovery; but if it
does occur, a format to guide a decision is a must.

Product problems can range in importance from simple, nonsignifi-
cant color changes to life-threatening hazards. Proper evaluation of
the problem can prevent a recall when one was not necessary, or can
prevent a company from not invoking a recall when one should have
been done. Straightforward objective thinking and clear evaluation
of problems with equanimity using well-established standards can lead
to a decision that will remain appropriate and free from criticism.

A. The Problem

A product problem may come to a firm's attention by:

1. Means of a consumer or user complaint
2. Notification from a distribution channel
3. Communication from a regulatory agency
4. Discovery from its own records that a product in distribution
   should be evaluated

B. Reasons

Although there are many reasons why products may present problems,
those most likely to occur include:

1. Deviation from an established standard, if the product represents that it complies with a particular standard
2. Contamination or decomposition of product
3. Leaking containers
4. Label mix-up or inadequate labeling
5. Faulty manufacturing practices, if product integrity is compromised
6. Legal reason, either discovery that the product fails to satisfy regulatory agency requirements, or by order of a regulatory agency

C. The Law

The federal Food, Drug, and Cosmetic Act (FDC Act) does not authorize the Food and Drug Administration (FDA) to order a recall of a cosmetic that is in violation of the act. In the preamble to the proposed
regulations of its recall policy (*Fed. Reg.*, *41*:26924, June 30, 1976),
the FDA noted that the Supreme Court has imposed a high standard
on the manufacturer, distributor, or processor of products governed
by the Act.

"The [FDC] Act imposes not only a positive duty to seek out and remedy violations when they occur but also, and primarily, a duty to implement measures that will insure that violations will not occur. The requirements of foresight and vigilance imposed on responsible corporate agents are beyond question demanding, and perhaps oner-ous, but they are no more stringent than the public has a right to expect of those who voluntarily assume positions of authority in busi-ness enterprises whose services and products affect the health and well-being of the public that supports them" (*U.S. v. Park*, U.S. Sup. Ct. 1975, 421 U.S. 658). In recognition of that obligation, most manufacturers have taken self-imposed measures to remove violative products from consumer availability. Present FDA policy on recalls [21 Code of Federal Regulations (C.F.R.) 7.40] is based on industry cooperation. However, the FDA's authority to seize and condemn vio-lative cosmetic products and to impose criminal sanctions encourages voluntary commencement of recalls upon the agency's request.

D. The Review

In establishing a recall program within your company, it is vital to identify clearly those areas where product problem information is most likely to occur. Those sources are usually quality assurance, market-ing, sales, public relations, consumer service, and legal. Ideally, a company recall coordinator should be designated in a written policy, and all information on product problems, irrespective of source, should be funneled to this coordinator. This individual should be thoroughly familiar with the regulations and company complaint re-porting procedures.

When the coordinator identifies a potential recall situation, this person quarantines current affected lots of product and calls together a group to evaluate the information. The group should consist of a broad spectrum of disciplines within the company, including market-ing/sales, regulatory, quality assurance, medical, operations, and legal personnel. As pointed out earlier, it is important to have gath-ered all pertinent information for a comprehensive evaluation of the problem. Meager information can lead to a wrong decision.

Once complete information is obtained and collated, the evaluation group will sift over the data and address several questions:

1. Does the product constitute a health hazard?
2. Does the product violate the law?
3. What is the cause of the problem?
4. What is happening in the field?
5. Where is the product in distribution?
6. What action, if any, should be taken?

E. The Decision

Once all pertinent information is obtained and answers to the above questions are available, the product recall committee should meet and decide on the recall question. Notification of the meeting is given by the coordinator. The company president chairs the meeting, with minutes taken by the coordinator. The product recall committee should be represented by the following areas:

1. President
2. Divisional manager
3. Operations
4. Information and control
5. Research
6. Quality assurance
7. Medical
8. Regulatory
9. Sales/marketing
10. Legal

During the meeting, the lot number(s) and product(s) to be recalled are identified. The product recall committee decides whether to:

1. Initiate a field correction
2. Initiate a field relabeling
3. Recall product immediately
4. Make a further investigation
5. Dismiss.

II. ACTIONS

For purposes of clarification, listed below are various actions a company could consider depending on the nature of the problem and the extent of distribution.

A. Stock Recovery

Products that have not left the direct control of the manufacturer or primary distributor, whether stored in their plant or in premises under their control, may be removed if no stocks have been released for distribution.

B. Market Withdrawal

Products from the market involving no violations or only minor violations that would not be subject to legal action under existing compliance policy may be removed from the market.

C. Product Withdrawals — Type A

This is a situation where none of the product has left the direct control of the manufacturer or primary distributor. This type of withdrawal would be a recall if in distribution channels. Normal FDA actions include the following:

1. Check on adequacy of withdrawal as for appropriate class of recall.
2. Not placed on public recall list.
3. No press release initiated. The FDA will respond to inquiries from the press and the public.

D. Product Withdrawals — Type B — Voluntary

This is a situation where no violations are involved or the violations are minor and not subject to seizure under current FDA policy and guidelines. Normal FDA action: none.

E. Recall

Recall is a firm's correction in the field or removal from the market of products which are subject to legal action under the FDA's existing compliance policy and present a threat or a potential threat to consumer safety and well-being, product adulteration, gross fraud or deception of consumers, or materially misleading, causing injury or damage.

Recall Criteria

A product recall must meet the following criteria:

1. Product has been in distribution channels.
2. Product is being removed from the market or is being repaired or relabeled, or
3. Product is adulterated or misbranded to the extent that it is subject to seizure under current FDA policy guidelines.

Recall Classification

Recalls can be classified into three categories.

Class I recall: A Class I recall refers to a situation in which there is a reasonable probability that the use of, or exposure to, a violative product will cause serious, adverse health consequences or death.

Examples of Class I recall situations include:

1. Label mix-up of potent drug
2. *C. botulinum* toxin in food product
3. Device with excessive electrical leakage or radiation hazard

Normal FDA actions for Class I recall include:

1. Product recall, to consumer level, as rapid and complete as possible
2. 100 percent effectiveness checks of known direct accounts and subdistribution points, including consumer
3. Placement on public recall lists as a Class I recall
4. Possible issuance of public warning

Class II recall: A Class II recall occurs when the use of, or exposure to, a violative product may cause temporary or medically reversible adverse health consequences or where the probability of serious adverse health consequences is remote.

Examples of Class II recall situations include:

1. Pathogenic micororganisms in food, exclusive of *C. botulinum*
2. Subpotent or superpotent drug, not life-saving in nature
3. Electrical shock hazard of small consequence, not life-threatening
4. Improperly calibrated thermometers

Normal FDA actions for Class II recall include:

1. Product recall to appropriate distribution level, completely and promptly
2. Effectiveness checks on a sliding scale depending on seriousness of hazard, ranging upward from 2 to 10 percent of known direct distribution points and upward from one to two subdistribution points for each direct distribution point checked
3. Placement on public recall list as a Class II recall
4. Possible issuance of press release as circumstances warrant

Class III recall: A Class III recall occurs when the use of, or exposure to, a violative product is not likely to cause adverse health consequences.

Examples of Class III recall situations include:

1. Product recall, generally to wholesale level
2. No effectiveness checks
3. Placement on public recall list as a Class III recall
4. Ordinarily, no press release initiated by the FDA, but the FDA will respond to inquiries from press and public.

III. DUTIES

A. General Responsibilities

If the decision is made to recall, the following orchestration ensues:

1. Communicate identified product and lot numbers to various operating levels.

2. Notify appropriate company officials and, if appropriate, the FDA district office.
3. Discontinue distribution of the affected lot(s) of product.
4. Determine geographic area and level of distribution of lot(s) of affected product.
5. Determine disposition of recalled product.
6. Institute a log of events and actions.
7. Prepare and issue, with FDA concurrence where appropriate, recall request (letters, telegrams, mailgrams, public statements) to known consignees, including necessary recall instructions. (Examples of recall letter, envelope, and consignee answer follow.)
8. Continue investigation of nature, extent, cause, and remedy of product.
9. Prepare information for company public relations for possible issue of news releases, as necessary.
10. Report progress of recall effectiveness to appropriate company officials as the recall progresses.
11. Prepare for FDA inspection.

B. Specific Responsibilities

Various areas of the company's operation have definite responsibilities during a recall. The following lists positions and responsibilities for a large company. The responsibilities can be tailored for small companies among those factions having appropriate control.

1. President
   The president of the company has overall responsibilities for the recall.
   a. Communicates the decision that a product is to be recalled to the following:
      (1) Chairman of the board
      (2) Law department
      (3) Director of personnel
      (4) Public relations department
   b. Reviews periodic status reports on the progress of the recall, received from quality assurance and regulatory affairs, to determine that all requirements of the recall are being met. If not, appropriate corrective action is to be taken.
   c. Reviews and approves all proposed communication to customers, consumers, the sales force, and the FDA.
   d. Decides the final disposition of returned, recalled product(s), i.e., to destroy the product, or with the approval of the FDA to reprocess.

2. Division manager
    a. Arranges that a control procedure is set up to monitor undelivered notifications.
    b. Reviews and approves all communications to customers, consumers, the press, and the FDA.
    c. Determines the level of involvement for members of the sales force, communicating relevant instructions to its members as the recall dictates.
    d. Develops a practical, smooth method of crediting customers for returns, either through replacement, refund, or credit.
    e. Helps to effect a smooth, rapid recovery of all stocks of affected product in the field.
    f. Participates in contacting customers and users of product and handles customer inquiries as received.
    g. Where appropriate, develops a policy relative to provide a refund for or replacement of the recalled product.
    h. Participates in the overall evaluation of the effectiveness of the recall, by assessing product remaining in the field.
3. Vice president — operations
    Initiates a detailed review of information relevant to the manufacture of the lot(s) of product being recalled to include:
    a. Date of manufacture
    b. Quantity manufactured
    c. Manufacturing deviations, if any
    d. Location of inventories
    e. Exact quantities in quarantine
    f. Other information, as appropriate, to investigate history of lot under scrutiny
4. Vice president — information and control
    a. Assigns a project number to the recall project so that all costs for time, supplies, returned goods, and other items may be appropriately charged.
    b. Evaluates cost impact of the entire recall project.
    c. Provides general direction to system's effort.
5. Vice president — research and development
    a. Assesses the product problem to determine if it is a unique (one-time) situation or one which represents a specific problem with additional implications.
    b. Assigns a team to evaluate:
        (1) The process of manufacture
        (2) The raw materials used and their specifications
        (3) The quality control testing of the product or its components

  (4) The container closure system of the product
  (5) Any other aspects of the product which may require
    evaluation
 c. Makes appropriate recommendations relative to the resump-
   tion of manufacturing and marketing of the affected product, and others which may be implicated.

6. Director — quality assurance
 a. Upon notification of a possible product problem, gathers
   all relevant information on which is necessary to evaluate
   the problem. Specifically this includes:
  (1) Identity of product, lot number(s) involved, and
    package sizes
  (2) Facts relating to the product, and if packages of
    the same lot are available in-house, testing of same
  (3) Name, address, and telephone number of person(s)
    reporting product problem
  (4) All pertinent information related to the problem
  (5) Information about any problem of medical consequence
    associated with product problem
  (6) History of lot(s) of product under scrutiny, includ-
    ing all previous received complaints
  (7) Processing records, retained samples, and quality
    control records, for review
  (8) Information about the testing of the lot(s) or pro-
    duct and summary of results
 b. Meets in order to evaluate this information with appropriate company personnel to verify that a product problem
   exists, and presents the investigative results.
 c. Upon verification of a product problem, immediately quarantines all in-house inventories of the lot(s) of product
   in question. The quarantined materials will also include
   recently pulled quantities of the lot(s) of product in
   question which may be in any stage of packing, loading,
   or ready for shipment, but which have not yet left company premises, warehouses, or distribution centers.
 d. If the evaluation indicates that a problem does exist, requests the president to call a meeting of the product recall committee.
 e. Confirms to appropriate personnel the suspension of distribution of all affected lot(s) of product under scrutiny.
 f. Collects processing records, retained samples, and quality control records for review.
 g. Obtains, on an ongoing basis, all packages of affected
   lot(s) of product and immediately removes these to quarantine, or quarantines them securely in a receiving area.

    h.   Issues to appropriate personnel periodic reports (at least weekly in frequency) as to:

        (1)   The quantity of recalled product received

        (2)   From how many affected customers

    i.   Assesses the analytical methods for the product in question and determines if these methods are adequate or need revision.

    j.   Monitors disposition of recall goods.

7. Manager — regulatory affairs

In the event of a recall situation, the manager of regulatory affairs will be responsible for coordinating the entire recall, through working with members of the product recall committee.

    a.   It is his responsibility to assure that:

        (1)   Each and every step of the recall procedure is understood and followed,

        (2)   All assignments relevant to the product recall are completed, and

        (3)   The overall objective, i.e., completing an effective recall, is accomplished. The effectiveness of the recall will be continually evaluated in order to demonstrate to the agency that an effective recall has been achieved, if appropriate.

    b.   This individual participates in both the preliminary meeting and the product recall meeting, acting as secretary for the latter.

    c.   Specifically, the manager of regulatory affairs is responsible for:

        (1)   Drafting recall letters and cards (See Figs. 1, 2, and 3)

        (2)   Drafting letters, telegrams, mailgrams, cards, and other communications to the FDA

        (3)   Internal communication of progress of recall

        (4)   If appropriate, notifying FDA in accordance with requirements, in consultation with the law department, and scheduling meetings with the FDA as required

    d.   Communicates written recall decision to appropriate company personnel.

    e.   Provides to the FDA the following information:

        (1)   Identity of the products involved by name, lot number(s), and package size(s).

        (2)   Description of the deficiency or possible deficiency in the product and the date and circumstances under which it is discovered.

John Doe Cosmetics
Somewhere, USA 12345

Date

URGENT: COSMETIC RECALL - Contamination

RE: Doe Facial Cleaner, Lot No. 4567

Recent tests show that the above lot number of this product is contaminated and, therefore, represents a potential public health hazard. Consequently, we are recalling this lot from the market. Other lot numbers are not involved.

Please examine your stocks immediately to determine if you have any of Lot No. 4567 on hand. If so, discontinue sale of the lot and promptly return via parcel post to our Somewhere Plant, ATTENTION: RETURNED GOODS.

IF A SUB-RECALL IS INDICATED, THE FOLLOWING PARAGRAPH SHOULD BE ADDED:
If you have distributed any of Lot No. 4567, please immediately contact your accounts, advise them of the recall situation and have them return their outstanding recalled stocks to you. Return these stocks as indicated above.

You will be reimbursed by check or credit memo for the returned goods and postage.

Please return the enclosed card immediately providing the requested information.

This recall is being made with the knowledge of the Food and Drug Administration (FDA). The FDA has classified this recall as class _____ (if classified).

We appreciate your assistance.

Sincerely yours,

John Doe
President

FIGURE 1 Model cosmetic recall letter.

489

PLEASE FILL OUT AND RETURN

We do not have any stock of Doe Facial Cleaner, Lot No. 4567

on hand. ☐

We have requested our accounts to return their stocks of this

merchandise to us. ☐

We are returning _____ jars of Doe Facial Cleaner, Lot

No. 4567.

Name_____

Address _____

_____

City _____ State _____ Zip_____

First Class
Permit No. 3
XXXXXX

BUSINESS REPLY MAIL
No Postage Stamp Necessary if Mailed in the USA

Postage will be paid by:

John Doe Cosmetics
Somewhere, USA 12345

ATTENTION: Mary Doe
            Manager-Regulatory

FIGURE 2 Model recall return postcard.

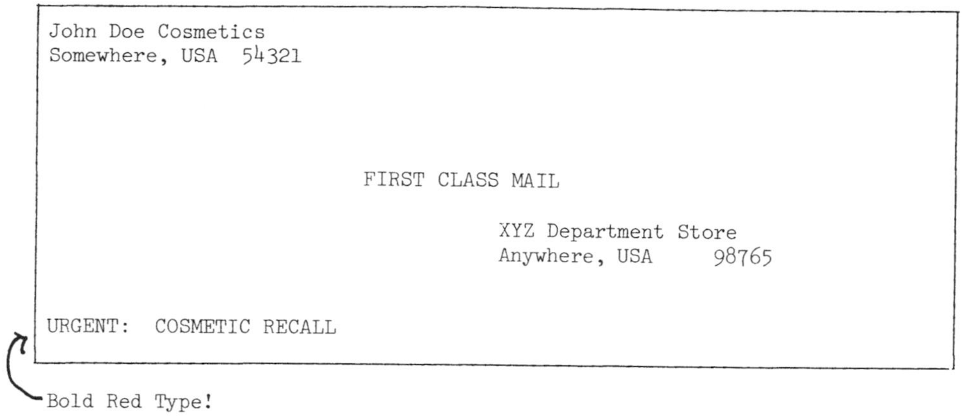

```
John Doe Cosmetics
Somewhere, USA  54321




                        FIRST CLASS MAIL

                             XYZ Department Store
                             Anywhere, USA      98765


URGENT:   COSMETIC RECALL
```

Bold Red Type!

FIGURE 3 Model cosmetic recall envelope.

(3) Evaluation of the risk associated with the deficiency or possible deficiency.
(4) Total number of such products produced and the time span of the production.
(5) Total amount of such products estimated to be in distribution channels, broken down into domestic and international distribution. Estimates will be based on distribution records.
(6) Distribution information, including identity of initial consignees, if known.
(7) Copy of recall communication or proposed communication if none has been issued.
(8) Proposed strategy for conducting the recall.
(9) Name and telephone number of regulatory personnel who should be contacted concerning the recall.

f. Receives from the FDA notice of the assigned recall classification, recommendations of any appropriate changes in company strategies for recall. Receives all FDA communications.
g. Maintains log of events and actions of the recall for documentation.

RECALL NUMBER:	DATE:		
	Date		
EVENT	IN PROGRESS	COMPLETED	REMARKS
1. Establish Product Problem & Gather Relevant Information			
2. Call Problem Meeting			
3. Call Product Recall Committee Meeting to Decide on Recall			
4. Institute Log of Recall Events			
5. Notify Company Officials			
6. Notify Corporate Officials			
7. Contact FDA (District Office)			
8. Contact Insurance Carrier			
9. Quarantine Inventories			Confirmed Yes/No
10. Define & Communicate to Director Distribution Services Product to be Recalled & Its Lot Number			
11. Discontinue Distribution			
12. Produce Customer Listings & Labels			
13. Mail Customer Notifications			
14. Notify Sales Force			
15. Establish Customer Listing with Amount of Product Distributed (Reconciling)			
16. Get Project Number for Recall Charges			
17. List Product as "Temporarily Unavailable" on all Current & Future Orders Until Problem Solved			
18. Prepare Bi-Weekly Status Report on Returned Goods			
19. Quarantine & Test Returned Product			
20. Collect Processing Records & Retained Samples, Quality Record for Affected Product			
21. Review Data Relative to the Manufacture of Lot Under Scrutiny			
22. Determine Replacement Product			
23. Assess Possible Causes for Product Problem			
24. Develop Remedy for Product Problem			
25. Decide on Disposition of Returned Goods			
26. Issue Credit for Returned Goods			
27. Prepare for FDA Inspection			
28. Certify Completion of Recall Process			

FIGURE 4 Checklist for coordinator of recall.

h. Coordinates activities of all individuals with responsibilities in the recall, by using as guide the Checklist for Coordinator (see Fig. 4), which will be filed as permanent record of the recall.

i. Escorts FDA inspectors to manufacturing facilities or to quality assurance as required.

j. Reports progress of recall effectiveness to company officials and the FDA district office.

k. Reports to the FDA the final recall status. This includes a review of the steps taken to:

(1) Dispose of the returned recalled product (as authorized by the president)

(2) Guard against a repetition of the same problem.

8. Vice president — personnel

a. Will be informed by the president to determine the intracompany publicity and policies which will be placed in effect.

b. Will further determine in case of work stoppage or layoffs that any program(s) which affects personnel will be instituted in accordance with that department's policies.

9. Manager — distribution

a. Receives the completed written recall authorization indicating which manufacturing lot number(s) of a specified product to recall.

b. Discontinues distribution of product(s) indicated.

c. Responsible for generating listings within reasonable time of all "definite recipients" and "potential recipients" of shipments from affected lot(s) of product. This is a "shipped to" listing, reflecting invoiced customers and includes both domestic and international.

d. Provides with these listings:

(1) Total number of customers involved in recall

(2) Total number of packages shipped

e. If the situation warrants broader coverage of customers to be notified, the system allows that additional *lot numbers* be included and the time period be expanded.

f. Provides the total reconciliation, showing:

(1) Number of customers per total recall

(2) Number of packages shipped per total recall

Note: Both numbers are by "shipping lot number." Depending on the type of recall, reconciliation could assure 90 percent completeness of all product shipped, including both domestic and international.

g. All return goods will be quarantined for inspection by quality assurance and disposition.

RECALL NUMBER	Date:		
EVENT	Date In Progress	Completed	Remarks
1. Receive Specifications Which Product & Lot Number(s) to be Recalled			
2. Contact Department Managers: Customer Service, Shipping Department, Product Planning			
3. Discontinue Distribution of Indicated Product(s)			
4. Run Recall System for Lot Control Numbers			
5. Establish Count of Packages Shipped for Total Recall Within 98% of Total Lot(s)			
6. Establish Count of Customers Per Total Recall			
7. Determine Exact Wording for Inquiry Responses			
8. Obtain Copy of Customer Notification (Letter to be Mailed)			
9. Research & Verify Potential Recall Shipments			
10. Communicate to Manager-Regulatory Affairs Both: "Number of Packages Shipped" and "Number of Customers Involved" (Total Recall Figure)			
11. Establish Number of Reply Cards Received.			
12. Receive Information Concerning Replacement Product(s)			
13. Develop Inventory Status and Forward to Director-Quality Assurance			
14. Review Returned Goods Reports, Prepare Weekly Status Reports and Assure Delivery of Same to Quality Assurance			
15. Receive a Project Number of the Recall From Vice President-Information & Control			
16. Arrange Replacement for Returned Goods			
17. Issue Credit to Customers for Returned Goods			
18. Provide Summary After Completion			

FIGURE 5 Checklist for manager — distribution.

    h. Provides general direction for all activities reporting to this function, using the checklist of events (see Fig. 5) as a guide and files the checklists as permanent record.

    i. Gets copy of customer notification and requests the exact wording for message to be used to answer inquiries.

     j. Reconciles the total quantity of packages of all lots in the recall with the total number of packages shipped, including the packages remaining in inventory.

     k. Provides number of reply cards received.

     l. Prepares a complete inventory status of lot(s) recalled and forwards to director of quality assurance.

     m. Directs the receiving department to channel returned goods through inventory control by logging in product (quantity, date received, from who received, condition upon receipt). As soon as all information relevant to returns is recorded, the returned goods will be transferred to the director of quality assurance for quarantine, until disposition is determined.

     n. Issues periodic (weekly) progress reports to manager of regulatory affairs concerning returned goods.

     o. Utilizing the guideline agreed to, determines from production planning when replacement material will be available to fill orders for replacement product.

     p. Arranges replacements, refunds, or credits for returned goods.

10. Product manager

     a. Participates in preliminary meeting to determine if a recall situation exists.

     b. Requests from vice president for information and control a project number for the recall. This number must be communicated to all individuals involved in the recall in order that all associated costs may be accumulated and properly charged.

     c. Works closely with the division manager to assist in accomplishing recall-related assignments or marketing.

     d. Supports customer service in the interest of accomplishing a smooth recall process and maintaining customer relations.

11. Director—operations

     a. Evaluates the product back order situation.

     b. Provides for manufacture of new lots of product. (If the product problem is more extensive than one lot, a delay in the manufacture of additional lots may be warranted.)

     c. Determines means for allocating and replacing recalled product.

     d. Revises product scheduling forecasts.

     e. Notifies suppliers of raw materials used in the affected product(s) whether any or all of the raw materials are below standards.

12. Legal representative

In the event of a recall situation, the legal representative will act as advisor on matters of applicable regulations, company

policy, and the governing statutes and appropriate amendments thereof. This may require planning the overall recall
strategy and reviewing proposed communications to customers,
users, the press, and the FDA. The legal representative will
also contact the insurance carriers.

13. Quality assurance
    a. Identifies the product in question by manufacturing lot
       number(s).
    b. Investigates batch samples of product in question to con-
       firm the complaint or deny further investigation as dis-
       missal of complaint.
    c. Investigates samples of complaint.
    d. Upon recall decision, stops shipments of and quarantines
       violative product.
    e. Monitors all returns.
    f. Issues periodic weekly reports to manager of regulatory
       affairs on the effectiveness of the recall.
    g. Makes all appropriate records of product available for FDA
       inspection.

14. Customer service
    a. Supports shipping department by researching files to pro-
       vide lot control number for customers on list.
    b. Prepares "total number of customers involved" for recon-
       ciling report.
    c. Provides general support as needed.
    d. Receives reply cards sent to customers indicating quantity
       on hand or quantity to be returned for credit or replace-
       ment.
    e. Prepares totals for reply cards received.

15. Receiving department
    a. Enters all recall shipments upon receipt at the dock in
       a designated logbook and assigns a sequential number
       to each entry, starting each book and recall with 0001.
    b. For each return shipment received, containing the pro-
       duct being recalled, fills out an individual receiving re-
       cord.
    c. Enters on this form:
       (1) Logbook number, as described above
       (2) Customer name and address
       (3) Lot number (shipping lot control number)
       (4) Freight (collect or prepaid) or common carrier
       (5) Date
       (6) Quantity
    d. Attaches a photocopy of the packaging slip from the ship-
       ment to each copy of the form and forwards.

> e. Provides scheduled reports to director of quality assurance (first daily and later weekly) to indicate the effectiveness of the recall.
> f. Collects the received recall material in a secured area.
> g. Through the vice president for operations, seeks decision on final disposition of recall material, decided by the president.
> h. Receives written authority from the president to destroy the specified goods of indicated lot(s) of product.
> i. Prepares a list of packages collected in the recall, indicating lot number(s) and number of packages and containers for each. Distributes this list in usual manner.
> j. In case a decision has been made to destroy all returned goods, the following three steps are to be followed. Otherwise, directions will be issued to reuse the material of the recalled product.
>> (1) Crushes containers *in-house*, which must be witnessed by a representative of quality assurance.
>> (2) When sufficient quantity of recalled material is accumulated, transports these crushed containers under security of two employees to designated landsite for final disposition.
>> (3) Witnesses destruction and burial. Records and issues written report of these last actions to appropriate company personnel.
> k. Files product recall logbook for future references.

C. Depth of Recall

The depth of the recall will depend on the nature of the problem and the extent of distribution. The depth is principally at three levels.

1. *Consumer or user level of recall*: Recall to the final consumer or users of the product, such as the individual consumer, physician, restaurant, hospital, etc.
2. *Retail level of recall*: Recall to the level immediately preceding the consumer or user level, such as retail establishments, as well as dispensing physicians, institutions, etc.
3. *Wholesale level of recall*: Level or levels between the manufacturer and the retailer. This level may not be encountered in every recall situation (i.e., the manufacturer may sell directly to the retail outlet).

D. Effectiveness Checks

Effectiveness checks of recalls are investigations by direct visits, telephone calls, letters, or other verified methods in order to assure that consignees have been notified of the recall and that they have taken appropriate action. Effectiveness checks can be at various levels and are dependent on the nature and depth of the recall.

1. *Level A*: 100 percent of known direct accounts and subaccounts and, if necessary, the consumers that are to be contacted.
2. *Level B*: Any percentage of direct accounts and subaccounts to be contacted which is greater than level C, but less than level A.
3. *Level C*: 10 percent of the total number of direct accounts or three per field office in whose area the direct accounts are located, whichever is greater, and two subaccounts of each direct account to be contacted.
4. *Level D*: 2 percent or less of the total number of direct account or per field office in whose area direct accounts are located, whichever is greater, and one subaccount for each direct account to be contacted in both cases.
5. *Level E*: No effectiveness checks will be done when this level is designated.

The FDA booklet, "Methods for Conducting Recall Effectiveness Checks" (6/16/78) is an excellent outline on the subject.

E. Completion

The recall is completed when the company has retrieved and impounded all outstanding recalled lots that could reasonably be expected to recover.

The recall can be closed at the point that satisfactory disposition of the recalled material has been reached. This includes corrective measures taken to eliminate the violative condition in current and future distribution of the product.

31

ESTABLISHING A CONSUMER RESPONSE PROGRAM

BETTY J. DIENER*

School of Business Administration, Old Dominion University Norfolk, Virginia

*Present affiliation: Secretary of Commerce and Resources, Commonwealth of Virginia, Richmond, Virginia.

I. INTRODUCTION

Each year, increasing numbers of consumers are communicating with personal care manufacturers—the marketers of cosmetic, toiletry, and fragrance products. More than two-thirds of these consumers are seeking information about personal care products or companies; fewer than one-third of the communications involve complaints.

Thus, well-planned consumer response programs can enable a manufacturer to respond to consumer information requests (whether to use, where to buy, how to buy, how to use) and also to the informal redress requests (refund, replacement, and recall) that may arise from product complaints and dissatisfactions.

Effective consumer response systems can also offer a way of reducing the feelings of frustration and impotence on the part of consumers who may feel that effective channels for information or redress do not exist. Finally, these systems can help to build consumer confidence in business—important in a period of declining confidence for many American institutions.

The purpose of this chapter is to assist companies in evaluating their consumer response systems already in place or to assist companies who are planning to establish such systems. The chapter will provide a brief background on the issues of consumerism that relate to consumer response programs and then will focus on the following issues:

What are the most important consumer information and redress
needs that are relevant to the personal care industry?
What are appropriate corporate responses to consumer information
and redress needs in the personal care industry?
How can a company be both consumer-*responsive* and cost-*efficient*
in establishing a consumer response system that responds to
consumer information and redress needs?
Are there organizational issues that affect either consumer
responsiveness or cost efficiency?

II. BACKGROUND

The personal care industry since the early 1960s has experienced increased levels of consumer communications. In addition, the industry has experienced increased pressure from consumers, consumer activists, business groups, regulatory agencies, and legislators with regard to the industry's methods of responding to consumer communications.

These pressures have probably been caused by a combination of heightened consumer expectations, increased marketplace complexity and confusion, and decreases in service and information at the retail level.

The phenomenon of the 1960s and 1970s that has come to be known as "consumerism" is simply an articulation of the public's desire for safe, reliable products, that perform as represented, for which redress is available if they break. It is the desire for product utility, durability, safety, and repairability. It is also the desire for adequate information about these goods or services so that an informed comparison of alternatives can aid the consumer in making a satisfactory purchase decision[1].

The basic consumer "rights," as first formally enumerated by President John F. Kennedy, include both the consumer's right to adequate information on which to make a reasoned purchase choice, and also the right to product safety, performance, and redress.

Consumers are becoming better educated, more sophisticated, and have higher expectations regarding product performance. This, combined with the pressures of inflation, probably leads consumers to demand more value of their products.

At the same time that consumers have heightened expectations, there are many more products and more complex products in the marketplace than ever before. The number of different items in a typical store has more than doubled in the last 10 years. Products which were once simple have become multifunctional and complex in composition and in use. This number and complexity of products can lead to consumer confusion and to an inability among consumers to evaluate product benefits or dangers.

There is also increasingly less information available at the retail level that consumers can use to make intelligent comparisons of alternatives and satisfactory purchase decisions. Food, drug, and department stores once featured sales help and in-store information almost on a one-to-one basis with the customer. The druggist, grocer, or salesperson was able to remember individual consumer probelms or preferences. The advent of self-service, however much it has aided purchasing convenience, has drastically reduced the quantity and quality of in-store information available. Concurrently, the sheer increase in the numbers of items available has also decreased the amount of information with which the few salespeople can be familiar.

Finally, labeling information which was intended to assist consumers in their purchase decisions has instead become long, meaningless lists of complex chemical names. Thus, consumers increasingly must turn to manufacturers for any information and redress needs which cannot be satisfied at the store level.

III. CONSUMER RESPONSE NEEDS

There are two basic types of consumer response needs in the personal care industry—information needs and redress needs.

Studies of almost a dozen companies over a period of 10 years have shown strong similarities in the types of consumer communications received. This has been true whether the company was large or small, made cosmetics or toiletries or fragrances, or directed their products to men or women.

Roughly two-thirds of all consumers who communicate with personal care manufacturers are seeking information—"how to use," "whether to use," "where to buy," or "what are the ingredients." The percentage of communications related to information needs can even rise above two-thirds for hair color or baby products or for companies that sponsor toll-free numbers for consumers to use.

The remaining one-third of all consumer communications seek redress——compensation or satisfaction for products that fail to live up to expectations, that perform poorly, whose package breaks, or whose use results in skin irritations or injury.

The National Business Council for Consumer Affairs (NBCCA) recommends:

> The handling of consumer [communications] should be swift, personalized, courteous, and as efficiently managed as any other function of prime importance to the company, including increased personnel training in the handling of consumer communications, complaint follow-up, and appropriate involvement by senior management[2].

With this in mind, there are two major problem areas that companies face in establishing consumer response programs to answer consumer communications regarding information and redress needs:

> How can the program's operations be made most responsive to consumer needs, via swift, personalized, yet standardized service and with courteous yet generous policies with regard to refund, replacement, and recalls?
> How can the program's operations be made most efficient with respect to organization (cost, location, personnel), automation, and response handling?

A. Information Needs

Consumer information needs can be defined as the information needed by a customer in order to make a reasoned or efficient purchase decision.

These needs might include questions of how to use a product, whether to use a product, where to buy a product, information about the way a product works, or information about the parent company that makes the product.

Consumers will seek out additional information in order to reduce the risk of making a wrong decision if they are inexperienced, if the product is really important to their ego needs, or if the needed information is not available at the place of purchase.

Where consumers do not know how to purchase a product, whether they should use it, or how to use the product correctly, they have a higher chance of being dissatisfied with its performance than if they have satisfactory information. Thus, an unsatisfied prepurchase information need which leads to a poor purchase choice can also lead to dissatisfaction on a post-purchase basis.

In the opposite direction, a consumer who can obtain adequate prepurchase information can then make a more reasoned purchase decision, and can be expected to experience fewer post-purchase disappointments. Thus, it is important for personal care companies to respond as effectively to information seekers as to complainers.

Studies of cosmetic industry files shows that consumer communications most frequently seek information on

Where to buy regular or discontinued products
Beauty advice—how to use/whether to use products
Product information—on ingredients and production processes
Company information
Free products—samples, donations, door prizes

B. Redress Needs

Redress needs of consumers involve post-purchase complaints which require some form of compensation or satisfaction.

Informal redress is that which can be accomplished without recourse to the legal system. Informal redress is available when a consumer can receive satisfaction from a manufacturer, a retail store, or from another informal channel. Such satisfaction usually involves refund, replacement, product recall, or information on how to use the product correctly.

The personal care industry's products are generally low-cost, widely available, and easily replaced if a problem arises.

Enough competition is present so that brand switching is frequently the easiest and least troublesome form of redress available to consumers.

In fact, because brand switching is so easy and replacements are available at relatively low cost, most industry customers can achieve redress without the bother of writing letters or returning products, either to the manufacturer or the retailer.

IV. APPROPRIATE CORPORATE RESPONSES

The following are suggested policies for handling each type of consumer information request and complaint. In each case an effort is made to balance the desire to be consumer-responsive and the need to be cost-efficient. This generally results in a choice between impersonal, yet efficient, prepared materials versus personalized, yet expensive, individually prepared responses. For every type of situation, responses need to be prompt, consistent, fair and, where possible, generous.

A. Consumer Information Requests

Where to buy regular or discontinued products

"I was shocked recently to hear that your product was discontinued six years ago. I realize I may be a little late, but do you have any of this left in your warehouse?"

Inquiries regarding where to buy products will be most frequent when products are in poor distribution or when they have recently been discontinued.

Customers who make these kinds of inquiries need to be advised if a product is available and, if so, where it can be found.

It is virtually impossible, however, for a consumer response unit to know which exact retail outlets carry which exact items. Salespeople sometimes have this information, but they are already overburdened with paperwork and would probably be slow in responding to an individual customer.

It is important, therefore, for current products, for a consumer response unit to be able to respond with a list of stores in the region who carry the general line of products. Customers themselves will still have to determine where a specific item is available. For discontinued products, it is useful for the consumer response unit to have a supply available for immediate free shipment to a customer. And, when stocks are completely gone, the units should be prepared to (1) recommend alternatives *and* (2) send a small or sample size of the recommended alternative. These *generous* responses hopefully build strong future customer relationships.

Beauty Advice

"I am sixteen, have long blond hair, gray eyes, light skin, weigh 125 pounds, and am 5'7" tall. Can you tell me what products to use?"

Consumers frequently seek advice on what cosmetics to use or how to use them. And their questions are frequently as complicated as the one shown above. Replies therefore are difficult to standardize, usually need to be prepared by experienced employees with technical backgrounds, and are therefore the most expensive type of reply to prepare.

Many companies have, however, been able to develop prepared advice materials. One example is the "100 Most Frequently Asked Questions About Hair Color."

Other companies have prepared general beauty advice booklets covering a wide range of topics.

For companies who do not have such materials on hand but would like to develop them, the topics that should be covered could be determined by evaluating all the beauty advice questions received in the last year. Usually three to five major categories of questions are involved. One company, for example, found that three booklets (the history of the company, how to care for your skin, how to use the products) cut by 70 percent the volume of beauty advice questions that had to be handled individually.

Where prepared materials are used, it is still important to retain a degree of personalization in the response. This can be done either by a cover letter or by the use of prewritten standardized paragraphs which are then combined in what appears to be a personalized letter. The response letter would appear to the consumer to be individually prepared although it is handled more efficiently than a letter that was completely individually prepared.

Another alternative which combines efficiency and responsiveness is the use of a telephone response. Such responses, where they are used, are considered ideal for highly technical or complicated problems. Surprisingly, the cost per response for a telephone response may be less than the cost for an individually handwritten response. This is due to the use of WATS lines and to a decrease in handling time. The primary problem with telephone responses, of course, is that it is difficult to identify and get answers at home phones during working hours.

Product Information Requests

"Please send me a complete list of ingredients for your product. I am allergic to it and wish to determine which ingredient is causing the allergy and thus avoid products containing it in the future."

Requests for product information will generally center around ingredients. Consumers either will ask for a list of all ingredients, or will ask whether products contain individual ingredients such as hexachlorophene, lanolin, or *PABA*. These requests are easily answered by having ingredient lists available for each product.

While it is easy to accommodate such requests, manufacturers should keep in mind that the answers may not satisfy the true consumer needs.

Ingredient lists are not very helpful, for example, to a consumer who wants to make a value comparison between brands. While the basic ingredients of a $1 and a $4 lipstick may appear similar, an average consumer does not have the technical knowledge necessary to evaluate the quality levels of the formulations. In addition, the names of the ingredients are complex and the functions of the ingredients are unintelligible.

Nor would ingredient listings, as presently developed, be very helpful to persons seeking to avoid allergic reactions. Fragrances are often the ingredients that cause such reactions, yet they cannot be completely identified in labeling or ingredient lists in a manner helpful to consumers.

With these problems in mind, manufacturers, in order to be consumer-responsive to unstated needs, must explore ways of clarifying ingredients lists—both for persons who seek value comparisons and for persons who wish to avoid allergic reactions. One way of doing this might be to have available, on request, an explanation of the ingredient lists—explanatory material, in simple language, that describes the functions and allergy potentials of the various ingredients within the overall formulations systems. This material, for example, might designate which of the ingredients are emulsifiers, and would then explain emulsifiers. Or it might identify one or two of the ingredients that are known to be potential allergens.

Request for Company Information

"I am writing a term paper and would be interested in any information you could send me concerning your company. Some areas I would like to cover are:

Company history
If you do importing or exporting
Production details (sampling, testing, ingredients and formulations)
Number of employees
A financial statement
How your products are distributed
What types of departments and executives you have
All the products you manufacture
Any business problems you have
I have a week to write my paper."

All companies receive requests for information on company histories, for lists of their full product lines, or for their marketing plans (including advertising and sales budgets)—although not usually in the same letter!

These requests are often associated with school assignments or projects.

Student requests for information about a company, a company's products, or its marketing plans represent an opportunity for a company to move responsively to improve student attitudes toward business. Most companies, however, find it impossible to respond individually to each request—either because of the length and complexity of the request, or because of the confidentiality of the information requested.

It is important therefore for a company to have promotional material available that explains how products are developed, tested, and taken to market, including examples of typical considerations that go into marketing plans. This together with prepared materials on company history and product lines should largely satisfy student needs.

Requests for Free Samples, Door Prizes, Donations

"I would like to have free samples for the 20 girls in my Scout troop."
"If it is convenient, I would like samples of all your products."
"Could you provide 200 samples as favors for our fashion show?"

Requests for free products come from a wide variety of individuals and organizations. Some will be for dozens of units, others for hundreds of units.

Some companies simply reject all such requests—frequently with an explanation that it is corporate policy not to provide free goods. Other companies view each request as a sampling opportunity and have plentiful sample-sized goods available to mail to customers. This is by far the most consumer-responsive method. The companies who use it have *not* found it to be abused.

Another approach that can be used is to offer one sampler of products that can be used as a door prize for charitable events.

The least efficient system that a company can use is to treat each request individually—some granted, others rejected. The problem with this approach is that it takes too much time and it doesn't treat consumers consistently. A company thus needs to develop a policy for the different types of requests.

Miscellaneous Information Requests

"How can you possibly justify the torture and suffering of untold numbers of civet cats to make your civet perfume? The whole process is disgusting and barbaric. Well, my money talks and I will never buy any of your produts until you stop torturing these animals."

In addition to the requests already discussed, there are also other, less frequent types of requests. One example is requests for information on careers available within the industry. Again, prepared promotional materials could be developed that outline typical industry jobs, educational preparation needed, etc.

Also, there are "topical" requests—requests which seem to increase, peak, and decline within rather short periods. Examples are ecological topics ("Do you have biodegradable plastic containers?) or questions about specific ingredients that have received media publicity (such as hexachlorophene when it was ordered removed from industry products).

As a consumer response unit notices a surge in a particular kind of letter, it is usually appropriate to involve marketing management (and often top management) in the development of a response policy for that situation. It should be noted, however, that the consumers in these problem areas are usually emotionally involved with the topics and are *very* hard to satisfy with any communications that they perceive to be manufacturer-generated and therefore biased.

B. Compliments

"I guess your company is so popular because you make your products not just to sell but to work."

Almost 5 percent of all consumer correspondence to companies will deal with unsolicited compliments.

These compliments usually mention how much the writer enjoys using a certain product. Sometimes the compliment relates to the way a previous complaint was handled—either with the speed or the generosity of the handling.

These letters are not only the easiest to answer, they are also the most gratifying for response personnel to handle. Generally, a standardized letter can be sent to a consumer thanking him or her for the compliment. In addition, a company can continue to build the relationship by also sending a sample of another product as a token of appreciation. For the companies that do this, there is no indication that this generosity is ever abused, nor is there any indication that the letters have been written solely in hopes of getting free goods from the manufacturers.

C. Complaints/Redress Needs

Redress needs have been defined as the needs created by some form of consumer dissatisfaction or complaint. According to the NBCCA guidelines, a company that wants to be consumer-responsive to complaints must have a system that handles these problems fairly, quickly, consistently, and generously.

The different basic types of complaints for the personal care industry include:

Promotional fulfillment—consumers send in by mail for free samples or other promotional offers and do not receive them.

Disappointment with product performance—consumers don't like the way the product works for them.

Packaging—caps, hinges, tubes, bottles, or other components break, aerosols clog, etc.

Unusual look/smell/feel of product—consumers are afraid of spoilage.

Irritation/rash/allergy—alleged by consumers to occur in connection with product use. (Usually fewer than 5 percent of total consumer communications to a company.)

Roughly one-third of the consumer correspondence for a company in this industry will deal with complaints and with redress needs for replacement or refund. While efficiency is also important, responsiveness is the primary consideration where redress needs to be effected.

Responsiveness, however, does not preclude the use of standardization (and thus efficiency) of response. In fact, in almost every problem category except alleged injury, a standardized response can be used to reply to 70 to 80 percent of the complaints. Standardization here means that while the responses are not identical on a word-for-word basis, the basic response formats for the response can be developed on a prepared basis. For example, "Thank you for writing about your problem with (the remainder to be filled in individually). We are sorry that this occurred, and feel it was caused by (to be filled in individually)."

Promotional Fulfillment

"You have taught me to distrust people—especially you! I am not easy to brush off and I will continue to complain! I sent in one measly dollar and you don't have the decency to fulfill my order."

Any company that has ever utilized mail-fulfilled promotions will have had significant numbers of complaints about those promotions. In fact, in large companies that run frequent promotions involving mail fulfillment, promotional fulfillment problems can easily constitute more than half the company's total complaints.

The fulfillment problem seems to have four basic causes:

The *manufacturer* underestimates the response volume or fails to have the promotional goods on hand at the fulfillment house on time.

The *consumer* writes illegibly, or fails to include the required information or payment, or moves away from the address given.

> The *fulfillment house*, with a labor-intensive system, uses lower-level personnel, resulting in a high level of clerical and handling errors.
>
> The *U.S. postal system* takes about a week to deliver the consumer order to the fulfillment location and up to three weeks to deliver the promotional material offered to the consumer.

It is estimated nationally that at least 10 percent of all promotional orders are not fulfilled as a result of some combination of the above reasons. When one considers that a good promotional offer (free trial size, or products for unusually low prices, for example) can pull more than 250,000 responses in this industry, the number of potentially unhappy consumers becomes apparent.

It should also be noted that these complaints are frequently among the most emotional complaints a company receives. One company, for example, offered a doll in exchange for proofs of purchase and a small sum of money. The doll, if everything had gone right, would have arrived just before Christmas. Unfortunately the company underestimated the response volume, and a large number of the consumers were unhappy at not getting the dolls they had intended to give as gifts. The doll was not related in any way to the quality of the company's products, but did result in consumer anger directed against the company and its products.

The major concern to the manufacturer should thus be that a promotional fulfillment problem can cause a consumer reaction that is quite emotional, is directed at the manufacturer, and is often caused by an offer for a product that the company doesn't even make.

Companies employ several different methods for dealing with this type of complaint.

Several companies simply write the consumer and advise him or her that another kit will be sent by the fulfillment house. This, however, can be seen by the consumer as an attempt to shift the blame. Also, it runs the risk of improper handling at the fulfillment house again. Several companies therefore mail the replacement item themselves (thus reducing extra correspondence and handling risk).

Another promotional fulfillment problem that occurs frequently is where the consumer remits an incorrect sum, remits nothing, or forgets to sign a check.

Most companies that run mail-fulfilled promotions anticipate these problems and have standard, prepared letters that can be used to request correction of the consumer oversight.

At the same time, manufacturers should realize the problems that are apparently inherent and insurmountable in the mail-fulfilled promotional system. They might even want to consider the elimination of this type of promotion, understanding that far more negative attitudes may be formed than are really worth it.

Disappointing Product Performance

"Your advertisement in a magazine convinced me to buy your product however I was not satisfied with it. Can I please have my money back."

This problem occurs when a product fails to live up to a consumer's expectations of its performance. The problem may be with just the one application and the consumer will want to continue using the product, or it may be with the brand in general and they will want to discontinue the brand's use completely.

If the problem was with the individual application, most of the companies will replace the same product. Some companies even have a policy of double replacement; if there is a problem with an 8-ounce bottle, then 16 ounces are sent to the customer.

If, however, the problem is with the brand in general ("I didn't think your product worked for me"), replacement with the same brand is obviously inappropriate. In that case, replacement will frequently be made with alternative products of equal or greater value. Or, where a refund is specifically requested, a refund for full retail value can be given.

Companies usually prefer, wherever possible, to replace with the same or an alternate product, rather than to refund at full retail value. With a relatively low cost of goods for personal care products, it can be both more generous (responsive) and less expensive (more efficient) to overwhelm the consumer with products than to refund the retail price. In addition, it keeps the company's goods with the consumer.

Even with the wide variety of performance problems that can occur within a company, given the width and variety of a product line, most companies can standardize the basic response to be used for responding to performance problems. One type of response can say:

Thank you for your letter regarding (fill in). We appreciate your bringing this to our attention because (fill in). In order to correct this problem, we are sending, under separate cover (fill in).

Such an approach is almost ideal—it is responsive by being fair and personalized and appearing individually prepared, and it is efficient because it is in large part pre-prepared and is highly standardized.

Packaging Complaints

"I have confidence in all your products except I sort of have lost all that confidence when I bought some of your mascara. The brush broke inside the tube and I can't get it out."

Persons who write to manufacturers complaining about packaging complaints generally are regular users of the products who want redress for their problem through refund or replacement. They like the products—their problem is with the broken packaging.

Packaging complaints range from gold flakes that come off caps, to aerosol cans that clog, to compact hinges that break, etc. In this area, the attitudes that consumers express in their correspondence will be quite mild. Often consumers won't even request refund or replacement—they'll just want the company to know or be aware of the packaging problem. Repurchase rates will also be higher for consumers with packaging problems. They probably assume that they just hit one bad package out of a million and could easily replace it, at low personal cost.

As was true of the performance complaint area, a high percentage of packaging complaints can be handled within a standardized letter format and with generous product replacement.

Unusual Look/Smell

"The scent in my bottle of perfume has changed. It smells quite bad now . . . it has the odor of a skunk."

This problem category is not a large one in personal care companies—only 1 to 3 percent of all communications—and involves complaints that deal with perceived physical problems with the products.

When such a problem occurs, it is usually important for a company to know whether the perceived problem is true and then whether it is an isolated problem or is true for an entire batch of products. In order to answer these questions it is often necessary for the company to recall the product from the consumer. The most responsive and efficient system for doing this is as follows:

A replacement product(s) is placed in a special mailing envelope.
Also in the envelope is a letter explaining the need for the return, the replacement, and requesting that the product be returned to the lab *in the same envelope* for testing, together with a return label and *return postage.*

Other systems, which are not as responsive or as efficient, might involve a letter to the consumer which

Questions the problem
Requests the return of the product before any consideration can be given to any form of redress
Gives neither mailing container, label, or stamps

This second kind of response can easily be considered insulting by the recipient. As an example of the insensitivity of this system, consider the case of the consumer who cannot use a 16-ounce container of a $10 cleanser because the odor seems to have turned. How can she easily return a pint-sized container?

The idea here is that companies should be selective in what they recall from consumers and should also make it as easy as possible for a consumer to return a product. This means supplying the consumer with a pre-addressed, pre-stamped mailer.

Companies who insist on product returns prior to redress decisions hear again from fewer than 20 percent of their customers. On the other hand, companies that ask that the products be returned, but include the replacement with the request, together with an addressed stamped return mailer, receive more than 50 percent of the products.

The first type of response inadvertently irritates a majority of its consumer—and this irritation is evidenced by the subsequent low rate of response to the recall demand. The second type of response, however, makes the recall as easy as possible. At the same time, it gives redress to the consumer with no questions asked. The consumer then has received satisfaction even if he or she is unable to or chooses not to return the product.

Alleged Personal Injury

Complaints alleging skin irritations, rashes, or allergies as a result of product usage constitute only 1 to 3 percent of the communications received by companies in the personal care industry. This translates, for most companies, into fewer than 100 complaints per year, and an alleged injury incidence rate of 2 or 3 per million units sold.

Complaints that relate to alleged personal injuries are usually handled on an individual basis. Consumers are generally contacted— by phone or by letter—to determine the alleged cause and extent of the injury, the treatment for the injury, and the compensation thought appropriate. If a settlement cannot be reached at this level, the matter is immediately referred to legal or insurance advisers.

Again, it should be emphasized that such complaints are normally only 1 to 3 percent of the communications received by a company in this industry.

V. OPERATIONAL GENERALIZATIONS

There are a number of generalizations that can be made about the ways in which companies operate their consumer response units. These include statements about the degree of standardization of response that is appropriate, the role of automation, appropriate response time, the use of telephone responses, generosity of redress, methods of product recall, the need for record keeping and management reports, and consumer follow up. Each of these is discussed in terms of the NBCCA objectives of fair, prompt, generous, and consistent responses which are both consumer-responsive and cost-efficient.

A. Standardization

One factor which determines, to some degree, the amount of volume that a company can handle is the degree of standardization and/or automation that is employed. Standardization, as used here, is defined as the use of prepared paragraphs of letters to respond to a request for customer service, a compliment, or a complaint. It does *not* imply the use of mimeographed material or letters where the typing of the address is different in appearance from the body of the letter. It does mean that much of the correspondence—estimated by many companies at around 70 percent—is repetitive and thus a relevant response can be prepared in advance.

Standardization usually involves the preparation of a standard book of responses. This is a major help in achieving speed, consistency, and efficiency of response.

A key question here is whether such prepared material can be truly responsive to the consumer—whether it can be completely relevant to an individual request. This of course depends on the quality of the response that is prepared. In most companies that use prepared responses, it appears that much of the customer information area (discontinued products, some beauty advice, free products, career information, product or company information) can be standardized without any loss in relevance of response. Some beauty or use advice, however, will always need to be handled individually.

The handling of complimentary letters is generally standardized, and can, at the same time, be courteous and responsive.

The majority of complaints, except for personal injury or property damage, can also be handled with a standardized format. Even in the area of product performance of technical products such as hair color, the development of informational materials can reduce the number of individualized responses necessary.

It should be noted that many companies have these letters/paragraphs prepared or at least revised by professional writers. This is done in order to get as much warmth and personalization as possible into the tone of the letters.

B. Automation

Automation refers to the use of electronic equipment such as MTST, magnetic cards, memory typewriters, and other word processing equipment. A regular typist normally can type 250 lines a day—an inexperienced person on an automated system can produce 400 lines per day—while an experienced operator can type up to 1,000 lines. Thus the machines improve productivity. Also, they reduce mistakes in the letters, thus improving efficiency and appearance.

The rental or purchase cost varies, depending on the equipment used and the flexibility of operations needed. Basically, a standardized letter or paragraph is placed in the system and is then available for automatic typing. In most systems, the operator individually types in the name, address, and date, and then, based on instructions from the consumer correspondence area, directs the machine to type a combination of paragraphs.

The automated machines can be found within the consumer response unit itself, or within centralized word processing centers.

C. Speed of Handling

The time that a piece of correspondence stays inside a company may be a key element in the overall consumer evaluation of the response.

Studies of actual consumer correspondence files have shown that letters average 5 days in the mails between date of postmark and date of receipt at a manufacturer. Thus, letters are in the mail system at least 10 days (incoming and outgoing), and consumers are therefore waiting 1-1/2 weeks for an answer, even without counting the time spent in the company. Even if a company generates a response in 4 days, the consumer is still left for two weeks without a response. Attempts therefore must be made to handle all routine correspondence within 1 to 2 days.

D. Telephone Responses

Telephone responses to consumers can be used as a way of getting back to a consumer quickly—within a week of the initial letter being written—and as a way of handling complicated problems (i.e., correspondence that would have been handled individually) more responsively.

Some companies even use the telephone for problems relating to personal injury and property damage. While there is some risk in handling these problems by phone, there are also benefits. Telephone calls provide an excellent method of getting more complete information on how a problem occurred. Also, if an injury complaint can be handled and settled by telephone, this avoids the delays (months) and negative attitudes generated when insurance or legal functions take over the problem.

It should be noted, however, that only about half of the attempts to reach consumers by phone can be completed. Many consumers work or have unlisted phone numbers. Obviously the cost increases for each unsuccessful attempt. Therefore, any company that uses a telephone response system heavily usually limits the number of attempts made to reach the consumer. Also, it is obvious from this that a telephone response system, while faster and more responsive, must still be combined with an efficient traditional correspondence function.

As was mentioned earlier, there is some risk of inconsistency or mistakes and there is also a problem of control in an "instant oral" response system. For major telephone users, this problem of consistency is overcome by initial and follow-up training for respondents and is supplemented by careful record-keeping systems.

E. Generosity of Redress

Mention was made earlier of companies in this industry that use every piece of correspondence as an opportunity to send a sample of their products to a consumer—and of companies that have a double replacement policy. Both of these policies could be considered generous since they give the consumer extra value over the original purchase cost.

It is recommended that a generous replacement program or sampling program be considered wherever possible.

Generous programs are recommended (and are preferred over refunds) because more value is given to the consumer with this procedure, at less cost to the company. Refunds have to be made for the *retail* cost of products—generally twice the factory sales value for this industry. Replacements, on the other hand, are done at the cost of goods plus mailing and handling. Thus, a company can even give the consumer twice the retail value of the problem product, usually still at less cost than would be true for a refund.

A generous replacement program should be considered for problems relating to packaging or product appearance, and even in some cases of product performance. In addition, all correspondence should be considered as an opportunity to cross-sample other products made by the company.

There is some fear that a generous replacement or sampling policy will lead to abuses of the system. There is no doubt that a close watch needs to be kept to prevent abuses, but abuses are usually obvious when they do occur, and abuses occur only rarely. Most companies can recall only one or two instances each year where obvious fraud or cheating has occurred.

F. Product Recall

As was mentioned earlier, some companies request that products be returned for a lab analysis if there is a problem with physical appearance, skin irritation, product performance, or packaging. Most of the products, however, are never returned, and frequently the lab analysis is felt to add little if anything to the response when a problem is already known, (aerosols clog, deodorant sticks evaporate, compact hinges break). In addition, the recall procedure usually adds little to the value of the consumer response or company quality control knowledge.

Recalls therefore should not be made unless a situation is new, or growing, or is otherwise unique and unusual. And, if an analysis would indeed be of value, the return of the product should be made as easy as possible for the consumer. In particular, a mailing container and a preaddressed prepaid label should be provided, together with a reasonable explanation for calling the product back.

G. Records

Records are necessary in order to retain a history of how correspondence has been handled and in order to provide information for summary reports to executives. Thus, it is necessary for a manufacturer to have file records of:

> The total experience (date, consumer name, product, problem, response, lab reports, etc.)
> Access to information by type of problem, consumer name, and name of product
> If injury, whether irritation, more serious personal injury, infection, or other
> Whether medical treatment necessary

A wide variety of filling systems are currently in use—most with serious deficiencies. For example, one company files correspondence alphabetically by consumer name—they cannot retrieve all the letters for a specific product. Another company files correspondence by product name—they cannot retrieve information by type of problem, such as allergy or packaging defects.

Some companies, however, have a system which keeps one basic file with all information and also keeps subfiles with information separated and retrievable by consumer name, product name, or type of problem.

H. Communications Reports

Summary reports vary widely in form, length, and frequency. Generally, monthly or quarterly reports should be prepared and directed to any or all of the following areas:

> General management
> All marketing levels
> Medical and legal or insurance
> Production
> Research and development

In general, managers in these areas should review the summaries by looking for trends, exceptions, or particular problems. Editorial comments or highlight comments are helpful in assisting these managers to understand consumer communications needs.

The reports should, on one page, emphasize overall trends for the period versus a year ago, and year-to-date versus a year ago.

I. Consumer Follow-up

Few companies ever do any follow-up research to see whether responses are well received by the consumers. Such information could
be helpful in determining ongoing consumer attitudes and also the
adequacy of the responses that are used. A simple system (such as
a postcard questionnaire, automatically sent one month later) could
give some quidance to the customer relations staff. It could also be
impressive to the consumers, who would be receiving extra attention
for their problems.

VI. ORGANIZATIONAL ISSUES

There are a number of organizational factors that can affect the
efficiency and responsiveness of a consumer response unit. These
include organizational location, duties and staffing of the consumer
response unit, the input relationship with marketing, and the degree
of managerial involvement in the unit's operations.

A. Location

Consumer response units in the personal care industry have, because
of the increasing volume of requests in recent years, become more
formally structured in most companies and have been enlarged in
terms of personnel.

This is reflected in the findlings of the Conference Board, which
found that fully 75 percent of the consumer affairs departments that
they had identified in 1973 had been formed in the last five years[3].

This is not to say, however, that the information service and
complain handling functions were not performed beofre. They were
simply performed by fewer people in different units of the organization.

In some companies, product managers or their secretaries formerly
answered consumer correspondence. In other companies, the secretary to the medical director handled complaints. In yet other companies, correspondence was handled by the sales department. There
were, of course, problems with all of these approaches. The medical
department might have been too clinical, the sales department too
busy with their primary function of selling, and the product managers
too busy with their main responsibilities of product management.

Howard and Julbert[4], for example, have indicated that brand
management is ineffective in handling complaints since their incentives
and evaluations fail to incorporate rewards for handling complaints
or even (since they cannot control production) preventing them.

With the increasing pressures of consumerism, however, almost
all manufacturers of personal care products are consolidating, reviewing, formalizing, and upgrading their consumer response units.

The concentration of operations in one department has been important in encouraging the development of professional, fast, and consistent responses to the consumer.

The Conference Board report found in 1973 that a majority (62 percent) of consumer affairs units reported directly to general management. This was considered to be the result of two factors—the newness of the units and the recent emergence of the importance of consumer affairs. As the units have matured, however, and as the newness of consumerism has faded, these units have usually been reorganized to fall under other staff or line departments. Some report to vice-presidents who report directly to top corporate levels, some report to a vice-president of public relations, and in other companies the units are part of broader marketing services departments.

B. Duties

Just as the organizational location of the response units are varied, so too are the duties of the units.

The Conference Board report indicated that all the consumer affairs units they had studied were directly concerned with one or more of the following:

Handling, resolution and analysis of customer complaints and inquiries

Developing and disseminating to consumers better information on the purchase and use of products or services sold by the company

Serving as an internal consumer "ombudsman" and consultant on consumer matters within the company

Providing liaison with consumer-interest organizations outside the company.

In some companies the consumer affairs departments even screen advertising in order to avoid possible conflicts with the consumer programs.

C. Staffing

A remarkably large volume of communications can usually be handled by very few people. From one to four full-time people can usually handle up to 10,000 pieces of communications annually. This would of course vary with the degree of standardization and automation systems in the company.

D. Professionalism

The degree of professionalism in the response units seems to be key to the efficiency of the units. This professionalism can be generated from experience in the area, from experience in the company, and also from outside sources such as the Society of Consumer Affairs Professionals (SOCAP) and the industry's trade association (CTFA). These organizations are increasingly active in the consumer response area.

E. Marketing Input

It is always important to review the response unit's relationship with line marketing, and to evaluate the degree to which the response unit is active or reactive within an organization.

Most consumer response units are reactive. They respond to correspondence or they handle traditional consumer relations programs. Some companies, however, are beginning to consider the possibility of incorporating assertive consumer responsiveness directly into their marketing strategy. As a Johnson & Johnson executive stated:

> "Our entire approach has changed over the last two years. At first, we wanted to get organized just to get the mail handled. Now we look upon our response function as providing an opportunity to be responsive to an individual consumer. We're questioning now whether consumer response can or should become a more integral part of our marketing strategy."

There should be no doubt that a consumer communications unit can provide key input to marketing strategy decisions. This input can be particularly valuable in the decisions regarding what products to market, at what prices, and with what promotional policies and executions. The response function can also provide a new idea resource for line marketing.

In addition, a manufacturer might want to stimulate additional consumer communications—either to gain a more complete picture of consumer needs, or to convey an additional openness and responsiveness to the consumer.

There is, however, a good deal of uncertainty about the resulting possible mixes of customer service and complaints. Such stimulation might, for example, result in:

A higher total level of correspondence, roughly divided by type as they are now, assuming that the current mix by type actually reflects consumer needs

A higher level of correspondence, with a heavier mix of informational and advice requests, assuming it is probably easier to stimulate the consumers who already truse the company

A higher level of complaints, assuming that increased stimulation would bring out the "submerged" portion of complaints.

Little or no increase in level, assuming that consumer inertia is to strong or overcome

Clearly, there is no "correct" answer to this dilemma, and no one can predict what the results of such consumer stimulation would be.

The uncertainties involved, and the cost involved in responding to increased levels of communication, might well make a manufacturer reluctant to stimulate increased levels of incoming communications. The point is that each manufacturer should still consider the dilemma, gauge the costs involved, and perhaps even test methods of consumer stimulation.

One alternative to consider in this area would be the institution of a toll-free number for consumers to use if they have questions, suggestions, or complaints.

Several companies in the personal care industry have instituted such a service and, to date, have found the results to be gratifying.

In one company, 20 percent of their letters involved complaints, but only 2 percent of the telephone calls were complaints. Thus there is some indication that consumers will respond to the ease of telephone service by expressing information and service needs, not their redress needs.

One other reason for this toll-free service, and a reason that other manufacturers should consider, is that the company saw a marketing advantage in stimulating consumer communications and in positioning themselves as a friend of the consumer.

VII. SUMMARY

In summary, well-designed consumer response programs can enable a manufacturer to deal efficiently and responsively to increasing volumes of consumer information requests and redress needs.

This chapter has identified the most important consumer information and redress needs for the personal care industry and has recommended appropriate corporate responses for each type of consumer need. In addition, operational suggestions have been made regarding ways in which standardization, automation, telephone responses, recall methods, increased speed of handling, generosity of redress, record keeping and summary reports, and consumer follow-up might contribute to more efficient and more responsive consumer response programs.

It is hoped that these observations and suggestions will be of use both to companies who already have consumer response programs or who wish to initiate such programs.

VIII. REFERENCES

1. David A. Aaker, and George S. Day, eds., *Consumerism: Search for the Consumer Interest*, The Free Press, Macmillan, New York 1971.
2. U.S. Department of Commerce, National Business Council for Consumer Affairs, *Responsive Approaches to Complaints and Remedies*, U.S. Government Printing Office, Washington, D.C., October 1972.

3. Conference Board, *The Consumer Affairs Department: Organization and Functions*, The Conference Board, New York, 1973.

4. John A. Howard, and James Hulbert, *Advertising and the Public Interest*, A Staff Report to the FTC, February, 1973.

32

SPECIAL PROBLEMS OF SMALL COMPANIES

LINDA R. MARSHALL

Elysee Scientific Cosmetics Inc., Madison, Wisconsin

I. INTRODUCTION

A few large corporations alone have not made the American business enterprise system the envy of the rest of the world. Thousands of individual independent entrepreneurs deserve the real credit for this remarkable achievement by building success for themselves—and tho nation—through determination and creativity.

The cosmetic industry has been built by entrepreneurs. Two-thirds of the companies in the cosmetic industry have annual sales under $500,000. Fifty percent have sales under $200,000. These include small cosmetic companies that manufacture their own products as well as those that distribute private-label products or work with a contract packager.

The entrepreneurs of yesterday were not controlled by the stringest and costly regulatory requirements with which American business must deal today. Will Rogers said "Thank God we don't get all the government we pay for." Today we get more government than we pay for, but it is clearly not a bargain for small businesses.

The costs of federal and state regulation of business in this country have taken a heavy toll on small business enterprises. In many heavily regulated businesses, small companies no longer exist. In others, regulatory requirements severely limit the areas of effective competition. Small over-the-counter drug companies, for example, must use private labelers because of the high cost of the good manufacturing practices required by the Food and Drug Administration (FDA) for drug companies.

Cosmetic products, however, are relatively harmless and have an enviable safety record. The cosmetic industry, through its trade association—The Cosmetic, Toiletry and Fragrance Association (CTFA)—has instituted a number of very successful programs of industry self-regulation. As a result, the cosmetic industry remains one area where government regulation has remained at a reasonably low level and thus where small business can still thrive.

II. WHAT IS REQUIRED TO GO INTO THE COSMETIC BUSINESS

The first decision for anyone who is considering entry into the cosmetic industry is whether to manufacture your own products or work with a private label house or contract packager. In the beginning, working with a private label cosmetic company is easiest and least expensive. The costs of personnel, equipment, research and testing, and packaging can all be reduced or eliminated. Your funds can be spent more effectively on labels for products, brochures, and marketing the line. There are a few large private label companies and numerious small private label companies. It is important to work with one in which you have confidence. Visit their plant, if possible, to learn about their facilities. Ask to see the product safety data. Make certain the company has product liability coverage. Check if the company is a member of the CTFA and complies with the FDA voluntary cosmetic establishment, registration, product listing, and product experience reporting programs. Involvement in these areas is not mandatory for a company to be considered legitimate, but serves as a very favorable indication of corporate knowledge and responsibility. If you have doubts, request FDA for all establishment inspection reports for the company under the Freedom of Information Act. Check other firms, given as references, for their experience.

A private label manufacturer should be able to advise you on all legal requirements for your labels. Ultimately, however, the legal responsibility for compliance with FDA and other requirements falls on you. The CTFA has prepared a labeling manual, available to both members and nonmembers, which is invaluable in helping resolve every labeling problem that may be encountered.

A private label manufacturer generally provides an already formulated product and suggests relatively standard labeling for it. If you are new in the cosmetic business, this is an enormous help. You can then concentrate primarily on building your business, without concern about the product itself.

A contract packager, in contrast, generally formulates the product according to your formulation and applies the label that you have designed. This demands greater knowledge and sophistication on your part, and may be more feasible after you have greater experience in the industry.

III. MANUFACTURING YOUR OWN COSMETICS

You may have a creative mind and know the type of products you want to develop and market. You want something different from the products presently on the market. What can you do?

You may wish to manufacture your own products. This is a more difficult and expensive approach and is likely to become increasingly more difficult in years to come because of increasing governmental regulation. When Calvin Coolidge declared that "The business of America is business," there was not much government regulation. Profit incentives were high, taxes modest, and regulation was minimal.

Before you decide to manufacture your own cosmetics, do extensive research on the technological and economic problems that will be encountered. You must explore fully the requirements for manufacturing and filling equipment, raw materials, packaging, product research, safety testing, overhead, marketing, and advertising. Perhaps you will need outside professional help to analyze all of these matters completely and objectively.

Inventory and cost control are major problems. The cosmetic business is a "glamor" industry, and packaging must be esthetically appealing. This presents a major problem because the small cosmetic manufacturer purchases raw materials and packaging in relatively small quantities. The end result is a larger cost per product unit, once again placing the small firm at a profit disadvantage.

For example, a small cosmetic company could never affort to have its own mold for a bottle or jar. Indeed, even minimum purchase requirements for stock packaging are too high for a small cosmetic company. A 10,000-unit minimum is not uncommon.

A small company just starting its own manufacturing must rely on standard stock packaging and labels applied by hand, unless the new

entrepreneur is well capitalized. Silk screening or hot stamping
methods, for example, create an impressive package, but a small company's volume may simply be two low to justify the cost. It is difficult
to create brand distinction under these circumstances, and even more
difficult to create elegant packaging.

IV. RESEARCH AND QUALITY CONTROL

In developing a line of cosmetics for a small company, it is imperative
to find suppliers who will work with you. They have the research
facilities and knowledge in areas where a small company cannot economically compete. Supply houses have basic formulas that are excellent
as "starting" formulas. They also have the necessary safety data on
raw materials. The more safety data that is compiled on each ingredient used, the less testing you will have to do on the final product.
As excellent source for safety data on individual cosmetic ingredients
is the Cosmetic Ingredient Review, sponsored by the CTFA. Another
major source is the series of monographs on over-the-counter (OTC)
drug ingredients, published by the FDA in the *Federal Register*. A
specification sheet and sample of each raw material will become your
standard.

A small cosmetic company often encounters serious problems in
dealing with large suppliers. Because of the poor state of the economy
and the high cost of labor, it is simply too costly for many suppliers
to work with small companies. The small company does not purchase
enough volume for suppliers to recoup costs, and they therefore prefer
to concentrate their efforts on large companies and higher profits. In
the past few years, small companies often have had to pay not only
higher costs for raw materials but also a repackaging fee for small
quantities.

Once the product is developed, an objective look at the product
must be made to determine how it is intended to be used. Safety data
must now be accumulated. A decision must be made regaring any ad-
titional safety testing to be done. Promotional claims must be analyzed.
Can claims be substantiated? Will they position the product as a "drug"
instead of a cosmetic? If a product is considered a drug because of the
claims made for it, the FDA has considerably greater control over it.
Thus, small cosmetics companies will want to avoid all drug claims except for products that clearly fall within OTC drug monographs
promulgated by the FDA.

V. QUALITY CONTROL IN MANUFACTURING

In developing quality control procedures, the *CTFA Technical Guidelines* are an excellent reference in outlining the procedures to follow.
A sound quality control program requires accurate records and

appropriate product testing. Representative microbiological testing is not expensive, and assures that products are free of contamination. Most small companies rely on an outside testing laboratory for whatever quality control testing is to be undertaken.

The CTFA has an excellent slide presentation on principles of microbiology relevant to the cosmetic industry, which can be used in training plant personnel. The CTFA also has a slide presentation on "Plant Housekeeping," which is very helpful in demonstrating principles of quality control.

Quality assurance is vital to the success and well-being of any cosmetic firm. It must be a major commitment from the president to the fillers and shippers. Small size is no excuse for poor quality.

VI. MARKETING BY THE SMALL COSMETIC COMPANY

Small cosmetic firms often encounter difficulty in determining the right market for their products. Department stores make costly demands. In addition to requiring a 40 to 50 percent discount, many department stores also require salary support for their salespersons and "push money," or p.m. as it is called in the trade—to support promotion of a name brand by a sales clerk. Department stores also expect manufacturers to provide cooperative advertising funds to help finance retail advertising.

Stores also routinely take between 90 and 120 days to pay their suppliers. Because cash flow is a major problem for the small cosmetic firm, such a lengthy delay in payment can be a serious problem. As the economy tightens, department stores rely more heavily on larger, well-known lines, resulting in the smaller lines being reduced or even losing their place in the store.

Small companies always operate at a disadvantage in department stores. They are allowed less selling space than the larger companies, which are in a stronger position to make demands for marketing prominence. The products and packaging of a small firm are always placed side by side with elegant, unique, and expensively packaged products, frequently from private molds. A larger cosmetic firm's advertising budget alone is many times higher than a small cosmetic firm's total sales volume. Thus, a department store will always receive more advertising dollars from the larger companies. The problem of persuading someone to introduce your line into the department stores or drug chains can be difficult. Most independent cosmetic industry representatives prefer to work with larger companies because their products are easier to sell. In the case of some small cosmetic firms, the entrepreneur has to handle the selling along with the responsibilities of running the company.

One often-unanticipated problem department stores create for small cosmetic firms is in the handling of "returns." Unreasonable

demands for credit on returned merchandise can quickly lead to the
demise of a small cosmetic firm.

This combination of forces can raise formidable challenges for a
small cosmetic firm with limited cash flow. Many small cosmetic companies have simply concluded not to join the fight to achieve the
prestige of major department store sales. The sales are there, but
the profits are not.

While this appears to be a very difficult marketing area for a small
cosmetic firm, it is not impossible. We all know of companies that have
succeeded. In earlier days, however, it was easier to succeed in department stores. Today, success for a new cosmetic firm in the department store field is the exception rather than the rule. Success
in this area is based on adequate capitalization; a creative, persistent,
and patient entrepreneur; and a great deal of luck thrown in for good
measure.

Small cosmetic firms must recognize that department store buyers
have their own problems, which lead them to emphasize well-established
national brands. These buyers must face the constant pressure of
sales quotas. They are forced to give the greatest prominence to lines
producing the greatest return on investment. The bottom line for
everyone must be profit.

As a result, the small cosmetic firm has a better chance to find an
adequate market by specializing in mail orders, direct sales, and
salons and boutiques. The investment is considerably less, and the
side-by-side competition is minimal. Marketing in these areas can also
be conducted on a cash basis.

Before entering any market area, some marketing research is
necessary. There are numerous booklets and articles written every
year on this subject. Know where you want to position your products
in the market. Know the customer you are targeting your line to
reach. What are the targeted consumer's needs? What is the targeted
consumer's income and lifestyle? How is your line designed to satisfy
the targeted consumer's needs, income, and lifestyle?

A small company has an advantage over a large company in this
area because it can "turn around" more quickly and develop truly
unique and exciting products on an overnight basis. A small company
is not hampered by a large bureaucracy of different groups working
with numerous products to gain final approval of top management.
Thus, a dynamic small company can carve out for itself a significant
and profitable market, if it studies the possible areas and targets the
right approach.

VII. MANAGEMENT PROBLEMS OF THE SMALL COSMETIC FIRM

A small cosmetic manufacturer or distributor of private label cosmetics
has other problem in addition to marketing. Typically, the company's
chief executive officer does everything. There are no executive vice

presidents and no vice presidents of finance or marketing. There is
no inside legal counsel and, in most cases, a small cosmetic manufacturer cannot afford an inside accountant. In face, the chief executive
of a small cosmetics firm usually survives by a combination of shrewd
instinct and practical experience.

Loose organization can, and often does, create a number of typical
management problems, basically centering around dealing with people
and developing resources for growth. Because the chief executive is
so intimately involved in operations and day-to-day survival, it is
sometimes difficult to evaluate the company objectively. A lack of
business experience—a common problem—may further aggravate the
company's problems.

Cash flow, for example, is a headache because it becomes an
emotional issue as well as a financial one for the small cosmetic entrepreneur. The pressure to meet a payroll can be tremendous. The
same chief executive who is hard-boiled when negotiating with suppliers
and resourceful when dealing with customers can be unbelievably
sentimental about employees.

Banks can also be troublesome. Contrary to popular perception,
they do not always view a balance sheet objectively. They often fail
to take into consideration the difference between a small and a large
firm. Profit and loss statements of owner-managed companies have a
different look from those of publicly held companies whose shareholders
are rarely involved in corporation operation.

Small cosmetic firms are also labor-intensive and tend to use production techniques that are less capital-extensive. Even if the small
firm uses its capital and its labor more efficiently and profitably, it
may still appear less productive than the capital-intensive corporation.

For example, small companies traditionally increase productivity
by improving employee attitudes, creating enthusiasm, and encouraging
a cohesive team spirit through good communications among employees.
Productivity also increases because small businesses offer new un-
skilled employees on-the-job training. Not only is the social and
economic value of this training overlooked by standard measures of
productivity, but the raw numbers continue to belie true productivity
in small firms.

As long as Congress and the government bureaucracy define
productivity as they do, small businesses will continue to have "tough
sledding" even if legislation is passed to help businesses with tax cuts,
accelerated depreciation, and similar financial assistance. Small
businesses cannot prosper when public policies do not reflect the
needs of small companies and thus are directed toward big business.
Government leaders must ignore the myths that "bigness" is inevitable
and that independent entrepreneurs are a "dying breed." If public
policies destroy the entrepreneural spirit of America, America will
cease being the great country it has become.

VIII. FEDERAL AND STATE REGULATIONS

Few small businesses can afford staff counsel. They hire attorneys who cannot be expected to know a client's operation as well as the company's president. The major cosmetic firms have inside legal staffs to handle everything from labeling to advertising, and the legal aspects of the FDA, the FTC, and other federal regulatory agencies. The outside attorney hired by the entrepreneur of a small cosmetic firm is probably not an expert in the legal areas that apply specifically to cosmetics. Therefore, the advice given may be general or require extensive research to supply the answer.

When a problem develops for a small cosmetic firm that relates to the FDA or the FTC, it would be advisable to hire an attorney experienced in such matters. The fee will probably be less than a corporate attorney in the long run, because the information is readily available and, more important, his or her information and judgment will be more accurate and reliable.

It would be advisable for the beginning cosmetic entrepreneur to have its attorney review all the labels and written product literature prior to launching the line. While it may cost more initially, it could save a great deal of time and effort by helping the small company avoid costly mistakes. The names of experts in FDA regulations are available from the CTFA.

Small companies manufacturing "treatment" products must conduct additional tests to substantiate product claims. In addition, claims must be carefully written to avoid drug claims that require a product to meet more stringent drug regulations. Making the wrong claims, whether because of a misunderstanding of the regulations or a desire to "out-claim" a competitor, can be a costly mistake for a small firm.

An FDA plant inspection becomes a fact of life for every cosmetic company at some time. Every company should operate in such a manner that it is prepared for an inspection. It is too late to prepare when the FDA inspector is at the door.

Every company should develop its own policies on handling an FDA inspector prior to an inspection. The CTFA has standard information and procedures to guide you. Following an inspection, you may never hear about the inspection again. If the FDA should conclude that it has uncovered a violation of the law or an FDA regulation, however, a "Notice of Adverse Findings" or a "Regulatory Letter" demanding corrections of the violation may arrive by mail. The FDA may also recommend to the Department of Justice that it institute a civil seizure action, an injunction, or a criminal prosecution, but this happens only in unusual situations. If any of these situations should arise, an attorney experienced in FDA matters should be consulted immediately. The legal staff at the CTFA can offer advice, which is very valuable, but cannot represent you.

Recently small business has made a regulatory gain. If the FDA or any other federal government agency takes a company to court and loses, the government agency is required to pay the company's legal fees. Prior to this, a small company could be forced out of business trying to pay legal fees even if the company ultimately prevailed in the courts. This ruling has caused government agencies to think twice about their position before taking a company to court.

The FDA's authority over cosmetics has increased throughout the years. The FDA first acquired the responsibility for the regulation of cosmetics with the enactment of the federal Food, Drug, and Cosmetic Act in 1938. Before then, cosmetics were not controlled by federal law. Since the 1938 act, there has been only one amendment affecting cosmetics. The Color Additive Amendments of 1960 were enacted to regulate all colorants used in cosmetics.

While the 1938 act has been amended only once, the FDA has gained considerably more authority over cosmetics through the years by promulgating implementing regulations. For example, cosmetic ingredient labeling began to be required by the FDA in the 1970s through regulations.

We are fortunate that cosmetics are not required to be approved by the FDA prior to marketing. Premarket approval would present a serious burden to the small cosmetic company. For example, the Drug Amendments of 1962 greatly increased the documentation needed to gain FDA approval of a "new drug." By 1976 it cost an average of $54 million to obtain FDA approval of a single new drug[1]. As a result, no small innovative drug company survived. Should the FDA gain this type of authority over cosmetics, I doubt that any small company would survive.

Premarket approval for cosmetics was discussed as early as 1948. Mr. Larrick, then associate commissioner and later commissioner of the FDA, felt strongly that more information was needed by the FDA regarding ingredients and safety substantiation of cosmetics. It appeared at this point that the cosmetic law might go in the same direction as the drug law. The CTFA prevailed, however, and has continued to prevail against premarket approval and other stringent statutory provisions that would have an effect similar to the new drug provisions of the law. Industry has stressed that the provisions of the 1938 act are adequate and that no additional legislation is necessary to protect the consumer.

In the *Federal Register* Of July 25, 1974, the FDA published the final regulations, and ingredient labeling became a "fact of life" for the cosmetic industry. Ingredient labeling supposedly helped in comparison shopping by the consumer and would also alert those consumers who have allergic reactions to certain ingredients. It has also proved to be one of the steps that has helped stay further federal legislation.

Unfortunately, it has helped "copycat" manufacturers imitate the formulas of successful products.

Product safety substantiation also became a reality for cosmetics during the 1970s. In 1975, the FDA required that each ingredient used in a cosmetic product and each finished product be adequately substantiated or bear the following warning: "The safety of this product has not been determined."[2].

To ward off further government regulation, the CTFA has initiated a number of voluntary self-regulation programs for the industry. Although cosmetic products have an enviable safety record, the CTFA petitioned the FDA to adopt a major voluntary program in 1972. This was a three-part program involving registration of cosmetic plants, submission of product formulas, and product experience reporting. While industry participation is not 100 percent, it has been very effective. The concept is excellent because it serves to confirm the cosmetic industry's credibility and concern for product safety. By voluntarily giving the FDA important product data, the industry has avoided regulatory requirements that other industries face.

In 1977, the CTFA petitioned the FDA to issue regulations governing cosmetic good manufacturing practices (GMP). Keeping in mind the needs of small companies, the CTFA petition avoided the more costly procedures mandated by the drug GMP regulations. To date, the FDA has taken no action on the CTFA petition, but has indicated general agreement with the less stringent approach.

The cost of compliance to industry overall is high, but it is proportionately even higher for small companies. Larger firms can afford professional staffs to handle compliance and can spread compliance overhead over a larger production volume. In the small cosmetic firm the responsibility for compliance falls on the chief executive.

Because of this competitive disadvantage in trying to comply with all regulatory standards, both mandatory and voluntary, the owner-manager may decide to live with the tension of partial compliance rather than pay the large costs of the voluntary program. One must weigh the cost of complete compliance against the need to demonstrate industry's ability to engage in useful self-regulation.

The Occupational Safety and Health Administration (OSHA) also creates problems for the small cosmetic firm. Its requirements are onerous, and they have strong enforcement authority. A fine of $1,000, or a $10,000 investment forced on a small company to meet OSHA requirements, can easily wipe out the profitability of a small firm while having little impact on a larger business.

State and local regulations are on the increase. These can become a nightmare for the small company since some are in conflict with F.D.A. federal requirements. The CTFA's legislative bulletins alert you to new federal and state legislation under consideration.

IX. WHERE SMALL COSMETIC FIRMS CAN SEEK HELP

There are numerous qualified cosmetic consultants who can prove valuable to the small company in helping it establish quality control procedures, develop new products, assist with labeling and packaging, and give advice on numerous other projects. It is also important to have both an attorney specializing in FDA matters and a regular corporate attorney. While the services of such expert consultants are definitely needed, the costs are often high. For the small company just getting started with limited cash flow, owner-managers should do as much as possible on their own and then get approval and advice from the experts. You should also develop a good working relationship with your suppliers, who often can supply the research answers needed at no additional cost.

I have found that one of the best resources available to the small cosmetic entrepreneur is the CTFA. The professional staff provides members—both large and small—with a comprehensive range of regulatory, legislative, and scientific services. For example, the first edition of the CTFA *Cosmetic Ingredient Dictionary* was published in 1973. Over the years, this dictionary has continued to grow, culminating in a recent publication of an extremely comprehensive third edition. The FDA has recognized the *Dictionary* as a source of uniform nomenclature to be used in cosmetic ingredient labeling.

The CTFA's color additive testing program is another effort to assure industry credibility and product safety. Through this program, the CTFA sponsors safety tests that the FDA requires before the agency can permanently list the color additives used in cosmetic products. The program costs millions of dollars that neither large nor small cosmetic companies could afford alone. Without this program, the FDA would be unable to approve any color as safe for use in cosmetics.

The CTFA has always strongly supported research on the safety of cosmetics and has established many programs to help industry meet its obligations to the consumer. It has formed scientific committees on microbiology, pharmacology and toxicology, quality assurance, hair coloring, ingredient nomenclature, and a wide variety of other subjects. The association regularly develops and disseminates information for raw material specifications, testing methods and ingredient descriptions, and technical guidelines to help ensure the quality and safety of the finished products. Through membership in the CTFA, small cosmetic firms have access to the assistance of all segments of the industry and information they could never afford to obtain on their own. All the CTFA's efforts have contributed to an increasingly positive impression of the cosmetic industry and have helped reduce the cost of safety substantiation for the small cosmetic firm.

The CTFA has developed a number of programs to assist the small entrepreneur over the regulatory stumbling blocks. It is up to each company, however, to use this association to its advantage when complying with federal requirements. It costs a little financially and takes a major commitment, but the return on the investment is high—a successful enterprise.

United by the CTFA, large cosmetic companies, as well as small, are determined to provide consumers with high-quality, safe cosmetics. To deviate from this pursuit would not only affect the individual company's profits, but would harm the reputation of the entire cosmetic industry. High-quality products take commitment from the top chief executive officer to the employee in the mailroom.

Whatever the problems of being an independent entrepreneur, the small cosmetic manufacturer will survive. He lives by the basic tenet expressed by Alexander Hamilton: "Here, Sirs, the people govern." Indeed, that is what a political democracy is all about. Small businesspeople, by nature, like to go their own way.

Although it may be fairly easy to enter the cosmetic industry, once you are in the competition is fierce. The chief executive officer must understand all the idiosyncrasies of the industry in order to succeed. He or she must learn to be selective, yet daring, when developing marketing programs, narrow when selecting the target audience, and discerning when choosing the best advertising media.

REFERENCES

1. Hansen, The Pharmaceutical Development Process: Estimate of Development Costs and Times and the Effect of Proposed Regulatory Changes, *Issues in Pharmaceutical Economics* (Chien, ed.), 1979, p. 151.
2. *Fed. Reg. 40*:8912 (March 2, 1975), 21 Code of Federal Regulations §740.10.

33

THE PRIVATE LABEL COMPANY—SPECIAL CONSIDERATIONS

JEROME K. MALBIN

Kolmar Laboratories, Inc., Port Jervis, New York

Private label manufacturer—contract manufacturer—co-packer. These are but a few of the identifiers that have been used, sometimes with a clear idea of what is intended, sometimes interchangeably, often with only an indefinite notion of what is or should be intended but always with the understanding that a function, vital to a product, is to be performed in an establishment outside of the organization and direct control and management of the marketer or distributor.

I. PRIVATE LABEL DEFINITION

It is appropriate to start with the following definition, which will
establish the common ground for our discussion and the phrase,
private label, or its equivalent: Private label manufacture is the
compounding, manufacturing, or filling of products under quality
controlled conditions to formulas or procedures or specifications provided or developed by the supplier or the customer, or both, utilizing
ingredients or bulk product or packaging components provided by the
supplier or the customer, or both, which products are to be marketed
by the customer under brand names owned or controlled by the
customer.

Note that the definition provides a considerable amount of versatility in all areas but two. The products are to be marketed and the
brand name is owned or controlled by the customer. The concept of
private label in the cosmetics and toiletries industry almost universally
recognizes that the private label manufacturer does not compete in the
consumer marketplace with his customers, the marketers or distributors, by selling the same products under his own brand name. He
may serve a variety of customers who compete with each other, but he
does not himself compete with them in the consumer marketplace.

Indeed, the fact that the private label manufacturer serves a
variety of competing customers may very well be the first step in assuring them of the quality they will be expecting and receiving. The
private label manufacturer may, perforce, be in a position to produce
to a standard that exceeds the industry mean in order to serve those
whose internal standards are higher than that mean. Thus, in the
foregoing definition, production has been identified as being conducted
under "quality controlled conditions." These standards embrace precision in manufacturing, care and awareness of plant and personnel
sanitation, an intimate knowledge of the properties and functions of
ingredients and raw materials that comprise the arsenal from which
formulas will be developed, a sense of esthetics which is a vital element
of product presentation in the industry, and a knowledge of the maze
of regulatory requirements pertaining to the products as well as their
manufacture and assembly.

A marketer may have a product formula that he has been marketing
for some time. He may require production beyond his in-house capacity. He may desire to transfer production out of his own establishment
for a variety of reasons. He may have in his possession a completely
new formula for a product that has not been tried in the marketplace.
The formula may have been developed within his own organization or
by an outside consultant, or it may simply have been obtained from a
source of standard formulas which are available from raw material
suppliers, fragrance houses, or trade publications. Indeed, the
marketer may have no formula at all but merely a concept for a product
or a line of products.

Each of these situations represents a case for considering the private label manufacturer. Each involves a joint venture of sorts between the marketer and the private label manufacturer, with each contributing functions and each assuming duties and responsibilities.

II. CONFIDENTIALITY

One cannot think of formulas that are in the possession of a customer without considering the issue of confidentiality. At first blush, the marketer may believe that the formula in his possession is unique and he will desire to discourage the world from duplicating it. He may desire the private label manufacturer to ally with him so that potential competitors will not be able to use the same private label house as a source of supply. Obviously, this is not realistic. However, the marketer may reasonably expect and demand a measure of confidentiality from the private label manufacturer, who must render that confidentiality without it being demanded, regardless of whether the point has been reduced to written contract form.

The nature and extent of the obligations of confidentiality between a private label manufacturer and the marketer merit some thought and discussion.

Current federal regulations require the declaration of ingredients on the labels of cosmetic products[1]. The devotion of time and space to any discussion concerning the confidentiality of the qualitative ingredients lists is beyond our present scope. However, absent patent protection for the quantitative formulas, once a cosmetic is exposed to the marketplace, it is not only in the public domain but is available for imitation and duplication. The mere fact that a product is in the public domain should not mean that it is not worthy of a measure of confidentiality in the relationship between the private label manufacturer and his customer. The public domain comprises a vast field of knowledge and information, of which some is widely circulated, known, and available to most, and of which other, though published someplace, is obscure and known to only a limited few. Though everything about a particular product may be technically in the public domain, someplace, its precise composition may be a matter of great interest and curiosity to competitors and this should be entitled to protection by contract between the private label manufacturer and the marketer who brought the formula to him. The private label manufacturer should not, via a license of exemption from confidentiality for matters in the public domain, be allowed to refer to the very piece of information given to him by his customer as a source on which to develop or produce products for his other customers.

A. Contractual Considerations

A suggested contractual arrangement to protect adequately the interests of both parties follows. Such an arrangement should include recognition that the information furnished by the customer is confidential in nature and will be used for the limited purpose of manufacture of the products for that customer. The arrangement should further provide that the contract manufacturer will not exploit his customer's success in the marketplace by soliciting business from others on the basis of making known to them that he is the manufacturer of this successful product under its successful name.

Generally, a marketer will desire to qualify the private label manufacturer who is to produce his cosmetic products and will want to tour his facilities. This presents a delicate problem to the supplier. On the one hand, he wants to demonstrate his capability and, on the other hand, he must preserve his integrity and the obligations of confidentiality he owes to his other customers. The most reasonable solution is to conduct the new customer on a cursory tour of the facilities designed to show only enough to demonstrate his capabilities but to avoid lingering at particular areas of activity where confidences may be revealed. Certainly no tour should permit access to areas that have been segregated and in which sensitive operations are being conducted. No tour should be prolonged to the point where a stop in a particular area will afford the opportunity for the visitor to learn the formulating, manufacturing, or other intimate details of products or procedures being made or conducted.

The contract manufacturer should not be so inhibited by a contractual arrangement that he will be proscribed from utilizing for other business his skills and the general storehouse of knowledge and information that he has accumulated during his years of experience. His ability to develop products for others, just as his competitors may, should be recognized, respected, and preserved, provided one is satisfied that he can successfully walk the tightrope of integrity that must be traversed by one who offers his services to many who compete with each other and about each of whom he knows numerous intimate business and technical details. The evaluation of the contract manufacturer's ability to walk this tightrope of integrity is beyond the scope of this chapter and is a matter of business judgment and business risk.

The following, in letter agreement form, should adequately cover the subject of confidentiality between a contract manufacturer and the marketer or distributor who has brought him a formula in light of the above discussion:

> We will disclose your formula only to those people in our organization who require it for the evaluation, pricing, and manufacture of the product. We will refrain from disclosing the formula to

others unless obliged to do so be federal, state, or municipal authorities and we will refrain from soliciting business from third parties based on the fact that we produce the product of the formula for you. Incidental observation of your products by business visitors to our plants shall not constitute disclosure within the intent of the foregoing.

Nothing contained in this agreement shall restrict us from the use of formulas or other technical knowledge or information which we either already possess, or may develop, learn, or acquire in the future from our own or other sources independent of you, nor shall we be restricted from producing for third parties pursuant to formulas which they may submit to use provided we have no reason to believe such third parties are in violation of an obligation of secrecy or confidentiality to you with respect to such formulas.

III. RESPONSIBILITIES

The most common situation involving the relationship between a cosmetic marketer or distributor and a private label manufacturer is one in which the marketer has a concept for a product or line of products but no formulas of his own. This will include a variety of facets that will include not only a method of selling and distribution but the price category and market position of the products. The selling methods may run the gamut from direct selling through house to house sales, mail order or party plan arrangements, sales through retail outlets such as boutiques, salons, department stores, franchise arrangements, or any number of other channels of distribution. The price position of the products may vary from popular-priced, standard packaged products to expensively packaged, prestige merchandise. The products themselves may vary from the color and fashion areas to treatment lines, fragrance items, toiletries, bath products, sun products, toiletries in general, or any combination of these and others.

The packaging may vary from standard selections available from numerous suppliers to uniquely designed containers and applicators requiring private molds and substantial investment. The products themselves may include special ingredients that form the heart of the promotional story that will be used in the marketing of the products or the involvement of a famous personality in connection with the sales effort.

The marketer frequently has no capability to develop his own formulas and is often not interested in taking the trouble to do so. He will normally approach the private label manufacturer for consultation and advice in translating his concepts into the products that he will be selling.

A. Choosing the Product

With this background, it is rare that the marketer will be willing to go
to market with the precise product, made to the precise formula and
packaged in the precise container, that the private label manufacturer
may happen to have on his shelf at the time. If indeed he were willing
to do so, the issues concerning responsibilities could be more readily
resolved.

More often than not, the first samples shown to the marketer or
distributor represent only the early stages of the product development
procedure. The products from the private label manufacturer's shelf
may represent examples of systems for products which have to be modified and customized as they evolve into the final form which the marketer determines to take to the market. The devlopment procedure
will generally involve sampling and resampling until the final version
is arrived at.

Apropos this development project, one must be aware of the regulatory considerations.

Drug or Cosmetic

Attention must be given to the status of a product as a cosmetic
or a drug depending on the label and labeling claims which will be made
by the marketer or distributor in connection with its sale. If the
claims are to be restricted to promotion of beauty and attractiveness,
the product is, by legal definition, a cosmetic[2]. If the claims are to
include representations that the product will cure, mitigate, treat, or
prevent disease or will affect the structure or any function of the
body, it will, by legal definition, be a drug[3]. Indeed, the product
may be both a cosmetic and a drug if its labeling claims make representations in both areas.

Consumer Rights

The entire spectrum of laws and regulations affecting the products
must also be given attention in the development of the product and in
getting it ready for market. These issues are not confined to situations involving the relationship between a marketer and a private label
manufacturer, but they do present the need to scrutinize and allocate
the responsibilities that each of the parties should assume. One must
consider the rights of the ultimate consumer as well as the respective
rights and duties of the marketer and the private label manufacture.
It is suggested that the latter should appropriately be the subject of a
contract.

It must be recognized at the outset that regardless of the contractual arrangement that may be concluded between the marketer and the
private label manufacturer, they cannot, in general, adversely affect
the rights of the consumer. The consumer may assert his rights in
situations involving adverse reactions, breaches of the various

warranties, or negligence against anyone in the chain of manufacture and distribution as he may be advised and as the total body of law giving him those rights provides. No contractual arrangement to which he is not a party can or should limit those rights.

Nonstandard Products

It has been stated that the questions concerning responsibilities between the marketer or distributor could be more readily resolved if the former were willing to purchase the precise product, formula, and container that the private label manufacturer may have on his shelf at the time. He would then be merely a reseller and would generally be able to assume that his vendor has in all respects substantiated the product that he is purchasing and reselling and has, as a responsible seller, taken the time and gone through all the neccessary background steps necessary before placing a product on the market.

However, the joint venture of this discussion does not fit that pattern. Regardless of whether the formula for a particular product originated with the customer or the private label manufacturer after the development procedure discussed earlier, it is submitted that the distinction should be immaterial in considering the responsibilities of the parties after having settled the question of confidentiality with regard to a customer's formula.

Manufacturer's standards of performance: Once the formula has been identified, the customer has the right to expect certain standards of performance from the private label manufacturer. The goal should be to establish, by contract, a set of objective criteria against which each party can measure performance. As an example, in a sale by sample, the obligation of the seller is to deliver to his customer a commercial duplicate of the agreed-upon sample, which then becomes a prototype against which performance may be measured. Consideration must be given to the mechanism for creating and identifying the sample.

Substantiation: In the regulatory climate of our times, and because of the responsibilities of the parties as manufacturers, marketers, and distributors, we know that before a product is presented to the marketplace it should be subjected to a program of substantiation. Note should be made that substantiation is a broad term that is not confined to safety substantiation, though it certainly includes that area.

Many marketers of cosmetics and toiletries go through at least months of devlopment and confirmatory work before a selected formula goes to market. The technical personnel at some point present a formula that should be put through certain testing protocols. These may include confirmation of whether the preservative system in the product is adequate to protect it against exposure to bacterial contamination in the manufacturing establishment as well as in the hands of the consumer; confirmation of whether the product and the packaging materials

with which it comes in contact are compatible; determination of the
stability and shelf life of the completed product; determination of the
safety of the total product as it is reasonably expected to be used by
the consumer; determination as to the functionality or efficacy of the
product and its bases for labeling and advertising claims; determination of the total esthetic effect of the product and its acceptability in
light of current fashions and trends; and any number of other areas.

Often, determinations such as these are in competition with marketing plans and deadlines. Contact during the course of product
development between a private label manufacturer and his customer is
frequently between the technical staff of the former and the marketing
personnel of the latter who have plans for a product or product line
that is to be invented, developed, and finalized and is expected to be
on retail shelves two weeks ago. Though these dilemmas are not peculiar to a private label situation, they present added pressures due to
the customer-supplier relationship between them which lacks a central
authority within a single organization who can determine when marketing plans are to be deferred in favor of the realities of the situation.

Thus, the articulation in writing of the respective duties and responsibilities of the parties will serve to avoid the misunderstandings
that are likely to occur if representatives of each party make erroneous
assumptions about what each expects of the other which the other did
not intend or understand. In identifying these duties and responsibilities, it is suggested that a distinction be recognized between the
responsibility for the inherent product and the responsibility for
duplication of the inherent product in day-to-day production.

B. The Inherent Product

The inherent product is the one which, if commercially duplicated in
production, the marketer is willing to attach his name to, invest in
inventory of, and advertise, promote, and distribute to the consumer.
If the consumer is dissatisfied for any reason, he will call on the person whose name is on the label. If the product falls under the scrutiny of regulatory agency, the marketer identified on the label will
invariably be the first to be approached. If the product succeeds in
the marketplace, the greatest gain will accrue to the marketer who has
really put it all together after committing his resources and accepting,
rejecting, modifying, or innovating on the counseling he has received
from the private label manufacturer and the others involved. Thus, it
is submitted, the responsibility for the inherent product should be
accepted by the marketer or distributor as his own, and it should make
little or no difference whether the formula for the product was developed within his own organization, was the result of the work of an
outside independent consultant retained by him, was gleaned from a
textbook, a trade publication, the catalog of a raw material supplier,

or was adapted and customized for him in the laboratories of a private label manufacturer.

Consulting the Manufacturer

Recognizing that in numerous instances the marketer looks to the private label manufacturer as his principal consultant in the product development process and in spite of the fact that he may be assuming the ultimate responsibility for the inherent product, it is submitted that he has a right to expect that the private label manufacturer will conduct himself in an aware and professional manner in preparing the various intermediate samples for consideration. Otherwise, he may find himself in an expensive game of trial and error whereby he chooses samples to be tested only to find that they have failed or that they may not be legally used. The marketer has the right to demand contractually that the private label manufacturer lives up to a high professional standard. There follows some suggested contractual language for an undertaking of this type:

> In formulating the various samples for your consideration, we [the private label manufacturer] agree that we will use our best efforts and judgment consistent with the then state of the art of the industry, to incorporate only raw materials and combinations thereof that are compatible with each other and that are neither known to possess unreasonable safety hazards, nor prohibited for use in the products. We further agree to be alert to any changes in the state of the art of the industry and in the law or regulations concerning the products which may have an affect on their safety or compliance. We agree to notify you with reasonable promptness after actually learning of such changes in order to afford you the opportunity to minimize the economic impact.

C. Manufacturer's Responsibility

What, then, one might ask, should be the area of responsibility of the private label manufacturer? This should be the area that involves the ongoing commercial duplication of the product in a reliable, timely manner to fill the inventory and sales requirements of the marketer. Examination of this area of responsibility is worthy of discussion.

Product-Package Compatibility

Generally, the marketer will purchase the containers and packaging materials directly from suppliers for shipment to the private label house. One must consider the question of compatibility of the packaging materials with the inherent product. Neither the packaging supplier nor the private label manufacturer will be willing, as a general rule, to guarantee that the two will be compatible.

Testing for such compatibility would seem, therefore, to be the logical mechanism. However, such testing involves the application of testing protocols over a period of time. Time is frequently a luxury that is not available in light of the deadlines that are ever present in the race to the marketplace. Many technical departments resort to accelerated testing procedures that involve the exposure of the products in their ultimate containers to cycles of heat or cold or ambient temperature and climate conditions. These accelerated cycles are often mistakenly interpreted as accurate syntheses of the longer periods that will be experienced in the channels of distribution. It is submitted that this is not necessarily the case and that the only way one can accurately determine that a packaged product will remain stable for, let us say, two years, is to observe a valid sampling for two years. under the conditions of exposure that are expected to be encountered.

What, then, is the value of the accelerated procedure of 30, 60, or 90 days? Its value is not so much for the positive results that will be disclosed, but rather for the negative results that may appear. Thus, if a product breaks down, splits, or otherwise fails, or if it proves to be incompatible with the packaging with which it comes in contact during the short term of the test, one can surmise that the longer-term exposure will also fail. Given such a short-term failure, the parties have not choice but to go back to the drawing board.

Who should assume the risk of failure if the accelerated tests do not produce negative results and there is neither time nor inclination to wait for the hypothetical two years to elapse? This may at any time be the subject of negotiation between the marketer and the private label manufacturer or packaging supplier. However, it would appear that the marketer must, in the end, assume these risks. He may minimize his exposure by the care he exercises in selecting suppliers whose experience in the field will enable them to counsel and guide him in the selection of materials and packaging in combinations that are least likely to fail. But the mere fact that the end product is to be a customized adaptation rather than an exact duplication of standard products and packages may fog predictability. The marketer would assume this risk in his own plant and will be hard-pressed to pass it on to his supplier in the private label situation.

Safety Evaluation

Much the same type of risk assessment and assumption must be addressed in regard to the safety evaluation of the inherent product. Having created a nonstandard product, one must determine its safety by testing. The first stage in the testing procedure will probably involve animal testing. Here, again, the question arises whether the conclusions of safety of the products on animals can be extrapolated to humans. Once again, the only definite conclusion that may be reached from animal testing of the product for safety is when there is a failure. It would be foolhardy to proceed with human testing if a product has produced irritations or other failures in laboratory animals.

If, however, the animal testing produces no untoward results, it is appropriate to consider going forward with testing on humans in appropriately designed tests on appropriately selected human panels. Consideration must be given to the cost of such testing and to the length of time in which such testing is to be conducted. The element of cost is, of course, one to be negotiated between the marketer and the private label house.

Since the product is to be marketed under the brand name of the marketer, these test results should be part of his dossier of the product. A prudent businessman should have this available in the event the question of substantiation should ever be directed against him either in the product liability arena or by regulatory agencies.

Selecting the samples: The time element once again becomes an area of focus when considering plans to get to the marketplace. In addition, it is essential, regardless of whether one is considering the total subject of substantiation for products produced in one's own facilities or that question where they are produced in a private label situation, that the marketer discipline himself to conduct his substantiation testing on the final version of the product. If the testing is conducted on an intermediate version which is subsequently changed, the entire value of what has been done may be frustrated because the changes may be so radical as to invalidate the tests.

Product Functionality

Finally, in considering the subject of substantiation, one should not overlook the subject of functionality mentioned earlier. The tests to determine functionality or efficacy of the product are not as scientifically sophisticated as are the tests for compatibility, stability, and safety. In making these determinations the marketer may have his own ideas and methods, and he certainly can call upon the private label manufacturer for guidance from his own experience. Oftentimes this determination involves as much subjective as objective evaluation. The same is true with regard to the determination of the esthetics of product. It is, therefore, an area within the province of the marketer.

Prototype Samples

The foregoing discussion has focused on the identification and substantiation of the inherent product on which the marketer has decided. To a very large extent this identification process would be substantially the same whether the marketer were to produce the products in his own facilities or through the vehicle of a private label house. It is appropriate to consider what the marketer may now require of the private label house in the repetitive production of the product to fill the marketer's supply requirements. Ideally, samples of the inherent product should be selected and retained by both the marketer and the private label house. The samples then become the prototype which the private label house is to duplicate commercially in production. This sample becomes the objective criterion against which performance is to be measured. A contract provision should incorporate an obligation by the private label house to produce and deliver commercial duplicates of these samples. The contract provision should recognize the fact that samples may be subject to change on the shelf and, therefore, with the passage of time cease to be representative of a newly manufactured batch. Therefore, the contract should provide for periodic selection of a new set of samples or prototypes that will become the sample for the following period of time.

   Standards and specifications: The private label house should be called upon to guarantee contractually that his production will be a commercial duplicate of the sample. In addition, a well thought out agreement between the marketer and the private label house should establish a set of standards and specifications against which production is to be measured. These specifications may include descriptions of confirmatory work that will be done as part of the production process. For example, the specifications should include requirements that the production batches be placed in quarantine while bacteriological work is conducted to ensure that the product is indeed not contaminated.

   There might be specifications for pH and viscosity of the product or its various phases in the process of production. In products involving colors, the specifications should set up a mechanism for determining whether the production batches match the master shade. This, of course, is often a subjective matter, but procedures can be articulated for submission of representative samples of production batches to a qualified person selected by the marketer.

   The total set of standards and specifications should be reduced to writing so that they will represent an objective criterion against which each party can measure the acceptability of the finished product. A well thought out contract should provide that the written specifications constitute all and the only criteria to which the product is to conform in order to determine its acceptability.

   Bearing in mind that the marketer usually supplies the packaging materials and labels and, sometimes, also supplies perfumes, key mixes, bulk or certain raw materials, the contract should set up a standard of allowable waste for each item and should provide that the private label manufacturer must pay for excess waste of the marketer's materials.

IV. CONTINUING FDA GUARANTEE

The Federal Food, Drug and Cosmetic (FDC) Act provides certain civil as well as criminal sanctions in the event a cosmetic shipped in interstate commerce is adulterated or misbranded[4]. In the area of criminal sanctions, a defense is provided if a party can establish that he has relied on a guarantee from his supplier that the product is not adulterated or misbranded[5]. In connection with this defense, the Food and Drug Administration (FDA) has published suggested forms of guarantee[6]. The general continuing form of guarantee suggested by the FDA has become widely used in the cosmetic industry as well as in the other industries under the jurisdiction of the FDA. This continuing guarantee deserves special attention in considering the relationship between the private label manufacturer and the marketer.

In light of the fact that the marketer is frequently the supplier of the packaging materials, the labels, and sometimes some of the ingredients in the product, it is inappropriate for the private label manufacturer to provide the general continuing guarantee for those items that he does not really supply. It is suggested, therefore, that an appropriate form of guarantee include mutual guarantees by each of the parties with regard to their respective functions and the materials each of them supplies. There follows, in letter form, a sample guarantee which should satisfy the responsibilities of each party:

The undersigned, private label manufacturer, whose principal office is at   , hereby guarantees that:

1. No product hereafter produced and/or filled and shipped or delivered by us for you will be, on the date of such shipment or delivery, adulterated within the meaning of the Federal Food, Drug and Cosmetic Act; and
2. No color additives contained in said products on the date of shipment or delivery thereof by us to you will be unsafe within the meaning of section 601(e) of the federal Food, Drug, and Cosmetic Act.

Nothing herein contained shall be deemed to extend the scope of this guarantee to cover components, labels, labeling, packaging materials, bulk product, or ingredients furnished by you in connection with products we may produce and ship to or for you. With regard to such items, you guarantee to us that same shall not, by virtue of their nature or condition when received by us, be or result in adulterated or misbranded products within the meaning of the Federal Food, Drug, and Cosmetic Act. All previous continuing guarantees of this nature are hereby revoked. This guarantee shall, upon becoming effective, take precedence over and be in lieu of any printed clauses in any purchase orders which may be issued by you to us. This guarantee shall become effective upon receipt by us from you of a counterpart hereof, and shall be a continuing guarantee and binding upon each of us with respect to all shipments made before the receipt of written notice from the other of its revocation.

V. INSURANCE

An arrangement between a marketer and a private label manufacturer
should also focus on the subject of insurance. Within that area consideration should be given to both property insurance and liability
insurance.

In dealing with a private label manufacturer, the marketer maintains an inventory of packaging materials, labels, finished products,
perfumes, raw materials, and other items of his property on the private label manufacturer's premises. One must consider the property
insurance that is to cover these items. Coverage for such hazards as
fire, the perils of extended coverage, vandalism, malicious mischief,
and crime are generally based on premiums applied to values. The
various packaging components may run the gamut from inexpensive
plastic cases or compacts to expensive pieces. The private label manufacturer frequently does not know the value and does not therefore
have a basis for reporting values to his insurance carrier. His position
is analogous to that of a public storage warehouse. It is customary
for business people who maintain inventories in various locations, including public storage warehouses, to maintain insurance of the type
described above through their own brokers or agents and under their
own multiple-location policies. These may be carried either on a
blanket basis or on a reporting form basis for the values that the marketer has at risk. The marketer should be willing to arrange for the
same insurance for his property in the hands of the private label
manufacturer.

The private label manufacturer's duty is no different than the
standard of performance imposed by the common law on any bailee for
hire. This is a duty of ordinary care. Absent negligence in that care
by the bailee, a prudent businessperson will want to insure that the
property at risk from various exposures will be covered by insurance.
By insuring for these risks through his own sources, he can assure
himself of the control and of the greatest economies.

A. Product Liability Insurance

The subject of product liability insurance requires special consideration. Marketers are frequently requested to provide to retailers,
through whom their products are sold to consumers, a broad-form
vendor's endorsement in connection with their product liability coverage. The gist of such endorsements is to protect the retailer, who
has not connection with the manufacture, packaging, design, or assembly of the product, from product liability claims that may be brought
against him by the customer. Marketers who deal with private label
manufacturers frequently request that they be supplied broad-form
vendors' endorsements as well. This is inappropriate. The standard

broad-form vendors' endorsement contains exclusionary language to the effect that if the customer (the marketer in this case) of the named insured (the private label manufacturer in this case) has provided any of the materials that go into the product, then the broad-form vendors' endorsement will have no validity in favor of the person to whom it is issued. Thus, as may be seen from the earlier discussion, because the marketer frequently provides the packaging materials, and sometimes provides some of the ingredients that go into the products, the broad-form vendors' endorsement that he has requested from the private label manufacturer would probably be of no value to him. It most assuredly would be of value to him if he were to buy the specific product off the shelf of the private label manufacturer in the packaging materials provided by the manufacturer. But this is not generally the case.

Furthermore, it is submitted that a prudent business person would be better advised to arrange for his own insurance coverage in all the areas of his exposure through his own insurance broker rather than try to pass the insurance placement responsibilities with an organization that may not be fully familiar with his activities and whose coverage is beyond the marketer's control.

The general subject of products liability requires discussion when considering the area of insurance. If the private label manufacturer fails to duplicate the products in manufacture or to conform to the specifications and, as a result of these omissions or shortcomings, if the product injures the ultimate consumer, the private label manufacturer should indeed indemnify the marketer for the results of his negligence. The marketer should insist that the private label manufacturer carry adequate insurance against these risks and demonstrate that he has such insurance by providing certificates of insurance. However, in most cases where consumer complaints arise that result in claims, they do not arise because the product was made wrong but, rather, because of an idiosyncratic problem of the consumer. It is submitted that the isolated allergic reaction problem of the consumer should fall more into the area of responsibility of the marketer than of the private label manufacturer.

VI. CONCLUSION

The foregoing discussion should demonstrate that there are indeed special considerations that deserve attention in the dealings between a marketer and a private label manufacturer. The relationship is certainly not as simple as selecting a product from a shelf. An awareness of the many facets of the dealings between the parties and a conscious resolution of each area will lay a firm foundation for a smooth and successful relationship between them.

REFERENCES

1. 21 Code of Federal Regulations (C.F.R.) 701.3.
2. 21 U.S. Code (U.S.C.) 201(i).
3. 21 U.S.C. 201(g).
4. 21 U.S.C. 301, 302, 303 and 304.
5. 21 U.S.C. 303(c)(2).
6. 21 C.F.R. 7.13.

34

INNOVATING IN A REGULATED ENVIRONMENT

MONROE LANZET

Research and Development, Max Factor and Company, Los Angeles, California

I. INTRODUCTION

Innovation lies at the heart of the cosmetics and toiletries business.
Past megagrowth in our industry has paralleled invention, but new
product forms and introduction of new raw materials have always involved some element of risk. For example, while most industry experts
agree that today's treatment cosmetics have entered the age of performance, more functional skin care products straddle the border
between cosmetics and drugs.

   The purpose of this chapter is to review those laws, regulations,
and other more voluntary limitations which affect the formulator in the
pursuit of new and demonstrably superior products. These may be
divided into compulsory and voluntary areas:

Compulsory	"Voluntary"
Food and Drug Administration	Network censors
Federal Trade Commission	Better Business Bureau/ National Advertising Division
Occupational Safety and Health Administration	Nader/Consumerists
International regulations	Ad agency review
State regulations	Cosmetic Ingredient Review
Consumer Product Safety Commission	Research Institute for Fragrance Materials

It is not my purpose to recover the ground of the other authors in
this book but to highlight some of the unexpected roadblocks which
might detour the creators of contemporary products and to show examples of how the various laws and regulations affect the free
exercise of pure creativity.

   The cosmetic scientist must be always mindful of the expense in
both time and money of stepping over the fine line that lies between
cosmetic and drug, the tried and true versus the unproven. The
burden is always the responsibility of the creative leader.

   One might apply a form of cost/benefit analysis to this process.
Being somewhat in advance of the industry allows an innovative company to command a better market share because of the extra price that
a value-added product can command. Market leadership also has a
halo effect on the other products in the line because retailers want to
give extra space, cooperative advertising, and special consideration
to the market leader. Apart from the educational advertising required

to launch really landmark new products, introduction of innovations may require less rather than more advertising dollars to ensure market penetration because a really demonstrable item speaks for itself. "Me too" commodities usually have the largest ad budgets. Toilet soap, laundry detergent, and the like are the most advertised.

While most of this chapter will be devoted to consideration of the ramifications of U.S. federal and state law, later in this chapter we will consider the additional complications of formulating for a multinational environment. Many of today's companies are developing products for simultaneous sale in the European Economic Community (EEC), South America, and the Orient as well as for North America.

II. HOW TO TURN A TRADITIONAL COSMETIC INTO A DRUG: ADD SUNSCREEN TO AN ORDINARY MAKEUP PRODUCT

Suppose your marketing department, taking note of the demonstrated wrinkle-accelerating effect of normal daily exposure to the sun and the proliferation of products which claim to protect against this, has required that your new facial moisturizer contain ultraviolet light absorbers. If you make a claim for protection against wrinkles in the United States, be aware that your product has, by current Food and Drug Administration (FDA) definitions, crossed the line between cosmetic and drug. This is true even if you specifically warn the consumer that the product is *not* intended to protect against sunburn.

The cosmetic industry prides itself on its ability to turn on a dime. For those weaned on this fast track, launching a makeup with a sunscreen will seem inordinately time-consuming. The final drug product must be tested for effectiveness by the method proposed in the tentative Over-the-Counter Drug Program monograph on sunscreens. The product must be assayed for active ingredients, and new analytical methods will often be required. After accelerated testing to simulate three years of shelf life, the assay and the sunprotective index must be successfully repeated. If this is not done, the product must be expiration dated.

All final testing must be conducted in the final package, so purchasing must bring in the exact package and closure nearly six months before the shipping date. Label text must contain all the mandated warnings, directions for use, and proscribed claims. Once the lab is through with its part, manufacturing has the problem of living with a drug good manufacturing practice (GMP)-regulated operation. All raw materials must be assayed, checked for conformity with the specifications, and then stored in a quarantine area. The same goes for the storage of components and finished goods. The factory must be registered as a drug manufacturing plant, with the high probability of strict FDA inspections at any time. The regulations on over-the-counter (OTC) drugs are far more rigorous than for cosmetics. If you have never played in this league before, at this point you may be wondering, along with the rest of us, if it is all worth it. A benefit/risk assessment on the part of management and an incremental cost-of-goods study would certainly be in order before blindly proceeding with such a project.

But there is a way to slip between the horns of the dilemma. If you add the sunscreen and know that it has a protective effect but say nearly nothing about it in the copy, then it is still a cosmetic.

Some firms have maintained a real cosmetic stance by including sunscreen but treating it as an ordinary ingredient in the normal ingredient listing. Moreover, the claims made are merely "contains sunscreen or ultraviolet absorber". No sun protection factor (SPF) or anti-skin cancer claim is included in the copy, and protection against premature aging (wrinkling) is alluded to rather than clearly stated. Your marketing department claims that it absolutely needs the "anti-aging" verbiage to promote the product effectively. Since what you say in most cases determines whether a product is a drug or cosmetic, you're now neatly hung on a cosmetic Catch-22. The much weaker final copy will clearly show the signs of a hard fought marketing/lab/legal battle, however, and no one really knows how the FDA will react once the tentative monograph on sunscreens ultimately becomes the final monograph.

A similar option exists for moisturizers and lipsticks as well, with similar potential for possible avoidance of drug status.

Recent Canadian rules are more reasonable. Use of sunscreen in cosmetics is not treated as a drug, provided that SPFs are not stated. Other protective claims are allowed.

Since most skin aging researchers would agree on the skin protecting benefits of regular use of sunscreens, even in non-sunbathing situations, it is ironic that the new U.S. regulations will actually discourage their use in makeup, lipsticks, and moisturizers.

Those who will continue to make the claims will undoubtedly pass the costs of the additional testing and separate handling on to the consumer.

III. HOW TO TURN A SOAP INTO A COSMETIC: MAKE A SKIN CARE CLAIM FOR AN ORDINARY SOAP BAR

Simple soap is excluded from regulation by the federal Food, Drug, and Cosmetic (FDC) Act, but is regulated under the Consumer Product Safety Act and labeling requirements of the Federal Trade Commission (FTC). Suppose your claims lab tells you that your ordinary alkali salt of fatty acids cleansing bar has been proven to accelerate cellular renewal by 19 percent. Marketing demands an exfoliating claim, and suddenly your cleansing bar is a cosmetic. You have changed nothing but the claim, but now you are in trouble. Adding ingredient labeling wasn't hard, but the nice stable pink color was made with noncertified colors permitted for use in soap but not allowed in cosmetics. Getting a stable pink with certified colors turns out to be a tough project. Adding foam stabilizers, emollients, synthetic surfactants, or proteins to an ordinary soap bar for improved performance claims also turns ordinary soap into a cosmetic, although the government has not imposed FDA regulation on soaps with small amounts of lanolin.

IV. HOW TO TURN A SOAP INTO A DRUG: ADD AN ANTIMICROBIAL OR BENZOYL PEROXIDE TO AN ORDINARY SOAP BAR

If you want to make a deodorant or medicated claim for an ordinary soap bar, you can add 1 to 2 percent triclocarbon, triclosan, or some other OTC Category No. 1 antimicrobial, but now the bar is a drug. Active ingredients, a full "other ingredients" list, and the appropriate warnings must be labeled, and all drug GMPs now apply to your plant. Moreover, you must be able to demonstrate that the bar is both safe and effective for the purpose claimed. Other active ingredients such as benzoyl peroxide fall into the same category.

V. WHEN IS A PRODUCT REGULATED AS BOTH A COSMETIC AND A DRUG?

Here again, what is claimed will determine the status of your product. Drugs are defined by section 201(g) of the FDC Act as "Articles intended for use in the . . . cure, mitigation, treatment or prevention of disease." As a consequence, cosmetics may also be regulated as drugs based on the intended use. Table 1 shows some typical examples.

Once more, the formulator must take care in choosing both ingredients and claims to avoid unintended drug status. While inclusion of some ingredients such as hydrocortisone, benzoyl peroxide, or sodium fluoride invariably confer drug status on the cosmetic base to which they are added, due to the implied claim, other nominal drug actives such as aluminum chlorhydrate and even zinc pyrithione have other purely cosmetic functions, such as astringency and preservation.

In these cases the claims made will largely determine how the product will be regulated. It will require the most skilled copywriters working closely with R&D, marketing, and the legal department to walk that tight line between effective cosmetic copy and verbiage which may be interpreted as a drug claim because you imply that you significantly affect the structure or function of the skin.

If, for example, you claimed that your nonmedicated shampoo was a "dandruff" shampoo by virtue of the simply mechanical removal of loose dandruff flakes as part of the cleansing process, then you have a clear cosmetic claim. Implication that the product loosened up adhering flakes or in some other fashion required less frequent

TABLE 1   Typical Cosmetic Drugs

Basic cosmetic Function	Intended drug use	Typical active ingredient(s)
Toothpaste	Prevention of tooth decay	Sodium fluoride-stannous fluoride
Deodorant	Inhibition of perspiration	Aluminum chlorhydrate, chloride, or chlorohydrex
Tanning product	Prevention of sunburn	OTC/Category No. 1 ultraviolet absorbers
Lip balm	Prevent chapping	Petrolatum, allantoin, or any other Category No. 1 skin protectant
Shampoo	Antidandruff action	Zinc pyrithione, colloidal sulfur, coal tar
Mouthwash	Treats mouth disease	Benzethonium chloride or other Category No. 1 antimicrobial accepted for use in the oral cavity
Oil-blotting astringents, exfolliants, or makeups	Treats acne	Benzoyl peroxide, sulfur, or Category No. 1 antimicrobial
Moisturizer	Antiinflamatory	Hydrocortisone

shampooing would be on the borderline between cosmetic and drug and would be open to government interpretation. Claims for "control" of dandruff would be a full drug claim, as would be calling the product an "antidandruff" shampoo.

Another example is in the skin protectants drug category. The OTC panel has concluded that certain skin protection product positioning involving claims for healing, helps heal, and prevents or protects against chapping are now drug claims although they have been applied to cosmetics for decades. This is independent of the ingredients used, since mineral oil, petrolatum, allantoin, and shark liver oil, all Category No. 1 protectants, have been used in cosmetics for may years, and can continue to be used as long as the specific healing claims are not applied.

Some new verbiage such as soothes or protects will have to be created to continue to make "cosmetic" skin protectant claims. If you wish to make healing claims and can handle OTC drug classification, you will have to formulate your skin protectant with Category No. 1 ingredients to make the claim.

Other elements of this issue related to moisturizers are further addressed below.

VI. DEFINING THE BORDERLINE BETWEEN COSMETIC AND DRUG IN CONTEMPORARY SKIN CARE

Cleansing, toning, firming, moisturizing, and protecting are all historic cosmetic claims associated with skin care. Some firms have avoided calling these products skin treatments although it is clear that the claims are solely cosmetic treatment verbiage. For the last 15 years, however, increasingly sophisticated research into the skin, how it functions, and how to affect its epidermal functioning, together with proportionately more sophisticated consumers, has ended the era of "hope in a bottle."

Out of this has come a new wave of claims, some of which are clearly substantiable and others which may not be as easily proven. New claims supporting tools such as autoradiography, scanning electron microscopy (SEM), oxygen electrode skin respirometry, skin moisturization by inductive impedence measurements, skin topography, and dansyl chloride cellular renewal techniques have come into wide use and acceptance.

Table 2 lists many of the claims currently being made for skin treatment together with some "active ingredients" either actually claimed or reported to be responsible for the newer claims.

TABLE 2   Contemporary Skin Care Claims

Claims	Ingredients associated with claim
Occlusive moisturizing	Waxes, mineral oil, petrolatum esters, proteins, fatty alcohols, glyceryl stearate
Protection	Same as above except proteins
Exfoliating and accelerated "cellular renewal"	Soap, synthetic detergents, allantoin, alcohol, polyethylene grains, pulverized fruit pit materials, oat flour, botanicals, sodium RNA
Nonocclusive moisturizing	Branched-chain esters, "natural moisture factor" ingredients, sodium PCA, amino acids, hyaluronic acid, lactic acid, sodium lactate, urea, soluble collagen, elastin, hydrolyzed proteins, humectant glycols, polyamino sugar condensates

TABLE 2.   (Continued)

Claims	Ingredients associated with claim
Toning and astringency	Ethanol, isopropanol, zinc salts, botanicals such as horse chestnut extract, tannin, tannic acid, aluminum salts, hydrolyzed yeast
Firming, smoothing, "age controlling"	Collagen, elastin and hydrolyzed proteins, sodium RNA and DNA, "live cell" extract, reticulin, astringent salts, botanicals, bovine and equine serum albumin, placental enzymes, human placental protein
Temporary wrinkle smoothing	Bovine serum albumin, human placental protein, "live cell" filtrate, soluble whole protein, hydrolyzed elastin, sodium silicate and other mineral film formers
"Nourishing" and nurturing	Sodium RNA, mucopolysaccharides, polyunsaturated fatty acids, glycogen, amino acids, simple sugars, polyamino sugar condensates, glycerin, glycerides, Revitalin, royal jelly, pollen, honey, low-molecular-weight polypeptide, hydrolyzed yeast, placental lipids, protein and enzymes, yeast extracts, certain vitamins such as A, D, E and B complex
Skin respiration	Sodium RNA, hydrolyzed yeast, bovine and equine serum albumin, cytochrome C oxidase and other oxidation promoting enzymes
Protection from ultraviolet light	Titanium dioxide, opacifying pigments, and ultraviolet absorbers
Repair of damage to skin by ultraviolet light and "night repair"	Chondrotin sulfate, hyaluronic acid, mucopolysaccharides, hyaluronidase, white nettle extract and other botanicals
Antiirritants and antiinflamatories, "desensitizers" and antiredness products	Allantoin, aluminum dihydroxyallantoinate and other allantoin complexes, matricaria oil, cammomile extract dimer and trimer acids, bisabolol, glycyrrhetinic acid, yeast extracts, guaiazulene, vitamins E, D, A, and B6 epinepherine sulfate, glycyrrhizic acid
Oil control	Talc, kaolin, silica, magnesium carbonate, calcium silicate, astringents and botanicals, quaternary ammonium compounds

VII. HOW TO TURN A COSMETIC INTO A NEW DRUG: ADD A NEW MEDICAMENT TO A COSMETIC BASE

Some of the most significant developments in cosmetics have come as a result of breakthrough discoveries of new active ingredients such as zinc pryrithione for dandruff control and pregnenolone acetate for cellular renewal and thickening the epidermal layer. Not only are these materials drugs but newly discovered ones as well.

When they were incorporated into ordinary cosmetic bases, the resulting product and the claims made for them required that the sponsoring firms process a full new drug application before the FDA would allow the products into the OTC market.

Subsequently other companies were able to market zinc pryrithione shampoos with abbreviated new drug applications (NDA) and more recently without any NDA at all. The same was true for fluoride toothpaste.

Although the NDA process is expensive and time-consuming, the marketing advantage for a significant breakthrough product is enormous and the approved NDA is virtually an exclusive license until some other firm takes the time to repeat the same demonstration of safety and efficacy for their version of the original. Innovation and market leadership may provide a company with six months to a year of lead time on the competition, but an approved NDA product grants two to three years of lead time before the competition can complete their studies.

VIII. HOW TO TURN AN OTC DRUG INTO A NEW DRUG: COMBINE TWO OR MORE OTC CATEGORY NO. 1 ACTIVES IN A NEW COMBINATION OR ADD AN ENTIRELY NEW ACTIVE INGREDIENT

Even if you have been involved in the complications of developing ordinary OTC drugs, there is nothing like the complexities of filing even a short-form NDA. FDA cautions that the combination of two Category No. 1 active ingredients not previously combined constitutes a new combination and hence require you to establish the safety and efficacy of the new blend. A combination of two Category No. 1 antimicrobials in a medicated facial cleanser or astringent would be a good example. The anticipated benefit of the new combination would have to be balanced against the time and cost of the NDA. For most OTCs the safest course is to rely on only one antimicrobial. This could get more complicated if you are contemplating addition of an antiinflammatory in a sunscreen or the like, where market studies have indicated that the new claims that could be made may warrant the expense in an increasingly crowded marketplace.

If you are not sure of the NDA consequences of some new combination, the wisest course of action is to get a preliminary opinion from the FDA before proceeding.

IX. HOW TO TURN A SPONGE INTO A MEDICAL DEVICE: MAKE DRUG CLAIMS FOR THE SPONGE

Suppose that marketing has proposed that your firm become involved with acne products. The cosmetic claim hedge usually employed is to make no specific claims about treating acne. A number of purely cosmetic cleansers and astringents are claimed to be suitable for oily and troubled skin. Any reference to the usual "troubles"—pimples, comedomes, or acne vulgaris—is carefully avoided. These preparations contain no antimicrobials or other medication for acne, but are useful nevertheless in keeping the skin clean and free of oils and sebum. One firm, however, ignored these precautions in describing the properties of a scrubbing sponge intended for us in conjunction with such an oily skin cleanser. Without consulting either the laboratory or its legal department, it proposed that the sponge be promoted as a gentle exfolliating scrubber for those with acne. Only at the last possible minute was the sleeve copy and ads for the scrub sponge changed, at considerable cost.

The lesson is quite clear. Devices have been regulated since the enactment of the Medical Devices Act in 1976. By the careful avoidance of medical claims for any cosmetic devices you intend to market, you can avoid the requirements of the rigorous device statute. Other possible examples include a toothbrush intended for use with certain oral cavity diseases such as gum disorders, or massaging devices, electrical or mechanical, which are claimed to be helpful for medical conditions such as arthritis or rheumatism.

Still another medical device issue that affects the cosmetic industry involves salons that provide cosmetic facials and other services. Certain devices used for massage and skin relaxation are regulated by the Medical Devices Act. They can be used only by a medical practitioner and cannot be used by a cosmetologist, licensed or not. The unwary cosmetics company might not be cognizant of this and so unwittingly step over the line to be regulated under the provisions of the act. Since 1976, several small companies have run afoul of this law.

X. INNOVATING IN A MULTINATIONAL REGULATORY ENVIRONMENT

Up to now we have concerned ourselves with the vagaries of U.S. law and regulation as it affects the formulator. While an extensive look at international regulations is beyond the scope of this chapter, it would be useful to look at how products are categorized in other major areas of the world. Many cosmetic drugs by U.S. definition are treated as pure cosmetics in other countries. On the other side of the coin, some OTC cosmetic drugs such as hydroquinone skin bleaching creams are prohibited in parts of Western Europe and in Japan.

Because of these differences, I find it convenient to divide the world into regulatory zones. Within limits, zone 1 consists of the United States together with Canada, Australia, and perhaps Venezuela. The second zone is the United Kingdom/EEC, which also includes many other European nations which are not in the common market but which have a similar regulatory philisophy. Greece, Sweden, and Israel are examples. Japan's regulations (zone 3) stand alone. Japan is probably one of the most regulated markets outside the United States, with an extensive list of ingredients and product categories that are prohibited. The fourth zone consists of Mexico, Brazil, Argentina, and many of the other smaller nations in South and Central America. They have borrowed some of the positive ingredient list regulatory apparatus of the EEC and Japan but adhere to US notions concerning GMPs and the like. The fifth and last zone is made up of those emerging nations where regulations are not yet sharply defined. We know from past history that it is just a matter of time until they are, however.

To avoid regulatory booby traps, it would be wise to know which ingredients are prohibited or limited in your major markets. Permitted colorants, preservatives, antimicrobials, and pharmaceuticals vary widely throughout the world. For formulations containing ultraviolet absorbers, it would be prudent to check the US OTC tentative monograph against the EEC and Japanese positive lists, although many of the same sunscreens are allowed. The difference is in the treatment of the category. These products are frequently considered quasi-drugs in Japan (similar to OTC drugs in the United States) but are not treated as drugs at all in Europe. Similarly, antimicrobials are definitely quasi-drugs in Japan, with elaborate safety and efficacy protocals, but they are not necessarily drugs in the EEC. Medicated scrubs and astringents are ordinary cosmetics.

Chlorhexidine is a good example of how ingredients differ. It is freely permitted in the EEC and Japan but would probably require a NDA for use as an antimicrobial in the United States. Treatment of hypoallergenic cosmetics differs as well. Even if one could find words that mean the same, such products would be classified as quasi-drugs in Japan. In Venezuela it is the expression itself that is prohibited. They can be marketed as "low allergy" cosmetics, however. In Argentina, the claim substantiation for hypoallergenic cosmetics is more rigorous. In Sweden and France they are sold only in pharmacies.

Many nations require formal safety substantiation for cosmetics, in contrast to the United States where submission of safety data is not required. Product registration with full safety testing and qualitative and quantitative formula disclosure is a Japanese Ministry of Health and Welfare requirement. Formula approval takes at least 120 days and even more for quasi-drugs. Europe prefers elaborate animal safety tests, and in some countries formula disclosure including specifications and manufacturing procedures is required even for ordinary cosmetics. Because of these differences, it would be prudent for the formulator to test any multinational product according to the requirements of the major markets in which it is to be sold. Launch dates for Japan, Argentina, and other nations that require preclearance should take the lengthy clearance times into consideration.

Requirements for support of cosmetic claims also vary widely. There is a heavy emphasis on botanical and biologically derived skin care in Europe. Claims are frequently made based on literature references and often circumstantial evidence. Much of what is considered "serious cosmetics" in Europe, based on the allowed claims, would undoubtedly be considered drugs in the United States. Even for those which would be true cosmetics, most of the claim substantiation would not be accepted by network censors or the NAD in the event there was a challenge of the copy claims by the Better Business Bureau or the like.

As a consequence, some U.S. multinational firms use different copy claims for their products in Europe, Japan, and the United States. A few larger companies have launched new products in Europe before the United States. This may be because the data on hand at the time were adequate to substantiate the claims in the EEC but further work was still required to meet US standards.

XI. CONCLUSION

Innovation in cosmetics and the health and beauty aids industry is a significant key to market expansion. I'm happy to report that it's alive and well despite encroaching regulation both in the United States and overseas. Those formulators who are familiar with the legal pitfalls will be best equipped to continue to innovate in these rapidly changing regulatory times.

Part III

CHALLENGES FOR TOMORROW

35

DEVELOPMENT OF PUBLIC POLICY ON CARCINOGENS

PETER BARTON HUTT

Covington & Burling, Washington, D.C.

Regulation of potential or proven carcinogens has increasingly been a major focus not just of the Food and Drug Administration (FDA) but indeed of several government agencies. Moreover, because cancer is our most feared disease, these regulatory activities have achieved widespread national interest. This chapter therefore can only highlight important landmarks in the development of public policy on cancer, and particularly those that involve cosmetic indredients.

I. BACKGROUND

Cancer is an old and well-established disease. It has been known for as long as medical records exist.

Because cancer is a disease that primarily strikes older people, its importance is directly proportional to the age of the population. In the early days of our country, life expectancy was quite low. In about 1845, for example, average life expectancy at birth in Boston and New York was 21.43 and 19.69 years, respectively.[1] Thus, cancer did not present a serious health problem at that time.

By 1900, as a result of the major public health measures undertaken during the last half of the nineteenth century, life expectancy
at birth in this country had risen to 47 years.[2] Since then, it has
steadily risen and in 1978, the most recent year for which information
is available, it reached 73 years. This represents a remarkable increase in national life expectancy of 26 years during the past 78 years.

This increase in life expectancy has been obtained at the expense
of the infectious diseases that formerly took a heavy toll in our population. The three leading causes of death in 1900 were pneumonia and
influenza, tuberculosis, and diarrhea.[3] Today, these diseases represent less than 2 percent of the problems they caused in 1900.[4] The
two leading causes of death today are heart disease and cancer.[5]

From all available information, it appears that the age-adjusted
incidence of cancer is not significantly increasing.[6] The total increase
in cancer has resulted from the substantial elimination of the infectious
diseases, not from any increase in the incidence of cancer itself. Current mortality statistics indicate that, on an age-adjusted basis, the
prevalence of cancer in this country remains remarkable steady. Some
forms of cancer have increased, and others have decreased, but overall the age-adjusted incidence has not changed perceptibly.[7]

Based on evidence of the variation of cancer among different populations throughout the world, epidemiologists have concluded that up
to 90 percent of all cancer is "environmental" rather than hereditary in
origin. Included within the "environmental" causes of cancer are such
widely diverse factors as dietary patterns, cigarette smoking, consumption of alcoholic beverages, and sexual promiscuity, as well as factors
such as air and water pollution that one would normally include within
this concept. In a conference held in 1979 at the American Health
Foundation, a group of cancer experts reach consensus on the risk
factors in human cancer. These factors on listed in Table 1.

The National Academy of Sciences has suggested that diet may be
responsible for 30 to 40 percent of cancers in men and 60 percent of
cancers in women.[8] While these figures represent judgment rather than
documented statistics, they nonetheless provide a very sound guide to
the relative risk provided by various sources of cancer in our country
today.

A 1981 report by the Office of Technology Assessment (OTA) of
the U.S. Congress states that skin cancer represents less than 2 percent of total cancer in the United States.[9] As Shimkin has pointed out
(Table 1), at least 50 percent of skin cancer is attributable to ultraviolet radiation, largely from the sun. The OTA report states that
consumer products, taken in their entirety, appear to be responsible
for less than 1 percent of U.S. mortality from cancer.[10] Thus, it appears unlikely that cosmetic products represent any significant cause
of cancer in this country.

TABLE 1   Factors in Cancer, as Summarized by the Conference on the Primary Promotion of Cancer, New York, 1979

	Percent of cancers involving factors	
	Men	Women
Smoking	25—35	5—10
Alcohol	7	2
Occupation	6	2
Nutrition	30	30—50
Food contaminants	0	0
Drugs	<1	<1
Air pollution	0	0
Ionizing radiation	<3	<3
Ultraviolet radiation	(skin 50)[a]	(skin 50)[a]
Heredity	10—25	10—25
Viruses	<1	<1
Immunodeficiency	<1	<1

[a]Excluded from total cancers.

Source: Shimkin, *Industrial and Life-Style Carcinogens,* 1980. p.10.

Indeed, it is likely that cosmetic products today prevent more cancer than they possibly might cause. In 1978, the FDA published the report of its advisory panel on over-the-counter (OTC) sunscreen drug products.[11] The panel concluded that the labeling for an effective sunscreen product could properly include the following claim:

Over exposure to the sun may lead to premature aging of the skin and skin cancer. The liberal and regular use over the years of this product may help reduce the chance of premature aging of the skin and skin cancer.

This has prompted a number of cosmetic manufacturers to include in their products effective sunscreen ingredients, whether or not a specific cancer prevention claim is made.

In spite of the fact that cosmetic products are highly unlikely to contribute significantly to cancer in the United States, the possibility that cosmetics may cause cancer remains a very emotional subject. The potential contribution of color additives to cancer risk was fully considered during enactment of the Color Additive Amendments of 1960 and the carcinogenicity of hair dye ingredients has been the subject of FDA regulatory concern during the last half of the 1970s. The remainder of this chapter will therefore trace the development of public policy on carcinogens in cosmetic products during the past 20 years.

II. THE DELANEY ANTICANCER CLAUSES IN THE FDC ACT

Legislation incorporating specific anticancer clauses resulted from the work of a Select Committee to Investigate the Use of Chemicals in Food Products, chaired by Representative James Delaney of New York, created by the House of Representatives in June 1950.[12] In October 1951, the Select Committee's jurisdiction was extended in include cosmetics. The report of the Select Committee, issued in 1952, recommended legislation to protect consumers from harmful chemicals in both foods and cosmetics.[13] As a result of the work of the Select Committee, the Federal Food, Drug, and Cosmetic (FDC) Act now contains three explicit anticancer clauses, the so-called Delaney clauses.

Section 409(c)(3)(A) of the act, as amended by the Food Additives Amendment of 1958 and the Drug Amendments of 1962, states:

Provided, That no additive shall be deemed to be safe if it is found to induce cancer when ingested by man or animal, or if it is found, after tests which are appropriate for the evaluation of the safety of food additives, to induce cancer in man or animal, except that this proviso shall not apply with respect to the use of a substance as an ingredient of feed for animals which are raised for food production, if the Secretary finds (i) that, under the conditions of use and feeding specified in proposed labeling and reasonably certain to be followed in practice, such additive will not adversely affect the animals for which such feed in intended, and (ii) that no residue of the additive will be found (by methods of examination prescribed or approved by the Secretary by regulations, which regulations shall not be subject to subsections (f) and (g)) in any edible portion of such animal after slaughter or in any food yielded by or derived from the living animal.

Section 706(b)(5)(B) of the act, as amended by the Color Additive Amendments of 1960 and the Drug Amendments of 1962, states:

A color additive (i) shall be deemed unsafe, and shall not be listed, for a use which will or may result in ingestion of all or part of such additive, if the additive is found by the Secretary to induce cancer when ingested by man or animal, or if it is found by the Secretary, after tests which are appropriate for the evaluation of safety additives for use in food, to induce cancer in man or animal, and (ii) shall be deemed unsafe, and shall not be listed for any use which will not result in ingestion of any part of such additive, if, after tests which are appropriate for the evaluation of the safety of additives for such use, or after other relevant exposure of man or animal to such additive, it is found by the Secretary to induce cancer in man or animal:

Provided, That clause (i) of this subparagraph (B) shall not apply with respect to the use of a color additive as an ingredient of feed for animals which are raised for food production, if the Secretary finds that, under the conditions of use and feeding specified in proposed labeling and reasonably certain to be followed in practice, such additive will not adversely affect the animals for which such feed is intended, and that no residue of the additive will be found (by methods of examination prescribed or approved by the Secretary by regulations, which regulations shall not be subject to subsection (d)) in any edible portion of such animals after slaughter or in any food yielded by or derived from the living animal.

Finally, section 512(d)(1)(H), as amended by the Animal Drug Amendments of 1968, states that a new animal drug application shall not be approved if:

such drug induces cancer when ingested by man or animal, or, after tests which are appropriate for the evaluation of the safety of such drug, induces cancer in man or animal, except that the foregoing provisions of this subparagraph shall not apply with respect to such drug if the Secretary finds that, under the conditions of use specified in proposed labeling and reasonably certain to be followed in practice (i) such drug will not adversely affect the animals for which it is intended, and (ii) no residue of such drug will be found (by methods of examination prescribed or approved by the Secretary by regulations, which regulations shall not be subject to subsections (c), (d), and (h)), in any edible portion of such animals after slaughter or in any food yielded by or derived from the living animals.

The Subcommittee on Health and Science of the House Interstate and Foreign Commerce Committee conducted hearings on food additives legislation in 1956—1958. The proposed legislation on which the 1956 hearings were held, including a bill introduced by Congressman James J. Delaney, contained no anticancer clause. In 1957, however, Congressman Delaney introduced a revised bill containing the following anticancer clause:

The Secretary shall not approve for use in food any chemical additive found to induce cancer in man, or, after tests, found to induce cancer in animals.

He testified in favor of that provision in 1957 and 1958. In its report on the bill, the Department of Health, Education, and Welfare initially objected to the provision on the following grounds:

> 1. H.R. 7798 specifies in the proposed section 409(b)(2) for the issuance of an order approving the use of a chemical additive shall contain reports of investigations to determine the additive's carcinogenicity. In section 409(b), it specifies that the Secretary "shall not approve for use in food any chemical additive found to induce cancer in man or, after tests, found to induce cancer in animals."
>
> We, of course, agree that no chemical should be permitted to be used in food if, as so used, it may cause cancer. We assume that this, and no more, is the aim of the sponsor. No specific reference to carcinogens is necessary for that purpose, however, since the general requirements of this bill give assurance that no chemical additive can be cleared if there is reasonable doubt about its safety in the respect.
>
> On the other hand, the above-quoted provisions are so broadly phrased that they could be read to bar an additive from the food supply even if it can induce cancer only when used on test animals in a way having no bearing on the question of carcinogenicity for its intended use. This, we think, would not be in the public interest. Scientists, I am advised, can produce cancer in test animals by injecting sugar in a certain manner, and they can produce cancers by injections into test animals of cottonseed oil, olive oil, or tannic acid (a component of many foods). Probably they can do the same thing with other naturally occurring food chemicals. We think that it would be unnecessary and undesirable to rule out of the food supply sugar, vegetable oils, or common table beverages simply because by an extraordinary method of application never encountered at the dining table, it is possible to induce cancer by injecting the substances into the muscles of test animals.

In July 1958, the Committee reported out a bill which required premarketing clearance of food additives, but which contained no anticancer clause.[14] Subsequent to the filing of the report, Congressman Delaney suggested the addition of the following anticancer provision to the bill:

> *Provided*, That no additive be deemed to be safe if it is found to induce cancer when ingested by man or animal, or if it is found, after tests which are appropriate for the evaluation of the safety of food additives, to induce cancer in man or animal.

Reportedly in order to assure enactment of the legislation, the Committee and the Department of Health, Education and Welfare (DHEW) agreed to this amendment. In a letter to the chairman of the committee, the Department accepted the concept of the amendment, again commenting that the anticancer amendment would not change the meaning of the bill:

> The widespread interest in cancer led to suggestions that the food additives legislation should mention disease by name and forbid the approval of any substance that is found upon test to cause cancer in animals. This Department is in complete accord with the intent of these suggestions—uses in food that might produce cancer in man. H.R. 13254 as approved by your committee, will accomplish this intent, since it specifically instructs the Secretary not to issue a regulation permitting use of an additive in food if a fair evaluation of the data before the Secretary fails to establish that the proposed use of the additive will be safe. The scientific tests that are adequate to establish the safety of an additive will give information about the tendency of an additive to produce cancer when it is present in food. Any indication that the additive may thus be carcinogenic would, under the terms of the bill, restrain the Secretary from approving the proposed use of the additive unless and until further testing shows to the point of reasonable certainty that the additive would not produce cancer and thus would be safe under the proposed conditions of use. This would afford good, strong public health protection.
>
> There are many serious conditions other than cancer that may be caused or aggravated by the improper use of chemicals. It is manifestly impracticable to itemize all of them in a bill. To single out one class of diseases for special mention would be anomalous and could be misinterpreted. Hence, in drafting the Department's bill (H.R. 6747) we chose general language that would restrain any use of an additive that would have any adverse effect on the public health. This approach has been followed in II.R. 13254.
>
> At the same time, if it would serve to allay any lingering apprehension on the part of those who desire an explicit statutory mandate on this point, the Department would interpose no objection to appropriate mention of cancer in food additive legislation. If the specific disease were referred to in the law, it would however, be important to everyone to have a clear understanding that this would in no way restrict the Department's freedom in guarding against other harmful effects from food additives.
>
> It would be important, also, to use language that would provide the intended safe-guards without creating unintended and unnecessary complications. For example, the language suggested by some to bar carcinogenic additives would, if read literally, forbid the approval for use in food of any substance that causes any type of cancer in any test animal by any route of administration. This could lead to undesirable results which obviously were not intended by those who suggested the language. Concentrated sugar solution,

lard, certain edible vegetable oils, and even cold water have been reported to cause a type of cancer at the site of injection when injected repeatedly by hypodermic needle into the same spot in a test animal. But scientists have not suggested that these same substances cause cancer when swallowed by mouth.

The enactment of a law which would seem to bar such common materials from the diet on the basis of the evidence described above, would place the agency that administered it in an untenable position. The agency would either have to try to enforce the law literally so as to keep these items out of the diet—evidently an impossible task—or it would have to read between the lines of the law an intent which would make the law workable, without a clear guide from Congress as to what was meant.

This difficulty could readily be avoided, if there is still a desire to make specific mention of cancer in the bill, by providing that "no additive shall be deemed to be safe if it is found to induce cancer when ingested by man or animal, or if it is found, after tests which are appropriate for the evaluation of the safety of food additives, to induce cancer in animals." Such language could be appropriately inserted as a proviso before the semicolon on page 24, line 16, of the bill as reported by your committee.

Thus, with both DHEW and Congressman Delaney proposing identical language, the bill was passed by the House with this amendment.

The bill then went to the Senate, where it was favorably reported without hearings, with the same anticancer clause as was added in the House debate. The Senate Report commented on the anticancer clause as follows:

We have no objections to that amendment whatsoever, but we would point out that in our opinion it is the intent and purpose of this bill even without the amendment, to assure our people that nothing shall be added to the foods they eat which can reasonably be expected to produce any type of cancer in humans or animals. . . . [W]e want the record to show that in our opinion the bill is aimed at preventing the addition to the food our people eat of any substances the ingestion of which reasonable people would expect to produce not just cancer but any disease or disability. In short, we believe the bill reads and means the same with or without the inclusion of the clause referred to. This is also the view of the Food and Drug Administration.[15]

The bill passed the Senate and, after a few minor Senate amendments were agreed to by the House, was signed into law as the Food Additives Amendment of 1958.

Thus, as originally enacted, the Delaney clause addressed the issue of the appropriateness of the evidence that a substance could cause cancer. Congress determined, on the basis of the best evidence available to it, that any substance that caused cancer when ingested, whether by people or by animals, is not to be permitted in the food supply. While scientific judgment is to be brought to bear on the decision whether or not a substance has been "found to induce cancer when ingested," once that determination is made the statute bans the use of the substance in food. Congress thus determined that ingestion is an appropriate test of a carcinogen as a matter of law.

But the Delaney clause also bars from the food supply substances which are found, "after tests which are appropriate for the evaluation of the safety of food additives," to induce cancer. This provision was added in response to the Secretary's concerns that scientists could cause cancers experimentally using common substances injected under the skin of animals. In these cases, a finding that a substance causes cancer does not require its elimination from the food supply. Instead, the Secretary must determine whether the test is "appropriate for the evaluation of the safety of food additives." A substance that is demonstrated to be a carcinogen by a route other than ingestion is not *per se* banned if the Secretary finds that the route of administration is not appropriate.

Not long after enactment of the Food Additives Amendment, the regulation of color additives came into question. Section 406(b) of the FDC Act as enacted in 1938 had permitted the use only of "harmless" food colors, and the Supreme Court ruled in 1958 that the FDA could not set tolerances for safe use of harmful coal-tar colors in specific foods.[16] Accordingly, legislation similar to the food additive legislation was sought to permit the establishment of tolerances for food colors.

The bill initially forwarded to Congress by DHEW contained no anticancer clause. It contained provisions comparable to the food additive legislation, requiring industry to prove the safety of a color before the FDA could permanently list it for approved uses. These provisions that were not controversial and were contained in the final version of the statute. The legislation was first considered by the Senate Committee on Labor and Public Welfare, which reported it out of Committee without inclusion of an anticancer clause.[17]

Just before Thanksgiving in 1959, the Secretary of Health, Education and Welfare concluded that a public statement should be made about the possible contamination of substantial quantities of cranberries with a pesticide, aminotriazol, which the FDA had just recently determined to be a carcinogen. Although the pesticide provisions of the law contained no anticancer clause, it was determined that use of this pesticide could be approved as safe, and that the public should bo warnod of tho potential hazard.

At the same time, the FDA determined that the previously approved use of diethylstilbestrol (DES) in poultry resulted in detectable residues in the liver and skin fat. Although the newly enacted anticancer

clause in the food additive provisions of the law was inapplicable, because DES was subject to a prior sanction and therefore was not a
"food additive," the FDA proposed to withdraw the new drug application for use of this drug in poultry on the ground that the finding of
residues showed that such use of the drug was unsafe.

In January 1960, just after these two major policy decisions relating
to carcinogenic substances, the Secretary appeared before the House
Committee on Interstate and Foreign Commerce to testify on the color
additive legislation. It was obvious that a major issue would be an
anticancer clause. Accordingly, the Secretary requested the National
Cancer Institute to prepare a statement summarizing the present scientific knowledge bearing on the causation of cancer. The conclusion of
this review was that:

> No one at this time can tell how much or how little of a carcinogen
> would be required to produce cancer in any human being, or how
> long it would take the cancer to develop.

After introducing the entire report into the record of the hearing on
the bill, the Secretary testified as follows:

> This is why we have no hesitancy in advocating the inclusion of
> the anticancer clause.
>
> Unless and until there is a sound scientific basis for the establish-
> ment of tolerances for carcinogens, I believe the Government has a
> duty to make clear—in law as well as in administrative policy—that
> it will do everything possible to put persons in a position where
> they will not unnecessarily be adding residues of carcinogens to
> their diet.[18]

The specific anticancer clause recommended by the Department was
basically identical to that included in the final legislation.

The Secretary went on to testify that this clause allowed for far
greater scientific judgment than its critics acknowledged:

> Some of the opposition, Mr. Chairman, to inclusion of an antican-
> cer provision in the proposed color additives amendment arises out
> of a misunderstanding of how this provision works in the food addi-
> tives amendment.
>
> It has been suggested that once a chemical is shown to induce a
> tumor in a single rat, this forecloses further research and forever
> forbids the use of the chemical in food. This is not true. The
> conclusion that an additive "is found to induce cancer when in-
> gested by man or animal" is a scientific one. The conclusion is
> reached by competent scientists using widely accepted scientific
> testing methods and critical judgment. An isolated and inexplicable
> tumor would not be a basis for concluding that the test substance
> produces cancer.[19]

The concept of appropriate tests was carried over to the color additives legislation, which, as introduced and as enacted, contains two parts to its Delaney clause. A color additive which will or may result in ingestion is deemed unsafe if it is found to cause cancer when ingested or if it is found to cause cancer after tests "which are appropriate for the evaluation of the safety of additives for use in food." A color additive that will not be ingested, on the other hand, falls within the proscription of the color additive Delaney clause only if it is found to cause cancer "after tests which are appropriate for the evaluation of the safety of additives for such use [i.e., noningestion], or after other relevant exposure of man or animal to such additive." The distinction between internal and external uses of color additives was recognized by the Secretary, whose department drafted the clause. In the section-by-section analysis of the bill accompanying the Secretary's transmittal letter, the provision was explained as follows:

> In determining whether the use of a color additive is safe, the Secretary is required to consider a broad range of factors. In particular, however, the color additive may not be listed if it has relevant carcinogenic potential.[20]

Because no hearings had been held prior to the addition of the Delaney clause to the Food Additives Amendment of 1958, the extensive House hearings on the color additives legislation focused primarily on the concept of the Delaney clause itself, and little attention was devolted to the language differences between the two different versions. At one point in the hearings, however, the Secretary did point up the difference:

> The Chairman. Otherwise these two bills, the House and Senate bills, are identical, with the exception of page 10 in the House bill, which has reference to the Delaney amendment—is that identical with the amendment in the food additive law, except in that respect?
> Secretary Flemming. It is not identical because of the colors that are applied to external parts.
> The Secretary. Except as to the colors?
> Secretary Flemming. That is right—except for the changes that have to be made in order to adapt it to the color situation—except for that, it is identical, so far as the policy issue is concerned.
> The Chairman. In other words, it gives identical authority, insofar as the color additive is concerned, as before?
> Secretary Flemming. That is correct.[21]

That the legislation distinguished between internal and external use of colors seemed to be well understood. Testifying in support of the legislation, Dr. Harold Aaron, Medical Advisor to Consumers Union,

said that Consumers Union believed that, to deal effectively with the
problem of color additives, a bill "should forbid the use of any color in
amount that causes cancer in man or animal, distinguishing, or course,
between the external and internal use of such colors.[22]

As enacted, then, the Delaney clause in the Color Additive Amendments establishes the following scheme regarding potential carcinogens.
A color additive which will or may be ingested is deemed unsafe if it
causes cancer when ingested, or if it causes cancer by other routes
appropriate for the evaluation of the safety of the use of color in food,
drugs, or cosmetics. A color additive that will not be ingested is
deemed unsafe only if it is found to cause cancer by tests appropriate
for the evaluation of its safety for its use, or by other relevant
exposure.

There is thus no *per se* bar to the use of a carcinogenic color in a
cosmetic product that will not be ingested. Such a color may be barred
under the Delaney clause only if the Secretary determines that the test
demonstrating carcinogenesis is "appropriate" for the evaluation of the
color additive for its external use. This important distinction between
the legal standard applicable to ingested and external colors is supported by the language of the statute and its legislative history, and
was clearly recognized by the FDA at the time the legislation was
enacted. FDA General Counsel William W. Goodrich, addressing an
industry group before the amendments were enacted, explained that
the Delaney clause would:

> Prohibit the approval of any color which may be ingested if the
> substance is found to induce cancer when ingested by man or
> animal or after other tests appropriate for the evaluation of the
> safety of food additives, to prohibit other exposures to color, if
> tests relevant to such exposure show that the additive will induce
> cancer in man or animal.[23]

In a speech delivered less than two weeks after the Color Additive
Amendments were signed into law, J. K. Kirk, Assistant to the FDA
Commissioner, explained the new provisions at an American Oil Chem-
ists Short Course. Mr. Kirk noted that:

> During its consideration by Congress the Color Additives Bill was
> quite a controversial subject because it includes the so-called
> Delaney Clause, which is essentially carried over from the Food
> Additives Amendment. In the case of the Food Additives Amend-
> ment, the Delaney Clause prohibits the establishment of any regu-
> lation for an additive which has been shown to induce cancer upon
> ingestion by man or animal or to induce cancer by other tests
> appropriate for the evaluation of food additives. The color bill
> carries a comparable provision taking into account, of course, that
> the tests will be appropriate for the proposed use of the particular
> color.[24]

Franklin D. Clark, Assistant to the FDA Deputy Commissioner, in a speech at the 1960 Joint National Conference of the Food and Drug Administration and the Food Law Institute, said the Delaney clause provided:

> Automatic disapproval of any color which may be ingested, if the substance is found to induce cancer when ingested by man or animal or after other tests appropriate for the evaluation of the safety of color additives. Automatic disapproval of color additives for other uses if tests relevant to such uses show that the additive will induce cancer in man or animal.[25]

In another speech, Mr. Clark stated that:

> The Color Additive Amendments also contain the Delaney Clause, which prohibits the listing of color additives which cause cancer when ingested by laboratory animals or, if the substance is not one which is ingested, if it is found to be carcinogenic by other appropriate tests.[26]

Perhaps the surest indication of the contemporary understanding of the meaning of the color additive Delaney clause is in the FDA's proposed regulations implementing the Color Additive Amendments.[27] Section 8.36 of the proposed regulations concerned application of the Delaney clause:

> (a) Color additives that may be ingested. Whenever the scientific data before the Commissioner (either the reports from the scientific literature or the results of biological testing) suggest the possibility that the color additive or any of its componets or impurities has induced cancer when ingested by man or animal, the Commissioner shall determine whether, based on the best judgment of appropriately qualified scientists, cancer has been induced and whether the color additive or any of its components or impurities was the causative substance. If it is his best judgment that the data do not establish these facts, the anticancer clause is not applicable; and if the data considered as a whole establish that the color additive will be safe under the conditions that can be specified in the regulation, it may be listed for such use. But if, in the best judgment of the Commissioner, based on information from qualified scientists, cancer has been induced by ingestion, no regulation may issue which permits its use.
>
> (b) Color additives that will not be ingested. Whenever the scientific data before the Commissioner suggest the possibility that the color additive or any of its components or impurities had induced cancer in man or animals by routes other than ingestion, the Commissioner shall determine whether, based on the best judgment of appropriately qualified scientists, the test suggesting the possibility of carcinogenesis is appropriate for the evaluation of the

color additive for a use which does not involve ingestion, cancer
has been induced, and the color additive or any of its components
or impurities was the causative substance. If it is his best judgment that the data do not establish these facts, the anticancer
clause is not applicable to preclude external drug and cosmetic
uses, and if the data as a whole establish that the color additive
will be safe under conditions which can be specified in the regulations, it may be listed for such use. But, if, in the best judgment of the Commissioner based on information from qualified
scientists the test is an appropriate one for the consideration of
safety for the proposed external use, and cancer has been induced
by the color additive or any of its components or impurities no
regulation may issue which permits its use in external drugs and
cosmetics.

Only minor editorial changes were made when proposed section 8.36
was published as part of the final color additive regulations.[28]
Although other sections of the regulations were successfully challenged,[29] the relevant paragraphs of section 8.36 appear today as
21 Code of Federal Regulations §70.50.

III. FDA ENFORCEMENT AGAINST CARCINOGENIC COSMETIC INGREDIENTS

As early as 1950, the FDA had adopted the administrative policy that
any substance shown to induce cancer in test animals would be regarded as a "poisonous or deleterious substance" that could not law-
fully be used in food. The agency's actions against carcinogenic
substances during 1950—1970 were, for the most part, taken under
the general safety provisions of the act rather than the specific anticancer clauses. Indeed, the FDA explicity invoked the anticancer
clauses during this period only to prohibit two little-known food
packaging materials.

Beginning in the early 1970s, however, the agency was forced
directly to address the issue of carcinogenic substances in the products
it regulates. Some years earlier the FDA had approved the carcinogenic animal drug, diethylstilbestrol (DES), for the use in animal feed
and implants as a growth promotant. With the increasing sensitivity
of detection methodology, residues of DES were inevitably found in the
liver of animals fed or implanted with DES. This raised very dramatically the possibility that virtually all carcinogenic animal drugs might,
at some point in the future, be found to leave trace residues in the
human food.

As already noted, the animal drug provisions of the FDC Act
provide that a carcinogenic animal drug may be used in food-producing
animals if "no residue" of the drug is found in the food derived from
that animal using the detection method prescribed by the FDA.

Obviously, the level of sensitivity of the detection method determines the likelihood of finding any such residue. If the detection method is very sensitive, it will pick up minute residues. If it is not very sensitive, it may well miss significant residues.

Faced with a number of important decisions respecting carcinogenic animal drugs in the early 1970s, the FDA determined to adopt quantitative risk assessment as a means of determining the required sensitivity of the detection method in order to justify approval of any carcinogenic animal drug. This approach, called the "sensitivity of the method" (or SOM) approach, has been pursued by the FDA since then for animal drugs and, as will be discussed below, has broader application to cosmetic products as well.

Under this approach, a quantitative risk assessment is undertaken for any carcinogenic animal drug, using a mathematical model to extrapolate from the high doses fed to test animals to the low level at which humans may be exposed. Compensating factors are included to account for the differences between animals and humans. FDA has concluded that a one in one million lifetime risk of cancer represents an insignificant risk and therefore is "safe." Thus, using this approach, it is possible to calculate a level of the animal drug which, if it were to occur in the food derived from the animal, would be regarded as safe. This is the level set as the required sensitivity for the detection method. If no residue is found at this level, the FDA reasons that the undetected presence of a residue at a lower level would present only an insignificant risk.[30]

This approach of regulating carcinogenic animal drugs is directly applicable to a number of other analogous situations that arise under the FDC Act. It could, for example, be used in determining insignificant levels of migration of carcinogenic packaging substances into food and cosmetics. It could also be used to determine insignificant levels of trace contaminants in food and cosmetics. Finally, it could be used to determine insignificant levels of carcinogenic substances in cosmetics that are only applied topically to humans and that are not sufficiently absorbed through the skin to present any significant risk. The FDA did not, however, immediately pursue the use of quantitative risk assessment for any of these purposes.

Instead, the agency continued to pursue carcinogenic substances in food and cosmetics in the same way it had during 1950—1970. In 1976, for example, the FDA summarily revoked the provisional listing for the color additive FD&C Red No. 2 on the basis of a highly controversial statistical reanalysis of a badly flawed animal study showing carcinogenicity. This action was upheld on court appeal.[31] The court quoted, in a footnote, Congressman Delaney's statement from the legislative history of the Color Additive Amendments that, because color additives "have no value at all, except so-called eye appeal," the FDA

should be particularly careful with them.[32] The court relied very
heavily on the potential carcinogenicity of the color to uphold the
agency's action:

> Courts have traditionally recognized a special judicial interest
> in protecting the public health, particularly where "the matter
> involved is as sensitive and fright laden as cancer." Where the
> harm envisaged is cancer, courts have recognized the need for action
> based upon lower standards of proof than otherwise acceptable.[33]

Similarly, the FDA banned the use of the carcinogenic packaging
ingredient, acrylonitrile, merely on the theoretical possibility that it
could migrate to food. On appeal in this instance, however, the court
held that the FDA must show more than a theoretical possibility to
justify action of this kind, and pointed out that the agency has inherent discretion under the FDC Act to ignore insignificant or de minimis
risks.[34] This has been widely interpreted as inviting the use of
quantitative risk assessment in determining the dividing line between
significant and insignificant migration of packaging ingredients to food
and cosmetics.

In January 1978, the FDA published a proposed regulation designed
to require warning consumers about coal-tar hair dyes containing 4-
methoxy-M-phenylenediamine (4-MMPD), also known as 2,4-diaminoan-
isole (2,4-DAA).[35] On the basis of an animal bioassay of 4-MMPD conducted for the National Cancer Institute, the FDA concluded that the
compound is carcinogenic in rats and mice. Because the agency determined that 4-MMPD "may be absorbed through the scalp during hair
dying, and pose a risk of cancer to users," it proposed to require hair
dyes containing 4-MMPD to bear the following label statement:

> Warning—Contains an ingredient that can penetrate your skin
> and has been determined to cause cancer in laboratory animals.

Because hair dyes are often applied in beauty salons, the FDA also
proposed requiring beauty salons to display a poster stating:

> Some hair dyes contain ingredients which may cause cancer.
> These hair dyes are required to bear a label warning. Ask to
> see the label of the product intended for hair.

The FDA proposed to warn consumers about the carcinogenic risk,
rather than to ban the ingredient outright, only because the FDC Act
exempts coal-tar hair dyes from the safety provisions otherwise applicable to cosmetics. An attempt by the agency to narrow this exemption
by administrative reinterpretation had been held unlawful a decade
earlier.[36]

The Cosmetic, Toiletry and Fragrance Association, Inc. (CTFA)
submitted extensive comments arguing against the proposed warnings.
The animal bioassay was shown to be seriously flawed, and the CTFA

contended that its results could not be accepted as scientifically valid. Even if the bioassay were to be accepted as valid, however, the maximum potential risk associated with hair dyes containing 4-MMPD was shown by the CTFA to be so small as not to be significant and, in any event, far smaller than carcinogenic risks for which the FDA and other agencies have never required warnings. The CTFA argued that to require a warning for 4-MMPD, and not for other comparable or greater risks, would be arbitrary and capricious.

To support its argument, the CTFA submitted data showing minimal absorption of 4-MMPD and a quantitative risk assessment performed by an outside consultant demonstrating only a very minor risk. The risk assessment showed that the overall individual lifetime risk to the population, using the most conservative method of extrapolation, could be no more than one in 4.3 million. Breaking the use of 4-MMPD down into various hair dye shades (because this substance is used at different concentrations in different shades), the risk ranged from one in 420,000 for the black shades to one in 62 million for the blond shades. The CTFA also obtained from the FDA various risk assessments conducted by the agency, which similarly showed a relatively low risk from this substance.

Finally, the CTFA argued that, if a warning is to be required, it should be revised in major respects. First, it should contain a complete and accurate statement of the maximum potential risk from 4-MMPD, including references to the high doses that were fed test animals, the uncertain relationship between animal feeding studies and potential human risk, and the small potential human risk that the substance was projected to present. Second, the CTFA contended that the proposed information about 4-MMPD should not be required to appear on posters distributed to beauty salons.

Eighteen months later, the FDA issued its final regulation on this matter, rejecting the CTFA position.[37] The FDA agreed that the bioassay was flawed in several respects, but concluded that those flaws were not sufficient to invalidate the test results. The agency thus determined that 4-MMPD is an animal carcinogen and poses a human risk.

The FDA did not dispute the CTFA quantitative risk assessment, but refused to take it into account in any way. The agency stated that any cancer risk at all, however small, from any deliverately added ingredient in food or cosmetics, must either be banned or, if the law prevents banning (as is true with coal-tar hair dyes), be the subject of a label warning. The FDA reiterated that it would have banned 4-MMPD altogether if it was not prevented from doing so by the hair dye exemption.

The FDA stated that the government can attack hazards seriatim, and is not required to deal with all of them at one time. The agency admitted that "many, perhaps most foods in a supermarket" today contain a carcinogenic constituent, but stated that it would not require a warning on them because it would be "confusing" to the public.

The FDA did agree to withdraw the proposed requirement of a warning poster in beauty salons. It declined, however, to change the label warning from the proposal. The agency also stated that industry would not be permitted to explain the warning to consumers, and in particular to inform consumers that hair dyes containing 4-MMPD present a much smaller risk of cancer than many other common products and activities.

Affected hair dye manufacturers then filed suit in the U.S. District Court for the Southern District of Georgia. Faced with this legal challenge to its position, the FDA consented to a court order remanding the matter to the agency for further consideration. The court order specifically required the use of quantitative risk assessment in determining the carcinogenic risk of 4-MMPD or any other hair dye in the future:

> 4. The regulation is remanded to FDA for reconsideration and further rulemaking, as necessary. In connection with any rulemaking undertaken pursuant to this remand, FDA agrees to propose the utilization of scientifically accepted procedures of risk assessment and to raise the issue as to whether, in view of those procedures, hair dyes containing 4-MMPD present a generaly recognized level of insignificant risk to human health.[38]

A similar situation had been developing for some years with another hair dye ingredient, lead acetate. Because this hair dye is not a coaltar ingredient, it is subject to the Color Additive Amendments. Substantial scientific work had been undertaken on the ingredient, and it was ready for approval by the FDA, when the agency concluded that lead acetate is an animal carcinogen and requested public comment on its regulatory status for use both as a hair dye ingredient and as a soldering agent in cans used to can food.[39]

On the basis of a skin penetration study conducted on humans, the principal affected manufacturer presented to the FDA a quantitative risk assessment demonstrating that lead acetate represents an insignificant human cancer risk. The agency agreed that the highest possible cancer risk from lead acetate in hair dyes was between one in 5 million and one in 18.5 million, and therefore approved the substance for hair dye use under the Color Additive Amendments.[40] In accordance with the legislative history and the interpretation of the Color Additive Amendments set out above, the FDA concluded that the animal carcinogenicity test, while entirely valid from a scientific standpoint, demonsuch an insignificant human risk that it could not be regarded as either "appropriate" or "relevant" within the meaning of those terms as used in the statute.

In April 1982, the FDA published an advance notice of proposed rule making to establish a policy for regulating carcinogenic chemicals in food and color additives.[41] The policy proposed by the FDA would differentiate between a functional "additive" and any nonfunctional "constituent" of that additive. The Delaney clause would be interpreted to apply only if the "additive" as a whole, including its "constituent," is tested and found to induce cancer in laboratory animals. The general safety provisions of the FDC Act, in contrast, would be interpreted to apply both to the "additive" as a whole, including its "constituents," and to each "constituent" viewed separately. Quantitative risk assessment would not be applied in determing whether an "additive" as a whole is safe under the Delaney clause, but would be applied in determining whether the "additive" as a whole, or any "constituent" viewed separately, is safe under the general safety provisions of the FDC Act.

Accompanying this advance notice of proposed rulemaking was a final order permanently listing D&C Green No. 6 for use in external cosmetics and drugs, in spite of the presence of the carcinogenic contaminant p-toluidine.[42] Using quantitative risk assessment, FDA determined that the carcinogenic risk to humans from the trace amount of p-toluidine is insignificant. Two months later, the FDA permanently listed D&C Green N0. 5 for use in external cosmetics and drugs, notwithstanding the presence of a trace amount of p-toluidine, in accordance with the same policy.[43] This action was challenged by a consumer activist in the courts and the FDA constituents policy and the approval of D&C Green N0. 5 were upheld as lawful.[44]

Thus, in recent situations, the agency has demonstrated its willingness to use quantitative risk assessment to determine the significance of any potential cancer risk, and to disregard any such risk that may properly be concluded to be *de minimis* or insignificant.

IV. FUTURE FDA POLICY

Without question, the regulation of potential and proven carcinogens by the FDA in the future will be undertaken using the principles of quantitative risk assessment set out above. As scientific knowledge progresses, moreover, it is inevitable that the agency will further refine its approach to regulation of carcinogens. It is likely, for example, that the FDA will begin to differentiate between primary (genotoxic) and secondary (epigenetic) carcinogens. Where metabolic and pharmacokinetic work is undertaken to demonstrate a secondary rather than a primary mechanism of action, the agency may well be able to conclude that a safe level exists for the use of the substance.

Even the use of quantitative risk assessment and recognition of a difference between primary and secondary carcinogens will not, however, be sufficient to allow the continued use of all carcinogenic cosmetic ingredients in the future. Some will be found to present a

important that the industry substantiate the safety of as many cosmetic ingredients, including color additives, as possible. If a few are
subsequently banned as significant risks, there will be a sufficient
number of acceptable substitutes to avoid any reduction in consumer
choice.

NOTES

1   Shattuck, *Report of the Sanitary Commission of Massachusetts*,
    1850, p. 104.
2   For the mortality statistics in this paragraph, see the annual
    publication of the Public Health Service, DHHS (formerly DHEW),
    Health—United States, 1975—1980.
3   Lilienfield, *Chronic Diseases*, in Maxcy and Rosenau, *Preventive
    Medicine and Public Health*, 1973, p. 497.
4   Fries, Aging, Natural Death, and the Compression of Morbidity,
    N. Engl. J. Med. 303:130, 132 (1980).
5   Public Health Service, *Health—United States*, DHHS Pub. No.
    (PHS) 81-1232, 1980, p. 26.
6   Doll and Peto, The Causes of Cancer: Qualitative Estimates of
    Avoidable Risks of Cancer in the United States Today, *J. Natl.
    Cancer Inst. 66*:1191, 1256 (June 1981), and National Academy
    of Sciences, *Diet, Nutrition, and Cancer*, 1982, p. 1-1.
7   See notes 2 and 6.
8   National Academy of Sciences, *Diet, Nutrition, and Cancer*, 1982,
    p.1-14.
9   Office of Technology Assessment, *Assessment of Technologies for
    Determining Cancer Risks from the Environment*, 1981, p. 4.
10 *Id.* at 109.
11 43 *Fed. Reg.* 38206 (August 25, 1978).
12 For a more complete legislative history of the anticancer clauses,
    see *Agriculture—Environmental, and Consumer Protection Appro-
    priations for 1975*, Hearings Before a Subcommittee of the Com-
    mittee on Appropriations, House of Representatives, 93rd Cong.,
    2nd Sess., part 8, at 180 (1974) and Hutt, Public Policy Issues
    in Regulating Carcinogens in Food, 33 *Food Drug Cosmet. L. J.*
    541 (1978).
13 H.R. Rep. No. 2182, 82nd Cong., 2nd Sess. (1952).
14 H.R. Rep. No. 2284, 85th Cong., 2nd Sess. (1958).
15 S. Rep. No. 2422, 85th Cong., 2nd Sess. (1958).
16 *Flemming v. Florida Citrus Exchange*, 358 U.S. 153 (1958).
17 S. Rep. No. 795, 86th Cong., 1st Sess. (1958).
18 *Color Additives*, Hearings Before the House Committee on Inter-
    state and Foreign Commerce, 86th Cong., 2nd Sess. 61 (1960).
19 *Id.* at 62.
20 H.R. Rep. No. 1761, 86th Cong., 2nd Sess. 71 (1960).
21 Note 18 *supra* at 102.

22 *Id.* at 335-336.
23 Goodrich, *From Bellyache to Headache—Is the Color-Additives Bill the Remedy?*, 15 *Food Drug Cosmet. L. J.* 297, 300 (1960).
24 Kirk, Address to American Oil Chemists Short Course (July 25, 1960).
25 Clark, Color Additives, 15 *Food Drug Cosmet. L. J.* 761, 764 (1960).
26 Clark, The Regulatory Functions of the Food and Drug Administration, 16 *Food Drug Cosmet. L. J.* 500, 505 (1961).
27 26 *Fed. Reg.* 679 (January 24, 1961).
28 28 *Fed. Reg.* 6439 (June 22, 1963).
29 *Toilet Goods Assn., Inc. v. Finch*, 419 F.2d 21 (2d Cir. 1969).
30 38 *Fed. Reg.* 19226 (July 19, 1973); 42 *Fed. Reg.* 10412 (February 22, 1977); 44 *Fed. Reg.* 17070 (March 20, 1979).
31 *Certified Color Manufacturers Association v. Matthews*, 543 F.2d 284 (D.C. Cir. 1976).
32 *Id.* at 196, n. 70.
33 *Id.* at 297-298.
34 *Monsanto Co. v. Kennedy*, 613 F.2d 947 (D.C. Cir. 1979).
35 43 *Fed. Reg.* 1101 (January 6, 1978).
36 *TGA v. Finch*, 419 F.2d 21 (2d Cir. 1969).
37 44 *Fed. Reg.* 59509 (October 16, 1979).
38 *Carson Products Co. v. Department of HHS*, Food Drug Cosmet. Law Rep. (CCH) ¶38,071 (S.D. Ga. 1980).
39 44 *Fed. Reg.* 51233 (August 31, 1979); 44 *Fed. Reg.* 12205 (March 6, 1979).
40 45 *Fed. Reg.* 72112 (October 31, 1980); 46 *Fed. Reg.* 15500 (March 6, 1981).
41 47 *Fed. Reg.* 14464 (April 2, 1982).
42 47 *Fed. Reg.* 14138 (April 2, 1982).
43 47 *Fed. Reg.* 24278 (June 4, 1982).
44 *Scott v. FDA*, Food Drug Cosmet. L. Rep. ¶38,260 (6th Cir., February 23, 1984).

36

RISK ASSESSMENT IN PERSPECTIVE

JOSEPH V. RODRICKS

Environ Corporation, Washington, D.C.

I. INTRODUCTION

It appears that the scientific, regulatory, business, and consumerist communities remain sharply divided on the issue of risk assessment and its role in regulatory decision making. Critics maintain that risk assessment cannot possibly fulfill the promises made on its behalf, and that its scientific underpinnings are either absent or extremely fragile. They also hold that currently used methods of risk assessment almost certainly misrepresent the truth about risk. The direction in which current methodologies are said to err often depends on the interests of the critic: Consumerists and environmentalists find that risks are routinely underestimated, while those whose products may be adversely affected find that risk projections are uniformly too high. Although advocates of the use of risk assessment are usually cautious about noting the uncertainties associated with current methodologies, they tend to forget or ignore these uncertainties when using the assessments for decision making. As a result, the critics of risk assessment find additional support for their skepticism.

This unhappy state of affairs is perhaps not surprising, given the relative immaturity of the subject and the degree of misunderstanding that exists regarding its conceptual basis and its practice. It is apparent, however, that risk assessment will continue to be used as a basis for regulatory decision making, and that its uses will increase. In this chapter I shall make an attempt to define and describe the subject of risk assessment and deal with its use in regulatory decision making; I shall also attempt to project its likely evolution.

II. WHAT RISK ASSESSMENT IS NOT

It is customary to begin a discussion of this type by defining the subject, but in this case there are compelling reasons to ensure at the outset that certain misconceptions are not allowed to linger.

The first of these arises when risk assessment is taken to be nothing but a problem of selecting and applying a mathematical model to predict risk at low levels of exposure from observations of risk at high exposure levels. To view risk assessment in such limited terms trivializes the subject and, moreover, promotes unwarranted criticism. Later in this chapter I shall attempt to show that this view of risk assessment is unduly narrow, and that excessive concern about the appropriateness of various extrapolation models will tend to divert attention from many other significant issues.

A second misconception concerns the pretense that the methodologies now in use yield estimates of actual human risk. Although they may do so, we have no readily definable method to know whether this is the case. I shall hope to show, however, that such methods need not be shown to yield the "truth" about human risk to be useful decision-making tools.

Finally, risk assessment should be seen to have only the goal of defining the likelihood of harm, and should not be considered to encompass questions of risk management. Thus, risk assessment is a scientific activity (even though science cannot yet supply all necessary answers), and is not concerned with the question of whether or not exposure to a substance shown to pose a risk should be limited or eliminated, nor with the means for achieving any such controls on exposure. Traditional methods of defining acceptable exposures have tended to obscure the difference between scientific and policy issues, and this has given rise to problems that can be overcome only if sharp distinctions are made between risk assessment and risk management. In the discussion to follow I shall further elaborate on these three issues.

III. THE PROBLEM

Risk is the probability of injury or death. For some activities we encounter no great difficulties in determining risk. Thus, it is possible to estimate quite accurately the risks of accidental death due to such activities as driving a car, working in a coal mine, riding a bicycle, hiking in the desert, or eating low-acid canned foods (botulism). Estimation of such risks is readily accomplished because historical statistical data are available, and because there is little difficulty in demonstrating the causal connections between injury and these types of activities. To estimate such risks is the work of actuaries, most of whom are employed by insurance companies.

Other risks cannot be so easily estimated because the necessary actuarial data do not exist and frequently cannot even be collected. Many of the potential risks from exposure to chemicals are in this second category. In addition to the absence of actuarial data relating to them, these risks tend to have the following characteristics:

1. Suspicion that exposure may lead to injury usually results from experimental observations, commonly involving animals.
2. Identifiable injury does not occur immediately following exposure, and may sometimes not occur for many years after initial exposure.
3. The conditions of exposure (level, frequency, duration, route) that give rise to experimentally-observed injury are frequently different (sometimes radically so) from the conditions of actual human exposure, which themselves may not be well defined.
4. The experimental environments in which information is collected on potential injury from a chemical exposure are usually free of the large number of factors in the human environment that may biologically or chemically interact with the chemical, and thus alter its capacity to cause injury.

5. Experiments used to collect data on chemical injury may involve several different species of test animals, and they may yield quantitatively, and sometimes qualitatively, different results. It is usually not possible to identify the species that best mimics human response.

6. Epidemiological investigations of chronic exposure or injury, while yielding data on the species of concern, are frequently limited because they cannot usually detect small but possibly important effects, and because they frequently cannot provide evidence of strict causation. Moreover, they can be conducted only after exposure has occurred and thus cannot be used to decide whether exposure to a newly introduced substance should be permitted.

Given the above, it would seem foolish to attempt to predict the human risks associated with exposures to chemicals. Many scientists faced with such a problem are not willing to attempt an answer, and proclaim the need for more research. Indeed, there are probably few greater research needs than those associated with this problem, and I shall have more to say about this subject later. However, if we fail to find workable approaches to the problem of assessing chemical risk, we shall indeed find ourselves in a serious predicament. We would be faced with the prospect of not being able to decide whether exposure to a chemical can or cannot be permitted, unless we make the decision completely unrelated to the question of risk. The latter course seems highly undesirable, although it has sometimes been taken.[1]

In the context of regulatory decision making, the difficulties of defining the nature and magnitude of chemical risk can be overcome (indeed, have been for years) by the application of certain *operational* schemes. Application of these schemes cannot be claimed to lead to true estimates of human risk, yet there are good reasons to believe that they meet the desirable criterion of being capable of distinguishing low from high risk exposures.

These operational schemes have been designed to address the following problem: Data are available suggesting the possibility of chemically induced injury under certain conditions of exposure, and we must decide whether and with what probability they are likely to occur under other conditions. This reduces to a problem of drawing inferences, and I shall now describe how the available operational schemes have approached its solution.

IV. SAFETY ASSESSMENT SCHEMES

The task of assigning safe exposure levels for chemicals has traditionally been assigned to toxicologists. During the first half of this century, this problem arose in connection with food additives, pesticides, drugs, and occupational exposures. Although toxicologists experimented with a variety of approaches, there emerged a scheme for assigning safe exposure levels that was based on the application of safety factors to experimental toxicity data, derived for the most part from studies in animals, but also from controlled studies involving humans. In general, toxicologists would divide experimentally determined "no observed effect levels" (NOELs) by a safety factor, the magnitude of which depended on the toxicologists' assessment of the quality of the underlying experimental data and on the extent to which the experimental exposure conditions mimicked those of the human population that was to be protected. The level of exposure arrived at by application of safety factors has never been claimed to be totally without risk, but it became widely accepted within the community of toxicologists that this type of scheme is appropriate for defining acceptable human exposure levels (except for carcinogens—see below). Thus arose the concepts of "acceptable daily intake" (ADI) for food and color additives and pesticides, and Threshold Limit Values (TLVs) for exposures in the workplace [1,2].

The central concept underlying this approach is that for most forms of toxicity, the production of effects requires a certain minimum dose (a threshold dose), and that unless the minimum dose is exceeded, no effect will occur. The essential problem is identifying this dose.

The experimental NOEL may approximate such a dose in the group of test animals studied (see below for further discussion of this point). However, there are plausible biological reasons as well as empirical evidence to engender the belief that the threshold dose is not fixed, and that it varies among individuals, and that in general, humans may be more susceptible than experimental animals to the toxic effects of chemicals. It thus became the practice to apply safety factors to NOELs in order to compensate for these possibilities.

This safety assessment scheme, which is still in wide use, has never been claimed to provide absolute safety (zero risk). There is, in fact, no scheme that could do so. But it does claim that any residual risk associated with exposures corresponding to an ADI is almot certainly very low. This is probably the case for most types of toxic agents, but we have no method to determine whether it is. But because the scheme claims to provide an estimate of low risk exposures, it is, at least implicitly, a risk assessment scheme that makes no attempt to characterize the risk that remains at exposures said to be "acceptable."

V. LIMITATIONS IN THE SAFETY ASSESSMENT SCHEME

The safety assessment scheme described above appears to have provided adequate public health protection, and will no doubt continue in use for some time to come. There are, however, certain limitations in the scheme that should be acknowledged, and although I shall now focus primarily on the application of this scheme to food and cosmetics exposures, the commentary has broader applicability.

First, the use of ADIs tends to give the impression that exposures to chemicals are either "safe" (below the ADI) or "unsafe" (above the ADI). Those who work in the area know that this is a false interpretation, because risk to a population does not simply "disappear" at a given dose. In fact there may be for some agents a range of doses well above their ADIs that fall well within the low or even zero risk category. On the other hand, risk may sometimes rise rapidly through and above an ADI. The point is that there are no sharp divisions in the continuum of dose-risk relations, at least insofar as we are concerned with population, not individual, risks.

It should be recognized that, no matter what risk assessment scheme is used, there will finally emerge an exposure level which will be said to be acceptable. There will probably always be a tendency to view such "official levels" as the dividing lines between "safe" and "unsafe" exposures. I suggest, however, that the use of a scheme that provides explicit estimates of risk, and from which policy makers decide on the risk that is tolerable in specific circumstances, is less likely to be misinterpreted as providing such sharp distinctions.

Procedures for estimating and using NOELs can be wasteful of data. The selection of the highest dose at which "no effect" is observed (the NOEL) ignores the possibility that the lack of observed effects could have been the result of chance variation about a true effect. If two experiments, identical except for sample size, yield identical NOELs, the larger experiment provides greater evidence of safety. The NOEL approach also does not fully utilize the experimental dose-response information. Dose-responses that decrease sharply with decreasing dose have different implications for risks at doses below the observed NOEL (i.e., the human dose) than do shallower dose responses. However, this difference is not accounted for in the setting of ADIs. In effect, the ADI scheme disregards the dose-response data.

Serious questions can also be raised about the use of specific "safety factors" to establish ADIs without scientific evidence to support the magnitude of such factors. In fact, there is nothing but custom to support the use of any specific safety factor. Because it can also be reasonably argued that the selection of specific safety factors is a matter of policy, not science, the safety assessment scheme can be seen as a blend of scientific and policy decisions that cannot be easily separated.

It appears, then, that some modification in the "NOEL-safety factor" approach is in order. There are difficulties that must be overcome before we can arrive at suitable alternative methods, but it is time to begin to move away from the concept that toxicologists can decide what is "safe" by simply selecting arbitrary "safety factors"; and to find ways to use the dose-response information in establishing ADIs.

Finally, the scheme has generally not been considered, even by its proponents, appropriate to apply to carcinogens. It is difficult to ascertain the reason for this. On the one hand, it may stem from the legal stricture (which exists in the United States in the form of the Delaney clause of the Food, Drug, and Cosmetic Act) that no ADI can be established for a carcinogenic additive, in which case no safety assessment scheme is needed. On the other hand, it may stem from a scientific view that the mode of action of carcinogens is such that exposure at a calculated ADI (experimental NOELs can be defined for many carcinogens) is almost assuredly going to pose a risk of cancer, regardless of the magnitude of the safety factor. Exposure to other types of toxic agents at a calculated ADI will, in many cases, also pose a finite risk. For both carcinogens and other types of toxicants, it is not possible to show that zero population risk is achieved at any finite dose. It is possible, however, to estimate low or even neglible risk doses for all forms of toxicants, including carcinogens, although I suggest that the traditional methods for establishing ADIs are probably not the best ways to accomplish these goals.

VI. THE PROBLEM OF CARCINOGENIC ADDITIVES

Directly introduced food and color additives are subject to the restrictions of the Delaney clause, and as long as this is the case, there is no need to estimate carcinogenic risk for such substances. However, food constituents of other types that may be carcinogenic are not so easily dealt with. The problem of carcinogenic residues of drugs used in food-producing animals provides one illustration of the need to develop a means for establishing tolerable or negligible doses for some carcinogens. The law governing the use of such drugs states that a carcinogenic drug may be used in food-producing animals if, under the conditions of its use, "no residue" of the drug is found in edible animal products when these products are examined with a method of analysis the Food and Drug Administration (FDA) finds acceptable [3]. This phraseology seems to answer the critical question of how much human exposure to a carcinogenic drug residue is acceptable. Of course, it does not provide a definitive answer because it does not establish the criteria the FDA must apply in order to judge the acceptability of a given method of analysis proposed to measure a given carcinogenic residue. The most important criterion—the requirement for limit of measurement (or "sensitivity") of such a method of analysis—is omitted from the language of the law. The criterion of method sensitivity is, of course, critical because it defines the maximum level of carcinogenic residue that shall be permitted to escape detection and possibly remain in food for human consumption. This dilemma hinges on the inescapable conclusion that we can never be certain that edible tissues contain "no residue" of any drug (in the absolute sense of the term) whenever a a drug has been administered to a food-producing animal. We can only be sure there is no residue at a concentration above the sensitivity limit of the analytical method used to examine tissues for the presence of drug residues.

An operational definition of "no residue" is therefore required, and the FDA has proposed that the concentration of a carcinogenic drug residue that corresponds to a maximum lifetime risk of cancer not greater than one in one million is a suitable definition [3]. This is a risk-based standard. It takes into account the fact that different carcinogens pose different risks at the same level of exposure; expressed in other terms, it holds that different levels of exposure to different carcinogens can present the same risk. Thus, applying a fixed risk standard results in varying operational definitions of "no residue"—in effect, "stronger" carcinogens will require more "sensitive" analytical methods (i.e., lower operational "zeros") than will "weaker" carcinogens.

This same problem applies in somewhat modified form to other classes of indirect food additives and to other classes of food constituents, including contaminants and natural components (see, for example, the FDA's "trace constituents" proposal, dealing with carcinogenic contaminants of food and color additives [4]).

Two decisions must be made before such a risk-based standard can be used: (1) How are risks to be assessed? (2) How much risk can be tolerated? These two decisions are linked insofar as the establishment of a tolerable level of risk will identify the problem we are up against in measuring the risk. If, to use an absurd example, we were to accept a risk of one in ten, then we could perform an animal test and directly measure the dose of the carcinogen that would produce such a disease incidence. Then the only major problem would be the one of extrapolating these observations from test animals to humans. Clearly, however, such a disease incidence, which is close to the lower limit of risk detection of most animal experiments, is of epidemic proportions and is outlandishly out of accord with the goal of protecting public health. Thus, to follow this approach to regulatory decision making requires both high-to-low dose and interspecies extrapolations. The methodology for carrying out such assessment, which is still in a state of development, is described in the next section. It will also be seen that these two extrapolation problems, while central to risk assessment, cannot be divorced from a host of other issues that arise during the conduct of risk assessment.

VII. CARCINOGEN RISK ASSESSMENT

Carcinogenic risk assessment has the following major components:

1. *Hazard evaluation* involves examining epidemiological, animal, and other types of experimental data to decide whether a

substance is carcinogenic and to characterize the strength of the evidence of carcinogenicity and the uncertainties in the evidence.

2. *Exposure evaluation* involves characterizing the level, duration, and frequency of human exposure, and the uncertainties in the estimates.

3. *Risk estimation* involves estimating the numerical probability that cancer will occur in humans under their conditions of exposure, and defining the uncertainties in the estimate.

Risk assessment involves bringing together all of the above information and analysis, the qualitative as well as the quantitative, to provide a complete characterization of the risks, and the associated uncertainties.

In the following, I shall describe some of the key features and uncertainties in each of these components of risk assessment.

A. Hazard Evaluation

Hazard evaluation is a multistep process. Its first step involves review of individual studies that contain information concerning the hazardous properties of the substance under review. Information may derive from clinical or epidemiological investigations, from animal toxicity and various types of *in vitro* studies, and may also be gained by examination of chemical and physical properties.

Hazard evaluation begins with critical and independent review of all pertinent studies. It is usually not satisfactory simply to accept the conclusions of the authors of a study, even if the study has been published in a scientific jornal, without some form of independent evaluation. Questions such as the following are critical:

1. Was the study design appropriate to the hypothesis under test?
2. Was the study properly conducted and reported?
3. Were the statistical methods appropriate and were they corrected applied?
4. Are the author's conclusions supported by the data? If not, what conclusions are appropriate?
5. What are the uncertainties in the conclusions?

The recommendation that studies be independently evaluated does not come from a view that scientists are not to be trusted. Rather, it stems from the fact that many studies not designed specifically for hazard evaluation may contain information highly useful for that purpose. Moreoever, some authors may overlook certain aspects of a study that may alter the conclusions, either because these aspects did not pertain to the central questions, or because they involved certain policy considerations that go beyond the authors' purview (e.g., how much statistical power should a "negative" animal or epidemiology study possess to be considered acceptable by a regulatory agency)?

After completing the review and evaluation of individual studies, the total body of data should be further evaluated. The further evaluation is designed to answer the following questions:

1. What is the total body of evidence suggesting that the substance under review is carcinogenic?
2. Have the observations of carcinogenicity been reproduced in a variety of systems, and under different conditions?
3. Are there any apparently contradictory observations, and is it possible to explain these apparent contradictions?
4. What is the *strength of the evidence* that the substance is carcinogenic?
5. What are the uncertainties in the characterization of carcinogenicity?

Even if some of these questions cannot be answered, or can be answered only with very low certainty, an adequate hazard evaluation should include discussions of all of them. Many times the discussion may reveal only that the answers are not known, but such a statement is important in that it summarizes the state of knowledge.

Part of this evaluation should focus on the possible interdependence of different sets of data. This sort of evaluation is best explained by use of an example. If a compound is shown to be carcinogenic in one system (e.g., an animal bioassay), many questions are raised, the answers to which might assist evaluation. The following are examples:

1. Is the compound carcinogenic under other conditions within the same bioassay (e.g., at different doses, or in different sexes)? If not, what are the possible reasons?
2. Is it similarly active in other systems? If not, what are the possible reasons?
3. Are there aspects of the toxicity or biological properties of the compound that might reveal its mode of carcinogenic action? Is it, for example, mutagenic? Is it active only at doses that also produce tissue damage? Does it require metabolic activation? Is there any evidence that a hormonally mediated mechanism is at work? How well are the answers to these questions known? If they are not known with sufficient certainty, what assumptions, if any, will be made about the mode of carcinogenic action?
4. What can the mechanistic information reveal about the nature of the dose-respose function at low doses?

In most cases, the answers to the third and fourth sets of questions will not be known, but a search for the answers may give rise to a number of interesting hypotheses for further research. And, of course, they may sometimes be answerable with sufficient certainty so that they can be incorporated into the risk assessment.

B. Exposure Assessment

Although the term "dose" is usually applied to drug administration, it is also appropriately applied to other substances to which we can be exposed. The usual method for expressing dose is to estimate the weight of the substance absorbed into the body per unit of body weight.

Estimation of human "dose" may be a relatively straightforward exercise for drugs, but for most other environmental agents it is beset with problems. The information needed and the way different pieces of information are put together to yield dose estimates are both easily defined, but what usually hampers the task is the absence of information. What is usually done when information is not available is, of course, to insert assumptions into the analysis. Whenever this is done, confidence in the estimates declines and the uncertainty in the ultimate risk assessment decreases. Nevertheless, when forced to make such assessments, and when assumptions are needed to complete them, it is appropriate to adhere to certain operating criteria: (1) assumptions should always be stated; (2) to the extent possible, assumptions should be consistent with knowledge available on related substances and conditions; (3) assessments should include, where possible, the likely population distribution of exposures (although "point" estimates are usually all that can be derived); and (4) assessments should probably include both "worst" and "best" case estimates, with both these terms defined.

Before detailing the elements of dose estimation, it is important to note that the population at risk should be identified and, where possible, its size and nature should be described. Are we concerned with the entire population, or only with certain subpopulations? Are we concerned primarily with one sex? Adults or children? Workers? Smokers or nonsmokers? Special subgroups that are at unusually high risk because of high exposure or susceptibility should be identified. This type of discussion is a critical feature of exposure assessment.

In may cases, the exposure assessment is complicated by the fact that individuals are exposed through several media and by different routes. Exposure to formaldehyde is a good example of this phenomenon. Many people are exposed in their homes and in other environments such as public buildings. Some people are also exposed in their places of work. Exposure may occur by inhalation, but it can also occur by the dermal and oral routes. It is not hard to imagine the difficulties this creates. If only one or a few types and routes of exposure out of many possible ones are to be assessed, it is important to point out that the assessment will acount for only a portion of the potential population risk.

In Table 1 are listed the data needed to estimate dose for a theoretical compound X. If all of the information in the table is known, it will be possible to estimate the total amount of compound X absorbed by a person/unit of body weight/unit time. Note that the total amount of compound X is derived by consideration of the amount coming from each of the media in which it may be present.

It is rare that all of the information in Table 1 is available. Availability of various types of information is usually a function of the medium and the source of the compound in the medium. For example, the amount of compound in food is very much better known if it is intentionally added to food than if it is an environmental contaminant.

The piece of information that is almost always absent is the fraction of the compound absorbed per exposure. Data on absorption of a compound from the lungs or gastrointestinal tract, or through the skin, are very meager. In the context of risk assessment, this is usually accommodated by assuming that the extent of absorption is the same in exposed humans as it is in the test animals from which the dose-response relationship derives. This assumption may not lead to serious error in those cases in which the route of exposure in test animals is identical to that of concern from the human population at risk, but this assumption is no doubt a source of some error.

Problems arise when the route of human exposure is not the same as that of the test animals. The most common situation of this sort arises when test animal data reflect oral exposure, and the major routes

TABLE 1   Exposure Assessment

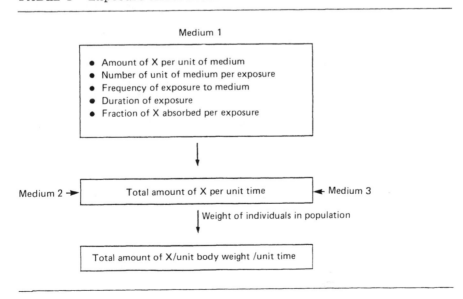

of human exposure are inhalation or dermal. Only if the extent of absorption in the test animals and that in humans are known can we achieve much confidence in the dose estimates. In the absence of such information, it becomes necessary to impose assumptions. The usual procedure is to use several sets of assumptions, including but not limited to the assumption that 100 percent absorption occurs by all routes. When possible, data on structurally related substances can be used to assist estimation of absorption rates.

In some cases, the expression of toxicity may be related to the route of exposure. Formaldehyde again provides an example. It may be that its effectiveness as a nasopharyngeal carcinogen is dependent on its inhalation, and the oral or dermal exposures (assuming that neither is accompanied by inhalation) do not carry this risk. In such cases, the exposure assessment should provide for separate consideration of population groups at risk because of their route of exposure.

There is a final aspect of exposure assessment that bears mentioning. The ideal goal is to provide estimates of the target site dose, perhaps the number of moles of a substance per unit weight of the ultimate biological target site. This ultimate target dose should provide the best measure of dose for purposes of risk assessment, assuming that information of this type exists for both the experimental subjects and the population groups of interest. To make such estimates requires careful metabolic and pharmacokinetic studies, both of which are almost impossible to perform using human subjects. If such data exist, however, it would likely be beneficial to incorporate them in the exposure assessment.

Exposure assessment is clearly a very complex subject, and because of its crucial role in risk assessment, it deserves as much critical evaluation and detailed presentation as other aspects of the subject. Of course, the uncertainties should be fully explicit.

C. Risk Estimation

After completing the hazard identification and exposure assessments, the task is to estimate risk—that is, the probability that cancer will occur in human populations under their conditions of exposure. Risk estimation is quantitative, so at this stage it becomes easy to feel that we have arrived at certainty. It should be kept in mind, however, that making these estimates involves a large number of choices and assumptions, not all of which can be defended on strictly scientific grounds (of course, this is true for any conceivable assessment scheme).

For example, if a single study, or set of data points from that study, is selected from a large body of data for use in risk estimation, it is sometimes for reasons that transcend purely scientific considerations. Often the data that represent the "most sensitive" of the various responding species are chosen for risk estimation. This is justified on the grounds that it is not possible to select the responding species "most biologically relevant" to humans, because of lack of knowledge; therefore, the "most sensitive" species is chosen in order to err on the side of public health protection. Making assumptions of this sort is frequently done, and, although it may be wise public policy, it should be recognized as extrascientific. Of course, the assumption that the "most sensitive" of responding species should be selected for purposes of risk estimation, if it is in fact done, should always be clearly stated, along with the uncertainties associated with such a selection. The same approach has to be taken with the many other selections and assumptions that have to be made to estimate risks (or, under the traditional safety assessment scheme, to arrive at an ADI).

The steps required to complete the risk estimation are the following:

1. Selection of appropriate studies and data points
2. Characterization of observed dose-response relations
3. Selection of appropriate dose-response data and application of a model for high-to-low dose, *intraspecies* extrapolation
4. Selection of appropriate interspecies scaling factors (milligrams per kilogram per day versus dietary concentration versus milligrams per square meter, etc.)
5. Consideration and incorporation of data bearing on interspecies differences in response
6. Selection of appropriate measures of human "dose" (average, "worst case," etc.)
7. Estimation of probabilty of effect, or ranges thereof, for each population group of concern
8. Presentation of results and uncertainties

Several of these steps require that certain subsets of data or certain models be selected from a larger body of data and models (steps 1, 3, 4, and 6). Such selections usually involve assumptions, some of which must be made on grounds other than scientific. There are, for example, several plausible models that could be used for high-to-low dose extrapolation. These various models yield different estimates of low dose risk (although those that are linear at low dose do not always differ substantially). It is not possible to select among the models on strictly scientific grounds, but this should not be an impediment to the use of one or more of these models for regulatory purposes. However, a science policy decision is needed to decide which of the several plausible models will be used. This is not unlike the policy decision that a 100-fold margin of safety is adequate to establish an ADI for chemicals producing nonneoplastic forms of toxicity. There is nothing improper about either policy decision, but both should be recognized as such. Selecting interspecies scaling factors (step 4) presents similar problems.[2] Again, all the uncertainties associated with such selections should be fully described.

D. Final Risk Assessment

Finally, all of the foregoing must be summarized in a succinct yet
scientifically accurate way, with no loss of the underlying uncertainties. The assessment should, in my view, never be reduced to a set
of numbers. The assessor must find a way to present the full richness
and complexity of a problem. Thus, the strength of evidence of carcinogenicity, and the extent to which mechanistic information may lead
one to believe that a risk has been overestimated or underestimated,
should also be discussed and incorporated into the final assessment.
Sensitivity analysis (i.e., examination of the effects of different
assumptions on the estimates) can also be a useful adjunct to the
assessment.

Such risk assessments are not especially desired by decision
makers, who naturally prefer simplicity. I think it is not possible to
justify reduction of scientific complexity to satisfy the needs of decision
makers. I do think, however, that the scientist has the responsibility
to summarize the assessment succinctly and in the clearest possible
language, and where possible, to point out which components of an
assessment have the strongest (albeit necessarily incomplete) scientific
support.

VIII. EXTENSION OF CARCINOGENESIS RISK ASSESSMENT METHODS TO OTHER TOXIC AGENTS

It would seem desirable to incorporate certain aspects of the risk
assessment methodologies now used for carcinogenesis into the schemes
used to establish ADIs. Most important among these is the use of dose-
response data rather than or perhaps together with NOELs to establish
an ADI. It may be possible to accomplish this goal with little modification in the present scheme.

IX. USE OF RISK ASSESSMENTS

Because most risk assessments are based on inferences, it is usually
not possible to state what relationship the assessed risk has to actual
human risk! Unfortunately, because the actual human risk is not
usually known at the time of the assessment and may never be known,
it is usually not possible to test empirically the results of an assessment. However, in those few cases in which quantitative human dose-
response data are available, it can be shown that risk assessments
based on animal data and involving the more cautious sets of assumptions (i.e., selection of "most sensitive" species and response, use of
models that are linear at low dose, use of milligram-per-square meter
scaling factors, etc.) tend to overestimate or approximate human risk
[5—7]. Furthermore, as currently practiced, risk assessments usually

involve assumptions intentionally designed to overstate the risk, in order to compensate for the possibility that animal responses are insufficiently sensitive measures of potential human response.

We need to make decisions about whether or not substances to which we are or might be exposed are too risky, and the methods outlined here are our only tools. Thus, the choice is either to make some attempt to assess risk, or to ignore the health and exposure data altogether, and make decisions entirely on other grounds. The latter course seems entirely unacceptable. Although these risk assessments schemes cannot be claimed to yield accurate measures of human risk,[3] they do provide an internally *consistent* and *operational* means for ranking risks and are therefore of enormous value in decision making. If the relative risks of regulated substances are known, and if there is some foundation for the belief that human risk is not likely to be higher than the estimated risk (as there is), then it becomes possible to develop operational definitions of safety (i.e., tolerable or negligible levels or risk) for various agents. It should be recalled, however, that it is not wise to reduce these definitions of safety to numbers—all the uncertainties need to be accommodated in the definition. This makes life for the decision maker exceedingly difficult, but it cannot be avoided.

X. CURRENT GAPS

It has been 12 years since the FDA first proposed the use of quantitative risk assessment in regulation [3]. Since that time the uses of risk assessment have expanded. The FDA has proposed its use in the regulation of certain food contaminants (e.g., aflatoxins and polychlorinated biphenyls), and to determine whether certain carcinogenic contaminants (trace constituents) of food and color additives pose a significant risk. Risk assessment also played a key role in the agency's resolution of the issue of risks posed by hair dyes containing lead acetate. The Environmental Protection Agency (EPA) has made more extensive use of risk assessment than has the FDA, and the Occupational Safety and Health Administration (OSHA), in reaction to the Supreme Court's decision on occupational exposure to benzene, has proposed to adopt quantitative methodologies of some type [8]. (OSHA's original cancer policy, published in 1979, rejected quantitative risk assessment for any purpose but priority setting. OSHA recently used such methods, for the first time, to define limits for arsenic exposure in the workplace [9].) Finally, several proposals for changes in our food safety laws [10—12] make explicit mention of risk assessment, and provide that the safety standard be redefined in terms of "insignificant risk."

As the uses of risk assessment have expanded, its practice has tended to remain relatively crude. There appear to be two major reasons for this. In the first place, the research community has not directed its attention to most of the crucial issues in risk assessment. Anyone who has worked through a full risk assessment in the manner I have described will come up against a large number of knowledge gaps that can now be filled only be the imposition of plausible, although untested, scientific assumptions. Thus, the practice of risk assessment reveals major research needs, and is therefore a superb guide to the systematic design of pertinent research programs. There is, for example, no way for biostatisticians to improve models for high-to-low dose extrapolation without new knowledge to guide them. Clearly, it will not be possible to gain the needed information on dose-response relations at low doses if we continue to rely on animal experiments in which the only measured end point is tumor incidence. Thus, new knowledge about dose-response relations will be acquired only if biostatisticians work together with biologists, toxicologists, biochemists, etc., to plan and conduct the types of experiments that can shed light on this question. A survey of programs in the research community reveals little activity of this type. The same could be said of many of the other issues in risk assessment (exposure assessment is probably the most seriously neglected area). We are thus continuing to make extremely important public health decisions under great uncertainty, and within a research vacuum. These same conclusions hold, and perhaps with greater force, for areas other than carcinogenesis. And, of course, they are true whether one supports traditional safety assessment schemes or newer risk assessment methods.

A second reason that the practice of risk assessment remains relatively crude is that many of its practioners fail to relate the full complexity of the subject. In may cases, crucial steps are either ignored or, more frequently, are presented with no explicit statement about the underlying assumptions and methods being used. Frequently, it is presented only as a problem in high-to-low dose extrapolation, and is hence trivialized. In fact, under the latter conditions, risk assessment is clearly nothing but a "numbers game," and as a result the ultimate decision-making process is itself in danger of trivialization. These dangers can surely be avoided, and it is likely that as we gain more experience and appreciation, the necessary improvements will follow.

XI. EVOLUTION IN PRACTICE AND USE OF RISK ASSESSMENT

It appears that current methods can also be improved if some of the long-standing assumptions and appraoches underlying the risk assessment methods are reexamined. Thus, some benefit might be gained by considering the following types of questions (limited here to carcinogenesis):

1. Is it possible to classify carcinogens according to their mechanisms of action? If so, what types of experimental evidence are required and what criteria should be used for classification?
2. If different types of carcinogens can be identified, what methods are appropriate for high-to-low dose extrapolation for each type? (How does the mechanism of action influence the functional relationship between dose and response?)
3. If several sets of carcinogenesis data are available, what criteria should be applied to select among them for risk estimation? Is it always necessary to use the "most sensitive" response? Is not the latter choice wasteful of data? How can all the data be used?
4. What are the proper interspecies scaling factors for interspecies extrapolation? How does the mechanism of action of a carcinogen influence the choice of scaling factors?
5. How can differences in timing of exposure between test animals and humans be taken into account?
6. Are we limited to "point estimates" of exposure? Need we always base risk on "worst case" exposure estimates? What are the alternatives?

This list is partial, but indicates the types of reexamination that might be undertaken. I suggest that the answers to many of these questions need not await new research, but can be found with careful analysis of existing data and methologies. Some of these questions are already being looked at, and improvements in current methods are likely to follow.

The proper use of risk assessment is an enormously complex subject and cannot be fully discussed here. There is, however, one crucial aspect of the subject that is closely linked to the preceding discussion and that is a useful conclusion to it. This concerns the notion of "insignificant risk."

There is not doubt that the significance of a risk is inextricably bound up with the question of benefits. I suggest, however, that it is possible to examine the question of "significance" purely as a matter of public health concern, without regard to benefits. Thus, we can examine the question of whether or not a given risk is an important risk to the public health (although such an analysis will hardly ever be sufficient for most regulatory decisions).

Assuming the risk assessment has presented the assessment in its full complexity, and with the appropriate description of uncertainties, resolution of the question of the public health importance of a risk requires consideration of *all of* at least the following:

1. The strength of the evidence that the substance under review is carcinogenic. It is presumed that as the evidence becomes

stronger, the potential public health importance of the risk also increases.

2. The extent to which the exposure assessment is likely to misrepresent the true exposure.

3. The extent to which all of the biological data have been incorporated into the analysis.

4. If information on mechanism has been used to characterize the dose-response relationship at low dose, the extent to which knowledge of the mechanism is uncertain.

5. The numerical estimates of risk, and the confidence limits thereon.

6. The sensitivity of the risk estimates to changes in assumptions.

There is no simple formula that permits incorporation of these and the other factors that may contribute to a determination of the public health importance of a given risk. Decision making of this type can, however, be made a more analytic process than it now is. The use of some of the methods for decision making under uncertainty that have been developed by decision analysts might usefully be applied to problems of risk significance. As in the case of risk assessment, it would probably be desirable that greater explicitness be incorporated into the processes of decision making than is now the case.

In closing, I suggest consideration of two different conceptual approaches and associated terms to deal with the question of risk management. In those instances in which it is found that the public health importance of a risk is clearly trivial, I suggest application of the phrase "negligible risk." This clearly denotes that the risk, of whatever type, is so small that it can truly be neglected, at least until we deal with more important risks. I suggest the use of "tolerable risk" in connection with risks that may not be clearly negligible, but which we must continue to live with because the social costs of reducing risk to negligible levels are too high. The phrase "tolerable risks" avoids the connotations of the phrase "acceptable risk," which should perhaps be applied only to those risks that are taken voluntarily.

NOTES ADDED IN PROOF

1 Thus, one approach to deciding how much exposure to a carcinogen can be permitted is to set limits at whatever the detection capability of available analytical methods happens to be. The latter has, of course, no relationship to risk. This is not to say that analytical capabilities as well as a host of other factors should not play a role in decision making. It is only to say that risk should not be ignored.

2   Such "science policy" decisions need to be sharply distinguished
    from those policy decisions that must be made concerning manage-
    ment of risks.
3   There may be some exceptions to this. For example, accurate esti-
    mates of human risk can sometimes be acquired if the dose-response
    data reflect human experience, and if the needed extrapolation is
    of limited magnitude.

37

TRACE CONTAMINANTS—THE INDUSTRY VIEW

JEROME H. HECKMAN

Keller and Heckman, Washington, D.C.

I. INTRODUCTION

In considering "challenges for tomorrow" that trace contaminants will pose for the cosmetic industry, the difficulties of prediction are complicated by the many dimensions of the problem. Concerns with trace contaminants involve the complex interaction of independent factors such as the vast improvements in analytical chemistry, the state of the art in toxicology, legal and legislative considerations, the dynamics of agency regulatory practice and policy, economics, and public perceptions. To make meaningful projections, it will be helpful to start by reviewing the role of each of these factors.

II. SCIENTIFIC CONTEXT

A. Analytical Chemistry

For many years, analytical chemistry in the cosmetic industry has been important in establishing specifications for raw materials and in assuring quality control so that finished products comply with applicable commercial standards. Chemical analysis assured the proper color, texture, aroma, functionality and, where indicated, appropriate standards of purity.

In the past, the safety of a cosmetic product was considered to depend almost exclusively on the safety of the major components that constituted the product. Over the last decade, however, and at an increasingly rapid pace, new analytical procedures or the application of new instrumentation to old procedures have made it possible for the chemist to isolate and identify the traces of impurities that inevitably accompany every raw material and every finished product. For example, the analytical chemist could examine a product that was deemed to be "chemically pure," perhaps 99.98 percent pure, and draw attention to the missing 0.02 percent, which was then characterized as 200 parts per million (ppm) or even 200,000 parts per billion (ppb). Chemists, electronic engineers, computer scientists, and instrument makers vied with each other. Now moderately equipped laboratories can routinely measure in the range of 100 ppb or less; well-equipped laboratories can often measure in the low parts-per-trillion (ppt) range.

With each new advance in analytical sensitivity, more trace contaminants were discovered at lower and lower levels. The product confidently and proudly acclaimed as very pure was now found to contain tens, scores, even hundreds of trace substances whose presence may never have been considered. Their discovery raised the questions: What does it mean? Is the product harmful? In looking for answers, industry, regulators, and the public turned to the science of toxicology.

B. Toxicology

Not too long ago, toxicologists who evaluated the safety of cosmetics conducted tests with animals that were considered to be appropriate surrogates for humans. The substance under investigation was applied to test animals and the effects noted. Because the toxicologists' test instruments, living animals, are so variable, toxicologists employed numbers of animals and observed both average and unusually sensitive responses. The mode of application was selected to simulate the mode of human exposure and additional tests were conducted to determine the effect of unanticipated or unintended human exposure. Thus, cosmetics not intended for ingestion or application to the eyes were often tested by these routes because it was possible that people might splash some into the eyes or might eat some as in the case of lipsticks or dentifrices.

The types of responses for which the toxicologist looked involved observable changes in blood chemistry, urine, rate of growth, adverse physiological effects on the organ the cosmetic contacted, and a study of many of the test animals' organ systems including both gross and microscopic examination of many tissues.

Because of the variability within the test animal species and the difficulty in predicting whether humans would be more or less sensitive than the selected test animals, toxicologists routinely exposed the test animals at much higher levels than those anticipated for humans. Frequently, toxicologists attempted to increase the dose until they forced an adverse response. Then the dose was lowered so that they could set a "no effect" level. Finally, they would routinely apply a safety factor of 10 to cover individual variations within a species and then another safety factor of 10 to cover the possible differences between the test species and humans. Moreover, if a relatively brief animal test was employed but long-term human exposure was anticipated, a third factor of 10 was utilized. In other words, having established a "no effect" level in the animals, the toxicologists would divide the exposure level by 100 or 1,000 to set a safe level for human exposure.

The concepts of a "no-effect" level and a safety factor were founded on the long-established observation that all living organisms have mechanisms to defend themselves against external insults. Bruises heal, animals recover from diseases, even frank poisons are not fatal unless the dose is too high. Thus, the observation of a "no effect" level in test animals was tacitly interpreted to mean that any effect that might have occurred upon exposure was reversed by the time the animal was examined.

But, like analytical chemistry, the science of toxicology has not been transfixed. In recent years, microbiologists have devoted increasing attention to DNA and its manipulation. They have developed a series of short-term tests on microorganisms or tissue cells to measure a substance's potential for causing genetic mutation and, therefore, possibly acting as a carcinogen. As these tests have been developed, toxicologists have added them to the battery of tests they perform. In contrast to the traditional animal study, a positive result at any level of exposure in the short-term test creates a presumption of potential adverse effect. There is no established procedure for either extrapolating this effect from bacteria to humans or for setting a safe level of human exposure even when a "no effect" concentration can be established with the test organisms.

Nor is this all. The traditional safety factor used by toxicologists has come under attack by statisticians. Although we are unaware of any case of a threat to human safety when the 100-fold safety factor was applied to an animal "no effect" level, statisticians have asked, "Why 100? Why not 73 or 114? Why any particular number absent a sound scientific basis for establishing it?" Furthermore, the statisticians claim that the traditional animal experiment conducted, for example, with even 50 test animals, cannot provide an adequate level of assurance that the chemical being tested is really without effect. If, the statistician speculates, the chemical could affect as much as 1 percent of an exposed animal population, a test with 50 animals would rarely detect the effect. Even if 100 animals were used, a test would more often miss the effect than observe it.

Although the criticism is statistically sound, it is questionable whether it is biologically sound. If an animal population truly exhibited an infinite range of sensitivity to a particular chemical, the statisticians' comments would be well taken. If the range of response is relatively narrow, however, the statistical criticism would not be based on biological facts. Nevertheless, this statistical criticism has made major inroads to the extent that no negative test can now be said to demonstrate the safety of a product. The impact has been greatest in the area of carcinogen testing, where traditional "no effect" concepts are routinely discredited by those who have embraced what might be called the "one molecule" theory. The theory posits that the intake of or exposure to one molecule of a putative carcinogen may ultimately initiate the disease.

Finally, mention should be made of the rapidly developing field of pharmacokinetics. Here, pathologists administer the test compound by an appropriate route and then measure its metabolic fate in the animal body. The safety, or lack thereof, is judged by the rate at which the administered compound is excreted, the quantity of the compound or its metabolites incorporated into "target" organs, and the nature of the compound-tissue interaction.

In each case, toxicology attempts to set a safe level of human exposure. Here, the scientific concept of safety interacts with the law.

III. LEGAL FRAMEWORK

A. Federal Food, Drug, and Cosmetic Act

The fundamental statute governing cosmetics is the federal Food, Drug, and Cosmetic Act (FDC Act). The Act's definition of "safe" is scarely illuminating. "The term 'safe'. . .has reference to the health of man or animal."[1] Better insight into the meaning of the term "safe" can be gleaned, in part, from the interpretative definition provided in the Food and Drug Administration's (FDA) Food Additive Regulations.

> "Safe" or "safety" means that there is a reasonable certainty in the minds of competent scientists that the substance is not harmful under the intended conditions of use. It is impossible in the present state of scientific knowledge to establish with complete certainty the absolute harmlessness of the use of any substance. Safety may be determined by scientific procedures or by general recognition of safety. In determining safety, the following factors shall be considered.
>
> (1)   The probable consumption of the substance and of any substance formed in or on food because of its use.
>
> (2)   The cumulative effect of the substance in the diet, taking into account any chemically or pharmacologically related substance or substances in such a diet.
>
> (3)   Safety factors which, in the opinion of experts qualified by scientific training and experience to evaluate the safety of food and food ingredients, are generally recognized as appropriate.[2]

Although written with respect to food ingredients, similar considerations should be applicable to cosmetics.

The safety of cosmetics *per se* (as well as foods, drugs, and medical devices) were not originally the concern of the Act. Rather, the fundamental presumption was that foods, drugs, devices, and cosmetics were safe unless they were adulterated. The early provisions of the Act dealt with the question of what constitutes adulteration.

In this vein, section 301 prohibits the adulteration or misbranding (mislabeling) of cosmetics. It also forbids the introduction, delivery for introduction, or receipt of adulterated or misbranded cosmetics in interstate commerce. Section 601 details the characteristics of an adulterated cosmetic.[3] It provides:

A cosmetic shall be deemed to be adulterated —
> (a)   If it bears or contains any poisonous or deleterious substance which may render it injurious to users under the conditions of use prescribed in the labeling thereof, or, under such conditions of use as are customary or usual. . .

(b)   If it consists in whole or in part of any filthy, putrid
      or decomposed substance.

(c)   If it has been prepared, packed, or held under insani-
      tary conditions whereby it may have become contaminated
      with filth, or whereby it may have been rendered in-
      jurious to health.

(d)   If its container is composed, in whole or part, of any
      poisonous or deleterious substance which may render
      the contents injurious to health.

(e)   If it is not a hair dye and it is, or it bears or contains
      a color additive which is unsafe within the meaning of
      section 706(a).

Trace contaminants are not directly addressed. Nevertheless,
a regulatory response to contaminants can readily be read into section
601. If a contaminant, at any level, is known to produce adverse effects, then the cosmetic containing this substance can be said to bear
or contain a poisonous or deleterious substance which may render it
injurious to users. At issue, of course, is whether the quantity of
contaminant in the product supports the conclusion that it may render
the cosmetic injurious under the usual or customary conditions of use.

A similar prohibition in section 402 of the Act dealing with foods
was legislatively modified. The earlier doctrine of toxic *per se* led
to the condemnation of foods containing any quantity of a poisonous
or deleterious substance, even though the food so contaminated was
not considered to pose any health hazard. Consequently, Congress
provided explicit recognition that added poisonous or deleterious substances in food could be permitted when they were required or where
they "cannot be avoided by good manufacturing practice. . .".[4]

A similar modification for cosmetics has never been incorporated
into the act, probably because such problems have not developed or
have not been publicized to any significant extent. Nevertheless,
it seems fair to assume that, since the FDA's Center for Food Safety
and Applied Nutrition also deals with the rare cosmetic ingredient issues
that arise, the unavoidable defect provision for food is probably
brought into play in the agency's enforcement approach to cosmetics.

In sum, the law assumes that cosmetics are safe unless adulterated, prohibits such adulteration, and relies on FDA regulations to
supply the details.

B. FDA Regulations

FDA regulations for cosmetics, codified in title 21 of the Code of Federal Regulations (C.F.R.), are silent on the question of contaminants
generally.[5] In a few instances, the regulations address specific substances. For example, 21 C.F.R. §700.11(a) states that bithionol

is a deleterious substance which may render any cosmetic product
that contains it injurious to users. Accordingly, any cosmetic
containing bithionol is deemed to be adulterated under section
601(a) of the Federal Food, Drug, and Cosmetic Act.

The intent was to ban the use of bithionol as a deliberately added
functional ingredient in cosmetic formulations. This section does not
make any provision for the possibility that a trace amount of bithionol
might be present in cosmetics and an unwanted and unavoidable contaminant in another ingredient.

Another provision governs the use of vinyl chloride monomer as
an ingredient in, or a propellant for cosmetic aerosol products. It
declares that vinyl chloride is a deleterious substance and that "any
cosmetic aerosol product containing vinyl chloride as an ingredient
is deemed to be adulterated."[6] Similarly, other regulations proscribe
the use of certain halogenated salicylanilides, zirconium, or chloroform as an ingredient in cosmetic products.[7] All these sections find
the named substances to be deleterious to health and declare that cosmetics containing them as ingredients are adulterated.

Only in considering mercurials do the regulations address the
question of trace contamination. Section 700.13 deals with the use
of mercury compounds in cosmetics, and notes that they are toxic,
pose environmental concerns, and are of questionable effectiveness
for some applications (skin bleaching agent). It also acknowledges
that in eye-area cosmetics the use of mercurial preservatives is considered warranted because of their exceptional effectiveness in preventing pseudomonas contamination. Consequently, the regulation
states that the FDA will regard as adulterated any cosmetic containing
mercury unless it contains no more than a trace amount, such trace
is unavoidable under conditions of good manufacturing practice, and
is less than 1 ppm. Alternatively, a cosmetic intended for use in the
eye area may not contain more than 65 ppm of mercury so long as
there is no effective, safe, nonmercurial substitute preservative available.

It is instructive to note that in this section where the presence
of mercury as a trace contaminant is explicitly permitted, this is so
only if the level of such trace contamination is sufficiently low and
is unavoidable under conditions of good manufacturing practice.

Except for these few substances, there are no regulations dealing
directly with cosmetics adulteration. Even in the cited cases, the substances are banned as intentional ingredients in cosmetics and the regulations are silent (except in the case of mercury) on how to deal
with the possible presence of traces of the banned chemicals.

This is not a matter of purely hypothetical concern. When vinyl
chloride monomer was found to be present as a residue in containers
manufactured from polyvinyl chloride (PVC), the potential migration
of the monomer into cosmetics packaged in the plastic received more

than passing attention, particularly as regards mouthwash. When the plastics manufacturing industry became aware of the phenomenon, polymerization and postpolymerization processes were modified to reduce the residual vinyl chloride to the lowest practicable levels. As a result, the potential for migration of vinyl chloride into cosmetics packaged in PVC has decreased so dramatically that migration is probably nonexistent by any reasonable definition.[8] However, the vinyl chloride experience, and other cases that might be cited, highlight the need for a clear policy governing the possible presence of trace contaminants in cosmetic products.

IV. PRACTICAL CONSTRAINTS

A. Economics

As a general principle, no contaminant can be completely removed from a product, although its level can be reduced to an insignificant concentration. The problem in effectuating this principle is that the cost generally increases disproportionately, often exponentially, as the level of contamination decreases. Consequently, products are purified to a degree that will assure their safety and effectiveness for their intended purpose at a cost that is acceptable to the consumer. This truism is especially applicable to the area of trace contaminants.

In some cases, a particular contaminant might be removed altogether by a change in a manufacturing process. Thus, if it were possible to synthesize a product by two completely different routes, the product would contain traces of different starting materials and intermediates depending on which route was selected. If the presence of a particular contaminant posed a problem, it might be possible to eliminate that contaminant if an altogether different synthetic sequence could be developed. In such a case, there might well be no economic penalty in eliminating the contaminant in question, but, of course, it should be recognized that the particular contaminant would be replaced by another or others deriving from the new starting materials and intermediates.

More generally, the level of specific contaminants can be reduced by the application of well-known techniques such as distillation, crystallization, and solvent extraction. The effectiveness of these procedures, however, varies inversely with the concentration of the product being removed, so that heroic measures become necessary as the "acceptable" level of contamination is decreased.

In addition, new sources of contamination are opened: traces of solvents, trace impurities in the solvents, and traces of the contacting surfaces. The more often a product is treated to remove certain specific trace contaminants, the more likely it is that new substances will appear in trace quantities.

Ultimately, the feasibility of extreme reductions becomes intimately entangled with the cost of further processing. The need for extreme reductions depends on all the factors discussed previously plus public perceptions as to the safety of the product.

B. Public Perceptions

Public perceptions of the utility, safety, and desirability of a cosmetic product play a critical role in marketing. If a public perception were to develop that a particular cosmetic product was unsafe because of the presence of a trace contaminant, that product could become unmarketable. Thus, a manufacturer's need to lower the content of particular trace contaminants to vanishingly small levels may be based on concerns that flow from the desire to maintain public confidence in the quality of a product rather than from true health or safety concerns. Evaluating possible public response to the presence of a trace contaminant that raises health concerns only at much higher levels has been one of the most difficult tasks facing manufacturers and others.

To a degree, some of the difficulty stems from the fact that there is no such thing as a monolithic "public" with a single perception or range of perceptions. Moreover, the mass media are likely to influence both public opinion and bureaucratic perceptions of public opinion.

There are some self-styled consumer organizations which are highly vocal and have ready access to the press but have no significant public membership; others are truly voluntary membership organizations and speak for a constituency. These organized groups are likely to have access to the media, but in almost no case do the self-proclaimed consumer spokesmen have any clear mandate from more than a miniscule portion of the population. Of course, this is usually irrelevant, since the lack of legitimacy has rarely been reflected in media coverage. Often, these "public" voices have been given great weight by legislators and regulators. Some legislative hearings and FDA regulatory actions that purported to protect the public health might more properly have been characterized as public relations activity taken when the major health issue was the public fear aroused by the press, radio, and television.

Despite the strident voices raised by its "public" champions, the public has, in more recent times, let the government know that it does not necessarily favor the banning of substances or products "at the drop of a rat." Thus, in the food area, the public consumption of saccharin-sweetened beverages, nitrite-cured meats, aflatoxin-containing produce, and a variety of other products containing minute amounts of substances known to be harmful to people or animals in large quantities has not been inhibited by regulatory decision making. Some view these cases as public recognition of the toxicological wisdom

that everything is harmful in excess and nothing is harmful when the
exposure is small enough.

V. RECENT DEVELOPMENTS AND FUTURE DIRECTIONS

A. Scientific Advances

In light of these considerations, what can we expect in the future in
the broad area of private- and public-sector regulation of trace contaminants in cosmetics?

It seems likely that continuing advances in analytical chemistry
will lead to the detection of more and more trace contaminants. This
will increase the pressure on toxicologists to develop methods for determining the effects, if any, of extremely small quantities of substances and assuring that these traces can be ignored without untoward health or public relations impact.

Toxicologists are also likely to discover even more subtle effects
in animal studies, such as neurological or even behavioral changes.
Hopefully, a more profound elucidation of the underlying mechanisms
whereby substances in cosmetics affect animals may also be anticipated. This should lead to a better understanding of body defenses
and an improved ability to set "no effect" levels on the basis of physiological principles rather than on the basis of informed empirical
judgment. We might then see a sounder set of bases for determining
what level of contamination is meaningful in terms of safety and what
level can and should be ignored.

B. Legislative Changes

The law, which broadly prohibits adulterated cosmetics, does not appear to require any fundamental changes in the near future. To the
extent that trace contaminants may be found to pose a public health
problem, the present statute is adequate. On the other hand, nothing
in the law requires adverse FDA action when the level of a trace contaminant is sufficiently low, as is almost always the case, to pose no
public health concerns. Inasmuch as the legal status quo is suitable,
there is little need for legislative change. To some extent, however,
this depends on continued reasonableness in the FDA's interpretation
and regulation.

In the food and food additives areas, legislation dealing with trace
contamination is probable. Unlike the situation with cosmetics, problems have arisen that need legislative attention, primarily because
the FDA, acting on a variety of stimuli (most of them irrelevant by
this author's standards), has been inconsistent as well as unreasonable in its day-to-day problem solving and policy determinations.
Congress may modify the basic law to give the FDA more flexibility
in dealing with the presence of traces of carcinogens or other contaminants.[9]

The same principles embodied in any food safety amendments should be applicable to cosmetics to the extent similar problems develop. In at least one respect, proposed legislative changes will affect cosmetics. All the proposals suggest amending the Act's current non-definition of safe. Some proposals refer to the "absence of significant risk" and others to "a reasonable certainty that the risks of a substance under the intended conditions of use are negligible."[10] If adopted, any of these various definitions would explicitly reject a search for absolute safety or purity. Instead, the proposals support the use of common sense and good science, including risk assessment when necessary. The same concept of safety should be applied to cosmetics.

C. Administrative Developments — The FDA's Carcinogen Policy

The rule of reason should play a role in all the FDA's regulatory activity, and there are signs that a realization of this fact is returning to the agency. This cautiously optimistic expectation is based on several influences now affecting cosmetic, food, and drug regulations and interpretations. If rationality is a contagious disease in the bureaucracy — and I think it is — any new age of reason at the FDA will be evidenced in its dealings with cosmetics and may well preface progress elsewhere. Indeed, in a few cases, precisely this effect is taking place.

One major area of FDA concern has been the presence of trace carcinogens in the food and drug supply. In particular, the so-called Delaney clause instructs the agency not to find a food additive safe

> if it is found to induce cancer when ingested by man or animal, or if it is found, after tests which are appropriate for the evaluation of the safety of food additives, to induce cancer in man or animal.[11]

Similar provisions are set forth for animal drugs and color additives.[12] In these cases, however, Congress provided that the Commissioner could promulgate a feed additive regulation, approve a new animal drug, or list a color additive for use in feed for animals which are raised for food production if the additive does not adversely affect the animals for which the feed was intended and if no residue of the additive could be found (by methods of examination prescribed or approved by regulation) in any edible portion of such animals after slaughter or in any food yielded by or derived from the living animal.

With the increasing sensitivity of analytical procedures, traces of carcinogenic animal drugs, feed additives, or their metabolites began to be found in edible tissues of treated animals, raising questions concerning additives previously considered suitable. In order to deal with this matter, the FDA devised a procedure for estimating the human risk posed by a given level of a carcinogenic residue in the edible tissue of animals. This procedure was based on dose-response data obtained at high levels in animal feeding tests and then extrapolated using conservative statistical procedures to yield a "virtually safe" dose. The agency then required that an analytical procedure be developed that would be sufficiently sensitive to detect the virtually safe level in the edible tissues of animals. This method would then be the method "of examination prescribed or approved by the Secretary by regulations" as appropriate for use to determine the absence of residues in animal tissues.[13] It is frequently referred to as the sensitivity of method (SOM) approach.

Obviously, this procedure does not guarantee that the food will not be contaminated by some level of a carcinogenic residue, only that the level will not exceed that defined by the sensitivity of the analytical method. In other words, traces of a carcinogen in food at levels less than that detectable by the prescribed procedure are deemed not to be food or color additives and, hence, outside the scope of the Delaney clause.

In November 1979, guidance with respect to trace contaminants in food was given by the U.S. Court of Appeals for the District of Columbia Circuit in connection with the possible migration of packaging components into food. The decision in *Monsanto Company v. Kennedy* recognized that there would usually be some migration from a package into food merely by virtue of contact between the two, but that this was not sufficient to make the migrant a food additive.[14] Rather, the court stated:

> For the component element of the definition to be satisfied, Congress must have intended the Commissioner [of Food and Drugs] to determine with a fair degree of confidence that a substance migrates into food *in more than insignificant amounts* [emphasis supplied].[15]

The court further explained:

> Thus, the Commissioner may determine. . .that the level of migration into food of a particular chemical is so negligible as to present no public health or safety concerns, even to assure a wide margin of safety. This authority derives from the administrative discretion, inherent in the statutory scheme, to deal appropriately with *de minimis* situations.[16]

In short, within a food additive framework, the court advised the FDA that it has the authority to deal with trace contaminants by ignoring them.

The FDA's application of these principles to cosmetics can be contrasted in two cases dealing with hair dyes that chronologically border the *Monsanto* decision. In the first, the FDA adopted a regulation in February 1979 that required "cancer" warning labels for coal-tar hair dyes containing 2,4-diaminoanisole regardless of the exposure level.[17] The FDA reasoned that some of the 2,4-diaminoanisole would be absorbed through the skin, since a related compound was absorbed.

During this rule-making proceeding, risk assessment data showed that any risk of cancer associated with even frequent use of the dye was extremely remote.[18] Nevertheless, the FDA explicitly stated that the same type of risk assessment proposed for animal drug residues discussed above was not applicable to the hair dye situation. As a result of a court challenge, the FDA agreed to stay the regulation's application and to apply risk assessment procedures to determine whether the level of the trace contaminants justified regulatory action.[19]

In the second case, the FDA also initially declined to use risk assessment when considering lead acetate as a component of hair dyes. Following judicial review, the agency changed its position, applied quantitative risk assessment and concluded that the amount of lead absorbed through the skin as a result of the use of such products presented a *de minimis* risk situation. Although viewed as a carcinogen, the FDA cleared the use of lead acetate as a color additive in 1980 and 1981 decisions.[20]

Undoubtedly the most significant development in this area has been the FDA's publication of a proposed carcinogen policy and its application of the policy to D&C Green No. 5 and Green No. 6. The policy, formally captioned the "Policy for Regulating Carcinogenic Chemicals in Food and Color Additives," was published as an advance notice of proposed rule-making (ANPRM) on April 2, 1982.[21] In the *Federal Register* notice, the FDA acknowledges two important regulatory principles. First, the agency recognizes that absolute purity is not attainable in any real or practical sense. It then analyzes the federal Food, Drug, and Cosmetic Act and concludes that safety refers to "a reasonable certainty that no harm will result," not "proof beyond any possible doubt that no harm will result under any conceivable circumstances."[22]

Having correctly concluded that Congress did not require any search for absolute — and hence unattainable — safety, the agency found that the science of risk assessment had developed to a stage where the FDA can "assess adequately the upper level of risk presented by the use of a noncarcinogenic additive that contains a carcinogenic chemical."[23] Because the assessment techniques employ multiple, conservative assumptions, the estimated risk will not be understated; on the contrary, the projected risk may always be expected to be significantly overstated. Therefore, if the use of a color with a trace amount of a carcinogen produces an insignificant risk under the FDA's procedures, the agency can conclude that there is a reasonable certainty that no harm will result and that the color, or other additive, is safe.

Continuing from the premises that the safety provisions of the
act impose a rule of reason and that risk assessment methodologies
are adequate, the FDA outlines the three elements of its proposed carcinogen policy. These elements are (1) clarifying the definition of
an "additive"; (2) interpreting the Delaney clause to apply only when
the additive as a whole has been shown to cause cancer (as opposed
to one of its components); and (3) using risk assessment as one of
the tools for determining whether the additive is safe.

In attempting to clarify the definition of an additive, the FDA
suggests three approaches. The constituents approach distinguishes
between the additive as a whole and unwanted trace substances. Under this approach, the term "color additive," for example, would include only those substances intended to be present in the dye or pigment. Any other chemical present in the additive would be a constituent. The constituents would include residual reactants, intermediates, manufacturing aids, and products of side reactions or chemical
degradation.

Another approach to the definition of an additive is the *de minimis*
concept. Without referring to it as such, this was the basis for the
FDA's lead acetate decision discussed earlier. Stemming from the
Monsanto decision, the *de minimis* approach would permit trace amounts
of a carcinogen where the potential human exposure and resulting risk
are so low as to be insignificant. Unfortunately, the FDA's recent
description of the *de minimis* approach might be interpreted as a quantitative approach triggered by certain fixed levels. In our view, the
sounder course is to use a qualitative approach that assesses the risk
resulting from various concentrations. It is the risk, not the quantity, that should be the agency's guide.

The final approach to defining an additive is the sensitivity of
method (SOM) approach. Under the SOM approach, the FDA could
decline to classify as an additive a carcinogenic chemical thought to
be present in the additive but undetectable by a specified method,
as long as the risk of cancer posed by the undetected concentration
of the carcinogen fell below a certain "acceptable" or "insignificant"
level. Using a specified analytical method might provide a needed
measure of guidance in itself. The sensitivity of the official analytical method would also mark a point beyond which the search for zero
need not proceed. Collectively, all three approaches to defining an
additive are a partial attempt to rationalize the operation of the Delaney
clauses in the act, which includes the anticancer clause governing
color additives in cosmetics.[24]

The FDA's first application of the policy was published together
with the proposal on April 2, 1982, and involved the agency's review
of D&C Green No. 6 as a color additive in externally applied drugs and
cosmetics.[25]

D&C Green No. 6 is formed by reacting p-toluidine and quinizarin.
Some p-toluidine remains in the purified color as an unwanted residual

reactant. Animal studies have demonstrated that p-toluidine is a carcinogen in mice.

Applying its new policy, the FDA found that p-toluidine is a constituent, not a color additive. Because there was no evidence that D&C Green No.6 as a whole caused cancer, Green No. 6 was removed from the operation of the color-additive Delaney clause. The FDA then focused the inquiry on the safety of the color.[26] The agency considered potential human exposure, determined the upper limit of individual lifetime risk from exposure to the contaminant, and held that "there is a reasonable certainty of no harm" from the color.[27]

A similar analysis was presented by the FDA when it "permanently" listed D&C Green No. 5.[28] D&C Green No. 5 is produced from the further reaction of D&C Green No. 6. Like Green No. 6, Green No. 5 also has trace quantities of p-toluidine as a residual reactant. Again, the FDA reviewed toxicological data and found that D&C Green No. 5 as a whole produced no significant adverse effects. Applying the carcinogen policy, the FDA determined that p-toluidine was a constituent and that the risk presented was insignificant.

In a sense, the application of the carcinogen policy to these colors was easy because the risk was so small and the potential exposure limited. More difficult situations will arise when the potential exposure is greater and the risk level increases. Although the FDA has suggested the use of one in one million as an acceptable risk level given the conservative bias of the risk assessment methodologies, and has used this figure in some regulatory actions, it is not clear that the FDA will employ this figure as a universal baseline.

The vast majority of comments filed in conjunction with the FDA's proposed carcinogen policy have voiced general support for the agency's efforts and encouraged the formulation of a broader carcinogen policy. Most commenters recommended a flexible combination of the constituents, *de minimis*, and SOM approaches. Others emphasized the use of one approach which, almost exclusively, was the constituents approach.

Critical comments were filed by the Health Research Group (HRG), a Nader organization, and Mr. Glenn M. W. Scott. These objections to the FDA's clearance of D&C Green No. 5 led to a significant appellate court decision upholding the validity of the FDA's proposed carcinogen policy and its application to color additives.[29]

From a broader perspective, indications are that neither the press nor television is as quick to give publicity to every new pronouncement of grave hazard discovered by the "public" spokesman. The frequent alarms about substances subsequently found to be inoffensive, the urgent pleading for action against substances generally found acceptable, and the petitioning to ban substances for which there is a clear desire have all made the media less eager to respond to discoveries of danger by those "public" protectors who have defended us against dangers that others have found difficult to perceive. Public good sense and practicality appear to be reasserting themselves so that actual problems can be dealt with realistically without the diversion of resources to deal with insignificant safety question.

VI. CONCLUSION

The challenge for tomorrow facing the cosmetics industry will be to reduce the level of trace contaminants to the extent necessary to assure the safety of the industry's products. At the same time, the industry should guard against unnecessary efforts and expenditures to remove trace contaminants of an inconsequential nature. This dual task will require the careful application of developing scientific realities in a changing regulatory environment.

NOTES

1. Section 201(u) of the act, 21 U.S.C. §321(u) states:

     The term "safe," as used in paragraph (s) of this section and in sections 409, 512, and 706, has reference to the health of man or animal.

     "Paragraph (s)" refers to the definition of the term "food additive." Section 409 deals with food additives; section 512 with new animal drugs; and section 706 with color additives.
2. 21 C.F.R. §170.3(i).
3. Section 402 of the act defines adulteration for food. Section 501 does the same for drugs and devices.
4. Section 406 of the act, 21 U.S.C. §346.
5. 21 C.F.R. §§700.3-740.18.
6. 21 C.F.R. §700.14(a).
7. 21 C.F.R. §§700.15, 700.16, and 700.18. The proscription on the use of zirconium is limited to aerosol cosmetic products.
8. *Monsanto Co. v. Kennedy*, 613 F.2d 947 (D.C. Cir. 1979).
9. At this time, pending legislation to amend the food safety laws includes: S. 1938 and H.R. 4121 [identical bills introduced by Senator Orrin G. Hatch (R, Utah) and Representative Edward R. Madigan (R, Ill.) in October 1983], and H. R. 5491 [introduced by Rep. Albert Gore, Jr. (D, Tenn.) in February 1982]. Other developments include the circulation of a revised working draft bill on April 20, 1982, by the Senate Committtee on Human and Labor Resources Staff. On May 5, 1982, the Sub-Cabinet Level Council Working Group on Food Safety sought comments on tentative draft language for statutory amendments as part of an effort to formulate an administration position on food safety legislation.

10. The quoted proposed amendments to section 201(u) of the act, 21 U.S.C.§321(u), are contained in section 101 of S. 1938 and H.R. 4121, and section 3 of H.R. 5491. The Senate Committee on Labor and Human Resources Staff Working Draft of April 20, 1982, refers, in section 5, to "a reasonable certainty that no harm will result from a substance under its intended conditions of use." The Sub-Cabinet Council Working Group on Food Safety suggested that the definition of safe in section 201(u) of the act be amended to mean "a reasonable certainty of no significant risk, based on adequate scientific data, under the intended conditions of use" (Tentative Draft Language of 10/30/81; released May 5, 1982).

11. Section 409(c) (3) (A) of the act, 21 U.S.C.§348(c) (3) (A).

12. Sections 512(d) (1) (H) and 706(b) (5) (B) of the act, 21 U.S.C. §§360b(d) (1) (H) and 376(b) (5) (B).

13. See, *Fed. Reg.*, 44:17070 (March 20, 1979) (Chemical Compounds in Food-Producing Animals; Criteria and Procedure for Evaluating Assays for Carcinogenic Residues) and *Fed. Reg.*, 47:4972 (February 2, 1982) (Chemical Compounds in Food-Producing Animals; Availability of New Threshold Assessment Criteria for Guideline).

14. 613 F.2d 947 (D.C. Cir. 1979).

15. 613 F.2d 955 (D.C. Cir. 1979).

16. 613 F.2d 955 (D.C. Cir. 1979).

17. 21 C.F.R. §740.18.

18. *Fed. Reg.*, 44:59509 (October 16, 1979).

19. The regulation, 21 C.F.R. §740.18, was stayed on September 18, 1980. See *Fed. Reg.*, 47:7829 (February 23, 1982).

20. *Fed. Reg.*, 45:72112 (October 31, 1980); *Fed. Reg.*, 46:15500 (March 6, 1981); see also *Fed. Reg.*, 44:59509 (October 16, 1979) (earlier decision declining to use quantitative risk assessment).

21. *Fed. Reg.*, 47:14464 (April 2, 1982).

22. *Fed. Reg.*, 47:14464 (April 2, 1982), quoting H.R. Rep. No. 2284, 85th Cong., 2d Sess. 1 (1958).

23. *Fed. Reg.*, 47:14466, col. 1 (April 2, 1982).

24. Section 706(b) (5) (B) of the act, 21 U.S.C. §376(b) (5) (B) provides in pertinent part:

> A color additive (i) shall be deemed unsafe, and shall not be listed, for any use which will or may result in ingestion of all or part of such additive, if the additive is found by the Secretary to induce cancer when ingested by man or animal, or if it is found by the Secretary, after tests which are appropriate for the evaluation of the safety of additives for use in food, to induce cancer in man or animal, and (ii) shall be deemed unsafe, and shall not be

listed, for any use which will not result in ingestion of
any part of such additive, if, after tests which are appropriate for the evaluation of the safety of additives for such
use, or after other relevant exposure of man or animal to
such additive, it is found by the Secretary to induce cancer in man or animal. . . .

25. *Fed. Reg.*, 47:14138 (April 2, 1982).
26. *Fed. Reg.*, 47:14140, col. 1, and 14142 (April 2, 1982).
27. *Fed. Reg.*, 47:14144, col. 3 (April 2, 1982).
28. *Fed. Reg.*, 4724278 (June 4, 1982).
29. Scott v. Food and Drug Administration, 728 F.2nd 322 (6th.
     cir. 1984).

38

THE FDA's RESEARCH PLAN FOR COSMETICS

SANFORD A. MILLER and PATRICIA THOMPSON

Center for Food Safety and Applied Nutrition, * *The Food and Drug Administration, U.S. Department of Health and Human Services, Washington, D.C.*

I. INTRODUCTION

A. The Role of Science in Regulatory Decision Making

From the beginning it was recognized that The Food and Drug Administration (FDA), to do its job well, had to have a science-based structure and that its decisions had to be developed on scientific grounds rather than in response to economic or political pressures. Thus, through the years, changes have occurred in the regulatory philosophy of the agency as a result of changing expectations of the Ameri-

---

*Formerly known as the Bureau of Foods

can people, and the agency has always attempted to develop the science necessary to meet those expectations.

The role of the scientific community and the role of the federal regulatory agency today are interwoven. In this situation, the FDA's services are only as good as the science supporting them. For science the choice is increasing participation in regulatory decisions or accepting the consequences – along with society in general – of emotionally inspired and scientifically deficient safeguards against the unwanted side effects of modern progress. The FDA, along with all of the federal and state agencies that combine science with regulation, today faces an enormously difficult task.

Although today regulatory agencies take for granted that their decisions regarding the substances they regulate are based on sound science, they are also frustrated by the gap between the scientific questions they are now able to ask and the ability of science to answer those questions. This nowhere more evident than in regard to toxicology.

Regulatory agencies must now cope with problems never dreamed of. Great skill and scientific sophistication are required to determine the hazards from chronic toxicity and to assess human risk. The need for scientific information on which to base regulatory decisions is rarely challenged. But it is often argued that the FDA itself, as primarily a regulatory agency, has no need to be involved directly in scientific research. This view is mistaken and has had a pernicious effect on the agency and the credibility of agency decisions. It is predicated upon misinterpretations of the agency's legislative mandate, the nature of cosmetics, and the degree of specification of contemporary science.

While there is general agreement on the need for more research and for the desirability of basing regulatory action on the most scientifically substantial base, nevertheless, regulatory agencies do not always have the luxury of waiting for science to take its normal evolutionary course. Often, the lack of a decision is a decision. As issues become more public, pressure increases for agency action; the less knowledge that exists, the greater the chance that the proposed regulatory course will be in the long run unfounded or counterproductive. Thus, there is much reason for the agency to await the results of science. However, it is probable that the necessary urgency and support for research on controversial issues will not be forthcoming unless the agency proposes regulation or threatens to do so. Only then will vested interests be mobilized to the point where pressure for substantial research investment will occur.

B. The Multiple Meanings of Research

The changing expectations of the American people, the increasing capabilities of analytical technology, the slower but still substantial advances in our understanding of life processes have served to raise substantial questions about the knowledge base on which future regulatory philosophy must be based. The development and refinement of a Bureau of Foods Research Plan [1] for foods and cosmetics has been an effort to anticipate these needs and prepare for the future. The plan is designed to provide a basis for reaching decisions on the safe use of chemical substances in foods and cosmetics and integrating research goals with regulatory responsibilities.

The task of developing the appropriate regulatory policies and strategies to meet these expectations has also become more complex as analytical chemistry has increased its ability to detect substances at concentrations of parts per billion or less. This increasing ability to detect extraordinarily low levels of substances has in turn raised substantive questions about the meaning of concepts such as "zero" or "without hazards" or "safe levels." For example, how do we regulate endogenous hazards associated with the natural components of our food supply or cosmetics, regardless of processing or other modifications? Since these issues generally have not been addressed by the FDA or other federal agencies, it appears clear that a change in the agency's view of the regulatory system is required if it is to fulfill these expectations and address the scientific questions involved. Today's need is for a system based not on the traditional view of the absolute determination of safety, "zero risk," but rather on relative safety within our environment.

The directions in which these regulatory activities are aimed should include factors that affect not only the quantity of life but also quality of life factors such as behavior. For these reasons, the Bureau of Foods has begun a process which will ultimately permit the development of the concepts, the models, and the tools that will enable the agency to evaluate hazard in terms not only of death but also of other issues which have arisen and will continue to arise over the last quarter of the twentieth century.

Among the most important initiatives in the agency's Bureau of Foods is the increased responsibility of the agency to do research in the development of new techniques and models to permit a new approach to these problems of risk assessment and regulation. The Agency's ability to fulfill any of its regulatory goals depends on its ability to develop both answers to specific questions about the safety of individual substances as well as appropriate models on which to build the systems needed to do the job.

II. THE ROLE OF THE FDA IN COSMETIC RESEARCH

Although research has been a fact of life in the Bureau of Foods almost since the foundation of the agency, little overall effort has been expended in its organization and its direction toward future societal needs. Lacking direct statutory authority to do research, this effort has been largely oriented toward immediate regulatory problems. Thus the pressure of daily crises has tended to direct these efforts toward solutions of these immediate problems while the chronic shortage of experienced and knowledgeable people has not permitted much speculation on future requirements. The result has been research directed toward testing and information acquisition rather than research pointed at the development of new concepts and understanding.

While similar arguments can be made for the impact of technology and science on many other areas overseen by the FDA, the problem is particularly acute for cosmetics. The general food safety provisions of the Food, Drug, and Cosmetic Act place the responsibility on the petitioner for proving the safety of substances added to food. For cosmetics, the act requires that the FDA demonstrate hazard, and given the practical limitations of resources this is a burdensome task. This puts more pressure on science and on methods of predicting and quantifying the hazards associated with cosmetic use. The cut in cosmetic program budgets in recent years has placed additional pressure on the FDA to develop more effective methods for identifying and evaluating possible cosmetic hazards. On the one hand, Congress has repeatedly maintained that the FDA has the responsibility to ensure the safety of cosmetics. On the other hand, it has just as consistently shown little enthusiasm for reforming the current statute to give the FDA the authority it needs to provide adequate protection for the consumer. The FDA's research activity may be the only practical way of communicating the agency's concern and commitment.

The largest share of federal responsibility for protection of the public from health hazards and deceptive practices associated with cosmetics rests with the Bureau of Foods of the Food and Drug Administration. The FDA has become the guardian of the transfer of cosmetic technology from science and industry to the public.

Adequate research provides the FDA with the information it needs to strike the balance. Since the turn of the century, scientific information has been indispensable as a necessary foundation for sound regulatory decisions. The proper use of what the agency calls "regulatory science" is a two-edged sword. For example, adequate toxicological screening methods can help the agency predict and prevent a possible hazard. Animal test data can exonerate a suspect substance or help the FDA prove in the final test in court that it may be injurious. Toxicological and metabolic data enable the establishment of tolerances, and these and other information enable the agency to permit the introduction of innovative and useful new additives while at the same time being assured that they will be used safely.

The bureau's capability to accomplish its formidable task is, however, hindered by information gaps in several scientific disciplines; by uncertainties in the interpretation of currently used toxicity testing methods (e.g., bioassays for carcinogenesis); by the absence of methods necessary to measure effects that are not lethal or severely debilitating but that nevertheless may produce impairment in more subtle ways (e.g., immunotoxic or behavioral effects); and by a still very limited ability to utilize animal toxicity information to assess probable human risk.

The FDA must make more discriminating judgments between those risks which are tolerable and those which are not. Better methods of human risk assessment are needed, including more useful models for dose and cross-species extrapolation. Unless the experimental methods used to study and evaluate hazards are improved and the uncertainties regarding those methods already in use are reduced, the public will continue to suffer the consequences of regulations based on incomplete and uncertain scientific information. For example, without more discriminating methods for testing carcinogens, and for evaluating their risks, we can expect to be forced to ban, with increasing frequency and possibly unnecessarily, technologically valuable materials. Replacement of these materials, when it can be done at all, can cost millions of dollars, a cost that is ultimately borne by society and the consumer. Moreover, the supply of these useful substances is not infinite, and where alternatives do not exist, the consumer can lose even more by decreased availability and convenience.

The acquisition of the knowledge and skills needed will not be easy or without cost, but this cost is a fraction of that required by regulation based on inadequate and uncertain information. The cost of faulty decisions may result in threats to the public health. The cost of excessive prudence can be high in terms of obligated expenditures and restrictions on freedom of choice. With proper information we can reduce these costs, reduce doubt as to safety, permit acceptable innovation, and improve both the safety of cosmetic products and the credibility of the bureau's decision makers. This last point is not trivial. The FDA's credibility with the public and the respect accorded it by the scientific community is a precious and indispensible resource. The degree to which regulatory decisions gain credibility and merit the support of the public and the scientific community depends on the extent to which they reflect the best available scientific knowledge.

A. Need for Relevant and Timely Studies

Since the scope of the research effort needed is large, the bureau must make effective use of the relevant knowledge developed by other institutions, both public and private. The problem of who will support the work is controversial. Some of this knowledge is being developed and will continue to be developed by the universities. The bureau need not and will not duplicate research done elsewhere. The goals of these and other institutions, however, are not the same as those of the bureau, and studies designed to meet the goals of one institution, even if they are in closely related areas, are not always certain to meet the goals of another. The overall goal is the achievement of maximum cosmetic safety. Contemporary science is unfortunately highly specialized, and no other institution can do the bulk of our research for us. The bureau cannot pick up a telephone and depend on any other governmental agency for necessary data to validate a proposed regulatory analytical method or to analyze products for potentially harmful substances. Such agencies and institutions can help, and much is to be gained by true collaboration with others in the planning and conduct of needed research. But the bureau can be sure that vital research programs will focus adequately on its needs only if some of them are conducted by or for the bureau and are supported by bureau funds. The bureau will, therefore, continue to explore areas of knowledge in which other agencies of the government are not performing, if these areas are essential to the regulatory mandate of the FDA.

Certain research activities are easily identifiable as directly supporting specific (and usually short-term) regulatory needs. The development of an analytical method to determine the amount of a toxic contaminant in a cosmetic product or the acquisition of information about use of the contaminated product are examples of such activities. It can often be a long way from a published analytical technique to a practical regulatory method capable of reliably monitoring a toxic substance in cosmetics. A strong and continuing analytical methods research activity is essential for the timely development and validation of these essential regulatory tools. The bureau has traditionally allotted a major share of its modest research resources to these kinds of activities, which are of immediate public health significance and are of unquestionable importance to the bureau's mission.

However, other kinds of research activity are equally essential but may appear to have less immediate ties to the objective of public health protection. Such research is currently needed to resolve uncertainties associated with the discipline of toxicology. Such research requires an expanded investment in specific projects which involve such scientific areas as: biochemistry, pharmacokinetics, endocrinology, molecular biology, immunology, neurotoxicology, pharmacology, behavioral toxicology, and epidemiology. It involves studies not only in whole animals from conception to death but also at the subcellular, cellular, tissue, and organ levels. Specifically directed investigations are needed by the bureau in these areas, not primarily because they increase understanding in a fundamental sense, which of course they may do, but because they are vital for the production of and reliable answers to regulatory questions. It is often argued that limited resources are best devoted to shorter-term research activities, proximately related to regulatory needs. These views are shortsighted and ignore several critical facts. For example, the current critical gaps in toxicological methodology are arguably fundamental or "basic" in nature, and yet neither academic science nor others have rushed to close them. Furthermore, many of these gaps are attributable to prior decisions that resulted in the neglect of "long-range" research.

Similarly, new parameters of safety are needed. In the past it was considered adequate to assess cosmetic safety by indices of gross human development. These indices are not adequate today, nor do they reflect the bureau's public health responsibilities with respect to the allergic, immunotoxic, behavioral, and other subtle debilitating effects possibly produced by cosmetic constituents, which cannot be adequately assessed by current methodologies.

The matter of timing is also of significance. The question of whether health hazards are or are not associated with cosmetics usually cannot be permitted to await a lengthy or uncertain resolution. We cannot wait for scientific issues to be resolved in time with the pendulum swings of interest of academic science; concern for safety does not permit us this prerogative. Only by being directly involved can we ensure that the research needs essential to a safe cosmetic supply will be met in a timely manner.

The bureau need not and should not conduct all this research itself, but it must clearly articulate its special needs and be in a position to assure that these vital research programs are carried out.

B. Need for Scientific Expertise and Leadership

As emphasized earlier, the FDA's credibility with the public and the respect accorded it by the scientific community is a precious and indispensable resource. Bureau scientists routinely deal face to face with many of the nation's best scientists from academia and industry. It is essential for the bureau's ultimate decision-making authority that these encounters take place in an atmosphere of mutual understanding and mutual respect. In the end, this depends on the quality and competence of its people and particularly its scientists.

The bureau must maintain a cadre of scientists who are close to developments at the frontiers of their disciplines. Top-quality scientists need to do top-quality research in order to maintain their expertise and self-respect, and much needs to be done that is of direct relevance to the bureau's mission. Recent budgetary constraints have begun to erode an already modest research program. Only through their own scientific publications can scientists establish and maintain their knowledge and reputations. An in-house bureau program of research is therefore indispensable in order to attract and retain competent scientists.

C. The FDA's Current Cosmetic Research Program

Cosmetics assume a singular position among consumer commodities. In the case of drugs, the health benefit may justify the assumption of substantial risk. Cosmetics, unlike drugs, are not essential to health. The assumption of any significant risk from cosmetics would therefore be difficult to defend.

Since the law does not require cosmetic manufacturers to submit data concerning the safety of their products and the ingredients in them, the agency frequently must develop necessary information or encourage others to do so.

So, the major thrust of the FDA's scientific activities for cosmetics centers on developing information about potentially harmful substances that may be components of cosmetic products. The cosmetic activities of recent years have focused primarily on studies to determine toxicological and microbiological health hazards, and development of analytical methods for the identification and determination of harmful cosmetic ingredients and microbial contaminants.

Scientific studies and knowledge are fundamental to the formulation of the FDA's cosmetic regulatory policy and to the operation of effective cosmetic surveillance and compliance programs. Cosmetic research enables the FDA to develop methodology to identify ingredients, and to determine the health significance of suspect chemicals found in cosmetic products.

The agency accomplishes its scientific activities in several ways: with its own personnel; jointly with other government agencies; with the use of advisory committees and outside consultants; and through the contract mechanism. Obviously, no regulatory agency can hope to perform effectively in the public interest if it lacks the data on which to base its actions or the brain power to evaluate and interpret the data in question. But that does not mean that the FDA or any other regulatory authority must bring on board a veritable army of experts whose know-how would presumably cover every possible question of fact or need for decision that might arise.

At the same time, however, the FDA is probably unique among federal regulatory agencies in terms of the scientific complexity of the decisions it has to make. Thus, while the agency cannot expect to build a staff of the foremost minds in such diverse fields as pathology, chemistry, and toxicology, to mention only a few, it does have to be able to draw on these capabilities. For without these capabilities, the FDA would be failing at a time when industry, the health profession, and the consumer insist more and more that it succeed.

Extramural efforts then are used to supplement the FDA's intramural capabilities and to assure that the best scientific expertise is available to support FDA executives in their regulatory decision making. The various categories of scientific activity conducted by the agency are listed below:

Sample Analysis: FDA laboratory scientists examine samples for possible contamination. If the samples are found to be contaminated, appropriate compliance and legal action may be taken to protect the public.

Methods development: The FDA must be able to detect and measure a harmful agent before the agency can enforce any standard or regulation. The FDA conducts research to develop testing methods which will allow it to determine the presence or absence of a harmful substance. A major thrust of this research is the development of more efficient and less costly testing procedures.

Research evaluation: The FDA must approve, within legally prescribed time limits, color additives (and food additives and new drugs) before they are sold in the United States. The FDA develops improved methodologies and protocols to assure proper scientific procedures, and reviews the studies submitted by industry in support of its petitions for these products.

Hazard determination: The FDA must establish the existence of potential health hazards before it can establish any standard or regulation. The agency conducts scientific studies to determine the biological effects of an agent and to document the potential hazard associated with its use or presence.

The agency's primary role in the actual performance of research is to conduct those studies that provide the critical information needed for sound regulatory decisions that others are not doing, cannot do, or cannot do in a timely fashion. The specific research areas span a considerable range -- all the way from information gathering to the development of toxicological test methods to efforts aimed at refining risk assessment techniques.

Several research studies are conducted for the FDA under contract, and others are being carried out within the agency. The studies range from the development of analytical methods to toxicological, microbiological, and epidemiological investigations related to cosmetic health hazards. Several concern issues of dermal, ocular, and respiratory toxicity. The results of these studies are generally published in the scientific literature.

The analytical research staff of the Bureau of Foods' Division of Cosmetics Technology is involved in the development of analytical methods for the qualitative and quantitative determination of antioxidants, preservatives, antibacterial agents, and certain perfume constituents. In addition, advanced analytical techniques are investigated in order to further the in-house capability of analyzing products by modern methodology. The Bureau's Division of Toxicology conducts animal research to develop suitable tests for predicting skin irritation and sensitization of cosmetic ingredients or cosmetic products in humans.

Many new issues of cosmetic product safety have arisen within the past few years; however, the FDA's budget for regulating cosmetics has not increased accordingly. The question of funding is particularly vexing and frustrating, since low priority is given to re-

search programs not essential to human health and for which there
is not a demonstrable, immediate societal need.

III. THE FDA'S BUREAU OF FOODS RESEARCH PLAN

Over a two-year period, mid-1979 through 1981, one of the Bureau
of Foods' major activities was the development of a research plan designed to help in predicting problem areas and then proposing solutions rather than reacting to emergencies as they arise. The principal approaches to this problem have been embodied in the Bureau of
Foods' Research Plan. The Research Plan outlines overall research
goals and objectives of the bureau for foods and cosmetics and the
research needs necessary to reach the objectives, and ultimately the
goals.

The Research Plan is a compendium of scientific information needs
that describes areas of knowledge the Bureau of Foods believes it
should acquire to improve the scientific basis for regulation. It
seemed clear, in developing the overall compendium, that large areas
of knowledge were not being explored in bureau research efforts, as
well as those in other agencies even though they were surely part of
the universe of bureau concerns.

There are several important characteristics of the plan that are
worthy of special note. For one, the plan emphasizes the need to expand knowledge in areas of concern to those members of society least
able to make individual risk judgments — for example, infants, children, and the unborn. Additionally, the plan emphasizes the need to
develop new measures of safety based on the changing expectations
of society. Thus, quality-of-life components such as behavior, immune response, and physical performance are targets for future exploration. Furthermore, this plan, for the first time, explicitly supports the need for research in statistical modeling as well as in the
areas of social and economic modeling. These latter areas are of concern in order to provide a basis for determining the future impact of
Bureau of Foods' actions on societal structures. Research on these
models also stresses the need for some means of predicting societal
changes that may require changes in bureau policy and activities.
Clearly, while the plan continues to stress the traditional needs of
the Bureau of Foods in a variety of areas of science, it also emphasizes the need for breaking new ground in support of the FDA's
changing mandate to protect the public health.

The need for improved planning and articulation of goals is nowhere more apparent than in the assessment of the safety and regulation of cosmetics. In the last 40 years, virtually no change has occurred in the statutory authority of the FDA to regulate cosmetics,
even though enormous strides have been made in cosmetic science and
technology.

In addition to the availability of new ingredients that have changed the technological foundation of the industry, the use of cosmetics has grown by more than an order of magnitude, reflecting both a myriad of new products and an expanded market. Concern for hazards arising from cosmetic use has also increased during this time. Science has provided more sensitive methods of analysis for the identification of toxic materials. More important, it has demonstrated that the skin is not an impervious barrier after all, but an active, functioning organ capable of responding to a large number of stimuli.

While similar arguments can be made for the impact of technology and science on many other areas overseen by the Bureau of Foods, the problem is particularly acute for cosmetics. The fact that the Food, Drug, and Cosmetic Act (FDC Act) requires that the FDA demonstrate hazard puts more pressure on science and on methods of predicting and quantifying the hazards associated with cosmetic use. Cosmetic program budget cuts have placed additional pressure on the bureau to develop more effective methods for identifying and evaluating possible cosmetic hazards.

For cosmetics then, the Research Plan serves several important purposes. In an area of limited statutory power, the research projects listed reflect an attempt to strengthen the scientific base and thus enable the FDA to implement better its limited regulatory options. In addition, exploration of these research areas will also strengthen those efforts devoted to anticipating and solving potential problems. Finally in an area that meets with mixed Congressional interest and enthusiasm, the Research Plan may be the only practical means of communicating FDA concern and commitment, and the listing of its priorities may be the agency's only way of pointing to deficiencies in its ability to deal with the problems of cosmetic safety.

A. Phases of the Plan

From the beginning, it was assumed that the approach to the problem would involve at least five phases. The first phase was devoted to the development of broad research goals, and the more specific research objectives, designed to define the boundaries of bureau research needs. The second phase was directed toward the development of the universe of research programs and projects required to fulfill these research goals, a universe constructed without reference to existing programs in the FDA or in other parts of the Public Health Service or federal government. Phase three was designed to determine which part of this program was appropriate for performance by the Bureau of Foods, either in-house or extramurally. This was accomplished by determining who was doing particular work or who had the programmatic responsibility for each of the areas defined in phase two. It was hoped in this way to define a comprehensive research matrix, defining the universe of knowledge required by the bureau to do its job. When the matrices are completed, it is hoped that deficiencies in bureau knowledge and in research support will be revealed. It is to these inadequacies in its research activity that bureau resources will be devoted. Phase four is concerned with putting into place a carefully constructed priority setting scheme. And finally, phase five will involve the planning and evaluation processes and procedures necessary to integrate these research needs with the bureau's umbrella planning and evaluation system.

B. The Research Goals

The bureau research goals include, in microcosm, the essential scientific and judgmental activities of the Bureau of Foods relevant to the technical evaluation and control of both food and cosmetic safety. They are listed more or less in order of their application to a typical regulatory problem. The examples provided are applicable primarily to cosmetics.

The bureau's approach of developing a research plan requires much more basic information for it to be useful than the bureau now has. Included in this information base are estimations of the natural level of toxic substances in the food and cosmetic environment to which people are exposed; better methods of risk assessment, involving both better understanding of the biology of the system and the mathematical models to make the assessment; exploration of the interaction between physiological state and response to such substance; methods of estimating the economic and social impact of FDA actions; development of methods to estimate functional hazards such as those involved in behavior, and immune response and physical performance.

To give perspective, to explain the justification of the research projects, and to help relate them to the bureau's research goals, summaries, where applicable to cosmetics, are provided.

Research goal A: To gather, analyze, and monitor published and unpublished information pertinent to the consumption, quality, and safety of foods and cosmetics.

It is essential that the bureau keep itself systematically informed; goal A reflects this need. Such information gathering not only is a prerequisite to the efficient conduct of research but also provides the major source of background information for all bureau activities, including informing the public. There are several major areas where information is currently inadequate and where an investment in appropriately developed and monitored data collection activities promises significant new information. Such areas of need include more complete identification of the composition of cosmetic ingredients and products; more detailed knowledge of cosmetic use patterns; estimates of human exposure to cosmetics; and surveys of chemical and toxicological data on cosmetic ingredients. Also needed is the development of more de-

tailed epidemiological information on human exposure to cosmetic ingredients, characteristics of populations that may be at risk, and the
consequences of such exposure. The foundation of all the bureau's
investigations and presently the weakest link in the research chain
is the ability to narrow this large number of substances to those that
are most significant from a public health standpoint. This effort requires expanded studies of production volumes and locations, markets,
uses, by-products, chemical and physical properties, and integration
of this information with relevant demographic and epidemiological data.

The bureau needs to strengthen its capability to anticipate future
problems created by rapid progress in cosmetic technology. An information resources center may be set up to provide centralized literature support to bureau professional staff and as a bureau-wide data
collection service.

Research goal B: To determine the nutritional needs of various
population groups and to develop strategies to improve the nutritional
status of those at risk.

This research goal is not generally applicable to cosmetics.

Research goal C: To isolate, purify, and identify potentially
hazardous food and cosmetic constituents and adulterants.

In order to evaluate the potential hazard from a constituent in
cosmetics, that substance must be identified and characterized. Because the number of substances to which humans are exposed through
cosmetics is large, this effort should be guided by knowledge (obtained from goal A) about the likelihood of exposure to potentially
harmful substances. Research under this goal encompasses the *qualitative* identification and characterization of potentially harmful substances in cosmetics, including chemical contaminants and microorganisms. Methods for the isolation, purification, and chemical or biological identification of harmful constituents will need to be developed.
Since the nature of a substance may be modified by environmental factors — e.g., heat, sunlight, air, water, ionizing radiation — and
other transforming influences including processing and handling, any
significant interactions of this kind will need to be determined for
their impact on safety.

Information on the occurrence of microorganisms in cosmetics must
be obtained. Credible regulatory action cannot be taken against
grossly contaminated products unless normal microbial levels are established. The vast majority of microorganisms may not be pathogenic
to humans; many bacteria are ubiquitous in nature and are pathogenic
to humans only if they possess virulence factors.

Virulence factors produced by proven human pathogens must be
isolated and purified in order to effectively determine the human risk.
The extent of contamination of cosmetic products with toxicogenic
fungi, toxic molds, and virus particles is largely unknown. Major

needs exist for better identification techniques for these substances
and also for other environmental contaminants.

Research goal D: To develop, validate, and apply quantitative
methods of analysis for potentially hazardous constituents, adulterants, and contaminants of foods and cosmetics.

This goal reflects the bureau's continuing need for accurate and
reliable quantitative chemical and biological assay methods. The preceeding goal and this one are virtually inseparable tasks; they both
are essential in the analysis of cosmetics for harmful constituents.
The qualitative identification and isolation activities under the preceeding goal provide the direction and necessary foundation for research under this goal.

Although analytical methods have improved greatly during the
last few decades, their application to complex mixtures typical of cosmetics is far more difficult and needs improvement. Such methods
are necessary in order to establish safety standards, set tolerances,
collect information for risk assessment, and permit regulatory monitoring and control.

Research goal E: To develop, improve, and validate biological
tests for various forms of human toxicity.

Vitally important to the quality of regulatory decisions in the cosmetic area is the reliability and practicability of toxicological assay
methods. Current methods are inadequate in several respects. A
single cancer animal bioassay costs approximately half a million dollars
and takes about three years to complete. The extrapolation of these
maximum tolerated dose (MTD) bioassays to predicting risks for humans exposed to much lower doses is currently necessary in the absence
of more discriminating assays. But the limitations of this approach,
including a significant disregard of the effects of dose on metabolism,
physiology, and pharmacokinetics, are painfully evident to toxicologists and are becoming obvious to the public as well. Insightful investigations that can lead to more critical, more cost-effective tests
are urgently needed.

Chronic toxicity: For some time to come the prediction of the long-
term or chronic effects of poisonous substances will continue to rely
heavily on animal bioassays. It is imperative that these animal models
be made more predictive of human risk and suitable for detecting toxicity from low levels of exposure. Investigations to establish the influence of the dose, route of administration, and length of exposure
to a substance are essential. Similar studies are needed to determine
the effect of age, stress, hormonal state, pregnancy, and nutritional
state on the toxic susceptibility of experimental animals. Studies in
comparative metabolism, biochemistry, and pharmacokinetics are likely
avenues of approach. Efficient use of animal studies requires the observation and selection of physical, physiological, biochemical, and

morphological indices that can afford early and reliable signs of chronic or irreversible disease. In designing toxicological methods, more emphasis needs to be placed on the detection of early signs of toxicity. Better statistical methods to optimize the design and analysis of chronic animal bioassays and more efficient use of mathematical models should be developed.

Fetal and neonatal toxicity: We are exposed to potentially toxic substances not only during the bulk of our lifetimes, but also during gestation and soon after birth when we are most vulnerable. Present methods for determining the susceptibility of the developing fetus to toxic substances taken in by the mother during pregnancy or of the neonate during lactation are inadequate. All of the questions about dosing and animal models apply here, but are further complicated because of incomplete knowledge regarding the uterine environment, transplacental migration, and infant metabolism.

Behavioral and immunotoxicity: There is growing awareness of the need to detect toxic manifestations that are not lethal or severely debilitating but that nevertheless impair human functionality in more subtle ways. Neurobehavioral effects are related directly to the problems of congenital mental retardation and abnormal behavior in children for which there is, in most cases, no known etiology. Many questions being addressed in this area relate to the sensitivity, feasibility and validity of presently used methods. The proposed research includes the development of *in vivo* and *in vitro* methods for assessing autoimmunity, allergies, and other immunological disorders.

Toxicological screens — short-term (*in vitro*) tests: The increasing number of substances that require testing, limitations of animal resources, and the length and cost of animal bioassays make it imperative to develop rapid, sensitive, and reliable predictors or screens of irreversible chronic toxicity (especially carcinogenicity and mutagenicity). These tests are needed to rank an increasing number of chemicals for further testing in whole animals and to provide unique information on the intrinsic toxicity of substances at a cellular or molecular level. In addition to the refinement and validation of tests already developed the use of tissue and cell culture systems as predictors of human toxicity needs to be investigated. Mammalian cell culture systems offer some real hope that safety evaluations in the future can make use of tests that combine speed and sensitivity with a more precise and credible extrapolation of their results to humans. *In vitro* assays for identifying carcinogenic promoters and other modifiers as well as those capable of detecting substances producing teratological defects (e.g., mammalian embryo culture), reproduction defects (e.g., oocyte development), and neurotoxins (e.g., mammalian neuronal cells) would be particularly valuable.

Dermal and ocular toxicity: Many acute or subchronic effects of chemicals can currently be best predicted through the use of animal

models or multicellular tissue systems. Ocular and skin toxicity are
two such examples. When these chemicals are absorbed through the
skin in sufficient quantity, they can produce both local or systemic
effects. The extent of skin penetration of suspect substances has
become an important determinant for regulatory action (e.g., hexachlorophene, N-nitrosodiethanolamine, hair dyes. Systematic studies
on skin penetration both in animals and humans are necessary in order
to predict the toxicological impact of cosmetic ingredients and to react
promptly to suspected hazards.

A newly recognized problem concerns potential photoinduced carcinogenesis following the application to the skin of substances which
are phototoxic, but not in themselves carcinogenic. It is important
to determine whether a readily demonstrable property, such as phototoxicity, can be used to screen chemicals that could enhance photo-
carcinogenesis. *In vitro* techniques that involve the chemical isolation
from the skin of DNA molecules, altered as a consequence of the effect
of sunlight (ultraviolet) and phototoxic substances, may permit rapid
screening of topically applied substances.

More useful methods for detecting and predicting phototoxicity,
photoallergy, skin irritation, cosmetic acne, and skin sensitization
are needed. The usual mildness and reversibility of these injuries
are not adequate reasons to neglect the discomfort they cause in hundreds of thousands of users annually, particularly since most of these
effects may be preventable.

Present tests for adverse ocular effects rely primarily on visual
observation. The subjective nature of these tests, such as the widely
used Draize rabbit eye tests introduced in the 1940s, makes establishment of ocular safety criteria difficult. Recent improvements in technology have made it likely that better techniques to detect and quantify adverse ocular effects can be developed. The application of the
specular microscope to measure corneal damage and the application
of liquid chromatographic techniques to measure accumulation of toxic
substances in the eye are two promising examples.

Other promising examples are tissue or cell culture techniques
for performing adverse reactions in the eye. Though currently they
may be reviewed only as scientific concepts, broad advances in toxicological, medical, and related scientific disciplines should bring about
significant advances that would bring them within the realm of useful
safety tests.

Research goal F: To apply toxicity tests and epidemiological studies to determine the adverse effects (hazards) of cosmetic ingredients,
singly or in combination.

The research effort highlighted in this goal is the critical need
for valid, relevant, and complete toxicological and epidemiological information for the large number of cosmetic chemicals the FDA must
gauge for safety or hazard. Some of the information developed here

may be useful for improving test systems (see goal E), but the major emphasis is on the substances under study, not the methodologies used. This activity depends on the availability of toxicological and epidemiological methods that have been validated and are predictive of adverse effects in humans, i.e., on progress under goal E.

Research goal G: To improve the accuracy of methods used to assess potential risks to human health associated with the constituents and contaminants of foods and cosmetics.

It has become increasingly evident that *absolute* assurance of safety cannot be provided for any product. Our responsibility in the bureau is to make some reasonable distinction between those risks that are tolerable and those that are not. The increased recognition of the ubiquity of carcinogenic chemicals in our diet and environment, coupled with the increased sensitivity of analytical instrumentation with which to detect them, make this a vital and urgent responsibility.

Risk assessment is the process of estimating human statistical risk from data on lower organisms, animals, or human subpopulations. This activity is not laboratory-oriented but requires integration and analysis of data derived from our own and other laboratories. The proposals in this area fall mainly into one of four categories: epidemiology, animal models, predictive *in vitro* tests, and theoretical models. The latter are necessary to extend laboratory dose-response data to the lower range of human exposure. In each case, risk assessment requires an understanding and a comparison of the dose-response behavior in the test system with the human population, together with an understanding and a comparison of the different susceptibilities of the two groups. There are major gaps in both science and methodology that currently preclude reliable risk assessments.

In the epidemiological area we need substantially to expand the usable data base on exposures, risk factors, and health outcomes by accessing and standardizing currently existing pertinent data bases in other agencies and by conducting agency-sponsored retrospective and prospective studies. It is important to develop more economical and more powerful statistical designs and analyses of human epidemiological studies to help us identify and minimize the effects of confounding factors. Definitive human epidemiological studies are especially few in number. There is a need to acquire data from specifically designed studies on a common set of known toxicants which can be used as a standard for developing future risk assessment methodologies.

Basic toxicological data are essential to risk assessment based on animal models. A fundamental problem is our lack of knowledge of the pharmacokinetics of absorption and metabolism of toxic compounds and their modifiers as well as the biomechanisms that determine a particular toxic response. We need to determine the proper definition of normal dosage, the proper toxicological indices for measuring response, and the effective dose reaching the blood, organ, or cellular site of action. Mathematical models relating applied dose to effective dose to response are needed. Such models based on the proper effective dose should be less variable than the less refined dose-response models used today. Because they are based on a less adequate data base, the proposals outlined in this area of research are of necessity more exploratory and less specific than those described for other bureau goals.

Short-term testing using lower organisms, organ systems, cell culture, or biochemical effects is becoming more important as the strain on our animal and economic resources mounts. Studies aimed at improving the relevance of these assays as predictors of human response should have the highest priority.

The assessment of risk is more credible when all available relevant information is utilized and integrated into a coherent "toxicity profile" for the chemical. Methods for correlating and integrating the dose-response relationships among human, animal, and short-term tests need to be developed.

Research goal H: To develop practical control technologies necessary to detect or prevent hazards resulting from the storage or processing of foods or cosmetics.

Processing, packaging, and storage can substantially transform cosmetics and cause them to become adulterated or inferior products. In order to enforce the provisions of the FDC Act, the FDA must monitor industry practices. Sensitive and rapid indicators of chemical and microbial contamination of cosmetics are needed. Rapid methods for isolating and identifying microorganisms in cosmetics and contact surfaces are required.

Research goal I: To assess social, economic, and environmental determinants and impacts of bureau actions.

The influence of bureau decisions on the fabric of daily life and commerce is immense. We have an obligation to take into account social, economic, and environmental consequences of bureau decisions, their costs and benefits, to ensure that the best possible decisions are made. While safety considerations have always been paramount and will remain so, the assessment of non-health-related costs and benefits have influenced bureau decisions, but usually on an ad hoc basis or in response to an Executive Order. The accuracy and timeliness of such analyses are often crucial to the development of the appropriate regulatory posture — recognizing that there is generally more than one way to achieve a given regulatory objective. In recent years the reporting of these assessments has become increasingly more systematic, explicit, and public for all regulatory agencies. Increasing demands for such analyses have created special needs for bureau resources in this area.

Unfortunately, scientific methods for these assessments are in states of relative infancy. Also, a major lack of information exists, particularly in the social and environmental areas. To carry out its responsibilities under Executive Order 12291 [2] (regulatory impact), the bureau needs to improve the data and the methods used in the area of cost analysis.

The FDA has a clear obligation under the National Environmental Policy Act and Council on Environmental Quality Regulations to assist applicants or petitioners in preparing environmental assessments. At present, there are no standard methods available for predicting the effects of chemical substances on the ecosystem. The development of a systematic and uniform assessment procedure would bring about needed improvement in the environmental review process by increasing both the reliability and consistency of environmental impact predictions and by assuring that such predictions are scientifically valid and legally defensible.

C. Setting Priorities: Reallocating Resources

The agency's cosmetics program has a history of being low in priority and limited in resources and personnel. It is well known that resources for all phases of the cosmetics program have been substantially reduced during the past few years. This requires the careful establishment of priorities in the technical area. The priority decisions are difficult ones. The program does not have the resources to permit it to deal with headline issues, let alone those that do not make the newspapers. The priorities are frequently modified to deal with emerging problems that require laboratory support for the determination of potentially harmful ingredients or contaminants reported to be in cosmetic products.

As a result of resource constraints, agency personnel must react to external decisions and events rather than acting on the basis of priorities. This forces a frequent reshuffling of priority decisions. With project rankings in such a state of flux, no priorities can be said to truly guide the program at all. Nevertheless, it is essential that priorities be established to guide the selection of the research to be performed. Whether the particular projects are undertaken by government, industry, or the academic community, selection from among the myriad of potential research projects must be made. Therefore, a ranking system for evaluating the importance and timeliness of the research projects is necessary regardless of who does the work. Research funding will be devoted primarily to priority research projects in direct support of the FDA's regulatory responsibilities. Consumers have a rising sense of awareness and expectations for redress. New questions of safety appear regularly in the press. Clearly, when the agency is forced to make resource allocation decisions, these decisions are likely to need defending before the public. If the public does not demand close scrutiny of decisions, industry will.

D. Funding

While the FDA should conduct research in some of the areas identified in the plan, institutions in the private sector as well as others in government also have a role and responsibility in these research areas. The magnitude of the plan is far beyond any resources of the FDA, current or future. Most of the work is expected to be done by others. Some of the work is already under way, and the bureau will continuously search available sources for program information. The plan was made available for public comment through a notice in the *Federal Register* [3], and comments were received, studied, and incorporated where appropriate. The bureau's hope is that by describing research goals and needs, it can focus the attention of others on these needs.

It is difficult to say at this point how much of the plan will be funded and what projects can be implemented by the Bureau of Foods. The pressure of reduced budgets resulting from both the general current attitude of fiscal restraint and the particular action of limiting the cosmetic budget makes difficult the prediction of how far the bureau can go in implementing this part of the plan. Nevertheless, the publication of the document itself is a clear expression of the determination of the bureau to continue to invest the maximum resources possible in this area. The bureau would also hope that some of the areas described in the plan will be explored by other federal agencies or, more important, by the industry.

Some indications to this effect are apparent, such as documented examples of industry responding to this challenge. For example, industry is involved in the development of methodology for the determination of nitrosamine and dioxane in cosmetic raw materials. Another example is the Cosmetic Ingredient Review Program, which has been a stimulating factor in toxicological research in the cosmetic area. Although not required by law, its sense of public responsibility should extend to industry's need to assure, with the best possible data and the most contemporary models, the safety of the products it sells. It is perhaps in this sense that the publication of the Research Plan will have its most important effect. The document may serve as bellwether and a common focus for a concerned, socially responsible industry and the FDA to join in a program of research in the interest of cosmetic product safety.

REFERENCES

1. Bureau of Food Research Plan, U.S. Food and Drug Administration, Washington, D.C., March 1980.
2. Executive Order 12291 of February 17, 1981, Federal Regulation.
3. *Fed. Reg.*, (March 21, 1980), Notice of Availability of the Bureau of Foods Research Plan.

39

THE NATIONAL TOXICOLOGY PROGRAM AND PREVENTIVE ONCOLOGY

JAMES E. HUFF, JOHN A. MOORE,* and DAVID P. RALL

National Toxicology Program, National Institute of Environmental Health Sciences, Research Triangle Park, North Carolina

I. INTRODUCTION

Chemicals cause cancer (*quod erat demonstrandum*). Fortunately, most do not. Certain chemicals cause cancer in humans [1,2,16,31]; and certain chemicals cause cancer in animals [5,13,21,30]. All chemicals known to induce cancer in humans cause cancer in laboratory animals (a possible exception is arsenic; yet Ivankovic et al. [17] diagnosed lung carcinomas in BD rats after a single intratracheal exposure to an arsenic-containing mixture used formerly in vineyards). Therefore, reducing exposure to chemicals known or suspected to cause cancer in humans or animals will reduce chemically induced cancer in humans. This public health stance of preventive oncology stands as one of the primary goals of the National Toxicology Program (NTP): to identify with certainty those chemicals most likely to be hazardous

---

*Present affiliation: Environmental Protection Agency, Washington, D.C.

to humans. From a preventive health view, the NTP underscores the concept that chemicals found to cause cancer in animals must be considered capable of causing cancer in humans [7,11,13,16,35,36]; and, thus, those chemicals so identified in well-conducted experiments should be controlled accordingly [13,41].

The NTP strives to accomplish this by evaluating those large-volume chemicals known to have a high index of human exposure. This scientific appraisal process comprises an integrated toxicological characterization approach: chemical disposition (absorption, distribution, metabolism, excretion), genetic toxicology (see Table 1), fertility and reproductive assessment, systemic toxicology (14-day and 90-120-day exposures), specific studies as needed (immunological, biochemical, and inhalation toxicology), clinical pathology where applicable (hematology, urinalysis, endocrine function, and clinical chemistry), and long-term (two-year) carcinogenesis bioassay.

This chapter concentrates mainly on the sequence of events surrounding the procedures and techniques of the carcinogenesis bioassay; for information about other research and testing areas within the NTP, consult the NTP Annual Plan [28,42].

II. BACKGROUND

The National Toxicology Program (NTP), established in 1978 [3], develops and evaluates scientific information about potentially toxic and hazardous chemicals. This knowledge can be used to protect the health of the American people and for the primary prevention of chemically induced disease. By bringing together the relevant programs, staff, and resources from the U.S. Public Health Service, Department of Health and Human Services (DHHS), the National Toxicology Program has centralized and strengthened activities relating to toxicology research, testing and test development/validation efforts, and has emphasized and increased the dissemination of toxicological information to the public and scientific communities and to the research and regulatory agencies [4,9,18,19,23-29,37-40,42].

The NTP is comprised of four charter DHHS agencies: the National Cancer Institute (NCI), National Institutes of Health; the National Institute of Environmental Health Sciences (NIEHS), National Institutes of Health; the National Center for Toxicological Research, Food and Drug Administration; and the National Institute for Occupational Safety and Health, Centers for Disease Control. In July 1981, the Carcinogenesis Bioassay Testing Program (NCI) was transferred to the NIEHS.

Also in July, Congressional hearings on the NTP were held by Representative Albert Gore, Jr., Chairman, Subcommittee on Investigations and Oversight of the Committee on Science and Technology [44].

In October 1981, the Secretary of DHHS removed the temporary status under which the NTP has functioned since 1978 and made the

NTP a permanent program. Specific goals have been early identified to (1) expand the toxicological profiles of the chemicals nominated, selected, and being tested; (2) increase the number and rate of chemicals under test, as funding permits; (3) develop, coordinate, and validate a series of tests/protocols more appropriate for regulatory needs; and (4) communicate program plans and results to governmental agencies, medical and scientific communities, and the public.

The NTP approach to testing emphasizes developing new and better test methods. This overture does not imply flaws in traditional toxicology and regulatory requirements, but reflects rapid advancements in testing methodology and expanding boundaries of scientific knowledge. Thus, the NTP plans to validate possible alternatives that may be performed more reliably, yield new toxicologic data, give results relevant to human disease, and develop a testing approach that produces equivalent results in a faster, more economical manner. Often, testing results affect regulatory or public health issues, and the NTP will meld these innovative techniques with "standard" methods to ensure results that are germane and of utility to regulatory and public health needs. When standard methods are used, the NTP will attempt to incorporate those standards presently advocated by regulatory agencies, such as the lifetime rodent bioassay [6,8,14,20,22,32,33,43 as examples].

III. TOXICOLOGY AND CARCINOGENESIS BIOASSAY

The "standard" two-year carcinogenesis bioassay remains the most definitive method for detecting chemical carcinogens in animals. The standard protocol as developed by the NCI (and frequently still used by the NTP) typically uses two rodent species (usually Fischer 344 rats and B6C3F1 mice), both sexes, and administration of multiple dose levels (concurrent controls, low dose, and high dose) of a chemical to groups of 50 animals, beginning at weaning and ending after two years. These series of experiments are designed primarily to determine if selected chemicals produce cancer in animals. (Excellent compendia are available that list those 8,252 chemicals already tested in carcinogenesis bioassays [34] and those 970 chemicals currently being tested [15].) NTP ordinarily uses four dose designs, with additional groups to be examined for pathology after 15 or 18 months exposure.

The results of the bioassay also serve as the reference base for the validation of short-term carcinogenesis assays. Two additional objectives have been identified as priority items: (1) to expand the bioassay experimental protocols to extend and better characterize the toxicologic profile of chemicals; and (2) to investigate, develop, and validate accurate, less costly, and more rapid methods for detecting carcinogenic potential.

Under the NTP, the carcinogenesis bioassay procedure(s) has been and continues to be changed to meet the objective of a broadened toxicologic characterization of chemicals, and, further, to lead or stay abreast of advancing scientific developments [10,12]. Prior to NTP involvement, the prechronic phases of the bioassay — which include single dose (acute), 14-day repeated dose, and 90- to 120-day repeated dose studies — were conducted to determine gross toxicity and general target organ effects at different dose levels as a primary basis for setting appropriate doses for the two-year bioassay studies. Now the NTP has begun to gather routinely other information related to target organ effects: Chemical disposition, fertility and reproduction, urinalysis, clinical chemistry, and hematology also are obtained from the prechronic studies — especially the 90-day study; certain other specific studies as applicable are included in the chronic two-year studies as well. Once those parameters that may be altered through exposure to the tested chemicals are identified, suspect chemicals are referred to specific organ system groups for more detailed study of the functional, biochemical, and morphologic effects of the test compounds. Also, wider analysis of the quantitative and comparative absorption, distribution, metabolism, and excretion patterns may be desired. For instance, 28 (or 70 percent) of the 40 chemical starts in fiscal year (FY) 1980 included specific toxicology studies in the prechronic testing phase. All chemicals started on test since then have had an expanded design including other select studies. Significantly, all chemicals selected for chronic bioassay will be profiled for chemical disposition patterns. The goal is to ensure that all major toxic effects will be identified for each chemical being considered for long-term bioassays.

Prior to commencing the actual long-term carcinogenesis bioassay, all chemicals undergo genetic toxicology testing in at least five *in vitro* short-term assays (Table 1). These data, together with other prechronic bioassay information, are used by the experimental design

TABLE 1 Cellular and Genetic Toxicology Program: Rapid In Vitro Test Systems

---

Gene mutations in bacteria
  Salmonella typhimurium/microsome

Gene mutations in mammalian cells
  Mouse lymphoma (L5178Y, thymidine kinase)

Chromosome damage in mammalian cells
  Cytogenic damage and sister chromatid exchange (*in vitro*, CHO)

A mammalian cell transformation assay
  (BALB/c-3T3)

A direct measure of DNA damage/repair (which does not necessarily result in mutation or transformation)
  Unscheduled DNA synthesis (rat hepatocytes)

---

groups to prepare appropriate study protocols and are used by staff
to assist in establishing priorities for chemicals queued into the long-
term carcinogenesis bioassay. A key decision that must be made at
this juncture between the completion of the prechronic phase and the
beginning of the chronic study centers directly on whether indeed the
lifetime bioassay should be done at all.

Thus, while the lifetime animal bioassay remains the best procedure for determining the carcinogenic potential of chemicals, the NTP
does not ordinarily use a standardized design. Rather, the design is
adapted to the special testing needs identified for the particular chemical. The NTP tailors its testing protocols to the particular chemicals,
based on the results from the prechronic testing phases, on available
literature, and on structure-activity relations. These new protocols
permit better, more specific information to be generated for the tested
compounds, which increases the effectiveness of the tests for potential
human risk estimations. Such protocols also will be useful as guidelines for testing undertaken by other agencies and by industry. As
examples, the NTP continues to pursue actively other design methodologies — increase the number of dose levels, utilize an "unbalanced"
distribution of animals among dose groups, introduce interim kills, reduce where appropriate the necessary pathology, and so on.

Carcinogenesis Bioassay Reports: These compilations receive considerable staff attention during generation, initial data analyses, and
draft report preparation; further intense focus is devoted to the draft
report by the NTP staff as an iterative review process. After the
draft receives internal approval, copies are sent for external peer review to the NTP Board of Scientific Counselors' Technical Reports Review Subcommittee (Peer Review Panel). At this stage draft reports
are made available to any individual seeking or requesting copies.
This ad hoc panel, comprised of nongovernment scientists (Table 2),
then reviews critically these draft reports and voices their opinions
and findings in sessions open to the public. When the draft reports
are considered by staff and by the Peer Review Panel to be scientifically acceptable and soundly based, the reports undergo a final technical and style edit prior to printing and distribution.

The Review Process: Meetings to perform peer review of NTP
draft bioassay technical reports are held approximately three times or
yearly in Research Triangle Park, N.C. (sometimes in Washington,
D.C.). A list of the reports to be reviewed, reviewer assignments,
review forms, and other information about a particular meeting are
sent to Peer Review Panel members at least one month prior to the
meeting data. At the same time notices about the meeting are published in the *Federal Register* and the *NTP Technical Bulletin*.
(Draft reports are also made available to anyone upon request.)

TABLE 2 The NTP Technical Reports Peer Review Panel Members

Louis S. Beliczky, M.S., M.P.H.
  (Occupational)
United Rubber Workers

Norman Breslow, Ph.D.
  (Biostatistics)
University of Washington

Devra L. Davis, Ph.D.
  (Epidemiology)
National Academy of Sciences

*Robert M. Elashoff, Ph.D.
  (Biostatistics)
Jonsson Comprehensive Cancer
  Center

Seymour L. Friess, Ph.D.
  (Toxicology)
Toxicology Consultant

Curtis Harper, Ph.D.
  (Pharmacology)
University of North Carolina

*Joseph H. Highland, Ph.D.
  (Chemistry)
Princeton University

*Margaret Hitchcock (Chair)
  (Toxicology)
J. B. Pierce Foundation Laboratory

*J. Michael Holland, D.V.M., Ph.D.
  (Pathology)
The Upjohn Company

Jerry B. Hook, Ph.D. (Chair)
  (Toxicology)
Smith Kline & French Laboratories

*Charles C. Irving, Ph.D.
  (Chemistry)
Veterans Administration Hospital

Thomas C. Jones, Ph.D.
  (Pathology)
Harvard Medical School

Richard J. Kociba, Ph.D.
  (Pathology)
Dow Chemical USA

David Kotelchuck, Ph.D.
  (Toxicology)
United Electrical, Radio and Mach-
  ine Workers of America

*Frank Mirer, Ph.D.
  (Toxicology)
United Auto Workers

*Sheldon D. Murphy, Ph.D.
  (Toxicology)
University of Texas

*Svend Nielson, Ph.D.
  (Pathology)
University of Connecticut

*Robert A. Scala, Ph.D.
  (Toxicology)
Exxon Corporation

*Bernard A. Schwetz, Ph.D.
  (Toxicology)
Dow Chemical USA (Now with NTP)

*Thomas H. Shephard, Ph.D.
  (Teratology)
University of Washington

*Roy Shore, Ph.D.
  (Biostatistics)
New York University Medical
  Center

Tom Slaga, Ph.D.
  (Carcinogenesis)
University of Texas

James Swenberg, Ph.D.
  (Pathology)
Chemical Industry Institute of
  Toxicology

TABLE 2 (Continued)

Steven R. Tannenbaum, Ph.D. (Toxicology) Massachusetts Institute of Technology	*Mary Vore, Ph.D. (Pharmacology) University of Kentucky
Bruce W. Turnbull, Ph.D. (Biostatistics) Cornell University	*Alice S. Whittemore, Ph.D. (Biostatistics) Stanford University
John R. Van Ryzin, Ph.D. (Biostatistics) Columbia University	*Gary M. Williams, Ph.D. (Pathology) Naylor Dana Institute of Disease Prevention
*Stan D. Vesselinovitch, Ph.D. (Pathology) University of Chicago	

*Term completed.

Members of the Peer Review Panel receive copies of *all* reports to be reviewed during the designated meeting. Two and sometimes three principal reviewers are assigned for each report, and each gives an oral critique at the meeting accompanied by written comments. These panel members are asked to provide the NTP with a critical review of each report in advance of the meeting. Deficiencies in design, conduct, or interpretation of the study should be identified, and errors or omissions in the draft report should be stated. Further, panel members are requested to read all reports scheduled for a particular meeting and to contribute their opinions and personal dissertation during the discussion period on each report. The recommendations of the reviewers and summary comments recorded at the meeting are incorporated where appropriate and relevant in the final revision of the report.

Following the meeting, draft summary minutes for each report review are prepared by the Executive Secretary and are sent to the reviewers for any necessary corrections and alterations. The edited minutes are then made available for distribution to any interested party. Likewise, immediately after the peer review, reports are readied for publication.

Carcinogenesis Studies Results: Since 1976 when the carcinogenesis testing program published the first in the series of Technical Reports, upwards of 300 studies have been designed and conducted to characterize and evaluate the toxicologic potential, including carcinogenic activity, of selected chemicals in laboratory animals — usually two species, rats and mice. Data taken from the 85 most recent car-

cinogenesis studies serve to illustrate the observed patterns of nonneoplastic and neoplastic responses. Most are two-year feeding or
gavage experiments involving Fischer 344/N inbred rats and B6C3F1
hybrid mice (C57BL/6N × C3H/HeN MTV⁻); control and two to three
treated groups contain 50 animals of each sex and species. Thus each
study consists of four separate experiments. Of the 85 studies, 42/85
(49%) were considered positive, 36/85 (42%) gave no evidence of carcinogenicity, 5/85 (6%) were equivocal, and 2/85 (2%) were inadequate.
[If one calculates the percentages by using the total number of experiments (one each for male rats, female rats, male mice, female mice)
then the figures become 188/310 (61%) no evidence, 98/310 (32%) positive, 17/310 (5%) equivocal, and 7/310 (2%) inadequate.]

Of the 42 studies with carcinogenic effects, rats and mice showed
similar sensitivity: 17 were positive in both species, 13 in mice only,
and 12 in rats only; 11 chemicals were positive in all four sex-species
groups. Results from these particular 85 studies showed that the use
of only male F344/N rats and female B6C3F1 mice would have detected
all 42 chemicals judged to be carcinogenic. Regarding exposure route,
19/23 (83%) gavage studies and 4/5 (80%) inhalation studies produced
carcinogenic responses. Obversely, 16/45 (36%) feed studies, 1/3
(33%) drinking water studies, and 1/6 (17%) dermal studies were considered positive. Among these 85 chemicals were 19 halogenated hy-
drocarbons: 15/19 (79%) produced carcinogenic responses, primarily
in the liver, and most were given by gavage. For these 15 studies
showing carcinogenic effects, the proportions were similar regardless
of exposure route: 11/13 (85%) positive studies by gavage compared
with 4/6 (67%) for all other routes.

Including the new chemical starts, there were 188 chemicals in the
prechronic (58), the chronic (45), or the histopathology (85), bioassay testing phases in July 1984. Within these 188 chemicals are 23 cosmetics, 50 dyes, 22 food additives or ingredients, and 84 pharmaceuticals (these numbers are not addable because most of these chemicals
fit in multiple categories). In each FY the NTP starts 20-30 new toxicology and carcinogenesis studies, all with expanded protocols, and
expects to complete (through peer review) 30-40 long-term bioassays
[10].

As examples, Table 3 lists 52 carcinogenesis bioassay technical reports that were evaluated and approved by the NTP staff and subsequently by the external peer review panel. These are grouped into
four results categories: (1) carcinogenic in rodents (26, 50.0 percent); (2) not carcinogenic in rodents (20, 38.5 percent); (3) equivocal in rodents (4, 7.7 percent); and (4) inadequate (2, 3.8 percent).

TABLE 3 Examples of 52 NTP/NCI Carcinogenesis Bioassay Results

I. Carcinogenic in Rodents		
Chemical technical report number (P = published)	Route/dose	Results under the conditions of these tests (* = statistically significant) (P < 0.05)
Allyl isothiocyanate, CAS No. 57-06-7 (TR-234)	Gavage (corn oil): 12 or 25 mg/kg 5 times/ week	Carcinogenic to male F 344 rats causing transitional-cell papilloma in the urinary bladder (0/49, 2/49, 4/49). (Additionally, epithelial hyperplasia of the urinary bladder occurred at 0/49, 1/49, 6/49 in male rats not having papillomas.); not carcinogenic to female F 344 rats or B 6C 3F 1 mice of either sex.
11-Aminoundecanoic acid, CAS No. 2432-99-7 (TR-216)	Feed: 7,500 or 15,000 ppm	Carcinogenic for male F 344 rats, inducing neoplastic nodules of the liver (1/50, 9/50,* 8/50*) and transitional cell carcinomas in the urinary bladder (0/48, 0/48, 7/49*); not carcinogenic for female F 344 rats. Not carcinogenic for B 6C 3F 1 mice of either sex, although the increase in male mice with malignant lymphoma (2/50, 9/50,* 4/50) may have been associated with the administration of 11-aminoundecanoic acid.
2-Biphenylamine hydrochloride (2-aminobiphenyl), CAS No. 2185-92-4 (TR-233)	Diet: 1,000 and 3,000 ppm	Not carcinogenic for F 344 rats of either sex. Equivocal for male B 6C 3F 1 mice causing hemangiosarcoma of the circulatory system 0/50, 2/50, 3/50 (however, poor survival); carcinogenic for female B 6C 3F 1 mice causing hemangiosarcomas of the circulatory system (0/49, 1/50, 7/50*).

TABLE 3 (Continued)

I. *Carcinogenic in Rodents*

Chemical technical report number (P = published)	Route/dose	Results under the conditions of these tests (* = statistically significant) (P < 0.05)
Bis(2-chloro-1-methylethyl) ether (BCMEE), CAS No. 108-60-1 (TR-239)	Gavage (corn oil) 5 times/week: mice — 100 or 200 mg/kg	Carcinogenic for B6C3F1 mice causing alveolar/bronchiolar adenomas in males (5/50, 13/50,* 11/50) and in females (1/50, 4/50, 8/50*) and inducing in males dose-related hepatocellular carcinomas (5/50, 13/50, 17/50*). Rare forestomach tumors were probably associated with the test chemical (low dose: 1 male; high dose: 2 males and 3 females).
C.I. disperse yellow 3, CAS No. 2832-40-8 (TR-222)	Feed: rats — 5,000 or 10,000 ppm; mice — 2,500 or 5,000 ppm	Carcinogenic for male F344 rats, causing neoplastic nodules of the liver (1/49, 15/50,* 10/50*); rare stomach tumors may have been associated with the test chemical (0/49, 4/50, 1/50). Not carcinogenic for female rats or male B6C3F1 mice. Carcinogenic for female mice causing hepatocellular adenomas (0/50, 6/50,* 12/50*).
C.I. solvent yellow 14, CAS No. 842-07-9 (TR-226)	Feed: rats — 250 or 500 ppm; mice — 500 or 1,000 ppm	Carcinogenic for F344 rats, inducing neoplastic nodules of the liver in both males (5/50, 10/50, 30/50*) and females (2/50, 3/49, 10/48*); not carcinogenic for B6C3F1 mice of either sex.
Cinnamyl anthranilate, CAS No. 87-29-6 (TR-196 P)	Feed: 15,000 or 30,000 ppm	Carcinogenic for male F344 rats, inducing low incidence of two rare tumors: acinar cell carcinomas or adenomas of the pancreas (0/42, 0/49, 3/45), and adenocarcinomas or adenomas of the renal cortex (0/48,

TABLE 3 (Continued)

I. Carcinogenic in Rodents

Chemical technical report number (P = published)	Route/dose	Results under the conditions of these tests (* = statistically significant) (P < 0.05)
		0/50, 4/49); not carcinogenic for female F344 rats. Carcinogenic for B6C3F1 mice, inducing hepatocellular adenomas in males (8/48, 23/50,* 29/47*; carcinomas were increased also 6/48, 7/50, 12/47) and in females causing hepatocellular carcinomas (1/50, 8/49,* 14/49*) and adenomas (2/50, 14/49,* 19/49*).
Cytembena, CAS No. 21739-91-3 (TR-207 P)	Intraperitoneal injection, 3 times/week: rats — 7 or 14 mg/kg; mice — 12 or 24 mg/kg	Carcinogenic for F344 rats (male: mesotheliomas in the tunica vaginalis 0/50, 11/50,* 10/50,* and in multiple organs 3/50, 26/50,* 26/50*; female: fibroadenomas in the mammary gland 13/49, 22/50, 36/50*); not carcinogenic for B6C3F1 mice of either sex.
D&C Red No. 9, CAS No. 5160-02-1 (TR-225)	Feed: rats — 1,000 or 3,000 ppm; mice — 1,000 or 2,000 ppm	Carcinogenic in male F344 rats, inducing sarcomas of the spleen (0/50, 0/50, 26/48*) and neoplastic nodules in the liver (0/50, 6/50,* 7/49*); not carcinogenic for female F344 rats, although the increased number of female rats with neoplastic nodules of the liver (1/50, 1/50, 5/50) may have been related to compound administration. Not carcinogenic for B6C3F1 mice of either sex.

TABLE 3 (Continued)

	I. *Carcinogenic in Rodents*	
Chemical technical report number (P = published)	 Route/dose	Results under the conditions of these tests (* = statistically significant) (P < 0.05)
1,2-Dibromo-3- chloropropane (DBCP), CAS No. 96-12-8 (TR-206 P)	Inhalation, 6 hours/day, 5 days/week: 0.6 or 3.0 ppm	Carcinogenic for F344 rats (nasal cavity tumors — male: 0/50, 40/50,* 39/49*; female: 1/50, 27/50,* 42/50*; and tumors of tongue — male: 0/ 50, 1/50, 11/49*; female: 0/ 50, 4/50, 9/50*; and cortical adenomas of the adrenal glands in females 0/50, 7/50,* 5/48*). Carcinogenic for B6C3F1 mice (nasal cavity tumors — male: 0/45, 1/42, 21/48*; female: 0/ 50, 11/50,* 38/50*; and alveo- lar/bronchiolar adenomas — male: 0/41, 1/40, 6/45*; fe- male: 3/49, 3/49, 10/47*).
1,2-Dibromoethane (ethylene dibro- mide), CAS No. 106-93-4 (TR-210 P)	Inhalation, 6 hours/day, 5 days/week: 10 or 40 ppm	Carcinogenic for F344 rats (both sexes — tumors of the nasal cavity and hemangiosarcomas of the circulatory system; male — mesotheliomas in the tunica vaginalis; female — al- veolar/bronchiolar tumors and mammary fibroadenomas). Car- cinogenic for B6C3F1 mice (both sexes — alveolar/bronch- iolar adenomas and carcinomas; female — hemangiosarcomas, subcutaneous fibrosarcomas, nasal cavity tumors, and mam- mary gland adenocarcinomas).
2,6-Dichloro-p- phenylenediamine, CAS No. 609-20-1 (TR-219 P)	Feed: rats — 1,000 or 2,000 ppm (males) and 2,000 or 6,000 ppm	Not carcinogenic for F344 rats. Carcinogenic for B6C3F1 mice (male: hepatocellular adenomas 4/50, 7/50, 15/50,* carcinomas not significant 12/50, 13/50, 17/50; female: hepatocellular

TABLE 3 (Continued)

	I. Carcinogenic in Rodents	
Chemical technical report number (P = published)	Route/dose	Results under the conditions of these tests (* = statistically significant) (P < 0.05)
		carcinomas 2/50, 2/50, 7/50, and adenomas 4/50, 4/50, 9/50, combined 6/50, 6/50, 16/50*).
Di(2-ethylhexyl) adipate, CAS No. 103-23-1 (TR-212 P)	Feed: 12,000 or 25,000 ppm	Not carcinogenic for F344 rats of either sex; carcinogenic for B6C3F1 mice (male: hepatocellular adenomas 6/50, 8/49, 15/49*; female: hepatocellular carcinomas 1/50, 14/50,* 12/49*).
Di(2-ethylhexyl) phthalate (DEHP), CAS No. 117-81-7 (TR-217 P)	Feed: rats – 6,000 or 12,000 ppm; mice – 3,000 or 6,000 ppm	Carcinogenic for F344 rats (male: hepatocellular carcinomas and neoplastic nodules combined 3/50, 6/49, 12/49*; female: hepatocellular carcinoma 0/50, 2/49, 8/50*). Carcinogenic for B6C3F1 mice (hepatocellular carcinomas – male: 9/50, 14/48, 19/50*; female: 0/50, 7/50,* 17/50*).
Hexachlorodibenzo-p-dioxins 1,2,3,6,7,8-/ and 1,2,3,7,8,9-, CAS No. 57653-85-7 (TR-198 P)	Gavage, 2 times/week: rats and male mice – 1.25, 2.5 or 5 mg/kg/wk; female mice – 2.5, 5.0, or 10 mg/kg/wk	Equivocal for male Osborne-Mendel rats (hepatocellular carcinomas or neoplastic nodules 0/74, 0/49, 1/50, 4/48*); carcinogenic for female Osborne-Mendel rats (neoplastic nodules of the liver 2/75, 10/50,* 12/50,* 30/50,* and hepatocellular carcinomas 0/75, 0/50, 0/50, 4/50*). Carcinogenic for B6C3F1 mice: male – hepatocellular adenoma (7/73, 5/50, 9/49, 15/48*); female – hepatocellular adenoma (2/73, 4/48, 4/47, 9/47*).

TABLE 3 (Continued)

I. *Carcinogenic in Rodents*

Chemical technical report number (P = published)	Route/dose	Results under the conditions of these tests (* = statistically significant) (P < 0.05)
4,4'-Oxydianiline, CAS no. 101-80-4 (TR-205 P)	Feed: rats — 200, 400, or 500 ppm; mice — 150, 300, or 800 ppm	Carcinogenic for F344 rats causing hepatocellular carcinoma (male — 0/50, 4/50, 23/50,* 22/50*; female — 0/50, 0/49, 4/50, 6/50*), neoplastic nodules of the liver (male — 1/50, 9/50,* 18/50,* 17/50*; female — 3/50, 0/49, 20/50,* 11/50*), follicular cell carcinoma of the thyroid (male — 0/46, 5/47,* 9/46,* 15/50*; female — 0/49, 2/48, 12/48,* 7/50*), follicular cell adenoma (male — 1/46, 1/47, 8/46,* 13/50*; female — 0/49, 2/48, 17/48,* 16/50*). Carcinogenic for B6C3F1 mice causing harderian gland adenoma (male — 1/50, 17/50,* 13/49,* 17/50*; female — 2/50, 15/50,* 14/50,* 12/50*); female — hepatocellular carcinoma (4/50, 7/49, 6/48, 15/50*), hepatocellular adenoma (4/50, 6/49, 9/48, 14/50*), and follicular cell adenomas of the thyroid (0/46, 0/43, 0/42, 7/48*).
Pentachloroethane, CAS No. 76-01-7 (TR-232)	Gavage (corn oil): rats — 75 and 150 mg/kg 5 times/week; mice — 250 and 500 mg/kg 5 times/week	Not carcinogenic for F344 rats of either sex (nephrotoxic for males: diffuse inflammation — 4/50, 14/49,* 33/50*; mineralization — 4/50, 29/49,* 29/50*); carcinogenic for B6C3F1 mice (male: hepatocellular carcinoma 4/48, 26/44,* 7/45; female: hepatocellular carcinoma 1/46, 28/42,* 13/45,* and hepatocellular adenoma 2/46, 8/42,* 19/45*).

TABLE 3   (Continued)

I. Carcinogenic in Rodents

Chemical technical report number (P = published)	Route/dose	Results under the conditions of these tests (* = statistically significant) (P < 0.05)
Polybrominated biphenyl mixture (Firemaster FF-1), CAS No. 67774-32-7 (TR-244)	Gavage (corn oil): 0, 0.1, 0.3, 1.0, 3.0, and 10 mg/kg 5 times/week for 6 months, held for lifetime observation	Carcinogenic for F344 rats (male: hepatocellular carcinoma 0/33, 2/39, 0/40, 1/33, 7/33,* 7/31*; female: neoplastic nodules 0/20, 2/21, 0/21, 2/11, 5/19,* 8/20,* and hepatocellular carcinoma 0/20, 0/21, 0/21, 0/11, 3/19, 7/20*); carcinogenic for B6C3F1 mice, causing hepatocellular carcinoma (male: 12/25, 8/27, 8/24, 12/25, 15/23, 21/22*; female: 0/13, 0/19, 2/15, 2/11, 3/17, 7/8*).
Reserpine, CAS 50-55-5 (TR-193 P)	Feed: 5 or 10 ppm	Carcinogenic for F344 male rats (adrenal medullary pheochromocytomas 3/48, 18/49,* 24/48*); not carcinogenic for female F344 rats (may have been able to tolerate higher doses). Carcinogenic for B6C3F1 mice: males — undifferentiated carcinomas of the seminal vesicles (0/50, 1/50, 5/49*); females — malignant tumors of the mammary gland (0/50, 7/49,* 7/48*).
Selenium sulfide, CAS No. 7446-34-6 (TR-194 P)	Gavage, 7 days/week: rats — 3 or 15 mg/kg/ day; mice — 20 or 100 mg/kg/day (suspended in 0.5% carboxymethylcellulose	Carcinogenic for F344 rats (hepatocellular carcinomas: males — 0/50, 0/50, 14/49*; females — 0/50, 0/50, 21/50,* and neoplastic nodules of the liver: males — 1/50, 0/50, 15/50*; females — 1/50, 0/50, 25/50*). Not carcinogenic for male B6C3F1 mice; carcinogenic for B6C3F1 female mice (hepatocellular carcinomas 0/49, 1/50,

TABLE 3 (Continued)

I. Carcinogenic in Rodents		
Chemical technical report number (P = published)	Route/dose	Results under the conditions of these tests (* = statistically significant) (P < 0.05)
		22/49,* hepatocellular adenoma 0/49, 1/50, 6/49,* alveolar/ bronchiolar carcinomas 0/49, 1/50, 4/49, and adenomas 0/49, 2/50, 8/49*).
2,3,7,8-Tetra-chlorodibenzo-p-dioxin (TCDD), CAS No. 1746-01-6 (TR-209 P)	Gavage, 2 days/week: rats and male mice − 0.01, 0.05, or 0.5 mg/kg/wk; female mice − 0.04, 0.2, or 2.0 mg/kg/wk	Carcinogenic for Osborne-Mendel rats (male: follicular cell thyroid adenomas 1/69, 5/48,* 6/50,* 10/50*; female: neoplastic nodules of the liver 5/75, 1/49, 3/50, 12/49*); carcinogenic for B6C3F1 mice (male: hepatocellular carcinomas 8/73, 9/49, 8/49, 17/50*; female: 1/73, 2/50, 2/48, 6/47*; female: follicular cell adenoma 0/69, 3/50, 1/47, 5/46*).
2,3,7,8-Tetra-chlorodibenzo-p-dioxin (TCDD), CAS No. 1746-01-6 (TR-201 P)	Dermal, 3 days/week: male mice − 0.001 mg; female mice − 0.005 mg	Not carcinogenic for male Swiss-Webster mice (increase in fibro-sarcomas of the integumentary system − 3/42, 7% versus 6/28, 21% − may have been associated with skin application of TCDD); carcinogenic for female Swiss-Webster mice (fibrosarcomas of the integumentary system − 2/41, 5% versus 8/27,* 30%).
1,1,1,2-Tetra-chloroethane, CAS No. 630-20-6 (TR-237)	Gavage (corn oil) 5 times/week: rats − 125 or 250 mg/kg; mice − 250 or 500 mg/kg	Not demonstrated carcinogenic for F344 rats (increased incidence of combined neoplastic nodules and hepatocellular carcinomas in males 0/49, 1/49, 3/48, and fibroadenomas of mammary gland in females 6/49, 15/49,* 7/46 may have been

TABLE 3 (Continued)

I. Carcinogenic in Rodents

Chemical technical report number (P = published)	Route/dose	Results under the conditions of these tests (* = statistically significant) (P < 0.05)
		associated with test chemical). Carcinogenic in B 6C 3F 1 mice (hepatocellular carcinomas — females: 1/49, 5/46, 6/48*; hepatocellular adenomas — males: 6/48, 14/46,* 21/50*; females: 4/49, 8/46, 24/48*).
Zearalenone, CAS no. 17924-92-4 (TR-235)	Feed: rats — 25 or 50 ppm; mice — 50 or 100 ppm	Not carcinogenic for F344 rats of either sex; should be considered carcinogenic for B 6C 3F 1 mice (male: pituitary adenoma 0/40, 4/45, 6/44*; female: pituitary adenoma 3/46, 2/43, 13/42*; hepatocellular adenoma 0/50, 2/49, 7/49*).
Ziram (zinc dimethyldithio-carbamate), CAS No. 137-30-4 (TR-238)	Feed: rats — 300 or 600 ppm; mice — 600 or 1200 ppm	Carcinogenic for male F344 rats causing C-cell carcinomas of the thyroid 0/50, 2/49, 7/49*; not carcinogenic for female F344 rats or for male B 6C 3F 1 mice; interpretation of the increased number of female B 6C 3F 1 mice with alveolar/bronchiolar adenoma (2/50, 5/49, 10/50*) is complicated because of a concomitant Sendai viral infection (in all groups of mice).

II. Not Carcinogenic in Rodents

Chemical technical report number (P = published)	Route/dose	Results under the conditions of these tests (* = statistically significant) (P < 0.05)
Agar, CAS No. 9002-18-0 (TR-230 P)	Feed: 25,000 or 50,000 ppm	Not carcinogenic for F344 rats or B 6C 3F 1 mice of either sex.

TABLE 3 (Continued)

II. Not Carcinogenic in Rodents		
Chemical technical report number (P = published)	Route/dose	Results under the conditions of these tests (* = statistically significant) (P < 0.05)
Arabic gum, CAS No. 9000-01-5 (TR-227)	Feed: 25,000 or 50,000 ppm	Not carcinogenic for F344 rats or B6C3F1 mice of either sex.
Asbestos, amosite, CAS No. 12172-73-5 (TR-249)	1% in diet	Not carcinogenic for Syrian golden hamsters of either sex.
Asbestos, chrysotile, CAS No. 12001-29-5 (TR-246)	1% in diet (short-range, SR; intermediate range, IR)	Not carcinogenic for Syrian golden hamsters of either sex. [Increased incidence of adrenal cortical adenomas in hamsters exposed to IR chrysotile when compared to pooled controls — males: 31/466 (7%) versus 29/244 (12%)*; females: 19/468 (4%) versus 23/234 (10%)*.]
Benzoin, CAS No. 119-53-9 (TR-204 P)	Feed: male rats — 125 or 250 ppm; female rats — 250 or 500 ppm; mice — 2,500 or 5,000 ppm	Not carcinogenic for F344 rats or B6C3F1 mice of either sex.
Bisphenol A, CAS No. 80-05-7 (TR-215 P)	Feed: rats — 1,000 or 2,000 ppm; mice — 1,000 or 5,000 (male) and 5,000 or 10,000 (female)	Not carcinogenic for F344 rats or B6C3F1 mice of either sex. (Increased incidences of leukemia in male rats 13/50, 12/50, 23/50,* and lymphoma in male mice 2/49, 8/50*, 3/50 may have been associated with the test chemical.)
Caprolactam, CAS No. 105-60-2 (TR-214 P)	Feed: rats — 3,750 or 7,500 ppm; mice — 7,500 or 15,000 ppm	Not carcinogenic for F344 rats or B6C3F1 mice of either sex.

TABLE 3 (Continued)

II. Not Carcinogenic in Rodents

Chemical technical report number (P = published)	Route/dose	Results under the conditions of these tests (* = statistically significant) (P < 0.05)
CI Acid Orange 10, CAS no. 1936-15-8 (TR-211)	Feed: rats — 1,000 or 3,000 ppm; mice — 3,000 or 6,000 ppm	Not carcinogenic for F 344 rats or B 6C 3F 1 mice of either sex.
CI Acid Red 14, CAS No. 3567-69-9 (TR-220 P)	Feed: rats — 6,000 or 12,000 ppm (males) and 12,000 or 25,000 ppm (females); mice — 3,000 or 6,000 ppm	Not carcinogenic for F344 rats or B 6C 3F 1 mice of either sex.
FD&C Yellow No. 6, CAS No. 2783-94-0 (TR-208 P)	Feed: 12,500 or 25,000 ppm	Not carcinogenic for F 344 rats or B 6C 3F 1 mice of either sex.
Guar gum, CAS No. 9000-30-0 (TR-229 P)	Feed: 25,000 or 50,000 ppm	Not carcinogenic for F 344 rats or B 6C 3F 1 mice of either sex.
Hexachlorodibenzo-p-dioxins 1,2,3,7,8,9- and 1,2,3,6,7,8-, CAS No. 19408-74-3 (TR-202 P)	Dermal, 3 times/week: 0.01 mg	Not carcinogenic for Swiss-Webster mice of either sex.
Locust bean gum (carob seed gum), CAS No. 9000-40-2 (TR-211 P)	Feed: 25,000 or 50,000 ppm	Not carcinogenic for F 344 rats or B 6C 3F 1 mice of either sex.
D-Mannitol, CAS No. 69-65-8 (TR-236)	Feed: 25,000 or 50,000 ppm	Not carcinogenic for F 344 rats or B 6C 3F 1 mice of either sex.

TABLE 3 (Continued)

II. Not Carcinogenic in Rodents

Chemical technical report number (P = published)	Route/dose	Results under the conditions of these tests (* = statistically significant) (P < 0.05)
Phenol, CAS No. 108-95-2 (TR-203 P)	Drinking water: 2,500 or 5,000 ppm	Not carcinogenic for F344 rats or B6C3F1 mice of either sex.
Selenium sulfide, CAS No. 7446-34-6 (TR-197 P)	Dermal, 3 times/week for 86 weeks: 0.5 or 1.0 mg/application	Not carcinogenic for ICR Swiss mice of either sex.
Selsun (trade name), no CAS No. (TR-199 P)	Dermal, 3 times/week for 88 weeks: 0.05 ml of 25% or 50%	Not carcinogenic for ICR Swiss mice of either sex.
Stannous chloride, CAS No. 7772-99-8 (TR-231)	Diet: 1,000 and 2,000 ppm	Not carcinogenic for F344 rats or B6C3F1 mice of either sex. (Increase in C-cell tumors of the thyroid gland in male rats may have been associated with the administration of the test chemical: adenomas 2/50, 9/49,* 5/50, or carcinomas 0/50, 4/49, 3/50, combined 2/50, 13/49,* 8/50*.)
Tara gum, CAS No. 39300-88-4 (TR-224 P)	Feed: 25,000 or 50,000 ppm	Not carcinogenic for F344 rats or B6C3F1 mice of either sex.
Toluene-2,6-diamine dihydrochloride, CAS No. 15481-70-6 (TR-200 P)	Feed: rats — 250 or 500 ppm; mice — 50 or 100 ppm	Not carcinogenic for F344 rats or B6C3F1 mice of either sex.

TABLE 3 (Continued)

	II. Not Carcinogenic in Rodents	
Chemical technical report number (P = published)	Route/dose	Results under the conditions of these tests (* = statistically significant) (P < 0.05)
Vinylidene chloride (1,1-dichloro-ethylene), CAS No. 75-35-4 (TR-228)	Gavage, 5 times/week: rats — 1 or 5 mg/kg; mice — 2 or 10 mg/kg	Not carcinogenic for F344 rats or B6C3F1 mice of either sex. (Increased incidence of liver necrosis in male mice 1/46, 3/46, 7/49*.)

	III. Equivocal	
Chemical technical report number (P = published)	Route/dose	Results under the conditions of these tests (* = statistically significant) (P < 0.05)
Butyl benzyl phthalate, CAS No. 85-68-7 (TR-213)	Feed: 6,000 or 12,000 ppm	Male F344 rat data inadequate for evaluation (early death caused by hemorrhage); probably carcinogenic for female F344 rats, causing an increased incidence of mononuclear cell leukemia (7/49, 7/49, 18/50*); not carcinogenic for B6C3F1 mice of either sex.
Eugenol, CAS No. 97-53-0 (TR-223)	Feed: female rats — 6,000 or 12,000 ppm; male rats and mice — 3,000 or 6,000 ppm	Not carcinogenic for F344 rats of either sex; equivocal for B6C3F1 mice (male: hepato-cellular adenomas 4/50, 13/50,* 10/49, or carcinomas 10/50, 20/50,* 9/49, combined 14/50, 28/50,* 18/49; female: hepato-cellular carcinomas 2/50, 3/49, 6/49, or adenomas 0/50, 4/49, 3/49, combined 2/50, 7/49, 9/49*).

TABLE 3 (Continued)

III. Equivocal

Chemical technical report number (P = published)	Route/dose	Results under the conditions of these tests (* = statistically significant) (P < 0.05)
Fluometuron, CAS No. 2164-17-2 (TR-195 P)	Feed: rats — 125 or 250 ppm; mice — 500 or 1,000 ppm	Not carcinogenic for F344 rats. Equivocal for male B6C3F1 mice (hepatocellular carcinomas 3/21, 8/47, 15/49, and adenomas 1/21, 5/47, 6/49, combined 4/21, 13/47, 21/49*). Not carcinogenic for female B6C3F1 mice. Rats and mice may have been able to tolerate higher doses.
Propyl gallate, CAS No. 121-79-9 (TR-240)	Feed: 6,000 or 12,000 ppm	Not considered carcinogenic for F344 rats, although there was evidence of an increased number of male rats with preputial gland tumors (1/50, 8/50,* 0/50), islet cell adenomas of the pancreas (0/50, 8/50*, 2/50), and pheochromocytomas of the adrenal glands (4/50, 13/48,* 8/50); rare tumors of the brain occurred in female rats (0/50, 3/50, 0/49). Not carcinogenic for B6C3F1 mice of either sex, although the increased number of male mice with malignant lymphoma (1/50, 3/49, 8/50*) may have been associated with the administration of propyl gallate.

TABLE 3 (Continued)

IV. *Inadequate*

Chemical technical report number (P = published)	Route/dose	Results under the conditions of these tests (* = statistically significant) (P < 0.05)
Asbestos, chryso-tile, and 1,2-dimethylhydrazine (CAS No. 540-73-8), CAS No. 12001-29-5 (TR-246)	1% in diet, (intermediate range, IR); gavage (4 mg/kg every other week for 5 doses)	Combination study in Syrian golden hamsters of either sex was considered inadequate because no increase in DMH-induced intestinal neoplasia (DMH is known to induce gastrointestinal tumors in animals).
2,3,7,8-TCDD and dimethylbenzan-thracene (DMBA), CAS No. 1746-01-6 (TR-201 P)	DMBA — 50 mg 1 week prior to TCDD	Initiation-promotion study considered inadequate because DMBA was not tested alone and DMBA-TCDD induced fibrosarcoma incidence not greater than that observed with TCDD alone.

Single copies of Technical Reports are available without charge from the NTP Public Information Office, P.O. Box 12233, Research Triangle Park, N.C. 27709.

Annual Report on Carcinogens [31]: A 1978 amendment to the Public Health Service Act (termed Public Law 95-622) requires that the DHHS publish an annual list of "all substances (1) which either are known to be carcinogens [in humans] and (2) to which a significant number of persons residing in the United States are exposed. . .". The Third Annual Report on Carcinogens lists those 117 selected chemicals, groups of chemicals, or industrial processes (Table 4) i) that have been evaluated by the International Agency for Research on Cancer [see, for example, 1,2,11,13,16] as being "carcinogenic for humans" or as having "sufficient evidence of carcinogenicity in experimental animals," ii) that have been regulated by the U.S. federal government as carcinogens, iii) that have been found to be carcinogenic in rodent bioassays by the National Cancer Institute or by the NTP, or iv) that have been suggested for inclusion by the agencies involved in preparing the Report.

These 117 entries do not constitute a complete compendium of all known or reasonably anticipated human carcinogens. Among others,

TABLE 4   Chemicals, Groups of Chemicals, and Manufacturing
Processes Included in the Third Annual Report on Carcinogens

1. 2-Acetylaminofluorene	37. Dibenzo(a,h)pyrene
2. Acrylonitrile	38. Dibenzo(a,i)pyrene
3. Aflatoxins	39. 1,2-Dibromo-3-chloropropane
4. 2-Aminoanthraquinone	40. 1,2-Dibromoethane
5. *4-Aminobiphenyl	41. 3,3'-Dichlorobenzidine
6. 1-Amino-2-Methylanthra-	42. 1,2-Dichloroethane
quinone	43. Diepoxybutane
7. *Amitrole	44. Di(2-ethylhexyl)phthalate
8. o-Anisidine Hydrochloride	45. *Diethylstilbestrol
9. Aramite	46. 3,3'-Dimethoxybenzidine
10. *Arsenic and certain arsenic	47. 3,3'-Dimethylbenzidine
compounds	48. 4-Dimethylaminoazobenzene
11. *Asbestos	49. Dimethyl carbamoyl chloride
12. *Auramine, manufacture of	50. Dimethyl sulfate
13. Benz(a)anthracene	51. 1,4-Dioxane
14. *Benzene	52. Direct Black 38
15. *Benzidine	53. Direct Blue 6
16. Benzo(b)fluoranthene	54. Ethylene Thiourea
17. Benzo(a)pyrene	55. Formaldehyde
18. Beryllium and certain beryl-	56. *Hematite underground mining
lium compounds	57. Hexachlorobenzene
19. *N,N-Bis(2-chloroethyl)-2-	58. Hydrazine and Hydrazine
naphthylamine	sulfate
20. *Bis(chloromethyl)ether and	59. Hydrazobenzene
chloromethyl methyl ether	60. Indeno(1,2,3-cd)pyrene
21. Cadmium and certain cad-	61. Iron dextran
mium compounds	62. *Isopropyl alcohol manufacture
22. Carbon tetrachloride	using strong acid process
23. *Chlorambucil	63. Kepone
24. Chloroform	64. Lead acetate and lead phos-
25. *Chromium and certain	phate
chromium compounds	65. Lindane and other hexachloro-
26. *Coke oven emissions	cyclohexane isomers
27. p-Cresidine	66. *Melphalan
28. Cupferron	67. 4,4'-Methylene bis(2-chloro-
29. Cycasin	aniline)
30. *Cyclophosphamide	68. 4,4'-Methylene bis(N,N-di-
31. 2,4-Diaminoanisole sulfate	methyl) benzenamine
32. 2,4-Diaminotoluene	69. Michler's Ketone
33. Dibenz(a,h)acridine	70. Mirex
34. Dibenz(a,j)acridine	71. *Mustard gas
35. Dibenz(a,h)anthracene	72. *2-Naphthylamine
36. 7H-Dibenzo(c,g)carbazole	73. *Nickel refining

TABLE 4 (Continued)

73. Nickel and certain nickel compounds	96. Polychlorinated biphenyls
74. Nitrilotriacetic acid	97. Procarbazine and procarbazine hydrochloride
75. Nitrofen	98. β-Propiolactone
76. 5-Nitro-o-anisidine	99. Reserpine
77. N-Nitrosodi-N-butylamine	100. Saccharin
78. N-Nitrosodiethanolamine	101. Safrole
79. N-Nitrosodiethylamine	102. Selenium sulfide
80. N-Nitrosodimethylamine	103. *Soots, tars, and mineral oils
81. p-Nitrosodiphenylamine	104. Streptozotocin
82. N-Nitrosodi-N-propylamine	105. Sulfallate
83. N-Nitroso-N-ethylurea	106. 2,3,7,8-Tetrachlorodibenzo-p-dioxin
84. N-Nitroso-N-methylurea	
85. N-Nitrosomethyl vinylamine	107. Thioacetamide
86. N-Nitrosomorpholine	108. Thiourea
87. N-Nitrosonornicotine	109. *Thorium dioxide
88. N-Nitrosopiperidine	110. o-Toluidine hydrochloride
89. N-Nitrosopyrrolidine	111. Toxaphene
90. N-Nitrososarcosine	112. 2,4,6-Trichlorophenol
91. Oxymethalone	113. Tris(aziridinyl)phosphine sulfide
92. Phenacetin	
93. Phenazopyridine hydrochloride	114. Tris(2,3-dibromopropyl) phosphate
94. Phenytoin	115. Urethane
95. Polybrominated biphenyls	116. *Vinyl chloride

Source: Ref. 31.
* = Established as a human carcinogen by epidemiology studies

tobacco smoke and alcohol are two known carcinogens not yet part of this listing. The available data are being examined on other substances that may represent potential carcinogenic hazards to the human population. These will be added only after careful evaluation of the scientific information. Single copies of this Report are available from the NTP.

IV. ACKNOWLEDGMENTS

We would like to thank Ms. Florence Jordan and Ms. Pamela Lemon for compiling and formatting Table 3, and Ms. Sharyn Wilkins and Ms. Pamela Chadwick for assistance in preparing this manuscript.

NOTES ADDED IN PROOF

1. The Ad Hoc Panel on Chemical Carcinogenesis Testing and Evaluation was established in response to a recommendation from the March 1983 meeting of the National Toxicology Program (NTP) Board of Scientific Counselors. The charge to the Panel was to review the basic biology and chemistry of chemical carcinogenesis and to recommend to the NTP Board of Scientific Counselors, methods that NTP should use for the detection and evaluation of chemical carcinogens. To accomplish this task, the 16-member Panel was divided into subpanels to deal with the general areas of Short Term Tests, Subchronic Studies and Related Issues, Chronic Studies, and Regulatory Aspects. A draft report of the Panel was released on February 15, 1984, with an invitation to interested parties for review, evaluation and comment. Since many comments were extensive and well documented and often included innovative and thoughtful suggestions, we have arranged to make all of the comments available through the National Technical Information Service repository service. The Ad Hoc Panel's Report is available upon request to the NTP.

2. Auditing of Data from Long-Term Carcinogenesis Studies. Data audits are an integral part of NTP's approach to quality assurance in the Toxicology Research and Testing Program. Whereas Good Laboratory Practices regulations establish a minimum level of management responsibility in the testing laboratory for the proper conduct of non-clinical studies, comprehensive data audits are the only means of ensuring the integrity and quality of the study data. Because of the impact of NTP technical Reports on public health issues relating to chemical toxicity and carcinogenicity, it is incumbent upon NTP to validate the data base upon which interpretive conclusions in Technical Reports are founded. During FY 1983, data discrepancies detected in the methylene chloride chronic study conducted by Gulf South Research Institute led to a Program decision not to issue a Technical Report on that study (Fed. Reg. 8-4-83, Vol. *48*, No. 151, p. 35508), or on any study not yet printed unless an audit was completed. Experimental data derived from chronic toxicological/carcinogenesis studies are audited for completeness, consistency, and accuracy and to determine if experimental procedures are consistent with Good Laboratory Practices. Audit procedures deal with eight major aspects of a study: (1) administrative information; (2) pretest animal data; (3) chemistry information; (4) dose preparation and administration; (5) environmental conditions − (temperature, relative humidity, lighting, air changes); (6) in-life observations; (7) pathology; and (8) Technical Report.

REFERENCES

1. R. Althouse, J. E. Huff, L. Tomatis, and J. D. Wilbourn, Chemicals and Industrial Processes Associated with Cancer in Humans (IARC Monographs, Volumes 1 to 20), *IARC Monographs on the Evaluation of the Carcinogenic Risk of Chemicals to Humans*, *IARC Monographs Supplement 1*, IARC, Lyon, France, September 1979.

2. R. Althouse, J. E. Huff, L. Tomatis, and J. D. Wilbourn, (Report of an IARC Working Group), An Evaluation of Chemicals and Industrial Processes Associated with Cancer in Humans Based on Human and Animal Data: IARC Monographs, Volumes 1 to 20, *Cancer Res.*, *40*:1-12 (1980).

3. J. A. Califano, Jr., Establishment of a National Toxicology Program, *Fed. Reg.*, *43*(221):53060-53061 (1978).

4. D. Canter, Role of the Regulatory Agencies in the Activities of the National Toxicology Program, *Reg. Toxicol. Pharmacol.*, *1*:8-18 (1981).

5. K. C. Chu, C. Cueto, Jr., and J. M. Ward, Factors in the Evaluation of 200 National Cancer Institute Carcinogen Bioassays, *J. Toxicol. Environ. Health*, *8*:251-280 (1981).

6. ECETOX, *A Contribution to the Strategy for the Identification and Control of Occupational Carcinogens*, European Chemical Industry Ecology and Toxicology Centre, Monograph No. 2, 1980.

7. E. J. Freireich, E. A. Geham, D. P. Rall, L. H. Schmidt, and H. E. Skipper, Quantitative Comparison of Toxicity of Anticancer Agents in Mouse, Rat, Hamster, Dog, Monkey, and Man, *Cancer Chemother. Rep.*, *50*:219-244 (1966).

8. Health Council of the Netherlands, *The Evaluation of the Carcinogenicity of Chemical Substances*, Leidschendam, The Netherlands, 1980.

9. L. G. Hart, J. E. Huff, J. A. Moore, and D. P. Rall, The National Toxicology Program's Research and Testing Activities, *Hazard Assessment of Chemicals. Current Developments*, *2*:191-244, J. Saxena, ed., Academic Press, New York, 361 pages (1983).

10. J. E. Huff, Carcinogenesis Bioassay Results from the National Toxicology Program, *Environ. Health Perspect.*, *45*:185-198 (1982).

11. J. E. Huff, Toxicology Data Evaluation Techniques and the International Agency for Research on Cancer, 55-78. *Proceedings of the Symposium on Information Transfer in Toxicology*, (G. J. Cosmides, ed.), PB82-220922, National Technical Information Service, Springfield VA. (1982). (Presented at the Symposium on Information Transfer in Toxicology, National Library of Medicine, Bethesda, MD., Sept. 16-17, 1981.)

12. J. E. Huff, J. A. Moore, J. Haseman, and E. E. McConnell, The National Toxicology Program, Toxicology Data Evaluation Techniques, and the Long-Term Carcinogenesis Bioassay, in *Safety Evaluation of Drugs and Chemicals*, Proceedings of a Symposium held at the Iowa State University on 1-3 June 1981, (1984).

13. *IARC Monographs on the Evaluation of the Carcinogenic Risk of Chemicals to Humans*, Volumes 1-33, International Agency for Research on Cancer, Lyon, France, 1972-1984.

14. IARC, Long-Term and Short-Term Screening Assays for Carcinogens: A Critical Appraisal, Supplement 2, *IARC Monographs on the Evaluation of the Carcinogenic Risk of Chemicals to Humans*, International Agency for Research on Cancer, Lyon, France, 1980.

15. *IARC Information Bulletin on the Survey of Chemicals Being Tested for Carcinogenicity*, Numbers 1-11, International Agency for Research on Cancer, Lyon, France, 1973-1984.

16. IARC Working Group Report, *International Agency for Research on Cancer Monographs on the Evaluation of the Carcinogenic Risk of Chemicals to Humans*, Supplement 4, Chemicals and Industries Associated with Cancer in Humans, IARC Monographs, Volumes 1-29, Lyon, France (1982).

17. S. Ivankovic, G. Eisenbrand, and R. Preussmann, Lung Carcinoma Induction in BD Rats After Single Intratracheal Instillatino of an Arsenic-Containing Pesticide Mixture Formerly Used in Vineyards, *Int. J. Cancer*, 24:786-788 (1979).

18. J. A. Moore, J. E. Huff, and J. H. Dean, The National Toxicology Program and Immunological Toxicology, *Pharmacol. Rev.*, 34:13-16 (1982).

19. J. A. Moore, J. E. Huff, L. Hart, and D. B. Walters, Overview of the National Toxicology Program, in *Environmental Health Chemistry* (J. D. McKinney, ed.), Ann Arbor Sci. Pub., Ann Arbor, Mich., 1980, pp. 555-574.

20. NCI/NTP, *Monitoring Guidelines for the Conduct of Carcinogen Bioassays*, National Toxicology Program Technical Report Series (TR 218), USDHHS, Washington, D.C., 1981.

21. National Cancer Institute, *Bioassay of "Chemical" for Possible Carcinogenicity*, Carcinogenesis Technical Report Series, Numbers 1-200, 1976-1980.

22. NIH/NTP, in *Bioassay Methodology: In Vivo, In Vitro and Mathematical Approaches*, Proceedings of the Symposium Held on 18-20 February 1981 in Washington, D.C., National Institutes of Health, Washington, D.C., 1981.

23. NTP, National Toxicology Program Fiscal Year 1979 Annual Plan: I — Research Plan; II — Review of Current DHEW Reearch Related to Toxicology, 1979.

24. NTP, National Toxicology Program Fiscal Year 1980 Annual Plan, 1979.
25. NTP, National Toxicology Program Fiscal Year 1980 Review of Current DHEW Research Related to Toxicology, 1979.
26. NTP, National Toxicology Program Fiscal Year 1981 Annual Plan, 1980.
27. NTP, National Toxicology Program Fiscal Year 1981 Review of Current DHHS, DOE, and EPA Research Related to Toxicology, 1981.
28. NTP Annual Plan for Fiscal Year 1982, National Toxicology Program, Research Triangle Park, N.C., April 1982; NTP Annual Plan for Fiscal Year 1983, January 1983; NTP Annual Plan for Fiscal Year 1984, February 1984.
29. NTP, National Toxicology Program Fiscal Year 1982 Review of Current DHHS, DOE, EPA Research Related to Toxicology, June 1982; Fiscal Year 1983 Review, January 1983; Fiscal Year 1984 Review, February 1984.
30. National Toxicology Program, *Carcinogenesis Bioassay of "Chemical" (CAS No.) in F344/N Rats and B6C3F1/N Mice (Dose Route)*, Technical Report Series, Numbers 201-300, 1981-1984.
31. NTP Third Annual Report on Carcinogens, National Toxicology Program, Research Triangle Park, N.C., December 1982.
32. OECD Guidelines for Testing of Chemicals, The Organization for Economic Co-operation and Development, Paris, France, 1981 draft.
33. OSHA, Identification, Classification and Regulation of Potential Occupational Carcinogens, Occupational Safety and Health Administration, *Fed. Reg.*, 45(15):5001-5296 (1980).
34. Public Health Service Publication No. 149, *Survey of Compounds Which Have Been Tested for Carcinogenic Activity*, U.S. Government Printing Office, Washington, D.C., J. L. Hartwell, Literature up to 1947 on 1329 compounds, 2nd ed., 1951; P. Shubik, and J. L. Hartwell, Supplement 1, Literature for the years 1948-1953 on 981 compounds, 1957; P. Shubik, and J. L. Hartwell (J. A. Peters, ed.), Supplement 2, Literature for the years 1954-1960 on 1048 compounds, 1969; National Cancer Institute, Literature for the years 1968-1969 on 882 compounds, 1971; National Cancer Institute, Literature for the years 1961-1967 on 1632 compounds, 1973; National Cancer Institute, Literature for the years 1970-1971 on 750 compounds, 1974; National Cancer Institute, Literature for the years 1972-1973 on 966 compounds, 1976; National Cancer Institute, Literature for the year 1978 on 664 compounds, 1980.
35. D. P. Rall, Validity of Extrapolation of Results of Animal Studies to Man, *Ann. N.Y. Acad. Sci.*, 329:85-91 (1979).

36. D. P. Rall, The Role of Laboratory Animal Studies in Estimating Carcinogenic Risks for Man, in *Carcinogenic Risks/Strategies for Intervention* (W. Davis and C. Rosenfeld, eds.), IARC Scientific Publications No. 25, International Agency for Research on Cancer, Lyon, France, 1979, pp. 179-189.

37. D. P. Rall, National Toxicology Program Fiscal Year 1979 Annual Plan, *Fed. Reg.*, *44*(143):43426-43435 (1979).

38. D. P. Rall, National Toxicology Program Fiscal Year 1980 Annual Plan, *Fed. Reg.*, *45*(28):8888-8918 (1980).

39. D. P. Rall, National Toxicology Program, 1-6, in *Toxic Control IV, Toxic Control in the 80's* (M. L. Miller, ed.), Government Institutes, Inc., Washington, D.C., 1980.

40. D. P. Rall, National Toxicology Program Fiscal Year 1981 Annual Plan, *Fed. Reg.*, *46*(39):14532-14625 (1981).

41. D. P. Rall, Issues in the Determination of Acceptable Risk, *Ann. N.Y. Acad. Sci.*, *363*:139-144 (1981).

42. D. P. Rall, NTP Annual Plan for Fiscal Year 1982, *Fed. Reg.*, *47*(94):14 May 1982; Fiscal Year 1983, *Fed. Reg.*, *48*(78): 21 April 1983; Fiscal Year 1984, *Fed. Reg.*, *49*(88): May 4, 1984.

43. J. M. Sontag, N. P. Page, and V. Saffioti, Guidelines for Carcinogen Bioassay in Small Rodents, NCI Carcinogenesis Technical Report (TR 1), DHEW, Washington, D.C., 1976.

44. U.S. House, National Toxicology Program, Hearing Before the Subcommittee on Investigations and Oversight of the Committee on Science and Technology, U.S. House of Representatives, 97th Cong., First Sess., July 15, 1981, No. 32, U.S. Government Printing Office, Washington, D.C., 1981.

40

ALTERNATIVES TO ANIMAL RESEARCH—AN ANIMAL WELFARE PERSPECTIVE

ANDREW N. ROWAN

School of Veterinary Medicine, Tufts University, Boston, Massachusetts

I. INTRODUCTION

A few years ago, the argument that animal rights would merit a place as a future challenge for the cosmetic industry would have been greeted with polite skepticism if not outright ridicule. However, recent events have changed that view, and it is clear that some vocal consumers would like to see the industry take constructive action on the animal testing issue. There are also a few studies, of varying rigor and depth, of consumer attitudes toward cosmetic testing on animals. At one end of the scale, there are surveys by the popular media such as *Glamour* magazine. In December 1981, *Glamour* published the results of a survey indicating that 84 percent of their respondents opposed cosmetic tests on animals. A majority of the respondents (59 percent) stated that they were even willing to use a drug which had not been tested on animals. In 1974, National Opinion Polls (England) reported that three-quarters of the survey sample expressed opposition to cosmetic tests on animals [1].

At the other end of the scale, two Australian sociologists investigated the attitudes of 302 undergraduates to various categories of animal use [2]. Strong and moderate disapproval of painful cosmetic tests (for eye irritancy) on mice, monkeys, and dogs was expressed by 75 percent, 87 percent, and 92 percent of the respondents, respectively. In questions concerning the use of animals in "nonmedical research," 46 percent expressed disapproval of the use of mice in painless experiments, with only 24 percent approving of such studies. When it came to painless tests on dogs, 76 percent disapproved while only 10 percent approved. It is clear that even the most innocuous experiments on animals by cosmetic companies are opposed, at least in theory, by a majority of the public. It is thus inevitable that organized protests by animal welfare groups will evoke widespread public support, and it is this feature which has raised the issues of animal rights and animal research significantly higher on the priority scale for the cosmetic industry.

The recent surge of interest in these topics is fueled by a number of factors, not least being the support of academic philosophers for the notion of animal rights.

II. ANIMAL RIGHTS

Animal rights is not a new rallying call. It has been promoted and ridiculed in various forms throughout the ages. The last decade has seen a resurgence of interest in the topic, beginning with the publication of an anthology called *Animals, Men and Morals* [3]. While this book did not have the impact that its authors hoped for, it did stimulate Singer, an Australian philsopher, to write *Animal Liberation* [4]. This has undoubtedly been a major landmark in the revitalization of the animal welfare movement. Many welfare advocates, young and old alike, have read the book and, emboldened by finding logical but simple arguments which support their own intuitive feelings about the treatment of animals, have become more vocal and more active. Other important books arising out of the early discussions of the Oxford Vegetarians [5] include Clark's *The Moral Status of Animals* [6] and Ryder's *Victims of Science* [7]. Ryder also introduced the term "speciesism" to describe unwarranted discrimination against other species, thus drawing an analogy with racism or sexism.

Singer's basic thesis was that all sentient beings have "interests" which matter to them and that these interests must be considered equally with like human interests. In line with classic Utilitarian thinking, Singer then argues that we should seek to maximize the satisfaction of those interests. It is important to note that equal consideration of like interests does not imply equal treatment. You may accept that your son and your neighbor's son have an equal interest in obtaining the best education possible, but that does not mean that you have to give your neighbor's son the same support you give your own. In addition, a chicken and a trout have an equal interest in an optimal environment, but a deep pool in a flowing mountain stream is not a good place for a chicken.

Singer accepts that his arguments are not new. Salt [8] enunciated many of them at the end of the nineteenth century, and Jeremy Bentham raised the issue of suffering and animal interests a hundred years earlier in his well-known and eloquent passage:

> The day may come when the rest of the animal creation may acquire those rights which never could have been witholden from them but by the hand of tyranny. . . . It may one day come to be recognized that the number of legs, the villosity of the skin, or the termination of the os sacrum, are reasons insufficient for abandoning a sensitive being to . . .torment. . . . The question is not, Can they reason? nor Can they talk? but, Can they suffer? [9].

The precise nature of the characteristics that make some or all animals worthy of moral concern is one of the critical questions in the modern debate. Singer argues that sentient animals should be included in the orbit of moral concern and defines sentience as a capacity to experience suffering or pleasure.

Regan [9] argues that a capacity to be harmed should be the distinguishing criterion and that even inanimate objects can be harmed. In this way, he attempts to unite the environmental and humane movements, which have traditionally been split by their respective concentration on the survival of populations of animals versus the suffering of individual animals.

The various criteria that have been put forward as justifying a distinction in kind between our moral concern for humans and our moral concern for animals include the following:

1. Only humans possess souls.
2. Humans have dominion and power over animals, whether or not this is granted by God.
3. Humans are rational and have language.
4. Only humans are moral agents (i.e., capable of distinguishing between right and wrong and acting on the distinction) who can, consequently, make social contracts to treat each other in a moral fashion.

This is not an exhaustive list, but it does include some of the more common arguments. However, the philosophical advocates of animal rights [cf. 10] do not consider the above distinctions to be morally relevant, in the same way that we do not now believe that skin color or religious belief is an adequate basis for relegating human beings to a lesser moral status.

The "soul" criterion is unsound since it is certainly not a universal belief that animals do not have souls. The Methodist church, for example, held that animals are admitted to the afterlife [11]. Even if one accepts that animals do not have souls, it does not necessarily mean that they are outside our orbit of moral concern. In fact, it has been argued that the proposed lack of a soul makes animal suffering worse than human suffering, because there is no afterlife in which the wrongs to the animal can be redressed.

The "dominion" argument is easily dismissed because otherwise it would require general acceptance that might makes right. Society does not accept that the strong can treat the weak as they like, and this must surely apply to animals as well as to humans. The claim that God has given us dominion over the animals is also balanced by many other scriptural messages that indicate that we have, at most, been placed on earth as stewards and not as domineering masters.

The "rationality and language" argument is probably the distinction most widely accepted by our secular society but it, too, has many flaws. First, it is by no means clear that animals are not rational. Also, we are only able to exclude them from linguistic capacities by a continuing escalation of what types of communication are actually admitted as indicating "true" linguistic ability. Second, there are humans who clearly lack either language or rationality (e.g., infants, the severely brain-damaged, the senile), but these individuals are included in our orbit of moral concern. Therefore, why do we include nonrational humans but exclude animals, at least some of whom show definite signs of what could be defined as rational behavior?

The "social contract" argument holds that one has to be able to enter into a contract (even if it is only understood and not explicit) in order to be included in the orbit of moral concern. However, there are humans who are unable to enter into such contracts and who are temporarily or permanently unable to act as moral agents, and yet we still include them in our moral concern. On what grounds are such humans preferred over animals?

The current broad consensus among advocates of animal rights is that animals should be brought fully into the moral universe because they have a life and interests, the satisfaction of which matters to them, and because they are capable of experiencing suffering. This capacity to suffer is cataloged under the broad term sentience and does not include the ability to sense and react in a reflex fashion to noxious stimuli. Sentience is put forward as a much better measure of moral considerability than rationality or linguistic ability.

There are obviously many practical difficulties in applying such a thesis. For a start, we are not sure where sentience begins and ends in the phylogenetic scale. Furthermore, surely there are grades of sentience; would this lead to a differential weighting of interests (does one human equal 100 rats, for example)? These issues and others are the subjects of heady debate in the academic periodicals, but they provide relatively little guidance for the overworked industrial toxicologist.

One of the benefits of these renewed philosophical stirrings is that we are beginning to move away from purely emotional responses when the topic of animal experimentation is raised. We know that intubating an animal with a toxic substance cannot be regarded as a kind act, but neither can it be regarded as a cruel act. The new philosophical arguments avoid the confusion raised by notions of kindness and cruelty and provide a basis for constructive dialogue.

III. ANIMAL RIGHTS AND ALTERNATIVES

Regan and Jamieson [12], two moral philosophers who advocate animal rights, have suggested that some animal research can be justified but only if the harm caused to the research animals will prevent greater harm being done to other living beings *and* if a conscientious effort has been made to explore the possibility of using alternatives. In fact, most philosophers who advocate granting animals certain moral or legal rights accept this position or something similar. Animal welfare groups are promoting alternatives extensively and have applied pressure on the cosmetic industry to support specific research to this end, particularly on the issue of irritancy testing.

The concept of alternatives is now fully integrated into the animal research campaign literature of the humane movement, but there are still semantic problems — different groups use the term "alternatives" to refer to different sets of techniques. However, the term "alternatives" usually refers to those techniques or methods which:

> *Replace* the use of laboratory animals altogether;
> *Reduce* the number of animals required;
> *Refine* an existing procedure or technique so as to minimize the
> level of stress endured by the animal.

Animal research techniques have undergone constant refinement in the past 150 years. Technical improvements have led to the replacement of animals in some instances and to reduced pain and suffering in others [13]. Nevertheless, most people date the concept of alternatives to 1959, when *The Principles of Humane Experimental Techniques* by Russell and Burch was published [14]. This book not only enunciated the three "R"s of Replacement, Reduction, and Refinement, it also introduced the important notion of fidelity and discrimination in animal models. For example, while a mammal is a high-fidelity model of the human, one may be able to conduct research or testing in a high-discrimination model instead, such as microorganisms for nutrition studies. Russell argued that progress toward replacement methods is "hindered by an insidious and widespread assumption"

which may be called the "hi-fi fallacy" [15]. In this fallacy, the major
premise is that high fidelity is desirable in general. The minor premise is that mammals are of exceptionally high fidelity as models of
the human organism and therefore should be used as much as possible.
As we learn more about mechanisms, it is becoming clear that a high-
fidelity model is not necessarily the most desirable, especially in detecting specific toxic effects.

By 1965, the idea of alternatives had attracted enough attention
that a Parliamentary committee of enquiry in England repeatedly questioned scientific witnesses about the prospects for the availability of
such techniques [16]. Seventeen years later, scientists still tend to
be suspicious of the concept. Some groups have gone to the lengths
of referring only to "adjunct" methods rather than "alternatives," because they feel that the term "alternatives to laboratory animals" gives
the public a false sense of the potential of such techniques. Others dismiss the idea altogether [17]. Nevertheless, industry is embracing
the concept, led by the Cosmetic, Toiletry and Fragrance Association,
which has established a center for alternatives research at The Johns
Hopkins School of Hygiene and Public Health.

This center and other recent initiatives have provoked a great
deal of interest in the idea of "alternatives." It is, perhaps, not unreasonable to credit Sir Peter Medawar with great prescience when,
13 years ago, he argued that the use of animals on the present scale
is a temporary episode in biomedical research and that the numbers
would start to decline in 10 years. He qualified this, however, by
further stating that, for the moment, we would have to live with the
paradox that only by doing research on animals could we one day develop the means which allow us to eliminate animal research altogether
[18]. In England, Medawar's prediction appears to be coming true,
although the decline in animal experiments may have more to do with
economics than the development and use of alternatives.

IV. ALTERNATIVES AS TECHNIQUES

In the sometimes heated debate on alternatives, we frequently forget
that the term "alternative" really refers only to a technique or method
and that the development of such techniques is a resource issue and
not one that will necessarily lead directly to exciting new concepts.
Funds and resources can therefore be applied with reasonably predict-
able results to the development of such new technology. Arguments
that "money cannot be thrown at the problem" are not as relevant in
this instance as, say, criticism of the "war on cancer."

Advancement in research depends on the creative act of formulating a theory, the self-critical process of testing that theory, and,
more often than not, the development of new or improved techniques.
The importance of the availability of new techniques cannot be over-

estimated. Nobel Laureate Hans Krebs argued for the importance of methodological papers and noted that "the frequent quotation of methods papers unequivocally demonstrates the usefulness of such papers" [19]. Thus paper chromatography and radioimmunoassay opened up exciting new research avenues when they were first introduced. The new monoclonal antibody techniques are having a similar impact. For the future, the answer to cancer and its prevention is more likely to lie in a thorough understanding of the growth requirements of cells, both normal and malignant, than in megamouse studies.

V. CHALLENGE FOR TOMORROW

Some time has been spent discussing various aspects of the alternatives concept since this appears to be one of the most constructive solutions to the animal welfare-cosmetic industry conflict. Animal welfare groups have protested the use of animals in toxicity testing because they believe that much of it is irrelevant or scientifically unsound and can point to scientific support for their views [20-22].

When such testing involves cosmetics, the concerns are aggravated by the belief that the production of a new cosmetic is not a sufficient reason to risk animal suffering. As a result, some animal welfare advocates have called for a prohibition on the development of new cosmetic products. There is unlikely to be any common ground over the ban on innovation but, fortunately, there is room for discussion and movement over the development of adequate alternative testing methods.

Steady progress toward the development of alternatives will be essential if the fragile truce between animal welfare groups and cosmetic companies is not to be broken. Some useful research has been done in the past, and there are a number of promising research projects underway at the moment [23], but the current initiatives must not be allowed to wither away. There are a number of possibilities for maintaining and even increasing momentum.

1. Companies should continue to support existing research programs and be responsive to new research proposals from both industry and outside laboratories. The Johns Hopkins Center has served as a focal point for industry research support and for interesting research ideas. However, the center has funds for only three years.* The industry will eventually have to provide funds for a much longer research program if it hopes to convince the animal welfare movement that it is really committed to solving the animal testing problem. In-house research by individual companies is another option, but critics will not be satisfied with mere assurances that such investigations are taking place. The company will have to provide more concrete evidence (e.g., published papers) that it is committed to the search.

---

*Funds have now been extended for two additional years.

2. Regulatory reform will be needed so that the alternatives concept can become an established element in safety evaluation dogma.
As long as the courts and the regulatory agencies look askance at alternative tests, it will be very difficult to make substantial progress
in reducing the number of animals used in safety testing. The cosmetic industry will have to play a role, perhaps in cooperation with
the animal welfare groups, to effect the necessary change in attitude
that will allow us to reduce animal testing without affecting human
safety standards [20].

3. Information on new research initiatives and procedural refinements must be rapidly distributed and put to use. There are several
possible outlets for such information, including The Johns Hopkins
Center for Alternatives. It is already acting as an informal information clearinghouse, and the Center's newsletter is expected to perform
this vital but neglected role. For example, on several occasions, data
from relevant in-house research projects was published only as a result of animal welfare pressure [24-26]. Procedural refinements, such
as those instituted by Revlon, which have reduced its rabbit testing
by 20 percent (Christian Science Monitor, July 15, 1982), should also
be publicized.

4. The industry trade associations should ensure that laboratory
practices are reviewed regularly (annually?) to ensure that alternatives are being employed where appropriate. It would help dialogue
and trust if a suitable animal welfare spokesperson was at least kept
informed of the results of the review, if not included in the review
process itself.

The above suggestions do not constitute an exhaustive list, nor
are they very detailed. However, they do provide some guidance as
to the type of initiatives that would be perceived by many in the animal welfare movement as being constructive.

REFERENCES

1. Report to Annual General Meeting of the Royal Society for the
    Prevention of Cruelty to Animals, National Opinion Polls, June
    28, 1974.
2. J. Braithwaite and V. Braithwaite, Attitudes Toward Animal
    Suffering: An Exploratory Study, Int. J. Study Animal Prob-
    lems, 3:42-49 (1982).
3. S. Godlovitch, R. Godlovitch, and J. Harris, eds., Animals,
    Men and Morals, Taplinger, New York, 1971.
4. P. Singer, Animal Liberation, New York Review/Random House,
    New York, 1975.
5. P. Singer, The Oxford Vegetarians — A Personal Account, Int.
    J. Study Animal Problems, 3:6-9 (1982).
6. S. R. L. Clark, The Moral Status of Animals, Oxford Univer-
    sity Press, Oxford, 1977.

7. R. D. Ryder, *Victims of Science*, Davis-Poynter, London, 1975.
8. H. S. Salt, *Animals' Rights*, Society for Animal Rights, Clarks Summit, Pa., 1980 (originally published in 1892).
9. J. Bentham, *An Introduction to the Principles of Morals and Legislation* (J. H. Burns and H. L. A. Hart, eds.), Athlone Press, London, 1970.
10. B. Rollin, *Animal Rights and Human Morality*, Prometheus, Buffalo, N.Y., 1981.
11. L. G. Stevenson, Religious Elements in the Background of the British Anti-Vivisection Movement, *Yale J. Biol. Med.*, 29:125-157 (1956).
12. T. Regan and D. Jamieson, The Case for Alternatives to Laboratory Animal Experimentation, *Lab Animal*, 11(1):21-23 (1982).
13. A. N. Rowan, Alternatives and Laboratory Animals, in *Animals in Research* (D. Sperlinger, ed.), J. Wiley & Sons, Chichester, England, 1981, pp. 257-283.
14. W. M. S. Russell and R. L. Burch, *The Principles of Humane Experimental Techniques*, Methuen, London, 1959.
15. W. M. S. Russell, The Increase of Humanity in Experimentation: Replacement, Reduction and Refinement, *Lab. Animals Bureau Collected Papers*, 6:23-25 (1957).
16. Littlewood Committee, Report of Departmental Committee on Experiments on Animals, Her Majesty's Stationery Office, London, 1965, p. 26.
17. M. B. Visscher, Animal Rights and Alternative Methods: Two New Twists in the Antivivisectionist Movement, *The Pharos Fall*:11-19 (1979).
18. P. B. Medawar, *The Hope of Progress*, Methuen, London, 1972, p. 86.
19. E. Garfield, To Remember Sir Hans Krebs: Nobelist, Friend and Adviser, *Current Contents*, 13(31):10 (1982).
20. I. Muul, A. F. Hegyeli, J. C. Dacre, and G. Woodard, Toxicological Testing Dilemma, *Science*, 193:834 (1976).
21. K. L. Melmon, The Clinical Pharmacologist and Scientifically Unsound Regulations for Drug Development, *Clin. Pharmacol. Ther.*, 20:125-129 (1976).
22. G. Zbinden, A Look at the World From Inside the Toxicologist's Cage, *Eur. J. Clin. Pharmacol.*, 9:333-338 (1976).
23. A. N. Rowan, The Draize Test: The Search for Alternatives, *Cosmet. Technol.*, 4(6):30-33 (1982).
24. P. J. Simons, An Alternative to the Draize Test, in *The Use of Alternatives in Drug Research* (A. N. Rowan and C. J. Stratmann, eds.), Macmillan, London, 1980, pp. 147-151.
25. A. W. Johnson, Use of Small Dosage and Corneal Anaesthetic for Eye Testing In Vivo, in *Proceedings of the CTFA Ocular Safety Testing Workshop: In Vivo and In Vitro Approaches, October 6 and 7, 1980*, Cosmetic, Toiletry and Fragrance Association, Washington, D.C., 1981.

26. J. McCormack, A Procedure for the *In Vitro* Evaluation of the
     Eye Irritation Potential of Surfactants, in *Trends in Bioassay
     Methodology: In Vivo, In Vitro and Mathematical Approaches*,
     National Institutes of Health (NIH Publications No. 82-2382),
     Washington, D.C., 1981, pp. 177-186.

INDEX